Multisensor Data Fusion and Machine Learning for Environmental Remote Sensing

Multisensor Data Fusion and Machine Learning for Environmental Remote Sensing

Ni-Bin Chang
Kaixu Bai

CRC Press

Taylor & Francis Group

Boca Raton London New York

CRC Press is an imprint of the
Taylor & Francis Group, an **informa** business

CRC Press
Taylor & Francis Group
6000 Broken Sound Parkway NW, Suite 300
Boca Raton, FL 33487-2742

First issued in paperback 2020

© 2018 by Taylor & Francis Group, LLC
CRC Press is an imprint of Taylor & Francis Group, an Informa business

No claim to original U.S. Government works

ISBN-13: 978-0-367-57197-9 (pbk)
ISBN-13: 978-1-4987-7433-8 (hbk)

Contents

PART I   Fundamental Principles of Remote Sensing

PART II   *Feature Extraction for Remote Sensing*

PART III Image and Data Fusion for Remote Sensing

PART IV   *Integrated Data Merging, Data Reconstruction, Data Fusion, and Machine Learning*

PART V Remote Sensing for Environmental Decision Analysis

Preface

Earth observation and environmental monitoring require data to be collected for assessing various types of natural systems and the man-made environment with varying scales. Such research enables us to deepen our understanding of a wealth of geophysical, geochemical, hydrological, meteorological, and ecological processes of interest. For the past few years, the scientific community has realized that obtaining a better understanding of interactions between natural systems and the man-made environment across different scales demands more research efforts in remote sensing. The key research questions include: (1) how to properly fuse the multisensor images with different spatial, temporal, and spectral resolution to minimize the data gaps and create a myriad of long-term consistent and cohesive observations for further feature extraction, and (2) how feature extraction can be adequately processed by these traditional algorithms and advanced computational intelligence methods to overcome barriers when complex features are embedded or constrained by heterogeneous images. From systems engineering perspectives, the method of integrating the latest forefronts of multisensor data fusion and feature extraction with the aid of advanced computational intelligence methods is of primary importance to achieve our common goal—"the whole is greater than the sum of its parts."

The aim of this book is thus to elucidate the essence of integrated multisensor data fusion and machine learning highlighted for promoting environmental sustainability. It emphasizes the concept of the "System of Systems Engineering" approach with implications for both art and science. Such an endeavor can accommodate an all-inclusive capability of sensing, monitoring, modeling, and decision making to help mitigate the natural and human-induced stresses on the environment.

On this foundation, many new techniques of remote sensing image processing tools and feature extraction methods in concert with existing space-borne, air-borne, and ground-based measurements have been collectively presented across five distinctive topical areas in this book. This initiative leads to a thorough discussion of possible future research with synergistic functionality across space and time at the end of the book. The book will be a useful reference for graduate students, academic scholars, and working professionals who are involved in the study of "earth systems science" and "environmental science and engineering" promoting environmental sustainability.

Ni-Bin Chang and Kaixu Bai

MATLAB® is a registered trademark of The MathWorks, Inc. For product information, please contact:

The MathWorks, Inc.
3 Apple Hill Drive
Natick, MA 01760-2098 USA Tel: 508 647 7000
Fax: 508-647-7001
E-mail: info@mathworks.com
Web: www.mathworks.com

Acknowledgments

This book grew out of a series of research grants and contracts as well as a wealth of international collaborative work. This book could not have been written without the valuable assistance of several people. The authors are indebted to Mr. Benjamin Vannah, Ms. Xiaoli Wei, Mr. Chandan Mostafiz, and Dr. Zhibin Sun for their collective contributions. Their helpful research and/or thesis work are gratefully acknowledged. We also extend our gratitude to Ms. Rachel Winter, who serves as the language editor of this book. Finally, without encouragement from Ms. Irma Shagla Britton, who is the senior editor of Environmental Sciences, Remote Sensing & GIS in the CRC Press/Taylor & Francis Group, we could have not made up our mind to complete this lengthy work. Special thanks are extended to her as well.

Authors

Ni-Bin Chang has been professor of Environmental Systems Engineering, having held this post in the United States of America since 2002. He received his BS in Civil Engineering from the National Chiao-Tung University in Taiwan in 1983, and MS and PhD in Environmental Systems Engineering from Cornell University in the United States of America in 1989 and 1991, respectively. Dr. Chang's highly interdisciplinary research lies at the intersection among "Environmental Sustainability, Green Engineering, and Systems Analysis." He is director of the Stormwater Management Academy and professor with the Department of Civil, Environmental, and Construction Engineering at the University of Central Florida in the United States of America. From August 2012 to August 2014, Professor Chang served as program director of the Hydrologic Sciences Program and Cyber-Innovated Sustainability Science and Engineering Program at the National Science Foundation in the United States of America. He was elevated to Fellow of the Institute of Electronics and Electrical Engineers (IEEE) in 2017, and he has been actively with the IEEE Geoscience and Remote Sensing Society, IEEE Systems, Man, and Cybernetics Society, and the IEEE Computational Intelligence Society. He also has distinctions which are selectively awarded titles, such as an inducted Fellow of the European Academy of Sciences in 2008, and an elected Fellow of American Society of Civil Engineers (F.ASCE) in 2009, the American Association for the Advancement of Science (F.AAAS) in 2011, the International Society of Optics and Photonics (F.SPIE) in 2014, and the Royal Society of Chemistry (the United Kingdom) (F.RSC) in 2015. He has been the editor-in-chief of *SPIE Journal of Applied Remote Sensing* since 2014. He is currently an editor, associated editor, or editorial board member of 20+ international journals.

Kaixu Bai has been an assistant professor of cartography and geographic information systems at School of Geographic Sciences of East China Normal University since 2016. He received his BS in urban and rural planning from Northeast Agricultural University in 2009, and PhD in geographic information system from East China Normal University in 2015, respectively, in China. His research interest focuses mainly on environmental remote sensing and modeling for earth system science by taking advantage of multisensor data merging, fusion, and mining as well as machine learning.

1 Introduction

1.1 BACKGROUND

Remote sensing is defined as the acquisition and analysis of remotely sensed images to gain information about the state and condition of an object through sensors that are not in physical contact with it and discover relevant knowledge for decision making. Remote sensing for environmental monitoring and Earth observations can be defined as:

> Remote sensing is the art and science of obtaining information about the surface or subsurface of Earth without needing to be in contact with it. This can be achieved by sensing and recording emitted or reflected energy toward processing, analyzing, and interpreting the retrieved information for decision-making.

The remote sensing process involves the use of various imaging systems where the following seven elements are involved for environmental monitoring and earth observations: (1) illumination by the sun or moon; (2) travel through the atmosphere; (3) interactions with the target; (4) recording of energy by the sensor; (5) transmission, absorption, reflection, and emission; (6) retrieval, interpretation, and analysis; and (7) decision making for applications.

Types of remote sensing technologies include air-borne, space-borne, ground-based, and sea-based remote sensing technologies with a wealth of sensors onboard different platforms. These sensors are designed to observe electromagnetic, acoustic, ultrasonic, seismic, and magnetic energy for environmental monitoring and earth observation. This book focuses on remote sensing sensors making use of the electromagnetic spectrum for environmental decision making. These sensors generally detect reflected and emitted energy wavelengths ranging from ultraviolet to optical, to infrared, to microwave remote sensing that can measure the electromagnetic energy.

Over the last few decades, satellite remote sensing that aims to observe solar radiation has become an invaluable tool for providing estimates of spatial and temporal time series variables with electromagnetic sensors. The traditional image-processing algorithms often involve image restoration, image enhancement, image segmentation, image transformation, image fusion, and data assimilation with feature extraction/classification models. With the availability of field observations, such image-processing efforts enable us to provide our society with an unprecedented learning capacity to observe, monitor, and quantify the fluxes of water, sediment, solutes, and heat through varying pathways at different scales on the surface of Earth. Environmental status and ecosystem state can then be assessed through a more lucid and objective approach. Yet this requires linking remote sensing image processing with change detection in a more innovative way.

In an attempt to enlarge the application potential, sensor and data fusion with improved spatial, temporal, and spectral resolution has become a precious decision support tool that helps observe complex and dynamic Earth systems at different scales. The need to build more comprehensive and predictive capabilities requires intercomparing earth observations across remote sensing platforms and *in situ* field sites, leading to cohesively explore multiscale earth observation from local up to a regional or global extent for scientific investigation (CUAHSI, 2011). Recent advancements in artificial intelligence techniques have motivated a significant initiative of advanced image processing for better feature extraction, information retrieval, classification, pattern recognition, and knowledge discovery. In concert with image and data fusion, the use of machine learning and an ensemble of classifiers to enhance such an initiative are gaining more attention. The progress in this regard will certainly help answer more sophisticated and difficult science questions as to how environmental, ecological, meteorological, hydrological, and geological components interact with each other in Earth systems.

Given the complexity of these technologies and the diversity of application domains, various ultraviolet, optical, infrared, and microwave remote sensing technologies must be used, either independently or collectively, in dealing with different challenges. To achieve a sound system design for remote sensing, the "system-of-systems engineering" (SoSE) approach must often be applied for tackling these complexities and challenges. SoSE is a set of developing processes, tools, and methods for designing, redesigning, and deploying solutions to System-of-Systems challenges (United States Department of Defense, 2008). In support of various types of missions of environmental monitoring and earth observations, the main components associated with a remote sensing system that the system design has to consider include (Figure 1.1): (1) a source of the electromagnetic radiation (i.e., the sun) to be detected by the sensors mounted on air-borne, space-borne, sea-based, or ground-based platforms; (2) the targets to be detected on Earth's surface which receive the incident energy from a source such as the sun and those targets which emit energy; (3) the quantity and type of sensor instruments to measure and record the reflected and/or emitted energy leaving Earth's surface; (4) the platforms supporting satellite sensors and payload operations such as altitude, gesture, orbital trajectory control, instrument calibration, power supply, data compression and transmission through digital telemetry, and communication between the receiving station and the satellite or among satellites in a satellite constellation; (5) the receiving station that collects raw digital data or compressed data through a wireless communication network and converts them into an appropriate format; (6) the processing unit in charge of pre-processing the raw data through a suite of correction procedures that are generally performed for generating remote sensing data products prior to the distribution of the products or the imageries; (7) the knowledge discovery or information retrieval to be carried out by professional analysts to convert the pre-processed image data into various thematic information of interest using a suite of image-processing techniques; and (8) the end-user communities who may utilize these thematic products for applications.

On many occasions, sensor networks onboard satellites and/or aircrafts are required to provide holistic viewpoints in concert with ground- or sea-based sensors. To further advance Earth system science, such sensor networks can be extended to cover sensors onboard unmanned aerial vehicles (UAVs) and autonomous underwater vehicles (AUVs) to overcome local or regional surveillance barriers. Such expanded sensor networks may be coordinated through proper sensing, networking,

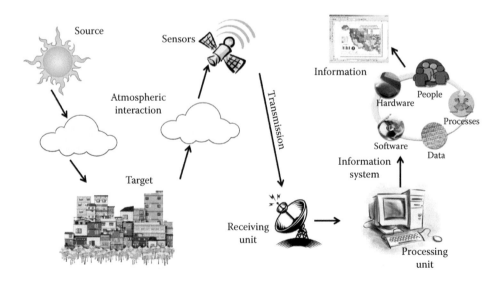

FIGURE 1.1   Basic structure of a remote sensing system.

and control with the aid of wired or wireless communication systems to produce multidimensional information and to monitor the presence of unique events. The SoSE approach may certainly provide sound routine monitoring, early warning, and emergency response capacity in our society, which is facing climate change, globalization, urbanization, economic development, population growth, and resource depletion.

An integrated Earth system observatory that merges surface-based, air-borne, space-borne, and even underground sensors with comprehensive and predictive capabilities indicates promise for revolutionizing the study of global water, energy, and carbon cycles as well as land use and land cover changes. This may especially be true if these multisensor data fusion and machine learning technologies are developed and deployed in a coordinated manner and the synergistic data are further screened, synthesized, analyzed, and assimilated into appropriate numerical simulation models for advanced decision analysis. Thus, the aim of this book is to present a suite of relevant concepts, tools, and methods of the integrated multisensor data fusion and machine learning technologies to promote environmental sustainability.

1.2 OBJECTIVES AND DEFINITIONS

The main objective of this book is to demonstrate the knowledge base for capacity building of integrated data fusion and machine learning with respect to a suite of satellite sensors and machine learning techniques to enhance the synergistic effect of multiscale remote sensing observations and intelligent image processing. The discussion of data fusion in this book is based on a definition derived from the recommendation of the United States Department of Defense Joint Directors of Laboratories (JDL) Data Fusion Subpanel:

> Data fusion is a multi-level, multifaceted process dealing with the automatic detection of the registration, detection, association, correlation, and combination of data and information from multiple sources to achieve refined state and identity estimation, and complete timely assessments of situation including both threats and opportunities.

The data fusion process proposed by JDL was classified into five processing levels, an associated database, and an information bus that connects the five levels (Castanedo, 2013). The five levels of processing are defined as (Figure 1.2):

- *Level 0—Source pre-processing*: Source pre-processing allocates data to suitable processes and performs data prescreening which maintains useful information for high-level processes (Hall and Llinas, 1997). It performs fusion at signal and pixel levels (Castanedo, 2013).
- *Level 1—Object refinement*: Object refinement transforms the processed data into consistent data structures. It combines features extracted from processed images and refines the identification of objects and classification (Hall and Llinas, 1997; Castanedo, 2013).
- *Level 2—Situation assessment*: Decision-level fusion starts at level 2. This level relates the events to the likely observed situations and helps us with data interpretation.
- *Level 3—Impact assessment*: The risk of future events is assessed at level 3 by evaluating current situations.
- *Level 4—Process refinement*: Finally, the last level monitors the data fusion process and identifies what additional information is required to improve the data fusion process (Castanedo, 2013).

In this book, machine learning or data mining algorithms are emphasized to help feature extraction of remote sensing images. The discussion of data mining in this book is based on the following definition:

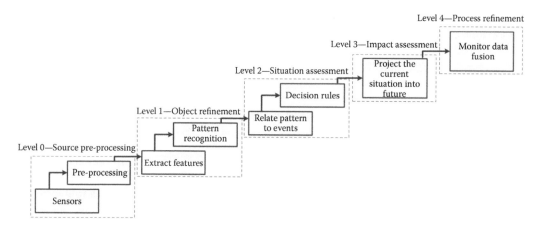

FIGURE 1.2   Structure of the JDL data fusion. (From Chang, N. B. et al., 2016. *IEEE Systems Journal*, 1–17.)

> Data mining, sometimes called knowledge discovery, is a big data analytic process designed to explore or investigate data from different perspectives in search of useful patterns, consistent rules, systematic relationships, and/or applicable information embedded among various types of flexibly grouped system variables of interests. It requires subsequent validation of the findings by applying these findings to new subsets of data.

The discussion of machine learning in this book is based on the following definition:

> Machine learning that was born from artificial intelligence is a computer science theory making computers learn something from different data analyses and even automate analytical or empirical model building without being coded to do so. By using various statistics, decision science, evolutionary computation, and optimization techniques to learn from data iteratively, machine learning allows computers to identify or discover hidden rules, inherent patterns, possible associations, and unknown interactions without being programmed what to search and where to look explicitly.

The major difference between data mining and machine learning is that the former has no clue about what the patterns or rules are in a system whereas the latter has some clues in advance about what the system looks like based on local or labeled samples. In image classification and feature extraction of remote sensing studies, such a distinction rests on whether or not the system of interest has some ground or sea truth data to infer. Since ground or sea truth data may or may not be available in different types of remote sensing studies, and collecting ground or sea truth data is favored for image classification and feature extraction toward much better prediction accuracy, the emphasis of this book is placed on machine learning rather than data mining although both cases are discussed in the context.

The process of data mining or machine learning consists of four general stages: (1) the initial exploration (i.e., data collection and/or sampling); (2) model building or pattern identification and recognition; (3) model verification and validation; and (4) application of the model to new data in order to generate predictions.

The niche for integrating data fusion and machine learning for remote sensing rests upon the creation of a new scientific architecture in remote sensing science that is designed to support numerical as well as symbolic data fusion managed by several cognitively oriented machine learning tasks. Whereas the former is represented by the JDL framework, the latter is driven by a series of image restorations, reconstructions, enhancements, segmentations, transformations, and fusions for intelligent image processing and knowledge discovery (Figure 1.3) (Chang et al., 2016). Well-known machine learning methods include but are not limited to genetic algorithms, genetic programming,

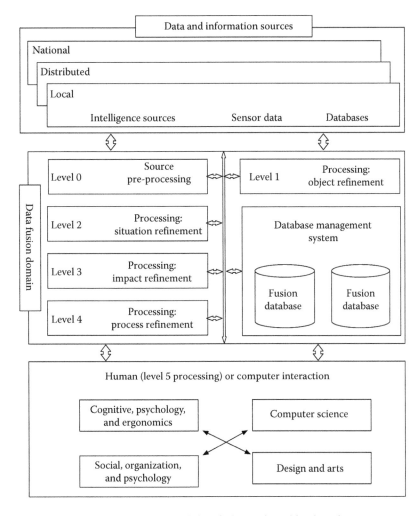

FIGURE 1.3   The new architecture of integrated data fusion and machine learning.

artificial neural networks, particle swarm optimization, support vector machines, and so on (Zilioli and Brivio, 1997; Volpe et al., 2007; Chen et al., 2009; Bai et al., 2015; Chang et al., 2015). They can be used for data mining as well if we do not have *in situ* observations.

1.3   FEATURED AREAS OF THE BOOK

The organization of this book will include most of the latest forefronts in the subject area streamlined by a logical sequence of multisensor data fusion and machine learning with a range of specific chapters associated with the following five featured areas:

- *Part I—Fundamental Principles of Remote Sensing*: This part of the discussion will demonstrate a contemporary coverage of the basic concepts, tools, and methods associated with remote sensing science. The relationship between electromagnetic radiation and remote sensing will be discussed in Chapter 2. Then the types of sensors and platforms that provide the image acquisition capacity will be delineated and emphasized in Chapter 3. Finally, the method of pre-processing raw images and how pre-processed images can be integrated with a geographical information system for different types of analyses via differing software packages will be introduced in Chapter 4.

- *Part II—Feature Extraction for Remote Sensing*: With the foundation of Part I, Part II aims to introduce basic feature extraction skills and discussion of the latest machine learning techniques for feature extraction. In this context, an overview of concepts and basics for feature extraction will be introduced for readers who do not have such a background (Chapter 5). The statistics or probability-based and machine-learning-based feature extraction methods will be described in sequence to entail how remote sensing products can be analyzed and interpreted for different purposes (Chapters 6 and 7). These machine learning analyses need to be validated with the aid of ground- or sea-based sensor networks that help collect reference data. Along this line, the final product can be integrated further with numerical simulation models in support of high-end research for environmental science and Earth system science.
- *Part III—Image and Data Fusion for Remote Sensing*: Following the streamlines of logic sequence in Parts I and II, Part III will focus on the concepts of image fusion with respect to current technology hubs (Chapter 8) toward an all-inclusive coverage of the most important image or data fusion algorithms (Chapter 9). These image and data fusion efforts may lead to the derivation of a series of new remote sensing data products with sound system design and the corresponding advancements must be specifically addressed and evaluated by a suite of performance-based evaluation metrics (Chapter 10). Situation refinement can be made possible with iterative work based on the knowledge developed in this part.
- *Part IV—Integrated Data Merging, Data Reconstruction, Data Fusion, and Machine Learning*: Starting at Part IV, focus is placed on large-scale, complex, and integrated data merging, fusion, image reconstruction, and machine learning scenarios. Chapter 11 offers intensive information about the concepts and tools of data merging with respect to multiple satellite sensors. Whether using data fusion or data merging, cloudy pixels cannot be fully recovered. Thus, cloudy pixel reconstruction at appropriate stages may come to elevate the accomplishment of data fusion and merging in support of machine learning (Chapter 12). With the inclusion of Chapter 12, regardless of intensity and spatial variability of cloudy pixels regionwide, the nature of cloudy pixel reconstruction with signal processing and machine learning techniques may greatly expand the general utility of fused or merged images. Chapter 13 presents a holistic discussion of integrated data fusion and machine learning for intelligent feature extraction. Chapter 14 demonstrates the highest level of synergy with a SoSE approach that enables readers to comprehend the sophistication of integrated cross-mission data merging, fusion, and machine learning algorithms flexibly toward better environmental surveillance. From various SoSE approaches, the integrated data fusion, merging, and machine learning processes may be evaluated by a set of indices for performance evaluation.
- *Part V—Remote Sensing for Environmental Decision Analysis*: Expanding upon the predictive capabilities from multisensor data fusion, merging, image reconstruction, and machine learning, the area of environmental applications may include but are not limited to air resources management (Chapter 15), water quality management (Chapter 16), ecosystem toxicity assessment (Chapter 17), land use and land cover change detection (Chapter 18), and air quality monitoring in support of public health assessment (Chapter 19). These case studies from Chapter 15 to Chapter 19 will be systematically organized in association with current state-of-the-art remote sensing sensors, platforms, tools, and methods applied for environmental decision making.

REFERENCES

Bai, K. X., Chang, N. B., and Chen, C. F., 2015. Spectral information adaptation and synthesis scheme for merging cross-mission consistent ocean color reflectance observations from MODIS and VIIRS. *IEEE Transactions on Geoscience and Remote Sensing*, 99, 1–19.

Castanedo, F., 2013. A review of data fusion techniques. *The Scientific World Journal*, 2013, 704504.

Chang, N. B., Bai, K. X., and Chen, C. F., 2015. Smart information reconstruction via time-space-spectrum continuum for cloud removal in satellite images. *IEEE Journal of Selected Topics in Applied Earth Observations*, 99, 1–19.

Chang, N. B., Bai, K. X., Imen, S., Chen, C. F., and Gao, W., 2016. Multi-sensor satellite image fusion, networking, and cloud removal for all-weather environmental monitoring. *IEEE Systems Journal*, 1–17, DOI: 10.1109/JSYST.2016.2565900.

Chen, H. W., Chang, N. B., Yu, R. F., and Huang, Y. W., 2009. Urban land use and land cover classification using the neural-fuzzy inference approach with Formosat-2 Data. *Journal of Applied Remote Sensing*, 3, 033558.

Consortium of Universities for the Advancement of Hydrologic Science, Inc. (CUAHSI), http://www.cuahsi.org/hos.html, accessed May 2011.

Hall, D. L. and Llinas, J., 1997. An introduction to multisensor data fusion. *Proceedings of the IEEE*, 85(1), 6–23.

United States Department of Defense, 2008. *Systems Engineering Guide for Systems of Systems*, http://www.acq.osd.mil/se/docs/SE-Guide-for-SoS.pdf, accessed February 13, 2016.

Volpe, G., Santoleri, R., Vellucci, V., d'Alcalà, M. R., Marullo, S., and D'Ortenzio, F., 2007. The colour of the Mediterranean Sea: Global versus regional bio-optical algorithms evaluation and implication for satellite chlorophyll estimates. *Remote Sensing of Environment*, 107, 625–638.

Zilioli, E. and Brivio, P. A., 1997. The satellite derived optical information for the comparative assessment of lacustrine water quality. *Science of the Total Environment*, 196, 229–245.

Part I

Fundamental Principles of Remote Sensing

2 Electromagnetic Radiation and Remote Sensing

2.1 INTRODUCTION

Identified by Einstein in 1905, quanta—or photons—stand for the energy packets that are particles of pure energy; such particles have no mass when they are at rest. While the German physicist Max Planck was developing the blackbody radiation law, he realized that the incorporation of the supposition that electromagnetic energy could be emitted only in "quantized" form was the key to smoothly interpreting the electromagnetic wave. This scientific discovery is the reason he was awarded the Nobel Prize in Physics in 1918. On the other hand, light must consist of bullet-like tiny particles, now known as photons, as Einstein pointed out in 1905. The photoelectric effect brought up by Einstein in 1905 successfully supplemented the quantization supposition that Planck proposed. This is also the reason Einstein was awarded the 1921 Nobel Prize in Physics. When possessing a certain quantity of energy, a photon is said to be quantized by that quantity of energy. Therefore, the well-known "wave-particle" duality entails the findings of Planck and Einstein that all forms of electromagnetic radiation (EMR) and light behave as waves and particles simultaneously in quantum mechanics. These findings imply that every quantic entity or elementary particle exhibits the properties of waves and particles, from which the properties of light may be characterized. Photons as quanta thus show a wide range of discrete energies, forming a basis for the spectrum of EMR. Quanta may travel in the form of electromagnetic waves, which provide remote sensing a classical basis for data collection.

Sunlight refers to the portion of the EMR spectrum given off by the sun, particularly in the range of infrared, visible, and ultraviolet light. On Earth, before the sunlight can reach ground level, sunlight is filtered by the atmosphere. The interactions among solar radiation, atmospheric scattering and reflections, and terrestrial absorption and emission play a key role in the ecosystem conditions at the surface of Earth. Atmospheric radiative transfer processes with the effect of transmission, absorption, reflection, and scattering have collectively affected the energy budget of the atmospheric system on Earth. For example, absorption by several gas-phase species in the atmosphere (e.g., water vapor, carbon dioxide, or methane) defines the so-called greenhouse effect and determines the general behavior of the atmosphere, which results in a surface temperature higher than zero degrees Celsius (273.15 K). In addition to the natural system, human activities have had a profound impact on the energy budget of the earth system. To some extent, air pollutants emitted by anthropogenic activities also affect the atmospheric radiative transfer processes and result in environmental effects and public health impact.

Following this deepened understanding of EMR, wavelength-dependent analyses for remote sensing data collection are often highlighted with respect to the given band specifications in the literature. Remote sensing sensor design based on specified bands and center wavelengths thus becomes feasible for collecting various images for processing, information extraction, and interpretation. Depending on the goals of each individual application, satellites onboard different sensors may be regarded as a cohesive task force to achieve a unique mission for earth observation and environmental monitoring. It is the aim of this chapter to establish a foundation by introducing a series of basic concepts and methods along this line.

2.2   PROPERTIES OF ELECTROMAGNETIC RADIATION

Quanta, or photons, travel as an electromagnetic (EM) wave at the speed of light, as shown in Figure 2.1. They have two components which are also coincident in time in an EM wave; one consists of the varying electric field and the other is the varying magnetic field. Such a wave can generate varying levels of EMR when the size and/or direction of the magnetic and electric field oscillating as sine waves mutually at right angles is changed over time at its source. The wave amplitudes of the two fields are a measure of radiation intensity, which is also called brightness. In Figure 2.1, the total number of peaks or crests at the top of the individual up-down or left-right curve passing by a reference point in one second is defined as that wave's frequency ν (read as nu), and the frequency is measured in units of cycles per second, for which the International System of Units (SI) version is Hertz [1 Hertz = s^{-1}]. The distance between two adjacent peaks on a wave is its wavelength λ and $\nu \sim 1/\lambda$. Note that c (speed of light) = $\lambda\nu$. Photons traveling at higher frequencies with smaller wavelengths are more energetic.

The level of energy that may characterize a photon can be determined with Planck's general equation. When an excited material has gone through a change from a higher energy level E_2 to a lower energy level E_1, we may use the formula below to characterize the radiant energy as:

$$E = h\nu = h(c/\lambda) \tag{2.1}$$

in which E is the radiant energy of a photon (in joules), h is Planck's constant (6.63×10^{-34} Joules-sec or Watt-sec^2), c is the speed of light ($=3 \times 10^8$ m $\cdot$ s^{-1}), λ is its wavelength (in meters), and ν represents frequency (in hertz).

2.3   SOLAR RADIATION

A beam of radiation that has photons of different energies, such as sunlight, is called a polychromatic light. The beam is monochromatic if only photons of one wavelength are involved. The EM spectrum is the range of all types of EM radiation associated with all photon energies. These types of EM radiation include gamma rays, X-rays, ultraviolet rays, visible, infrared rays, microwaves, and radio waves, which together make up the entire EM spectrum (Figure 2.2). The EM spectrum covers a wide range of photon energies, wavelengths, and regions of the EM spectrum, as summarized in Table 2.1.

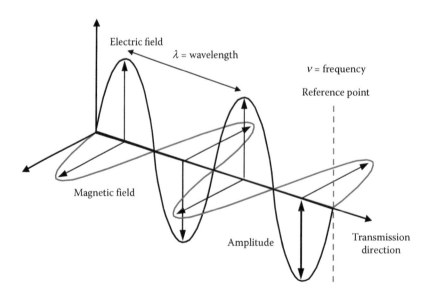

FIGURE 2.1   Structure of an electromagnetic wave.

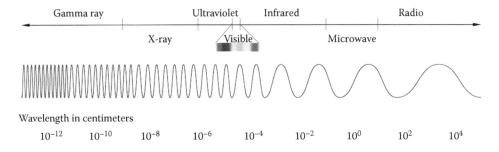

FIGURE 2.2 Comparison of wavelength, frequency, and energy for the electromagnetic spectrum. (NASA's Imagine the Universe, https://imagine.gsfc.nasa.gov/science/toolbox/emspectrum1.html, accessed March, 2013.)

TABLE 2.1

Approximate Frequency, Wavelength, and Energy Limits of the Various Regions of the EM spectrum

	Wavelength (m)	Frequency (Hz)	Energy (J)
Radio Waves	$>1 \times 10^{-1}$	$<3 \times 10^{9}$	$<2 \times 10^{-24}$
Microwave	$1 \times 10^{-3} \sim 1 \times 10^{-1}$	$3 \times 10^{9} \sim 3 \times 10^{11}$	$2 \times 10^{-24} \sim 2 \times 10^{-22}$
Infrared	$7 \times 10^{-7} \sim 1 \times 10^{-3}$	$3 \times 10^{11} \sim 4 \times 10^{14}$	$2 \times 10^{-22} \sim 3 \times 10^{-19}$
Optical	$4 \times 10^{-7} \sim 7 \times 10^{-7}$	$4 \times 10^{14} \sim 7.5 \times 10^{14}$	$3 \times 10^{-19} \sim 5 \times 10^{-19}$
Ultra Violet	$1 \times 10^{-8} \sim 4 \times 10^{-7}$	$7.5 \times 10^{14} \sim 3 \times 10^{16}$	$5 \times 10^{-19} \sim 2 \times 10^{-17}$
X-ray	$1 \times 10^{-11} \sim 1 \times 10^{-8}$	$3 \times 10^{16} \sim 3 \times 10^{19}$	$2 \times 10^{-17} \sim 2 \times 10^{-14}$

Source: NASA Goddard Space Flight Center-Science Toolbox, 2016.

All parts of the EM spectrum consist of EM radiation produced through different processes. They are detected in different ways by remote sensing, although they are not fundamentally different relative to the nature of EM radiation.

2.4 ATMOSPHERIC RADIATIVE TRANSFER

2.4.1 PRINCIPLES OF RADIATIVE TRANSFER

During the transfer of radiant energy through a medium, such as Earth's atmosphere, four types of radiative transfer processes may occur at the same time: transmission, absorption, reflection, and scattering. Scattering may be combined into the reflection category for simplicity. When considering a beam of photons from a source passing through medium 1 that impinges upon medium 2, the beam will experience some reactions of transmission, absorption, and reflection, as explained below (Figure 2.3):

- *Transmission*—Some fraction of the total radiation energy in a beam of photons may penetrate certain surface materials.
- *Absorption*—Some fraction of the total radiation energy in a beam of photons may be absorbed through molecular or electronic reactions within the medium; a fraction of this absorbed radiation energy is re-emitted at longer wavelengths as thermal emissions, and the rest of it remains while heating the target.
- *Reflection*—Some fractions of the total radiation energy in a beam of photons reflect at specific angles moving or scattering away from the target at different angles, depending on the incidence angle of the beam and the surface roughness.

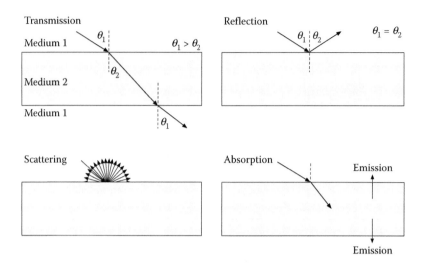

FIGURE 2.3   Radiative transfer processes through a medium.

For energy conservation, the summation of the fraction of the total radiation energy associated with transmission, absorption, and reflection must be equal to 1.

Similarly, the atmosphere filters the energy delivered by the sun and emitted from Earth while performing radiative transfer. A series of radiative transfer processes may collectively describe the interaction between matter and radiation; such interactions might involve matter such as gases, aerosols, and cloud droplets in the atmosphere and the four key processes of absorption, reflection, emission, and scattering. Whereas scattering of an incident radiation by the atmospheric matter results in a redistribution of the radiative energy in all directions, absorption of an incident radiation by the atmospheric matter results in a decrease of radiative energy in the incident direction. Figure 2.4 conceptually illustrates the radiative transfer processes through an atmospheric layer. In such

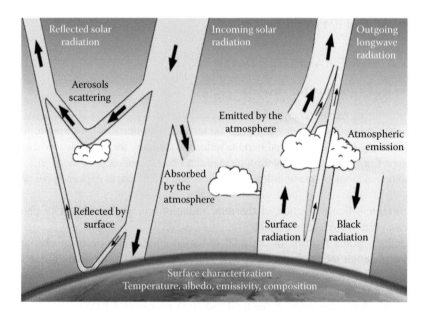

FIGURE 2.4   Radiative transfer processes through an atmospheric layer. (NASA Goddard Space Flight Center, PSG, 2017. https://psg.gsfc.nasa.gov/helpmodel.php, accessed December, 2017.)

an environment, whether radiation is absorbed or transmitted depends on the wavelength and the surface properties of the matter in the atmospheric environment. When an air-borne or a space-borne sensor views Earth, reflection and refraction are two key radiative transfer processes; these two processes are discussed in detail below.

2.4.2 REFLECTION

Our ability to see luminous objects with our eyes depends on the reflective properties of light, as does an air-borne or a space-borne sensor. In the earth system, types of reflections at the surface of Earth include specular and diffuse reflection (Figure 2.5). Factors affecting surface reflectance on Earth include absorption features (e.g., water, pigments, and minerals) at ground level, surface roughness, and observation and illumination angles. Specular reflection occurs on smooth surfaces (Figure 2.5a) whereas varying surface roughness may result in Lambertian or diffuse reflectance (Figure 2.5b). Note that diffuse reflection that is also termed Lambertian bidirectional reflection distribution function (BRDF) occurs on rough surfaces (Figure 2.5b) such as forest or agricultural fields. In diffuse reflection, the roughness of the surface results in variations of the normals along the surface; however, all of the reflected rays still behave according to the law of reflection, which states that the incident ray, the reflected ray, and the normal to the surface of the mirror all lie in the same plane. The BRDF depends on wavelength and the BRDF function is based on illumination geometry and viewing geometry, which is determined by the optical and structural properties of the surface. These properties include but are not limited to: multiple scattering, facet orientation distribution, facet density, shadow casting, mutual shadowing, reflection, absorption, transmission, and emission by surface objects. Hence, BRDF is related to Lambertian reflection, which defines how light reflected at an opaque surface differs from what we may see with our eyes with respect to the same scene when Earth moves over different positions relative to the sun.

2.4.3 REFRACTION

Refraction is a light movement effect that happens between transparent media of different densities in which the transparent media can be air, water, or even snow on dearth (Robinson, 1997). The bending effect of light in association with the refraction media is a physical representation of the longer time it takes for light to move through the denser of two media (Figure 2.6). The level of refraction is dependent on the incident angle with which the ray of light strikes the surface of the medium. Given that the temperature of the atmospheric layers may vary with height, this effect could affect the density of air, and bias of the signals collected by remote sensing sensors could impact the measurement accuracy.

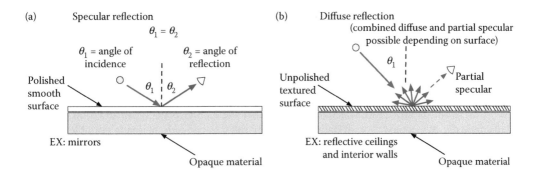

FIGURE 2.5 Types of reflections.

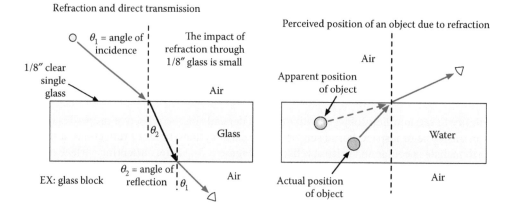

FIGURE 2.6   Types of refraction layers.

2.5   REMOTE SENSING DATA COLLECTION

2.5.1   ATMOSPHERIC WINDOWS FOR REMOTE SENSING

In earth system science, the interactions among solar radiation, atmospheric matter, and terrestrial objects play a critical role for conditions at Earth's surface. Line-of-sight is defined as an imaginary line from the sensor to a perceived object. When a satellite sensor has a line-of-sight with an object that is reflecting solar energy, the sensor collects that reflected energy and stores the observation for further analysis. In summary, the radiative transfer processes that account for the reflected energy include: (1) reflection, (2) emission, and (3) emission-reflection. As shown in Figure 2.4, emission is associated with outgoing longwave radiation that is EMR emitted by the Earth-atmosphere system in the form of thermal radiation. As mentioned before, EM propagation of the electromagnetic radiation in an atmosphere is affected by the state of the atmospheric conditions and the atmospheric composition. Whereas the former is related to temperature, pressure, and air density, the latter is associated with gases and particulates in the air. Gases can have a strong scattering effect on incoming radiation. Gases can absorb (and/or emit) a fraction of this incoming radiation energy depending on their molecular structure. In addition, particulates including aerosol and clouds in the air can also scatter radiation energy, and some of them can absorb (and/or emit) radiation energy depending on their refractive properties, which are driven by composition. Consequently, reflected solar radiation plus emission and emission-reflection radiation as measured by satellite sensors can produce remote sensing data and tell us about the physical, chemical, and biological properties of objects in the air or at ground level.

Remote sensing data are collected by multiple channels or bands which divide the EMR ranging from ultraviolet to radio waves (Table 2.1). These remote sensing data are called multichannel data, multiband data or multispectral data. Remote sensing data are then digitalized through a process of sampling and quantization of the electromagnetic energy detected and gathered by space-borne, air-borne, sea-based, or ground-based sensors. To save the information embedded over different wavelengths, remote sensing data use different quantization levels in the process of data gathering, digitization, and storage. Remote sensing data are always compared with other data collected by the same or different satellite platforms to decide whether the quality is good or bad based on selected quality indexes. Yet comparison of data with quality indexes can only be made possible when both use the same quantization levels.

Remote sensing data collection is limited by atmospheric windows, however. This implies that Earth's atmosphere prevents several types of solar EM from reaching Earth's surface.

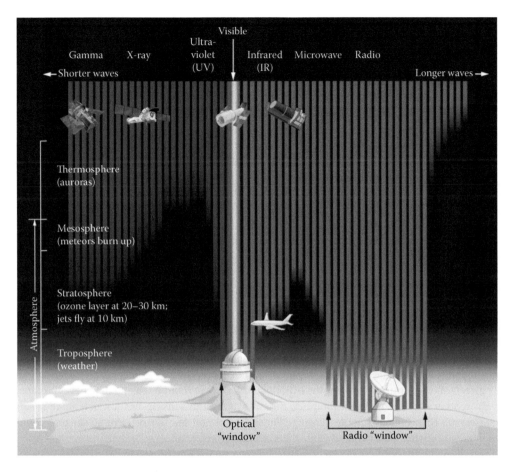

FIGURE 2.7 Atmospheric windows and radiative transfer. (Modification of work by STScI/JHU/NASA.)

The illustration of Figure 2.7 shows how far different portions of the EM spectrum can move forward before being absorbed in the atmosphere. It is notable that only portions of visible light, infrared, and some ultraviolet light can reach Earth's ground surface or make it to sea level. EM radiation from space that is able to reach the surface of Earth through the atmosphere window provides a wealth of ground-leaving or water-leaving reflectance data for remote sensing sensors to collect.

Such atmospheric windows deeply affect the assessment of the environmental sustainability of the earth system. For instance, the stratosphere, located in the upper atmosphere, and the troposphere, located in the lower atmosphere, are chemically identical in terms of ozone molecules. However, these ozone molecules have very different roles in these two layers and very different effects on life systems on the surface of Earth. As shown in Figure 2.7, stratospheric ozone filters most of the solar ultraviolet radiation, playing a beneficial role by absorbing most of the biologically damaging ultraviolet sunlight, known as UV-B. Ozone near the ground surface in the tropospheric layer not only lacks the filtering action of the ozone layer, but is also toxic to life on Earth. In addition, the change of water-leaving reflectance associated with different wavelengths of visible light can be regarded as a surrogate index for monitoring water pollution. Terrestrial thermal emissions are correlated with the evapotranspiration process through all plant species, which may be monitored by remote sensing to understand the ecosystem status.

2.5.2 Specific Spectral Region for Remote Sensing

Light can be used to "see" an object at the Earth's surface if and only if light has a wavelength about the same order of magnitude as or smaller than the object. Given the allowable penetration of the EM waves through limited atmospheric windows, the approximate scale of the wavelength at the objective level can be associated with a radiation-type wavelength for remote sensing design. For example, visible light and infrared regions of the spectrum may span the wavelengths suitable for studying human society and the ecosystem, which can then be used to support a myriad of earth science applications. Many satellite sensors are designed based on this range of the EM spectrum. However, remote sensing associated with the visible light and infrared regions of the spectrum works with two different types of EMR (Eismann, 2012). One works with the reflective spectral region, and the other works with the emissive spectral region (i.e., thermal radiation) (Figure 2.8) (Eismann, 2012). This is due to the nature of the thermal equilibrium at the ground level. In fact, remote sensing sensors can measure these two types of radiation sources from the reflective spectral and/or emissive spectral region in terms of spectral radiance at the sensor (Figure 2.8).

2.5.3 Band Distribution for Remote Sensing

According to Figure 2.7, remote sensing sensors may be designed to measure reflected radiation over the microwave, visible light, and infrared regions of the spectrum throughout the appropriate atmospheric windows. Microwave radiation with longer wavelengths can penetrate through cloud, haze, and dust because the longer wavelengths are not susceptible to atmospheric scattering, which only affects shorter optical wavelengths. Although microwave remote sensing may have the capacity for all-weather and day or night imaging, microwave remote sensing has a much smaller signal-to-noise ratio than optical remote sensing. Thus, optical and microwave remote sensing could be complementary when observing objects at the ground or sea level.

Overall, band distribution in specific spectral regions for sensor design over these atmospheric windows can be a challenging task. While the band distribution in visible light and infrared regions of the spectrum may be designed to meet the needs of various hyperspectral and multispectral remote sensing applications flexible to any central wavelength, microwave remote sensing is normally designed based on the fixed band distribution (Table 2.2).

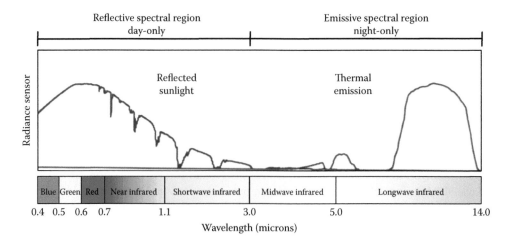

FIGURE 2.8 Definition of specific spectral regions and visible and infrared radiation.

TABLE 2.2

Band Distribution of Microwave Remote Sensing and Related Properties

Designation	Wavelength Range	Frequency (GHz)	Applications
Ka	5.0–11.3 mm	26.5–40	Earth observation, satellite communications
K	11.3–16.7 mm	18–26.5	Earth observation, radar, satellite communications
Ku	16.7–25 mm	12–18	Earth observation, satellite communications
X	25–37.5 mm	8–12	Satellite communications, radar, terrestrial broadband
C	3.75–7.5 cm	4–8	Earth observation, long-distance radio telecommunications
S	7.5–15 cm	2–4	Weather radar, satellite communications, surface ship radar
L	15–30 cm	1–2	Earth observation, global positioning system, mobile phones
P	30–100 cm	>0.39	Earth exploration for various applications

2.6 RATIONALE OF THERMAL REMOTE SENSING

2.6.1 Thermal Radiation

Some types of radiation processes in remote sensing may account for thermal emissions. Blackbody temperature plays a critical role in better understanding thermal emissions. As mentioned before, Planck's major research back in the 1910s dealt with blackbody temperature. A blackbody that is a perfect absorber (emitter) of radiation is one that absorbs all the EMR falling on it at all wavelengths, and a blackbody emits more radiation than any other type of object at the same temperature. To stay in a thermal equilibrium, it must emit radiation at the same rate as it absorbs. Blackbody radiation at a given wavelength depends on the temperature only. Radiation emitted by a blackbody is homogeneous, isotropic, and unpolarized. Any two blackbodies emit the same radiation at the same temperature. In this context, thermodynamic equilibrium is defined as the state of radiation and matter inside an enclosure under isolated constant temperature. However, the atmosphere is not in thermodynamic equilibrium because its pressure and temperature vary in terms of position. Hence, the atmosphere is usually subdivided into small subsystems, assuming that each subsystem holds the isobaric and isothermal condition generally referred to as local thermodynamic equilibrium.

In fact, every object emits electromagnetic radiation based on its temperature. In Figure 2.9, the temperature of objects at which this radiation emits the most intense wavelength is highlighted. Emitted radiation can be reflected by the particles in the air. However, objects around room temperature radiate mainly in the infrared, as seen in Figure 2.9, in which the graph shows examples for three objects including the sun, Earth, and a candle, with widely different temperatures given in Kelvin (K = C + 273.15). Solar irradiance is defined as the power per unit area received from the Sun in the form of EMR in the wavelength range of the measuring instrument. The solar irradiance integrated over time is called solar irradiation, insolation, or solar exposure. Total irradiance at the Top of Atmosphere (TOA) level on Earth in this case is defined as the sum of solar irradiance, direct irradiance, and diffuse irradiance. For example, the solar radiation spectrum is described by the blackbody radiation curve on the top, which has the temperature of the sun (close to 6,000 K). The peak of the sun's energy emission is exactly in the range of wavelengths of visible light, although it also emits a portion of infrared and ultraviolet. Conversely, as shown in the green curve in the same graph, the peak of Earth's energy emission is in the range of wavelengths of infrared only. Given that the green curve stands for Earth's condition, it is indicative that Earth emits orders of magnitude less radiation and that radiation peaks are mainly in the infrared region. This will differentiate the type of emission from the same object experiencing differing temperatures.

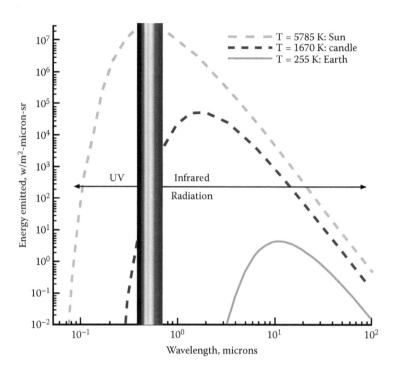

FIGURE 2.9 Comparisons of the spectral emittance from a 6,000 K Blackbody with a candle and total irradiance at Earth's surface at the TOA level. (NASA Data Center, 2016)

2.6.2 ENERGY BUDGET AND EARTH'S NET RADIATION

The solar constant is the portion of total radiation that is emitted by the sun and reaches the TOA on Earth. In fact, it is not a constant, but is a function of several factors such as sun spots, sun activity, and the distance between sun and Earth. As discussed in Section 2.4, sunlight will pass through the atmospheric layers and reach the ground level after experiencing the scattering and reflection effects due to gaseous species and particles in the air. After reaching the ground level, a fraction of the incident sunlight reflected from Earth's surface will go back to the atmosphere. The residual sun radiation not reflected is absorbed by the surface. This part of the energy that evaporates water as well as sublimates/melts snow and ice with latent heat raises the surface temperature with sensible heat, and energizes the heat exchange in the turbulent boundary layer, which affects the weather conditions. The surface albedo that accounts for the reflection of radiation that is a function of both surface and atmospheric properties is a key attribute in remote sensing from space.

In general, the state of the atmosphere is determined by its energy budget, which is comprised of the solar and terrestrial radiation fluxes as well as the latent and sensible heat fluxes, all of which are related to EMR. Note that flux can be used to delineate longwave radiation, shortwave radiation or the total of shortwave and longwave radiation that is moving up or down in the earth system. While latent heat is related to changes in phase between gases, liquids, and solids, sensible heat is related to changes in temperature of an object or a gas without any change in phase. Broadly speaking, sensible heat is the radiative flux caused by conduction and convection. For instance, heat will be conducted into the atmosphere and then convection will follow when a warm front meets a cold front in the planetary boundary layer. Latent heat is also regarded as the heat moved up into the atmosphere by water evaporation and condensation processes. The rate of heat penetration into the soil depends on the thermal diffusivity, which is defined as the thermal conductivity divided by the heat capacity. In fact, the thermal conductivity stands for how well heat is conducted through the soil layer, and the heat capacity is related to how much heat it takes to increase the unitary temperature of the soil layer. The lower the value of the thermal diffusivity, the less the temperature rises further into the soil, and the higher the reflected radiation into the atmosphere. Net radiation in this context is the amount of energy actually added to the earth system.

Earth's net radiation is the balance between outgoing and incoming energy at the TOA level. The solar energy arriving at the surface can vary from 550 W/m² with cirrus clouds to 1025 W/m² with a clear sky (Krivova et al., 2011). Earth and the atmosphere absorb 341 W/m² of solar radiation on average annually (Johnson, 1954). In view of the energy budget delineation in Figure 2.4, outgoing longwave radiation is EMR emitted from Earth and its atmosphere out to space in the form of thermal radiation through both soil layers and atmospheric layers. Most of the outgoing longwave radiation has wavelengths (from 4 to 100 μm) in the thermal infrared part of the electromagnetic spectrum. In fact, our planet's climate is driven by absorption, reflection, shortwave or longwave emission, and scattering of radiation within the atmosphere due to the presence of thin clouds, aerosol, and some gases. Cases can be seen in some extreme weather events such as tropical storms and hurricane assessment.

2.7 BASIC TERMINOLOGIES OF REMOTE SENSING

Measurements of solar radiation are based on a few parameters in remote sensing defined below (Figure 2.10):

- *Azimuth*: Geographic orientation of a line given as an angle measured in degrees clockwise from the north.
- *Azimuth direction*: This is the direction of a celestial object, measured clockwise around the observer's horizon from the north. In air-borne or space-borne radar imaging, the direction in which the aircraft or satellite is heading is the azimuth direction, which is the same as the flight direction.
- *Zenith angle*: Zenith angle is the angle between the sun and the vertical and it is a complementary angle to solar elevation or altitude.
- *Nadir point*: Nadir point is the direction at the ground level directly in line with the remote sensing system and the center of the earth. In other words, the nadir at a given point is the local vertical direction pointing in the direction of the force of gravity at that location.
- *Off-Nadir*: Any point not directly beneath a scanner's detectors, but rather off at an angle.
- *Solar azimuth angle*: The azimuth angle of the sun.

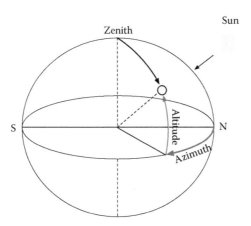

FIGURE 2.10   Basic terminologies of remote sensing for earth observation.

- *Top of the atmosphere (TOA)*: TOA is defined as the outermost layer of Earth's atmosphere, which is the upper limit of the atmosphere—the boundary of Earth that receives sun radiation.
- *Albedo*: Albedo means whiteness in Latin, and refers to the fraction of the incident sunlight that the surface reflects. The residual radiation not reflected is then absorbed by the surface.
- *Spherical albedo of the atmosphere*: Spherical albedo is the average of the plane albedo over all sun angles. Spherical albedo of the atmosphere is the effective albedo of an entire planet that is the average of the plane albedo over all sun angles at TOA.

2.8 SUMMARY

In this chapter, some basic properties of light and concepts of EMR are introduced sequentially to support the basic understanding of remote sensing. The discussion is followed by remote sensing data collection conditional to atmospheric windows and specified band regions. In addition, the global energy budget in relation to thermal radiation is presented to provide a complementary view of thermal emission relative to sun light reflection. The chapter ends by including the basic terminologies of remote sensing for environmental monitoring and earth observation.

REFERENCES

Eismann, M. T., 2012. *Hyperspectral Remote Sensing*. SPIE Press, Bellingham, Washington, USA.

Johnson, F. S., 1954. The solar constant. *Journal of Meteorology*, 11, 431–439.

Krivova, N. A., Solanki, S. K., and Unruh, Y. C., 2011. Towards a long-term record of solar total and spectral irradiance. *Journal of Atmospheric and Solar-Terrestrial Physics*, 73, 223–234.

Robinson, D. A., 1997. Hemispheric snow cover and surface albedo for model validation. *Annals of Glaciology*, 25, 241–245.

National Aeronautics and Space Administration (NASA) Data Center, https://mynasadata.larc.nasa.gov/radiation-energy-transfer/, accessed June, 2016.

National Aeronautics and Space Administration (NASA) Goddard Space Flight Center, http://imagine.gsfc.nasa.gov/science/toolbox/emspectrum1.html, accessed June, 2016.

National Aeronautics and Space Administration (NASA) Goddard Space Flight Center, PSG, 2017. https://psg.gsfc.nasa.gov/helpmodel.php, accessed December, 2017.

National Aeronautics and Space Administration (NASA) Goddard Space Flight Center, http://imagine.gsfc.nasa.gov/science/toolbox/emspectrum1.html, accessed August, 2016.

3 Remote Sensing Sensors and Platforms

3.1 INTRODUCTION

Solar energy moves from the sun to Earth and finally to the satellite sensors onboard a variety of platforms for measurement. Remote sensing images and data provide critical information about how the natural system can be sustained over time. Such radiative transfer processes reveal how the solar energy is partitioned into different compartments in the natural system. Remote sensing images and data with different spatial, spectral, radiometric, and temporal resolution need to be pre-processed, retrieved, analyzed, interpreted, and mapped in an iterative and holistic way to support various types of decision analysis for sustainable development. In cases that require involving more than one satellite for applications, automated data merging and/or fusion processes for dealing with challenging problems are critical for supporting human decision making, which requires linking data with information, knowledge discovery, and decision analysis to achieve timely and reliable projections of a given situation in a system (Figure 3.1), such as climate change impact.

Consequently, as mentioned in Chapter 2, the following energy partition terminologies in the radiative transfer processes in the natural environment deeply influence the system design of both sensors and platforms and deserve our attention:

- *Transmitted energy*—The energy that passes through a medium with a change in the velocity of the light as determined by the refraction index for two adjacent media of interest.
- *Absorbed energy*—The energy that is surrendered to the target through electron or even molecular reactions.
- *Reflected energy*—The energy bounced back with an angle of incidence equal to the angle of reflection.
- *Scattered energy*—The energy that is diffused into the air with directions of energy propagation in a randomly changing condition. Rayleigh and Mie scattering are the two major types of scattering in the atmosphere.
- *Emitted energy*—The energy that is first absorbed, then re-emitted as thermal emissions at longer wavelengths while the target, such as the ground level, heats up.

As mentioned, remote sensing images and data play a critical role in understanding the solar energy paths and extracting features associated with targets. Before conducting data merging and/or fusion, which fuses or merges images of different spatial, temporal, and spectral resolution, there is a need to understand important functionalities of different sensors and platforms individually or collectively. This chapter thus aims to investigate quality sensors and platforms capable of supporting multisensor data merging and/or fusion based on synthetic aperture radar, infrared, and optical remote sensing images and data. The following sections present different classification principles of sensors and platforms to establish a fundamental understanding of remote sensing systems as well as the current, historic, and future missions of remote sensing with inherent connections. These sensors and platforms are regarded as enabling technologies for monitoring solar radiation and improving our comprehension of the interactions between solar energy and materials over different wavelengths at the ground level or in atmospheric layers. The sensor, bands, spatial resolution, swath width, spectral range, and temporal resolution associated with a suite of major multispectral remote

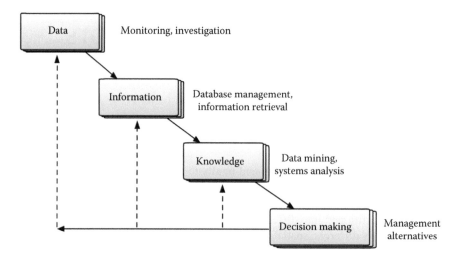

FIGURE 3.1   Contribution of remote sensing images and data to sustainable development.

sensing platforms are highlighted for demonstration. An understanding of this relevant knowledge may lead to optimizing the system planning and design of data merging and/or fusion, meeting the overarching goal of the system of systems engineering, when more sensors and platforms with different features become available and join the synergistic endeavor.

3.2   REMOTE SENSING PLATFORMS

A remote sensing system is an integrated software and hardware system structured by sensors, instruments, platforms, receiving stations, data storage centers, and a control room. Platforms in such systems have normally included various mounted sensors and grouped instruments such as global positioning systems. The platform design, which is geared toward the optimal payload distribution, determines a number of attributes. The payload for remote sensing satellites, for example, can typically include photographic systems, electro-optical sensors, and microwave systems. More than one sensing system can be housed on a single satellite to increase the benefit of simultaneous coverage by different sensors. There are auxiliary devices for recording, pre-processing, compressing, and transmitting the data in addition to these sensing systems. These attributes include periodicity and timing of image acquisition, orbiting geometry or flight path, types of sensors and instruments for the viewing object of interest, the longitudinal extent of the swath, and even the trade-off between swath extent (coverage) and resolution.

There are three broadly based categories of remote sensing platforms: ground- or sea-based, airborne, and space-borne platforms. They are discussed below.

3.2.1   SPACE-BORNE PLATFORMS

The most popular platform for remote sensing aloft is a space-borne satellite. Over three-thousand remote sensing satellites have been launched since 1957 at which Russia launched the first man-made satellite of Sputnik 1. In addition, the space shuttle, which functions as a remote sensing satellite, belongs to this category. However, the space shuttle can be reused for multiple missions, unlike satellites. The path of a satellite in space is referred to as its orbit. Satellites can be classified based on either orbital geometry or timing for image acquisition. Two types of orbits, including geostationary/equatorial and polar/Sun synchronous, are commonly used as a broad guideline for the classification of remote sensing satellites (Natural Resources Canada, 2017). These orbits are fixed after launch and can be only slightly adjusted to maintain their anticipated position for environmental monitoring and earth observation over time. The type of orbit that affects the design of the sensor onboard determines its altitude with respect to Earth and the limit of its instantaneous field of view (i.e., the area on Earth which can be viewed at any moment in time).

In general, geostationary or equatorial satellites are designed to have a period of rotation equal to 24 hours, the same as that of the Earth, making these satellites consistently stay over the same location on the top of Earth. These geostationary satellites must be placed at a very high altitude (~36,000 km) to maintain an orbital period equal to that of Earth's rotation and appear to be stationary with respect to Earth, as illustrated in Figure 3.2. Due to the stationarity, any sensor onboard these satellites can only view the same area of Earth over a very large area because of the high altitude. Such satellites normally circle Earth at a low inclination in an equatorial orbit (i.e., inclination is defined as the angle between the orbital plane and the equatorial plane). This type of system design of geostationary orbits can meet the needs for communications and weather monitoring, hence many of them are located over the equator. However, the space shuttle chose an equatorial orbit with an inclination of 57 degrees. The space shuttle has a low orbital altitude of 300 km, whereas other common polar satellites typically maintain orbits ranging from 200 to 1,000 km.

Polar-orbiting or sun-synchronous satellites are designed to pass above (i.e., polar) or nearly above (i.e., sun-synchronous or near-polar orbits) each of Earth's poles periodically. Polar or sun-synchronous orbits are thus the most common orbits for remote sensing due to the need to provide illumination for passive sensors. Note that although active sensors such as LiDAR and radar do not need the Sun's illumination for image acquisition, passive sensors count on solar energy as a source of power. We will define active and passive sensors later in this chapter. Both types of polar-orbiting satellites with similar polar orbits can pass over the equator at a different longitude at the same local sun time on each revolution, as illustrated in Figure 3.2. The satellite revisit time (or revisit interval or revisit period) is the time elapsed between two successive observations of the same point on Earth, and this time interval is called the repeat cycle of the satellite. Each repeat cycle enables a polar-orbiting satellite to eventually see every part of Earth's surface. A satellite with a near-polar orbit that passes close to the poles can cover nearly the whole earth surface in a repeat cycle depending on sensor and orbital characteristics. For most polar-orbiting or sun-synchronous satellites, the repeat cycle ranges from twice a day to once every 16 days. Real-world examples include Landsat and the well-known Earth Observing System (EOS) series satellites such as Terra and Aqua. Such an attribute of global coverage is often required for holistic earth observation.

Data collected by most remote sensing satellites can be transmitted to ground receiving stations immediately or can be temporarily stored on the satellite in a compressed form. This option depends on whether the receiving station has a line of sight to the satellite when the satellite wishes to

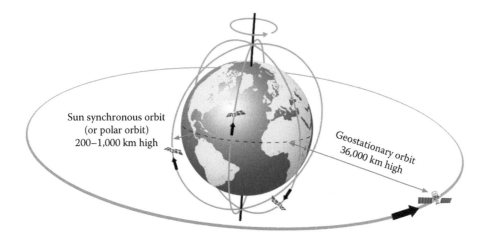

FIGURE 3.2 Orbital geometry of space-borne satellites.

transmit the data. If there are not enough designated receiving stations around the world to be in line with the satellite, data can be temporarily stored onboard the satellite until acquiring direct contact with the ground-level receiving station. Nowadays, there is a network of geosynchronous (geostationary) communications satellites deployed to relay data from satellites to ground receiving stations, and they are called the Tracking and Data Relay Satellite System (TDRSS). With the availability of TDRSS, data may be relayed from the TDRSS toward the nearest receiving stations without needing to be stored temporarily onboard the satellite.

3.2.2 Air-Borne Platforms

Air-borne platforms collect aerial images with cameras or sensors; airplanes are currently the most common air-borne platforms. When altitude and stability requirements are not limiting factors for a sensor, simple, low-cost aircraft such as an Unmanned Aerial Vehicle (UAV) can be used as platforms, too. If instrument stability and/or higher altitude requirements become essential, more sophisticated high-altitude aircraft platforms that can fly at altitudes greater than 10,000 meters above sea level must be employed for remote sensing applications. Included in this case are the fixed-wing, propeller-driven planes handled by pilots. Although they are limited to a relatively small area, all air-borne platforms are more suitable for acquiring high spatial resolution data. Remote sensing instruments may be either outfitted in the underside of the airplane or simply hung out the door of the airplane using simple mounts. On the other hand, mid-altitude aircraft have an altitude limit under 10,000 meters above sea level, and are used when stability is demanded and when it is essential for acquiring imagery remotely that would not be available from low-altitude aircraft such as helicopters or UAV. A real-world example includes the C-130 air-borne platform, owned by the National Center for Atmospheric Research (NCAR) and the National Science Foundation (NSF) in the United States.

3.2.3 Ground- or Sea-Based Platforms

Ground- or sea-based platforms with instruments can be used to measure indirect illumination, which is also known as diffuse illumination at the surface of Earth, from the sky as opposed to direct sunlight. They can be handheld devices, tripods, towers, buoys or cranes on which the instruments are mounted. They are used for monitoring not only atmospheric phenomenon, but also long-term terrestrial or oceanic features for close-range characterization of objects. Real-world examples include: (1) the NSF-owned ground-based, polarimetric weather radars that provide us with measurements of precipitating cloud systems with different movement, structure, and severity; and (2) the United States Department of Agriculture-owned Ultraviolet-B Monitoring Climatological and Research Network.

3.3 REMOTE SENSING SENSORS

Sensors can be onboard satellites or airplanes or mounted at ground- or sea-based stations for measuring electromagnetic radiation at specific wavelength ranges, usually named bands. Remote sensing systems measuring energy distribution that is naturally available through radiative transfer processes are called passive sensors (Figure 3.3a). Because the sun's energy is either reflected, as it is for visible wavelengths, or absorbed and then re-emitted, as it is for thermal infrared wavelengths, the energy being measured by passive sensors may be associated with visible light, infrared, and/ or the microwave spectrum. In remote sensing systems, these measurements are quantized and converted into a digital image, where each pixel (i.e., picture element) has a discrete value in units of Digital Number (DN). These space-borne passive satellite sensors operate in the predesigned frequency bands allocated to the satellites, and these bands are driven by fixed physical properties such as molecular resonance of the object being monitored. Once designed, these frequencies do not change and information embedded in these bands is unique and cannot be duplicated in other

(a)   (b)

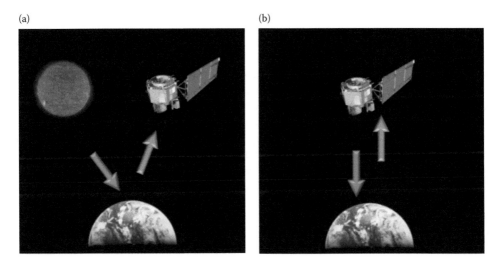

FIGURE 3.3   The difference between (a) passive sensors and (b) active sensors. (National Aeronautics and Space Administration (NASA), 2012. https://www.nasa.gov/directorates/heo/scan/communications/outreach/funfacts/txt_passive_active.html, accessed May 2017)

frequency bands. On the other hand, active sensors send out their own energy for illumination. This means the sensors can emit their own radiation directed toward the target of interest. The reflected radiation from that target of interest is detected and measured by the sensor (Figure 3.3b).

Regardless of the type of sensors that are radiation-detection imagers, the resulting images embedded with these DN values associated with each band have different resolutions, as summarized below.

- *Spatial resolution*: Spatial resolution is usually measured in pixel size, depending on focal length, detector size, and sensor altitude. Spatial resolution is a key factor required for the discrimination of essential features.
- *Spectral resolution*: Spectral resolution is the density of the spectral bands in the electromagnetic spectrum of multispectral or hyperspectral sensors; each band corresponds to an image.
- *Radiometric resolution*: Radiometric resolution, usually measured in binary digits (bits), is the range of available brightness values corresponding to the maximum range of DNs in the image, specifying the ability of a sensor to distinguish the differences in brightness (or grey-scale values) while acquiring an image. For example, an image with 8-bit resolution has 256 levels of brightness (Richards and Jia, 2006).
- *Temporal resolution*: Temporal resolution is the time required for revisiting the same area of Earth (NASA, 2013).

Most of the current satellite platforms for remote sensing follow polar or near-polar (i.e., sun-synchronous) orbits. These satellites travel to the northern pole on one side of Earth (ascending passes) and then proceed toward the southern pole on the second half of their orbital paths (descending passes). If the orbit is a sun-synchronous orbit rather than a pure polar orbit, the descending mode is on the sunlit side, while the ascending mode is normally on the shadowed side of Earth. Passive optical sensors onboard these satellites may record reflected images from the surface on a descending pass, when solar illumination is available. Unlike active sensors, which count on their own illumination, passive sensors only record emitted radiation (e.g., thermal radiation) and can also image the surface of Earth on ascending passes. Active and passive sensors are further classified in detail below.

3.3.1 PASSIVE SENSORS

Passive sensors used in remote sensing systems measure the radiance corresponding to the brightness along the direction toward the sensor. Due to the radiative transfer processes, sensors are used to measure the sum of direct and indirect reflectance, defined as the ratio of reflected versus total power energy, through a set of spectral reflectance curves for different targets (Figure 3.4). The spectral signature is the reflectance that is a function of wavelength, in which each target material has a unique signature useful for feature extraction and classification. Passive sensors have a variety of applications related to ecology, urban planning, geology, hydrology, meteorology, environmental science, and atmospheric science. For example, multispectral remote sensing images have been used widely to monitor water quality (Arenz et al., 1995; Agren et al., 2008; Chang et al., 2013; Chang et al., 2014) and air quality (King and Byrne, 1976; King et al., 1999; Lee et al., 2012; Li et al., 2012; Li et al., 2015) issues in different regions.

Space-borne passive satellite sensors provide the capability to acquire global observations of Earth. Most passive sensors can detect reflectance (i.e., visible light), emission (i.e., infrared and thermal infrared), and/or microwave portions of the electromagnetic spectrum by using different types of radiometers and spectrometers. These radiometers may include, but are not limited to:

- *Radiometer*: A remote sensing instrument that measures the intensity of electromagnetic radiation in some visible, infrared or microwave bands within the spectrum.
- *Hyperspectral radiometer*: An advanced remote sensing instrument that can detect hundreds of narrow spectral bands throughout the visible, near-infrared, and mid-infrared portions of the electromagnetic spectrum. This type of sensor, which has very high spectral resolution, can conduct fine discrimination among different targets based on their spectral response in association with each of the narrow bands.
- *Imaging radiometer*: A remote sensing instrument that has a scanning capability to generate a two-dimensional array of pixels from which an image of targets may be produced. Scanning can be carried out electronically or mechanically by using an array of detectors.

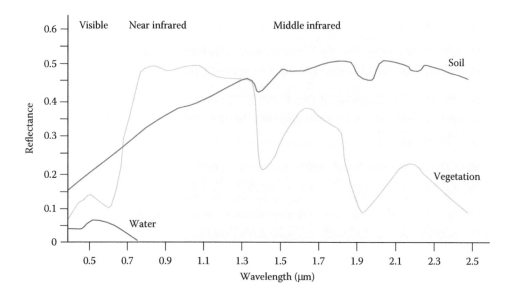

FIGURE 3.4   Reflectance spectra of soil, water, and vegetation.

In addition, these spectrometers may include, but are not limited to:

- *Sounder*: A remote sensing instrument that is designed to measure vertical distributions of specific atmospheric parameters such as composition, temperature, humidity, and pressure from multispectral information.
- *Spectrometer*: A remote sensing instrument that is designed to detect, measure, and analyze the spectral content of incident electromagnetic radiation. Traditional imaging spectrometers count on prisms or gratings to disperse the radiation for image discrimination analysis of spectral content.
- *Spectroradiometer*: A radiometer that is designed to detect and measure the intensity of radiation in multispectral bands or hyperspectral bands. Multispectral remote sensing has only limited bands available.

The difference between multispectral and hyperspectral imaging is illustrated in the diagram shown below (Figure 3.5). Broadband sensors typically produce panchromatic images with very wide bandwidths, typically 400–500 nanometers. For example, WorldView-1 produced panchromatic images with a high spatial resolution of 50 centimeters. Most multispectral imagers have four basic spectral bands, including blue, green, red, and near-infrared bands. Some multispectral imaging satellites, such as Landsats 7 and 8, have additional spectral bands in the shortwave infrared (SWIR) region of the spectrum. Hyperspectral imaging systems are designed to obtain imagery over hundreds of narrow, continuous spectral bands with typical bandwidths of 10 nanometers or less. For example, the NASA JPL AVIRIS air-borne hyperspectral imaging sensor obtains spectral data over 224 continuous channels, each with a bandwidth of 10 nm over a spectral range from 400 to 2,500 nanometers. Ultraspectral sensors represent the future design of hyperspectral imaging technology.

3.3.2 Active Sensors

An active sensor in remote sensing systems is a radar, laser or acronym for light detection and ranging (LiDAR) instrument used for detecting, measuring, and analyzing signals transmitted by the sensor,

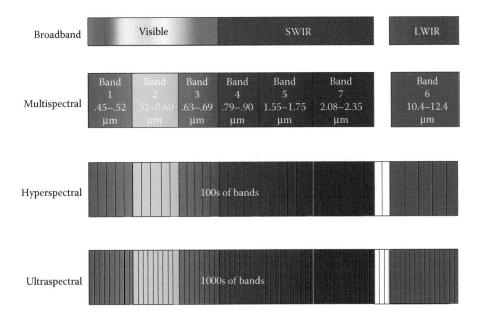

FIGURE 3.5 The comparison among broadband, multispectral, hyperspectral, and ultraspectral remote sensing.

which are reflected, refracted or scattered back by the surface of Earth and/or its atmosphere. The majority of active sensors operate in the microwave portion of the electromagnetic spectrum, and the frequency allocations of an active sensor from Ka band to L band (Table 2.2) are common to other radar systems. Some active sensors are specifically designed to detect precipitation, aerosol, and clouds, simply based on the radar echos. Active sensors have a variety of applications related to hydrology, meteorology, ecology, environmental science, and atmospheric science. For example, precipitation radars, either ground-based (Zrnic and Ryzhkov, 1999; Wood et al., 2001), air-borne (Atlas and Matejka, 1985), or space-borne (Anagnostou and Kummerow, 1997), are designed to measure the radar echo intensity (reflectivity) measured in dBZ (decibels) from rainfall droplets in all directions to determine the holistic rainfall rate over the surface of Earth. These radar, LiDAR, and laser sensors may include, but are not limited to:

- *LiDAR*: A Light Detection And Ranging active sensor (LiDAR) is designed to measure the distance to a target by illuminating that target with a pulsed light amplification through stimulated emission of radiation (laser) light and measuring the reflected pulses (the backscattered or reflected light) using a receiver with sensitive detectors. Distance is equal to velocity multiplied by time, so the distance to the object is calculated by recording the time between transmitted and backscattered pulses multiplied by the speed of light.
- *Laser altimeter*: Mounted on a spacecraft or aircraft, a laser altimeter is a remote sensing instrument that is designed to use a LiDAR to measure the height of Earth's surface (either sea level or ground level). It works by emitting short flashes of laser light toward the surface of Earth. The height of the sea level or ground level with respect to the mean surface of Earth is then calculated by the time spent between emitted and reflected pulses multiplied by the speed of light to produce the topography of the underlying surface.
- *Radar*: An active radar sensor, whether air-borne or space-borne, emits microwave radiation in a series of pulses from an antenna based on its own source of electromagnetic energy. When the energy hits the target in the air or at the ground/sea level, some of the energy is reflected back toward the sensor. This backscattered or reflected microwave radiation is detected, measured, and analyzed. The time required for the energy to travel to the target and return back to the sensor multiplied by the speed of light determines the distance or range to the target. Therefore, a two-dimensional image of the surface can be produced by calculating the distance of all targets as the remote sensing system passes through.

Microwave instruments that are designed for finer environmental monitoring and earth observation may include, but are not limited to:

- *Scatterometer*: A high-frequency microwave radar that is designed to measure backscattered radiation over ocean surfaces to derive maps of surface wind speed and direction.
- *Sounder*: An active sensor that specifically measures vertical distribution of atmospheric characteristics such as humidity, temperature, and cloud composition, as well as precipitation.

3.4   REAL-WORLD REMOTE SENSING SYSTEMS

The advancement of new space, communication, and sensor technologies developed in the 1950s triggered the world's first man-made satellite when Sputnik-1 was launched on October 4, 1957, by Russia. It was the first man-made satellite launched into space with scientific instruments onboard, albeit it had only a 90-day lifetime. After a silence of three years, in 1960, the term "remote sensing" was formally introduced. On April 1, 1960, the Television and Infrared Observation Satellite (TIROS), the first true weather satellite, was launched by the United States of America (USA). TIROS carried special television cameras used for Earth's cloud cover observation from a 720 km (450 mile) orbit, and was the first experimental attempt of the National Aeronautics and Space Administration (NASA) to study Earth with satellite instruments. Since then, satellites have become the primary platform utilized to carry remotely sensed instruments for earth observation and environmental monitoring. These space-borne instruments take advantage of large spatial coverage and regular revisiting periods. From 1964 to 1970, a series of four meteorological research satellites, named after Nimbus, were launched into space and had profound impacts due to their synoptic views of Earth; they provided information on issues such as weather dynamics and vegetation patterns.

With the rapid progress of new remote sensing technologies in each successive generation of new satellites, remotely sensed instruments and platforms became increasingly sophisticated and professional, generating finer temporal, spectral, and spatial resolution imagery on a routine basis. There have been abundant remotely sensed instruments onboard different platforms since the passive remote sensing techniques were first proposed for satellite Earth observation in the 1970s. The chronological history of a set of famous satellite remote sensing platforms is illustrated in Figure 3.6.

One of the world's best-known families of remote sensing satellites is Landsat, which is operated by the USA and has evolved over the past 40 years. Landsat 1, which was also called the "Earth Resources Technology Satellite" until 1975, is the first satellite in the Landsat family and was launched in 1972, dedicated to periodic environmental monitoring. The Landsat 1 satellite had two sensors, called the Return Beam Vidicon (RBV) and the Multispectral Scanner (MSS). The MSS sensor was designed to capture images in the red, blue, and green spectra at 60 m resampled resolution over four separate spectral bands between 500 and 1,100 nm. The other two successive satellites (Landsat 2-3), were launched in 1975 and 1978, respectively. The same sensors were deployed onboard the Landsat 2, while the spectral capability of the MSS sensor on Landsat 3 was extended to measure radiation between 1,050 and 1,240 nm. Following the success of Landsat 1-3, the Landsat 4 was launched in 1982 with improved spectral and spatial resolution. The RBV was replaced with the thematic mapper (TM) sensor, providing seven bands from 450 to 2,350 nm with 30-m resolution pixels. In addition, the revisiting time of the satellite was improved from 18 days to

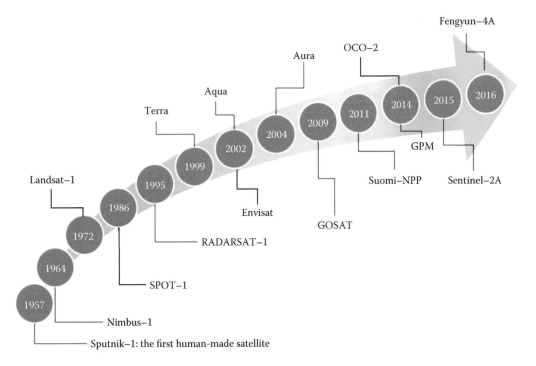

FIGURE 3.6 The timeline of some well-known Earth science satellite systems since the 1950s.

16 days. Launched in 1985, Landsat 5 is a duplicate of Landsat 4, and its TM sensor remains active 25 years beyond its designated lifetime. The next two satellites (Landsat 6-7) were launched in 1993 and 1999, respectively. However, Landsat 6 did not reach its orbit due to launch failure. These two satellites were equipped with the Panchromatic (PAN), Enhanced Thematic Mapper (ETM), and Enhanced Thematic Mapper Plus (ETM+) sensors, providing a spatial resolution of 15-m panchromatic and 30-m multispectral images. In addition, the latest generation Landsat satellite, Landsat 8, was launched in 2013 with a two-sensor payload, including the Operational Land Imager (OLI) and the Thermal InfraRed Sensor (TIRS). Landsat 8 OLI and TIRS images are comprised of nine spectral bands with a spatial resolution of 30 m for bands 1 to 7 and 9 (Table 3.1) (Barsi et al., 2014). The ultra-blue band 1 is useful for coastal and aerosol studies and band 9 is useful for cirrus cloud detection (Barsi et al., 2014). Thermal bands 10 and 11 are useful for providing more accurate surface temperatures and are collected at 100 m (Barsi et al., 2014).

Developed by NASA in the USA since the early 1970s, the Landsat program is the longest remote sensing program in the world, providing over 40 years of calibrated data about Earth's surface with moderate resolution to a broad user community. In summary, Landsats 1–3 images are comprised of four spectral bands with 60-m spatial resolution, and the approximate scene size is 170 km north-south by 185 km east-west (USGS, 2017). Specific band designations differ from Landsats 1, 2, and 3 to Landsats 4 and 5 (Tables 3.2 and 3.3). Landsats 4–5 images are comprised of seven spectral bands with a spatial resolution of 30 m for bands 1 to 5 and 7 (Table 3.3). The approximate scene size is 170 km north-south by 183 km east-west (USGS, 2017). The images of ETM+ are comprised of eight spectral bands with a spatial resolution of 30 m for bands 1 to 7 (Table 3.4). Yet the resolution for band 8 (panchromatic) is 15 m, which provides a niche for data fusion. The approximate scene size of the ETM+ images is 170 km north-south by 183 km east-west (USGS, 2017). The resolution for band 8 (panchromatic) is 15 m. The approximate scene size is 170 km north-south by 183 km east-west (USGS, 2017). Overlapped bands provide a critical basis for information consistency, which is essential for cross-checking the continuity of the multispectral data coverage provided by Landsat missions (Figure 3.7).

Besides Landsat satellites, the second important remote sensing satellite family, SPOT (Satellite Pour l'Observation de la Terre), was designed and subsequently launched by a French–Belgian–Swedish

TABLE 3.1

Comparison of Corresponding Basic Properties of Landsat 8 OLI and TIRS Images

Landsat 8 Bands	Wavelength (μm)	Resolution (m)
Band 1—Ultra Blue (coastal/aerosol)	0.435–0.451	30
Band 2—Blue	0.452–0.512	30
Band 3—Green	0.533–0.590	30
Band 4—Red	0.636–0.673	30
Band 5—Near Infrared (NIR)	0.851–0.879	30
Band 6—Shortwave Infrared (SWIR) 1	1.566–1.651	30
Band 7—Shortwave Infrared (SWIR) 2	2.107–2.294	30
Band 8—Panchromatic	0.503–0.676	15
Band 9—Cirrus	1.363–1.384	30
Band 10—Thermal Infrared (TIRS) 1	10.60–11.19	100 * (30)
Band 11—Thermal Infrared (TIRS) 2	11.50–12.51	100 * (30)

Source: United States Geological Survey (USGS), 2017. https://landsat.usgs.gov/what-are-band-designations-landsat-satellites, accessed December 2017.

* TIRS bands are acquired at 100 meter resolution, but are resampled to 30 meter in delivered data product.

TABLE 3.2

Comparison of Corresponding Basic Properties of Landsats 1–3 Multispectral Scanner (MSS) Images

Landsat 1–3 MSS Bands	Wavelength (μm)	Resolution (m)
Band 4—Green	0.5–0.6	60[a]
Band 5—Red	0.6–0.7	60[a]
Band 6—Near Infrared (NIR)	0.7–0.8	60[a]
Band 7—Near Infrared (NIR)	0.8–1.1	60[a]

Source: United States Geological Survey (USGS), 2017. https://landsat.usgs.gov/what-are-band-designations-landsat-satellites, accessed December 2017.

[a] Original MSS pixel size was 79×57 meters; production systems now resample the data to 60 meters.

TABLE 3.3

Comparison of Corresponding Basic Properties of Landsats 4–5 Thematic Mapper (TM) Images

Landsat 4–5 TM Bands	Wavelength (μm)	Resolution (m)
Band 1—Blue	0.45–0.52	30
Band 2—Green	0.52–0.60	30
Band 3—Red	0.63–0.69	30
Band 4—Near Infrared (NIR)	0.76–0.90	30
Band 5—Shortwave Infrared (SWIR) 1	1.55–1.75	30
Band 6—Thermal	10.40–12.50	120 (resampled to 30)
Band 7—Shortwave Infrared (SWIR) 2	2.08–2.35	30

Source: United States Geological Survey (USGS), 2017. https://landsat.usgs.gov/what-are-band-designations-landsat-satellites, accessed December 2017.

TABLE 3.4

Comparison of Corresponding Basic Properties of Landsat 7 Enhanced Thematic Mapper Plus (ETM+) Images

Landsat 7 ETM+ Bands	Wavelength (μm)	Resolution (m)
Band 1—Blue	0.45–0.52	30
Band 2—Green	0.52–0.60	30
Band 3—Red	0.63–0.69	30
Band 4—Near Infrared (NIR)	0.77–0.90	30
Band 5—Shortwave Infrared (SWIR) 1	1.57–1.75	30
Band 6—Thermal	10.40–12.50	60 (resampled to 30)
Band 7—Shortwave Infrared (SWIR) 2	2.09–2.35	30
Band 8—Panchromatic	0.52–0.90	15

Source: United States Geological Survey (USGS), 2017. https://landsat.usgs.gov/what-are-band-designations-landsat-satellites, accessed December 2017.

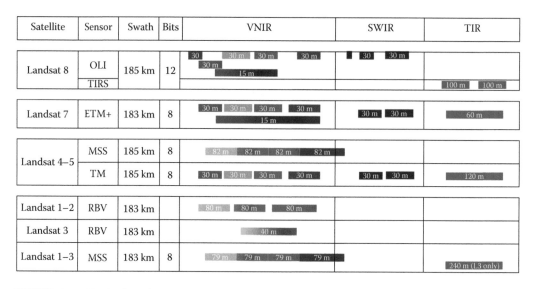

FIGURE 3.7 Continuity of multispectral data coverage provided by Landsat missions. (United States Geological Survey (USGS), 2014. Landsat 8 (L8) Data Users Handbook. https://landsat.usgs.gov/landsat-8-l8-data-users-handbook-section-1, accessed May 2017.)

joint program beginning in 1986. In 1986, the SPOT 1 was launched, equipped with a high-resolution visible (HRV) sensor that offered 10-m panchromatic and 20-m multispectral images with a 26-day revisit interval (i.e., repeat cycle). Improvements in the spatial resolution of SPOT 1 provided more accurate data for better understanding and monitoring the surface of Earth. The SPOT 2, SPOT 3, and SPOT 4 were launched with the same instruments in 1990, 1993, and 1998, respectively. SPOT 5, the latest SPOT in the sky, was launched in 2002 with 2.5- or 5-m panchromatic and 10-m multispectral image resolution, with the same 26-day revisit interval.

IKONOS, the first commercialized remote sensing satellite launched by a private entity, was launched in 1999 and was capable of providing a high spatial resolution (1 m) and high temporal resolution (1.5 to 3 days) imagery. During the same year, Terra, the flagship satellite for the Earth Observing System (EOS), was launched with five different instruments aboard, including the Clouds and the Earth's Radiant Energy System (CERES), the Multi-angle Imaging SpectroRadiometer (MISR), the Moderate-Resolution Imaging Spectroradiometer (MODIS), the Measurements of Pollution in the Troposphere (MOPITT), and the Advanced Spaceborne Thermal Emission and Reflection Radiometer (ASTER). These sensors were designed to monitor the state of Earth's environment and ongoing changes in its climate system with a spatial resolution of 250–1,000 m around the globe. Aqua is another EOS satellite similar to Terra but with a different equator-crossing time. Both sensors of MODIS onboard Aqua and Terra provide us with the ability of near real-time environmental monitoring on a daily basis (i.e., Aqua passes the equator daily at the local time of 1:30 p.m. as it heads north (ascending mode) in contrast to Terra, which passes the equator daily at the local time of 10:30 a.m. (descending mode)).

In addition to remotely sensed sensors such as MOPITT, other specific instruments were utilized to provide information regarding the atmospheric compositions and air quality. Total Ozone Mapping Spectrometer (TOMS), first deployed onboard NASA's Nimbus-7 satellite in 1978, was the first instrument designed to monitor the total column ozone at the global scale in order to track the ozone depletion from space. The following instruments, such as the Ozone Monitoring Instrument (OMI) onboard Aura launched in 2004, the Total Ozone Unit (TOU) onboard the Chinese FY-3 series satellite since 2008, and the Ozone Monitoring and Profiler Suite (OMPS) onboard the Suomi-NPP satellite launched in 2011, are all dedicated to monitoring ozone variability at the global scale. Similarly, the Greenhouse Gases Observing SATellite (GOSAT), launched in

2009 by Japan, was designed to monitor global carbon dioxide (CO_2) and methane (CH_4) variability. The latest Orbiting Carbon Observatory-2 satellite, NASA's first dedicated Earth remote sensing satellite used to study atmospheric carbon dioxide from space, was launched in 2014. With the aid of these satellite platforms and instruments, more detailed information about Earth is provided to help us better understand our changing world.

The third family of remote sensing satellites is the National Oceanic and Atmospheric Administration (NOAA) family of polar-orbiting platforms (POES). The Coastal Zone Color Scanner (CZCS) was launched in 1978 for measuring ocean color from space. Following the success of the CZCS, other similar ocean color sensors were also launched. They include the Moderate Optoelectrical Scanner (MOS), the Ocean Color Temperature Scanner (OCTS), Polarization and Directionality of the Earth's Reflectances (POLDER), and the Sea-Viewing Wide Field-of-View Sensor (SeaWiFS). The most popular space-borne sensor, Advanced Very High Resolution Radiometer (AVHRR), embarked on this series of satellites for remotely determining cloud cover and the surface temperature. With the technological advances of the late 20th century, satellite sensors with improved spatiotemporal and spectral resolutions were designed and utilized for different Earth observation purposes.

In addition, the Medium Resolution Imaging Spectrometer (MERIS), one of the main payloads onboard Europe's Environmental Satellite (ENVISAT-1), provided hyperspectral rather than multispectral remote sensing images with a relatively high spatial resolution (300 m). Although the ENVISAT-1 was lost in space, those onboard instruments such as MERIS had provided an ample record of Earth's environment. Meanwhile, more and more commercial satellites managed by private sectors in the USA, such as Quickbird 2, Worldview 1, and Worldview 2, provided remotely sensed optical imagery with enhanced spatial and spectral details.

In the 1990s, microwave sensors, such as synthetic aperture radar (SAR), were deployed onboard a series of satellites as the fourth family of remote sensing satellites; they were designed to provide microwave sensing capacity with higher resolution in all weather conditions. They include JERS-1 (Japan), ERS-1 and ERS-2 (Europe), and RADARSAT-1 (Canada). With increased spatial, spectral, and temporal resolutions, more detailed information on Earth's changing environment and climate can be provided. As the importance of microwave remote sensing was generally recognized, new generations of satellites of this kind continued the advancement in SAR remote sensing with varying polarization (HH+HV+VH+VV) modes, including Japan (ALOS), Canada (RADARSAT-2), and Germany (TerraSAR-X).

3.5   CURRENT, HISTORICAL, AND FUTURE IMPORTANT MISSIONS

3.5.1   Current Important Missions

Through NASA's EOS program, a variety of remotely sensed data has become available in the Earth Observing System Data Information System (EOSDIS) from diversified operational satellites in the USA since the 1980s. Visible Infrared Imaging Radiometer Suite (VIIRS), a subsequent mission to MODIS, is a scanning radiometer that collects visible and infrared imagery and radiometric measurements rendering complementary support for date merging and fusion to Soil Moisture Active Passive (SMAP) (L-band SAR), which was launched in late 2014 to provide global soil moisture data. In Europe, Copernicus, which was previously known as Global Monitoring for Environment and Security (GMES), is a major program in the European Space Agency (ESA) in the European Union (EU) for the establishment of a European capacity for Earth observation. Many of the active and passive sensors onboard different platforms whose data are supported by EOSDIS (USA) and Copernicus (EU) are highlighted in this section. In addition, the German Aerospace Center (Deutsches Zentrum für Luft- und Raumfahrt; DLR) has been very successful in producing and managing TerraSAR-X, TanDEM-X, and Sentinel series satellites. France's Centre National d'Etudes Spatiales (CNES) has been successful in the SPOT satellite series. In addition, the Canadian Space Agency (CSA) helps fund the construction and launch of the RADARSAT satellite and recovers this investment through the supply of RADARSAT-1 and RADARSAT-2 data to the Government of Canada and various user communities during the lifetime of the mission. The Japan Aerospace Exploration Agency (JAXA) also performs various space activities related to Earth observation. JAXA handles the Advanced Land Observation Satellite, known as ALOS.

The following summary tables (Tables 3.5 through 3.8) present most of the current satellites as of July 2017 relevant for environmental monitoring and Earth observation. These remote sensing systems are operated mainly by space agencies in many countries such as NASA (USA), ESA (European Union), DLR (Germany), CNES (France), JAXA (Japan), and CSA (Canada). In addition, IKONOS and GeoEye/RapidEye are commercial optical-NIR (near infrared) providing high-resolution satellite imageries.

3.5.2 Historic Important Missions

There were some important programs that provided valuable space-time observations of selected parameters for various environmental applications in the past three decades. They are summarized in Table 3.9. As old satellites stepped down at the end of their product life cycle, new satellites with more advanced functionalities were sent into space for carrying out even more complex missions. From microwave to optical remote sensing, a few important successions in this regard deserve our attention; for instance, ALOS-2 (L-band SAR) was launched in 2014 to replace ALOS-1 in Japan and RADARSAT-2 was launched in 2007 to replace RADARSAT-1. Landsat 8 was launched in

---

TABLE 3.5

Current Important Missions Using Passive Spectrometers for Environmental Applications

Platform	Sensor	Type	Feature
Aircraft	Airborne Visible/ Infrared Imaging Spectrometer (AVIRIS)	Imaging Spectrometer (Passive Sensor)	• AVIRIS has 224 contiguous channels. • Measurements are used for studying water vapor, ocean color, vegetation classification, mineral mapping, and snow and ice cover.
Suomi National Polar-orbiting Partnership (Suomi-NPP)	Cross-Track Infrared Sounder (CrIS)	Spectrometer (Passive Sensor)	• CrIS produces high-resolution, three-dimensional temperature, pressure, and moisture profiles.
Suomi National Polar-orbiting Partnership (Suomi-NPP)	Ozone Mapping Profiler Suite (OMPS)	Spectrometer (Passive Sensor)	• OMPS is an advanced suite of two hyperspectral instruments. • OMPS extends the 25+ year total-ozone and ozone-profile records.
Terra	Multi-angle Imaging SpectroRadiometer (MISR)	Imaging Spectrometer (Passive Sensor)	• MISR obtains images in four spectral bands at nine different angles. • MISR provides aerosol, cloud, and land surface data.
Sentinel-2	MultiSpectral Imager (MSI)	Imaging Spectrometer (Passive Sensor)	• Sentinel-2A and 2B provide satellite image data to support generic land cover, land use and change detection, leaf area index, leaf chlorophyll content, and leaf water content.

Source: National Aeronautics and Space Administration (NASA), 2017. EOSDIS – Remote Sensors. https://earthdata.nasa.gov/user-resources/remote-sensors, accessed May 2017.

TABLE 3.6

Current Important Missions Using Passive Multispectral Radiometers for Environmental Applications

Platform	Sensor	Type	Feature
Aqua	Advanced Microwave Scanning Radiometer (AMSR-E)	Multichannel Microwave Radiometer (Passive Sensor)	• AMSR-E measures precipitation, oceanic water vapor, cloud water, near-surface wind speed, sea and land surface temperature, soil moisture, snow cover, and sea ice.
Aqua	Moderate-Resolution Imaging Spectroradiometer (MODIS)	Imaging Spectroradiometer (Passive Sensor)	• MODIS measures ocean and land surface properties, surface reflectance and emissivity, and air properties.
Landsat 7	Enhanced Thermatic Mapper Plus (ETM+)	Scanning Radiometer (Passive Sensor)	• The ETM+ instrument provides high-resolution imaging information of Earth's surface.
Landsat 8	The Operational Land Imager (OLI) and the Thermal Infrared Sensor (TIRS)	Radiometer (Passive Sensor)	• OLI and TIRS are designed similarly to Landsat 7 for the same purpose in applications.
Soil Moisture Active Passive (SMAP)	L-Band Radiometer (LBR)	Radiometer (Passive Sensor)	• SMAP-LBR radiometer chooses an advanced radiometer to monitor water and energy fluxes and improve flood predictions and drought monitoring.
Suomi National Polar-orbiting Partnership (Suomi-NPP)	Visible Infrared Imaging Radiometer Suite (VIIRS)	Radiometer (Passive Sensor)	• VIIRS collects water-leaving reflectance and land-reflective data.
Terra	Advanced Spaceborne Thermal Emission and Reflection Radiometer (ASTER)	Multispectral Radiometer (Passive Sensor)	• ASTER measures surface radiance, reflectance, emissivity, and temperature. Provides spatial resolutions of 15 m, 30 m, and 90 m.
Terra	Clouds and the Earth's Radiant Energy System (CERES)	Broadband Scanning Radiometer (Passive Sensor)	• CERES measures atmospheric and surface energy fluxes.
Terra	Moderate-Resolution Imaging Spectroradiometer (MODIS)	Imaging Spectroradiometer (Passive Sensor)	• The same as MODIS Aqua
Aura	Ozone Monitoring Instrument (OMI)	Multispectral Radiometer (Passive Sensor)	• OMI collects 740 wavelength bands in the visible and ultraviolet electromagnetic spectrum. • OMI measures total ozone and profiles of ozone, N_2O, SO_2, and several other chemical species.
SPOT	High Resolution Visible (HRV) Imaging Spectroradiometer	Multispectral Radiometer (Passive Sensor)	• SPOT provides high-resolution maps for change detections of Earth's surface.
IKONOS	High Resolution Visible (HRV) Imaging Spectroradiometer	Multispectral and Panchromatic Radiometer (Passive Sensor)	• IKONOS provides high-resolution maps for change detections of Earth's surface.

(Continued)

TABLE 3.6 (*Continued*)

Current Important Missions Using Passive Multispectral Radiometers for Environmental Applications

Platform	Sensor	Type	Feature
GeoEye	High-Resolution Visible (HRV) Imaging Spectroradiometer	Multispectral and Panchromatic Radiometer (Passive Sensor)	• GeoEye provides high-resolution maps for change detections of Earth's surface.
WorldView	High-Resolution Visible (HRV) Imaging Spectroradiometer	Panchromatic Radiometer (Passive Sensor)	• WorldView provides extra-fine resolution (less than 1 m) panchromatic maps of Earth's surface.
QuickBird	High-Resolution Visible (HRV) Imaging Spectroradiometer	Multispectral and Panchromatic Radiometer (Passive Sensor)	• QuickBird provides high-resolution maps for change detections of Earth's surface.

Source: National Aeronautics and Space Administration (NASA), 2017. EOSDIS – Remote Sensors. https://earthdata.nasa. gov/user-resources/remote-sensors, accessed May 2017, CNES, and private sectors' web sites.

February 2013, although Landsat 7 is still in operation. This will certainly continue the previous Earth observation Landsat mission in the optical and infrared remote sensing regime over the past four decades. GOES-R was deployed as GOES1-12 retired from its geostationary orbit. The mission of TOMS aboard Nimbus 7 for total ozone mapping was continued by OMI aboard Aura and later by OMPS aboard Suomi-NPP. The new Sentinel 1 satellite continued the mission of ERS-1 and ERS-2 for SAR imaging from space.

3.5.3 Future Important Missions

In view of future important missions, it is critical to stress the Copernicus program, which is an initiative headed by the European Commission (EC) in partnership with the ESA. The Copernicus program, known as GMES, is one of the most ambitious Earth observation programs to date. The new family of satellites planned in GMES, called Sentinels, will meet the operational needs of the Copernicus program. The family of Sentinels will provide a unique set of Earth observations, starting with the all-weather, day and night radar images from Sentinel-1A and -1B launched in April 2014 and April 2016, respectively, to Sentinel-3A, launched on 16 February 2016 to provide data for services relevant to the ocean and land. In the future, Sentinel-4 and -5 will provide data for atmospheric composition monitoring from geostationary and polar orbits, respectively (ESA, 2017). Sentinel-6 will carry a radar altimeter to measure global sea surface height, primarily for operational oceanography and climate studies (ESA, 2017). In addition, a Sentinel-5 precursor mission is being developed to reduce data gaps between ENVISAT, in particular the SCIAMACHY instrument, and the launch of Sentinel-5 (ESA, 2017). For forest management, the ESA Biomass Earth Explorer mission will be the first satellite P-band SAR, with a potential launch in 2019. All of these efforts will lead to the collection of accurate, timely, and easily accessible information to improve the management of the environment, understand and mitigate the effects of climate change, and ensure civil security.

Parallel to the thrust of the Copernicus program in Europe, in North America the RADARSAT Constellation Mission is scheduled for launch in 2018 to collect the C-band radar data. SWOT (Surface Water Ocean Topography), designed by the Jet Propulsion Laboratory in the USA, is a Ka-band SAR altimeter and radar interferometer with a potential launch in 2020. This will provide an opportunity to create a breakthrough in hydrological science, leading to closure of the loop of water-balance calculation in the natural environment. In addition, the Ice, Cloud, and land Elevation Satellite-2 (ICESat-2) that is the benchmark EOS mission is the second generation of the orbiting

TABLE 3.7

Current Important Missions Using Radar, LiDAR, Gravimeter, and Laser for Environmental Applications

Platform	Sensor	Type	Mission
Airborne Microwave Observatory of Subcanopy and Subsurface (AirMOSS)	Synthetic Aperture Radar (SAR)	Radar (Active Sensor)	• P-band SAR provides calibrated polarimetric measurements to retrieve root-zone soil moisture.
Ice, Cloud, and land Elevation Satellite (ICESat)	Geoscience Laser Altimeter System (GLAS)	Radar (Active Sensor)	• ICESat measures ice sheet elevations and changes in elevation through time in addition to the measurement of cloud and aerosol height profiles, land elevation and vegetation cover, and sea ice thickness.
Cloud-Aerosol LiDAR and Infrared Pathfinder Satellite Observations (CALIPSO)	Cloud-Aerosol LiDAR with Orthogonal Polarization (CALIOP)	Cloud and Aerosol LiDAR (Active Sensor)	• CALIOP is a two-wavelength polarization-sensitive LiDAR that provides high-resolution vertical profiles of aerosols and clouds.
Cloud-Aerosol Transport System on the International Space Station (CATS)	Light Detection and Ranging (LiDAR)	LiDAR (Active Sensor)	• LiDAR provides range-resolved profile measurements of atmospheric aerosols and clouds.
Global Precipitation Measurement (GPM)	Dual-Frequency Precipitation Radar (DPR)	Radar (Active Sensor)	• DPR provides information regarding rain and snow worldwide.
Ocean Surface Topography Mission/Jason-2 (OSTM/Jason-2)	Poseidon-3 Altimeter (PA)	Altimeter (Active Sensor)	• PA provides sea surface heights for determining ocean circulation, climate change, and sea level rise.
Sentinel-1	Synthetic Aperture Radar (SAR)	Radar (Active Sensor)	• Sentinel-1 SAR provides land and ocean monitoring regardless of the weather.
Sentinel 3	Synthetic Aperture Radar (SAR)	Radar (Active Sensor)	• Sentinel 3 supports marine observation, and will study sea-surface topography, sea and land surface temperature, ocean and land color.
Soil Moisture Active Passive (SMAP)	L-Band Radar (LBR)	Radar (Active Sensor)	• SMAP-LBR radar measures the amount of water in the top 5 cm of soil everywhere on Earth's surface.
Advanced Land Observation Satellite (ALOS)	L-band ALOS PALSAR	Phased Array L-band Synthetic Aperture Radar (Active Sensor)	• ALOS expands SAR data utilization by enhancing its performance.
TerraSAR-X	X-band SAR sensor	Radar (Active Sensor)	• TerraSAR-X provides SAR images with high resolution.
TanDEM-X	X-band SAR sensor	Radar (Active Sensor)	• TanDEM-X provides land subsidence, digital elevation model and other land cover conditions.
Gravity Recovery and Climate Experiment (GRACE) Satellite	Low-earth orbit satellite gravimetry	Passive Sensor	• GRACE measures gravity changes to infer the water storage at the surface of Earth.

Source: National Aeronautics and Space Administration (NASA), 2017. EOSDIS – Remote Sensors https://earthdata.nasa.gov/user-resources/remote-sensors, accessed May 2017.

TABLE 3.8

Current Important Missions of Scatterometers and Sounding Instruments for Environmental Applications

Platform	Sensor	Type	Feature
Cyclone Global Navigation Satellite System (CYGNSS)	Delay Doppler Mapping Instrument (DDMI)	Scatterometer (Active Sensor)	• DDMI measures ocean surface wind speed in all precipitating conditions.
Aqua	Atmospheric Infrared Sounder (AIRS)	Sounder (Passive Sensor)	• AIRS measures air temperature, humidity, clouds, and surface temperature.
Aqua	Advanced Microwave Sounding Unit (AMSU)	Sounder (Passive Sensor)	• AMSU measures temperature profiles in the upper atmosphere.
Aura	High-Resolution Dynamics Limb Sounder (HIRDLS)	Sounder (Passive Sensor)	• HIRDLS measures profiles of temperature, ozone, CFCs, and various other gases affecting ozone chemistry.
Aura	Microwave Limb Sounder (MLS)	Sounder (Passive Sensor)	• MLS derives profiles of ozone, SO_2, N_2O, OH and other atmospheric gases, temperature, pressure, and cloud ice.
Suomi-National Polar-orbiting Partnership (Suomi-NPP)	Ozone Mapping Profiler Suite (OMPS)	Sounder (Passive Sensor)	• OMPS provides operational ozone measurements.
Terra	Measurements of Pollution in the Troposphere (MOPITT)	Sounder (Passive Sensor)	• MOPITT measures carbon monoxide and methane in the troposphere.

Source: National Aeronautics and Space Administration (NASA), 2017. EOSDIS – Remote Sensors https://earthdata.nasa.gov/user-resources/remote-sensors, accessed May 2017.

laser altimeter ICESat, and is scheduled for launch in 2018. The ICESat-2 mission will provide multi-year elevation data needed to determine ice sheet mass balance as well as cloud property information, especially for stratospheric clouds common over polar areas. The Gravity Recovery and Climate Experiment Follow-on (GRACE-FO) (a.k.a. GFO) mission that is part of the US-German GRACE consortium (NASA/Jet Propulsion Laboratory, Center for Space Research/University of Texas, DLR, GFZ Helmholtz Center Potsdam) is heavily focused on maintaining data continuity from GRACE and minimizing any data gap after GRACE. In concert with the functionality of SWOT, this effort leads to the final closure of the water balance in the hydrological system.

Specifically, the future German satellite mission EnMAP (Environmental Mapping and Analysis Program) addresses the future need for hyperspectral remote sensing (Stuffler et al., 2007). It aims to measure, derive, and analyze diagnostic parameters for the vital processes on Earth's land and water surfaces. The EnMAP hyperspectral remote sensing products take images of 1024×1024 pixels ($\sim 30 \times 30$ km^2), which are generated by the processing system on demand and delivered to the user community.

3.6   SYSTEM PLANNING OF REMOTE SENSING APPLICATIONS

Satellite imagery demonstrates the same physical environment in different ways as a result of different spectral, spatial, and temporal resolutions. Various software dealing with image processing may become essential for different missions (Zhang, 2010). In the broadest sense, data fusion

TABLE 3.9

Historic Important Missions for Environmental Applications

Platform	Sensor	Type	Mission
Advanced Land Observing Satellite (ALOS)	Phased Array L-band Synthetic Aperture Radar (PALSAR)	Radar (Active Sensor)	• PALSAR provided mapping of regional land coverage, disaster monitoring, and resource surveying.
Advanced Land Observing Satellite (ALOS)	Panchromatic Remote Sensing Instrument for Stereo Mapping (PRISM)	Spectrometer (Passive Sensor)	• PRISM provided panchromatic images with 2.5-m spatial resolution that digital surface model (DSM).
Radar Satellite (RADARSAT-1)	Synthetic Aperture Radar (SAR)	Radar (Active Sensor)	• RADARSAT-1 collected data on resource management, ice, ocean and environmental monitoring, and Arctic and off-shore surveillance.
Nimbus-7	Coastal Zone Color Scanner (CZCS)	Radiometer (Passive Sensor)	• CZCS attempted to discriminate between organic and inorganic materials in the water.
Nimbus-7	Earth Radiation Budget Experiment (ERBE)	Radiometer (Passive Sensor)	• ERBE attempted to test infrared limb scanning radiometry to sound the composition and structure of the middle atmosphere.
Nimbus-7	Stratospheric Aerosol Measurement II (SAM II)	Photometer (Passive Sensor)	• SAM II attempted to measure stratospheric aerosol measurement and provided vertical profiles of aerosol extinction in both the Arctic and Antarctic polar regions.
Nimbus-7	Solar Backscatter Ultraviolet (SBUV), Total Ozone Mapping Spectrometer II (TOMS II)	Spectrometer (Passive Sensor)	• SBUV and TOMS sensors provided first-hand data of UV-B and total column ozone.
Nimbus-7	Scanning Multichannel Microwave Radiometer (SMMR)	Multispectral Microwave Radiometer (Passive Sensor)	• SMMR measured sea surface temperatures, ocean near-surface winds, water vapor and cloud liquid water content, sea ice extent, sea ice concentration, snow cover, snow moisture, rainfall rates, and differential of ice types.
Ice, Cloud, and land Elevation Satellite (ICESat)	Geoscience Laser Altimeter System (GLAS)	Laser Altimeter (Active Sensor)	• GLAS measured ice sheet elevations and changes in elevation through time. • GLAS measured cloud and aerosol height profiles, land elevation and vegetation cover, and sea ice thickness.
European Remote Sensing Satellite (ERS-1, ERS-2)	Synthetic Aperture Radar (SAR)	Radar (Active Sensor)	• ERS SAR emitted a radar pulse with a spherical wavefront which reflected from the surface.
Cosmo/SkyMed 1, 2, 3, 4	Synthetic Aperture Radar (SAR)	SAR 2000 (Active Sensor)	• Cosmo/SkyMed SAR emitted a radar pulse.
European Remote Sensing Satellite (ERS-1, ERS-2)	Active Microwave Instrument (AMI-WIND)	Microwave (Active Sensor)	• ERS AMI-WIND emitted a radar pulse with a spherical wavefront which reflected from the surface.
European Remote Sensing satellite (ERS-1, ERS-2)	Radar Altimetry (RA)	Radar (Active Sensor)	• ERS RA emitted a radar pulse with a spherical wavefront which reflected from the surface.
Geostationary Operational Environmental Satellite (GOES 1-12)	Advanced Very High Resolution Radiometer (AVHRR)	Radiometer (Passive Sensor)	• AVHRR can be used for remotely determining cloud cover and the surface temperature.

Source: National Aeronautics and Space Administration (NASA), 2017. EOSDIS – Remote Sensors https://earthdata.nasa.gov/user-resources/remote-sensors, accessed May 2017. CNES, DLR, European Space Agency (ESA), 2017. The Copernicus Programme, http://www.esa.int/Our_Activities/ Observing_the_Earth/Copernicus/Overview3, CSA (2017).

TABLE 3.10

Future Important Missions for Environmental Applications

Platform	Sensor	Type	Mission
Surface Water Ocean Topography (SWOT)	Advanced Microwave Radiometer (AMR)	Radiometer (Passive Sensor)	• SWOT will provide sea surface heights and terrestrial water heights over a 120-km-wide swath with a ±10-km gap at the nadir track.
Sentinel 5P	Synthetic Aperture Radar (SAR)	Radar (Active Sensor)	• Sentinel 5P will aim to fill in the data gap and provide data continuity between the retirement of the ENVISAT satellite and NASA's Aura mission and the launch of Sentinel-5.
Biomass	Imaging Spectroradiometer	Multispectral Radiometer (Passive Sensor)	• Biomass will address the status and dynamics of tropical forests.
EnMap	Imaging Spectroradiometer	Hyperspectral Radiometer (Passive Sensor)	• EnMap will address the dynamics of the land and water surface.

Source: National Aeronautics and Space Administration (NASA), 2017. EOSDIS – Remote Sensors https://earthdata.nasa.gov/user-resources/remote-sensors, accessed May 2017. CNES, DLR, European Space Agency (ESA), 2017. The Copernicus Programme, http://www.esa.int/Our_Activities/Observing_the_Earth/Copernicus/Overview3, CSA (2017).

involves combining information to estimate or predict the state of some aspect of the system. This can be geared toward much better Earth observations and to tackling a few challenging problems. For instance, due to the impact of global climate change, data fusion missions can be organized based on the existing and future satellites groupwise, such as ESA Sentinels 1 and 2 and Landsat-8, gravimetry missions, TerraSAR-X, SWOT, SMAP, GRACE-FO, GOES-R and TanDEM-X, to produce data fusion products. Data fusion, with respect to the different remote sensing sensors above, can be carried out to blend different modalities of satellite imagery into a single image for various Earth observation applications over temporal and spatial scales, leading to better environmental decision making. However, a satellite constellation such as the A-train program, which is a joint program between NASA, CNES, and JAXA, may group several satellites by design, providing some insightful and complementary support to this type of profound research. Note that the A-train (from Afternoon Train) is a satellite constellation of six Earth observation satellites of varied nationalities in sun-synchronous orbit at an altitude of 705 km above Earth (Figure 3.8); they include OCO-2, GCOM-W1 SHIZUKU, Aqua, CloudSat, CALIPSO, and Aura as of July 2014.

In addition, to fill in different data gaps of space-borne remote sensing and to facilitate the system planning goals of providing low-cost and full-coverage images, the community has further adopted a standard dubbed as CubeSat. (Heidt et al., 2000). A CubeSat is a spacecraft sized in units, or Us, typically up to 12U (a unit is defined as a volume of about 10 cm × 10 cm × 10 cm) that is launched fully enclosed in a container, enabling ease of launch vehicle system integration, thus easing access to space (National Academies of Sciences, Engineering, and Medicine, 2016). Continuous creation of customized nano-satellites and cube-satellites in the optical, microwave, and radio frequency domains has become a big initiative and a giant niche for different types of environmental applications. This fast evolution in Earth observation will possibly disrupt the conventional ways of environmental monitoring.

Possible data merging and data fusion opportunities to tackle permafrost remote sensing studies are shown in Figure 3.9. This system planning diagram exhibits the possible space-time plot of selected near-term (2013–2020) satellite sensor observations with potential relevance for permafrost. These parameters are linked to ALOS-2 (L-band SAR), Biomass Earth Explorer mission, Landsat 8, RADARSAT (C-band radar data), IKONOS and GeoEye/RapidEye, SWOT,

FIGURE 3.8 The A-Train that consists of six satellites in the constellation as of 2014. (National Aeronautics and Space Administration (NASA), 2017. NASA A-Train portal, https://atrain.gsfc.nasa.gov/ accessed May 2017.)

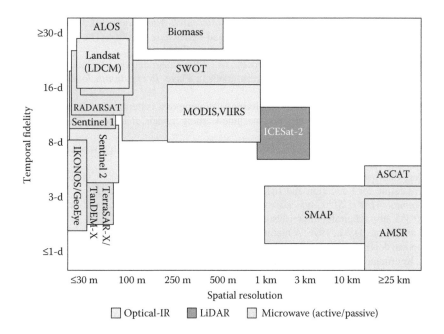

FIGURE 3.9 Possible data fusion opportunities to tackle permafrost remote sensing studies. (Adapted from The National Academies, 2014. *Opportunities to Use Remote Sensing in Understanding Permafrost and Related Ecological Characteristics: Report of a Workshop.* ISBN 978-0-309-30121-3.)

VIIRS, ICESat-2 (LiDAR), ASCAT (Advanced SCATterometer), SMAP (L-band SAR), AMSR (Advanced Microwave Scanning Radiometer), ESA Sentinels 1 and 2, GRACE-FO, X-band SAR, TerraSAR-X, and TanDEM-X.

3.7 SUMMARY

Many remote sensing studies have been carried out on national and international levels by different government agencies, academia, research institutions, and industries to investigate new techniques for observing Earth from space or sky. The close-knit relationships across many satellite missions have welded several families of satellites that may help perform data merging or create some hidden niches to promote data fusion, albeit not necessarily through a satellite constellation. The connection between the EOS and SPOT families lives on as well from past missions to the current ones and to the future missions—such as the Copernicus program—that spin off even more potential for long-term data fusion and machine learning research. Future missions will influence cross-mission ties from which the past, the current, and the future missions can be tailored cohesively for large-scale research dealing with specific earth system science problems. These kinds of linkages have glorified the principle of system engineering—"the whole is greater than the sum of its parts". Various data merging, data fusion, and machine learning algorithms have played a key behind-the-scenes role in helping to expand the sum further. These algorithms will be introduced in subsequent chapters.

REFERENCES

Agren, A., Jansson, M., Ivarsson, H., Biship, K., and Seibert, J., 2008. Seasonal and runoff-related changes in total organic carbon concentrations in the River Ore, Northern Sweden. *Aquatic Sciences*, 70(1), 21–29.

Anagnostou, E. N. and C. Kummerow, 1997. Stratiform and convective classification of rainfall using SSM/I 85-GHz brightness temperature observations. *Journal of Atmospheric and Oceanic Technology*, 14, 570–575.

Arenz, R., Lewis, W., and Saunders III, J., 1995. Determination of chlorophyll and dissolved organic carbon from reflectance data for colorado reservoirs. *International Journal of Remote Sensing*, 17(8), 1547–1566.

Atlas, D. and Matejka, T. J., 1985. Airborne Doppler radar velocity measurements of precipitation seen in ocean surface reflection. *Journal of Geophysical Research-Atmosphere*, 90, 5820–5828.

Barsi, J. A., Lee, K., Kvaran, G., Markham, B. L., and Pedelty, J. A., 2014. The spectral response of the landsat-8 operational land imager. *Remote Sensing*, 6, 10232–10251.

Chang, N. B., Vannah, B., Yang, Y. J., and Elovitz, M., 2014. Integrated data fusion and mining techniques for monitoring total organic carbon concentrations in a lake. *International Journal of Remote Sensing*, 35, 1064–1093.

Chang, N., Xuan, Z., and Yang, Y., 2013. Exploring spatiotemporal patterns of phosphorus concentrations in a coastal bay with MODIS images and machine learning models. *Remote Sensing of Environment*, 134, 100–110.

European Space Agency (ESA), 2017. The Copernicus Programme, http://www.esa.int/Our_Activities/Observing_the_Earth/Copernicus/Overview3

Heidt, H., Puig-Suari, J., Moore, A. S., Nakasuka, S., and Twiggs, R. J., 2000. CubeSat: A New Generation of Picosatellite for Education and Industry Low-Cost Space Experimentation. In: *Proceedings of the 14th Annual AIAA/USU Conference on Small Satellites, Lessons Learned-In Success and Failure, SSC00-V-5.* http://digitalcommons.usu.edu/smallsat/2000/All2000/32/

King, M. D. and Byrne, D. M., 1976. A method for inferring total ozone content from the spectral variation of total optical depth obtained with a solar radiometer. *Journal of the Atmospheric Sciences*, 33, 2242–2251.

King, M. D., Kaufman, Y. J., Tanré, D., and Nakajima, T., 1999. Remote sensing of tropospheric aerosols from space: Past, present, and future. *Bulletin of the American Meteorological Society*, 80, 2229–2259.

Lee, H. J., Coull, B. A., Bell, M. L., and Koutrakis, P., 2012. Use of satellite-based aerosol optical depth and spatial clustering to predict ambient PM2.5 concentrations. *Environmental Research*, 118, 8–15.

Li, J., Carlson, B. E., and Lacis, A. A., 2015. How well do satellite AOD observations represent the spatial and temporal variability of PM2.5 concentration for the United States? *Atmospheric Environment*, 102, 260–273.

Li, Q., Li, C., and Mao, J., 2012. Evaluation of atmospheric aerosol optical depth products at ultraviolet bands derived from MODIS products. *Aerosol Science and Technology*, 46, 1025–1034.

National Academies of Sciences, Engineering, and Medicine. 2016. *Achieving Science With CubeSats: Thinking Inside the Box.* The National Academies Press, Washington, DC. doi:10.17226/23503

National Aeronautics and Space Administration (NASA), 2012. https://www.nasa.gov/directorates/heo/scan/communications/outreach/funfacts/txt_passive_active.html, accessed May 2017.

National Aeronautics and Space Administration (NASA), 2013. Landsat 7 Science Data User's Handbook. Available at http://landsathandbook.gsfc.nasa.gov

National Aeronautics and Space Administration (NASA), 2017. EOSDIS - Remote Sensors. https://earthdata.nasa.gov/user-resources/remote-sensors, accessed May 2017.

National Aeronautics and Space Administration (NASA), 2017. NASA A-Train portal. https://atrain.gsfc.nasa.gov/ accessed May 2017.

Natural Resources Canada, 2017. In the *"Fundamentals of Remote Sensing" tutorial, by the Canada Centre for Remote Sensing (CCRS)*, Natural Resources Canada. http://www.nrcan.gc.ca/earth-sciences/geomatics/satellite-imagery-air-photos/satellite-imagery-products/educational-resources/9283, accessed May 2017.

Richards, J. A. and Jia, X., 2006. *Remote Sensing Digital Image Analysis: An Introduction.* Springer, Berlin, Germany.

Stuffler, T., Kaufmann, C., Hofer, S., Förster, K.-P., Schreier, G., Mueller, A., and Eckardt, A. et al., 2007. The EnMAP hyperspectral imager—An advanced optical payload for future applications in Earth observation programmes. *Acta Astronautica*, 61, 115120.

The National Academies, 2014. *Opportunities to Use Remote Sensing in Understanding Permafrost and Related Ecological Characteristics: Report of a Workshop.* ISBN 978-0-309-30121-3

United States Geological Survey (USGS), 2017. https://landsat.usgs.gov/what-are-band-designations-landsat-satellites, accessed December 2017.

United States Geological Survey (USGS), 2014. Landsat 8 (L8) Data Users Handbook. https://landsat.usgs.gov/landsat-8-l8-data-users-handbook-section-1, accessed May 2017.

Wood, V. T., Brown, R. A., and Sirmans, D., 2001. Technique for improving detection of WSR-88D mesocyclone signatures by increasing angular sampling. *Weather and Forecasting*, 16, 177–184.

Zhang, Y., 2010. Ten years of remote sensing advancement & the research outcome of the CRC-AGIP Lab. *Geomatica*, 64, 173–189.

Zrnic, D. S. and Ryzhkov, A. V., 1999. Polarimetry for weather surveillance radars. *Bulletin of American Meteorology Society*, 80, 389–406.

4 Image Processing Techniques in Remote Sensing

4.1 INTRODUCTION

Remote sensing involves the collection of data from different sensors far from the target (e.g., space-borne instruments onboard satellites); the information is collected without making any physical contact with the object. Data collected from remotely sensed instruments can be recorded either in analog (e.g., audios) or digital format. Compared to the digital format, the conventional analog format is confined to several drawbacks, such as a limitation on the size of the data that can be transmitted at any given time, and inconveniences with manipulation as well. Therefore, the digital format is commonly applied to archive remotely sensed data, especially in the form of images.

In remote sensing, data recorded by remotely sensed sensors are commonly archived in different formats convenient for information storage and transformation, such as the Hierarchical Data Format (HDF), network Common Data Format (netCDF), and so forth. Data archived in such formats are always hard to be manipulated for visualization and interpretation without the aid of professional image processing software and tools. In addition, remotely sensed data may contain noise or other deficiencies resulting from various reasons, such as abnormal vibration of the observing systems. Therefore, further processing procedures should be conducted to deal with these flaws. Since the remotely sensed data are always archived in two-dimensional image forms, any further processing procedures performed on the raw remotely sensed images can be generally interpreted as image processing.

Toward establishing a definition of image processing, two classic definitions are provided below:

- Image processing is a method to perform some operations on an image in order to get an enhanced image or to extract some useful information from it. It is a type of signal processing in which input is an image and output may be an image or characteristics/features associated with that image (Acharya and Ray, 2005).
- Image processing is the processing of images with mathematical operations by using any form of signal processing for which the input is an image or a series of images (Gonzalez and Woods, 2008).

As suggested therein, the overarching goal of image processing is to produce a better image to aid in visualization or information extraction by performing specific operations on raw images. In general, image processing is intended to make the raw image more interpretable for better understanding of the object of interest.

With the development of remote sensing and computer sciences, various image processing techniques have been developed to aid in the interpretation or information extraction from remotely sensed images. Although the choice of image processing techniques depends on the goals of each individual application, some basic techniques are still common in most remote sensing applications. In this chapter, a set of basic image processing techniques, as well as several software and programming languages that are commonly applied for image processing and analysis in remote sensing will be introduced.

4.2 IMAGE PROCESSING TECHNIQUES

Numerous image processing techniques have been developed in the past decades, and these techniques can be categorized into the following four broad categories: pre-processing, enhancement, transformation, and classification. In this section, techniques associated with each type of application to process remotely sensed images will be introduced.

4.2.1 PRE-PROCESSING TECHNIQUES

As remote sensing systems fly high above Earth's surface, remotely sensed images may contain various distortions due to the vibration of imaging systems, the rotation and curvature of Earth, and atmospheric contaminations. Therefore, it is essential to correct possible flaws in remotely sensed images prior to post-processing and analysis. Forms of this image preparation step are referred to as image pre-processing, which involves a set of initial processing operations that can be used to correct or minimize distortions due to problems resulting from imaging systems and observing conditions (Khorram et al., 2013). In this chapter, typical image pre-processing operations including atmospheric correction, radiometric correction, geometric correction, registration, resampling, mosaic, and gap filling will be discussed.

4.2.1.1 Atmospheric Correction

In remote sensing, radiation from Earth's surface undergoes significant interaction with the atmosphere before it reaches remote sensing sensors. Therefore, the measured radiation typically contains at least two kinds of energy: part of the measured radiance from the reflection of the target of interest and the remnant from the atmosphere itself (Hadjimitsis et al., 2010). For most terrestrial applications, remotely sensed images should be collected under clear scenes without other contaminants. However, it is hard to ensure that the observing scenes are fully clear, as thin clouds, haze, and aerosols in the atmosphere are pervasive. Due to atmospheric effects such as attenuation and scattering of clouds and aerosol particles, radiation measured within these scenes contain atmospheric noise, which in turn affects the interpretation accuracy of the terrestrial targets, especially in visible and near-infrared remote sensing (Herman et al., 1971; Duggin and Piwinski, 1984; Kaufman and Sendra, 1988; Kaufman and Tanré, 1996).

Broadly, atmospheric correction algorithms fall into two categories, absolute and relative correction methods.

- *Absolute correction method*: Absolute correction method involves radiative transfer calculations (e.g., Rayleigh and Mie scattering) by considering many time-dependent parameters, such as sensor viewing geometry, irradiance at the top of the atmosphere, solar zenith angle at the time of satellite overpass, and total optical depth associated with aerosols and other molecular gases (Gonzalez and Woods, 2008). The physics-based absolute correction methods are accurate in atmospheric correction, and have been widely applied in real world applications, such as the commonly used dark object method (e.g., Chavez, 1988; Song et al., 2001; Hadjimitsis and Clayton, 2009). Common models that can be used for absolute atmospheric correction include 6S (Second Simulation of the Satellite Signal in the Solar Spectrum) (Vermote et al., 1997), MODTRAN (MODerate resolution atmospheric TRANsmission model) (Berk et al., 1989), and LOWTRAN (Kneizys et al., 1981).
- *Relative correction method*: Relative correction method works to minimize the variations within a scene through normalization of multiple images collected on different dates compared with a reference of the same scene (Hall et al., 1991; Giri, 2012). A variety of relative correction methods have been proposed, such as pseudo-invariant features (Schott et al., 1988), multivariate alteration detection (Nielsen et al., 1998), and histogram matching (Richter, 1996a,b). Compared to the absolute correction methods involving complex radiative transfer calculations, the relative correction methods are simpler and more efficient.

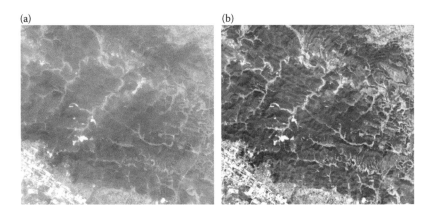

FIGURE 4.1   Comparison of Landsat 8 Operational Land Imager (OLI) true color image (RGB composite of bands 4, 3, 2) on October 6, 2014, (a) before and (b) after atmospheric correction by using Fast Line-of-sight Atmospheric Analysis of Hypercube (FLAASH).

Despite atmospheric correction methods providing a way toward accurate remotely sensed images, atmospheric correction should be carefully conducted as many factors must be estimated. If these estimations are not properly derived, the atmospheric correction might result in even larger bias than the atmosphere itself. An illustrative example of atmospheric correction is demonstrated in Figure 4.1. As suggested therein, significant atmospheric effects are observed within the Landsat 8 Operational Land Imager (OLI) scene on October 6, 2014, due to smog and aerosols, and these effects were predominantly removed after performing atmospheric correction with the aid of the Fast Line-of-sight Atmospheric Analysis of Hypercube (FLAASH) method.

4.2.1.2 Radiometric Correction

Due to radiometric distortions arising from differences in illumination conditions and sensor characteristics, the radiation observed by the imaging system may not coincide with the emitted or reflected energy from the objects of interest (e.g., Duggin and Piwinski, 1984; Schott et al., 1988; Song et al., 2001; Du et al., 2002; Teillet et al., 2007). In order to attain radiometric consistency among different remotely sensed images, radiometric inconsistencies and distortions should be accurately addressed prior to further image interpretation and analysis; processes to remove these radiometric distortions are referred to as radiometric corrections (Pons and Solé-Sugrañes, 1994; Yang and Lo, 2000; Janzen et al., 2006; Pons et al., 2014).

Generally, radiometric correction aims at calibrating the remotely sensed data in order to improve the fidelity of reflected or emitted radiation from objects by removing radiometric distortions due to sensors' irregularities, topographic effects, and sun angles (e.g., Vicente-Serrano et al., 2008; Chander et al., 2009; Gopalan et al., 2009). Considering different origins of radiometric noises, radiometric corrections can be classified into the following two categories associated with sensor irregularities as well as sun angle and topography:

- *Radiometric correction of noise due to sensor irregularities*: Sensor-related radiometric noise may arise from changes in sensor sensitivity, such as degradation of remote sensing sensors after operating in orbit for a long time period (Wang et al., 2012; Kim et al., 2014). The measured signal from a degraded sensor will contain radiometric errors due to changes in the sensor response function. To address these radiometric errors, new relationships between calibrated irradiance measurement and sensor output signal should be established; this process is also termed as calibration (Thorne et al., 1997; Obata et al., 2015).

(a)   (b)

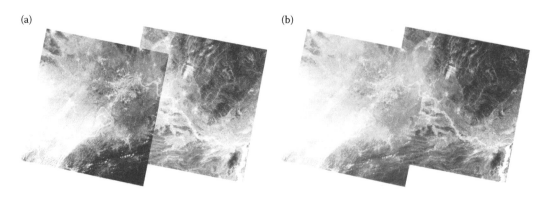

FIGURE 4.2   Landsat OLI scenes (a) before and (b) after radiometric correction.

- *Radiometric correction of noise due to sun angle and topography*: In remote sensing, particularly over the water surface, the observed scenes may be contaminated by the diffusion of the sunlight, resulting in lighter areas in an image (i.e., sun glint). This effect can be corrected by estimating a shading curve which is determined by Fourier analysis to extract a low-frequency component (Kay et al., 2009). In addition, due to topographic effects, especially in mountainous areas, shading effects could result in another kind of radiometric distortion, making the shaded area darker than normal. To remove shading effects, corrections can be conducted using the angle between the solar radiation direction and the normal vector to the ground surface (Dozier and Frew, 1990; Essery and Marks, 2007) (Figure 4.2).

4.2.1.3   Geometric Correction

In remote sensing, radiometric characteristics of objects of interest are always collected along with geospatial (i.e., geographic location) information. Ideally, one pixel with the same geolocation information in two images should precisely coincide to the same grid point on the ground. In other words, pixels with the same geolocation should spatially overlay each other in the geographic domain. However, remotely sensed images are always recorded along with the motion of imaging instruments and platforms, hence geometric distortions may arise from changes in the sensor-Earth geometry and altitude variations, such as movement of imaging platforms, variations in platform velocity and scanning rate, and distortions due to the curvature of Earth (Khorram et al., 2013, 2016). Pixels recorded in different images may not be projected onto the same geographic grid due to geometric distortions. Therefore, essential corrections or registrations should be performed to improve the spatial coincidence among images.

Processes to adjust the coordinate system of images to reduce possible geometric distortions and make pixels in different images coincide to the same grid points are referred to as geometric correction (Toutin, 2004). Typically, geometric correction can be achieved through image-to-map rectification or image-to-image registration (coregistration) by establishing affine relationships between image coordinate system and geographic coordinate system with ground control points (Toutin, 2004; Baboo and Devi, 2011; Khorram et al., 2013). The control points are used to build a transformation relationship (e.g., polynomial) that will shift the raster dataset from its existing location to the spatially correct location. This process also includes image registration, which is an integrated image processing technique commonly used to align multiple scenes into a single integrated image by transforming different sets of data into one identical coordinate system. Registration is essential to image overlaying, because it helps to resolve common issues arising from image rotation and scale, especially for overlaying two or more images of the same scene taken at different times, from different viewpoints, and/or by different sensors for further grid scale calculations (Zitová and Flusser, 2003). Image registration is considered a crucial step in all image analysis tasks, in particular for the final knowledge gained through the integration of information from multiple data sources. Thus, it is widely used in remote sensing, medical imaging, computer vision, and cartography.

The traditional approach for geometric correction mainly depends on the manual identification of many ground control points to align the raster data, which is labor-intensive and time-consuming (Goshtasby, 1987). In addition, the number of remotely sensed images has grown tremendously, which has reinforced the need for highly efficient and automatic correction methods. With the development of computer sciences and remote sensing technologies, a variety of methods has been developed to advance automatic geometric correction, such as the automated ground control points extraction technique (Gianinetto and Scaioni, 2008), scale-invariant feature transform (Deng et al., 2013), and contour-based image matching (Eugenio et al., 2002).

With advancements in remote sensing technologies, high resolution satellite imagery (aerial imagery as well) has become popular in real-world applications. Due to the high resolution of each pixel, horizontal accuracy is of critical importance, because a tiny geometric variation from either systematic sensors or terrain-related errors could result in significant distortions in the observed imagery. Orthorectification, a process of removing inaccuracies caused by sensor, satellite/aircraft motion, and terrain-related geometric distortions from raw imagery to improve the horizontal accuracy, is also essential in geometric correction. Orthorectified imagery is required for most applications involving multiple image analyses, especially for tasks when overlaying images with existing data sets and maps, such as data fusion, change detection, and map updating. Compared to the original imagery, the resulting orthorectified imagery is planimetric at every location with consistent scale across all parts of the image so that features are represented in their "true" positions, allowing for accurate direct measurement of distances, angles, and areas.

4.2.1.4   Geometric Transformation

Geometric transformation is the process of changing the geometry of a raster dataset from one "coordinate" space to another. Since this "coordinate" space can be either the image coordinate or map coordinate, we have to be very careful when talking about geometric transformations as the object being transformed can be either the geometric objects or the coordinate system. Typically, geometric transformation can be divided into three classes: Euclidean transformations, affine transformations, and projective transformations.

- *Euclidean transformations*: The Euclidean transformations are the most commonly used transformations, and can be a translation, a rotation, or a reflection. Essentially, Euclidean transformations do not change length and angle measures. Moreover, Euclidean transformations preserve the shape of a geometric object (e.g., lines transform to lines, and circles transform to circles). In other words, only position and orientation of the object will be changed.
- *Affine transformations*: Affine transformations are considered generalizations of Euclidean transformations. An affine transformation (or affinity) refers to any transformation that preserves collinearity (i.e., all points lying on a line initially still lie on a line after transformation) and ratios of distances (e.g., the midpoint of a line segment remains the midpoint after transformation) (Weisstein, 2017). In general, affine is a type of linear mapping method. Hence, operations such as scaling, resampling, shearing, and rotation are all affine transformations. The difference between Euclidean and affine is that all Euclidean spaces are affine whereas affine spaces can be non-Euclidean.
- *Projective transformations*: Projective transformations are commonly applied to remotely sensed imagery to transform the observed data from one coordinate system to another by given projection information. Certain properties remain invariant after projective transformations, which include collinearity, concurrency, tangency, and incidence.

In performing geometric transformations on remotely sensed imagery, in particular the affine transformations and projective transformations, resampling is one of the most commonly applied techniques to manipulate one digital image and transform it into a more consistent form by changing its spatial resolution or orientation for visualization or data analysis (Gurjar and Padmanabhan, 2005). Due to limitations imposed by imaging systems, remotely sensed images captured from different remote sensing instruments may have different spatial resolutions. However, in real-world applications, the initial spatial resolution of a remotely sensed image may not be sufficient, or may need to be consistent with other images. In such cases, resampling should be applied to transform the original image into another form to satisfy application needs.

4.2.1.5   Resampling

Mathematically, resampling involves interpolation and sampling to produce new estimates for pixels at different grids (Parker et al., 1983; Baboo and Devi, 2010). To date, a variety of methods have been developed for resampling, and the choice of resampling kernels is highly application-dependent. The three most common resampling kernels are nearest neighbor, bilinear interpolation, and cubic convolution.

- *Nearest neighbor*: Nearest neighbor is a method frequently used for resampling in remote sensing, which estimates a new value for each "corrected" pixel (i.e., new grid) using data values from the nearest "uncorrected" pixels (i.e., original grids). The advantages of nearest neighbor are its simplicity and capability to preserve original values in the unaltered scene. Nevertheless, the disadvantages of nearest neighbor are also significant, in particularly its blocky effects (Baboo and Devi, 2010). An example of image resampling with nearest neighbor is shown in Figure 4.3.
- *Bilinear interpolation*: Bilinear interpolation is an image smoothing method which uses only values from the four nearest pixels that are located in diagonal directions from a given pixel to estimate appropriate values of that pixel (Parker et al., 1983; Baboo and Devi, 2010). In general, bilinear interpolation takes a weighted average of the closest 2×2 neighborhood of known pixel values surrounding the corresponding pixel to produce an interpolated value. Weights assigned to the four pixel values are normally based on the computed pixel's distance (in 2D space) from each of the known points.
- *Cubic convolution*: Cubic convolution is conducted through a weighted average of 16 pixels nearby the corresponding input pixel through a cubic function. Compared to bilinear interpolation, cubic convolution performs better and the result does not have a disjointed appearance like nearest neighbor (Keys, 1981; Reichenbach and Geng, 2003). However, computational times required by cubic convolution are about 10 times more than those required by the nearest neighbor method (Baboo and Devi, 2010).

(a)   (b)

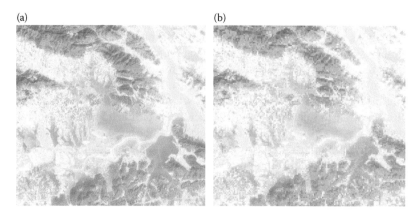

FIGURE 4.3   Comparison of Landsat Thematic Mapper (TM) image (RGB composite of bands 4, 3, 2) on October 17, 2009, at (a) 30-meter and (b) 200-meter (resampled) spatial resolution.

In addition to the three aforementioned commonly used resampling kernels, there are some other methods for resampling, such as the fast Fourier transformation resampling (Li, 2014) and quadratic interpolation (Dodgson, 1997).

4.2.1.6 Mosaicking

Due to constraints of imaging systems, observations within a single scene may be incapable of providing a full coverage of targets of interest. Therefore, an assemblage of different images together to form one image with larger spatial coverage is desirable. In image processing, such a blending process is referred to as mosaicking (Inampudi, 1998; Abraham and Simon, 2013).

Generally, mosaicking of images relies on the identification of controlled points or features in different images, and then blends these images based on the overlap of these extracted common controlled points or features (Inampudi, 1998). The most straightforward mosaicking is to blend images collected from the same or adjacent satellite paths because of minimal radiometric differences between these images (e.g., Figure 4.2). However, when images are collected from different paths with significant time-elapsing differences, radiometric corrections should be conducted prior to mosaicking. Otherwise, new radiometric distortions might be introduced to the blended images.

4.2.1.7 Gap Filling

Due to the presence of thick clouds or instrumental malfunctions, missing data values are commonly observed in remotely sensed images. The most well-known example is the unscanned gaps presenting in the Landsat 7 Enhanced Thermal Mapper Plus (ETM+) images, which result in roughly 22% information loss in each ETM+ image (e.g., Figure 4.4a) due to the failure of the scan-line corrector (SLC). Gaps in these remotely sensed images significantly reduce their utility for environmental monitoring applications. In order to apply these valuable data sources, gaps in these remotely sensed images should be removed by recovering or reconstructing all value-missing pixels through a process known as gap filling (e.g., Maxwell et al., 2007; Zhang et al., 2007; Chen et al., 2011; Zhu et al., 2012).

To date, numerous gap-filling approaches have been developed and used in real-world applications. In general, these approaches can be broadly classified into three categories based on spatial information, temporal information, or both.

- *Gap-filling methods relying on spatial information*: Methods relying on spatial information to fill data gaps are mainly based on spatial interpolation, which use the non-gap neighboring pixels to estimate values for value-missing pixels. Methods relying on this theory can be found in numerous literatures, especially for removing gaps presenting in EMT+ images (Addink, 1999; Maxwell et al., 2007; Zhang et al., 2007; Chen et al., 2011; Zhu et al., 2012).

(a) (b)

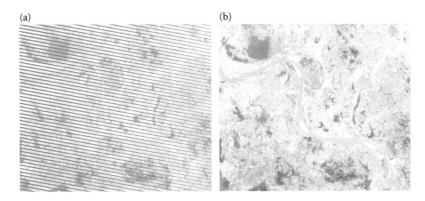

FIGURE 4.4 An example of Landsat 7 ETM+ SLC-off image (RGB composite of bands 4, 3, 2) on February 12, 2015 (a) before and (b) after gap filling.

- *Gap-filling methods relying on temporal information*: Such gap-filling methods mainly take advantage of historical information in time series to recover missing pixel values through memory effects (Roerink et al., 2000; Gao et al., 2008; Kandasamy et al., 2013; Verger et al., 2013). These methods are commonly used to fill gaps in time series such as Leaf Area Index (LAI) and Normalized Difference Vegetation Index (NDVI), which have prominent seasonal recurrence along the time horizon.
- *Gap-filling methods using both spatial and temporal information*: To achieve higher accuracy, in some cases both spatial and temporal information are used to model the missing pixel values. A simple example is the method proposed by Kang et al. (2005), who proposed an approach for gap-filling ecosystem metrics using simple spatial interpolation within land cover classes. If no cloud-free pixels were found within a 5×5 pixel window, the algorithm used temporal interpolation to fill the pixel using data from earlier and later dates. More complicated methods can also be found in the literature, such as the smart information reconstruction algorithm (Chang et al., 2015) and other hybrid approaches (Weiss et al., 2014).

4.2.2 ADVANCED PROCESSING TECHNIQUES

In addition to those common pre-processing techniques mentioned above, certain advanced processing techniques can also be applied to the processed imagery to further enhance the property or quality of the imagery. This type of effort leads to the advancement of our understanding in regard to how to manipulate those valuable image data sources. The following four techniques are typically used to further manipulate remotely sensed imagery.

4.2.2.1 Image Enhancement

Basically, image enhancement modifies attributes of objects in an image with the aim of improving the appearance or perception of these objects in order to aid the visual interpretability and analysis (Maini and Aggarwal, 2010). In comparison to other manipulations performed on images, the enhancement changes only the dynamic range of the chosen features for better visualization and interpretation of objects instead of increasing the inherent information content of the original data (Hummel, 1977; Starck et al., 2003). Techniques developed for image enhancement can be generally divided into two categories including spatial domain methods and frequency domain methods.

- *Spatial domain methods*: Spatial domain methods directly deal with image pixels through different operations, such as histogram equalization (Hummel, 1977) and contrast stretching (Yang, 2006). An overview of spatial domain methods can be found in the literature (Maini and Aggarwal, 2010; Bedi and Khandelwal, 2013). An example of image enhancement through histogram equalization is shown in Figure 4.5. Contrast is the difference in visual properties that makes an object distinguishable from other objects and the background. In visual perception, contrast is determined by the difference in the color and brightness of the object compared to other objects. Methods such as contrast stretching for image enhancement are oftentimes used to increase contrast between different objects in order to make the objects of interest distinguishable (Starck et al., 2003; Yang, 2006).
- *Frequency domain methods*: Frequency domain methods operate on Fourier transform of an image. This means that enhancement operations are performed on the Fourier transform of the image, and the final output image is obtained by using the inverse Fourier transform. Filtering is the commonly applied method for image enhancement; filtering out unnecessary information (or noise) highlights certain frequency components (Chen et al., 1994; Silva Centeno and Haertel, 1997).

(a)   (b)

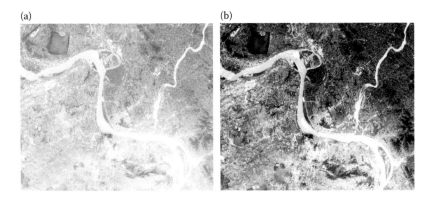

FIGURE 4.5 Landsat TM image (RGB composite of bands 3, 2, 1) on October 17, 2009 (a) before and (b) after enhancement by performing histogram equalization.

Nevertheless, there is no general theory for determining the quality of image enhancement, which means that most enhancements are empirical and require interactive procedures to obtain satisfactory results.

4.2.2.2 Image Restoration

Similar to image enhancement, the purpose of image restoration is also to improve the quality of image. However, image restoration attempts to reconstruct or recover an image that has been degraded or corrupted. Corruptions and degradations can come in many forms, for instance, motion blur, noise, illumination, and color imperfections. Image restoration works in principle by modeling the degradation using *a priori* knowledge and applying the inverse process to restore the lost image information (Lagendijk and Biemond, 1999). The ultimate goal is to reduce noise/corruption and recover information loss. Because image processing is commonly performed either in image domain or frequency domain, the most straightforward way for image restoration is to perform filtering and deconvolution (Gunturk and Li, 2013). To date, a wide range of techniques have been developed to perform image restoration, such as Weiner filtering (Lagendijk and Biemond, 1999), Fourier transform (Lagendijk and Biemond, 1999), wavelet transform (Figueiredo and Nowak, 2003), and blind deconvolution (Figueiredo and Nowak, 2003).

4.2.2.3 Image Transformation

In image processing, remotely sensed images can be converted from one representation or form to another by applying simple arithmetic or complex mathematic operations; these processes are referred to as image transformations (Gonzalez and Woods, 2008). In general, image transformations include broad categories as forms of images are changed through operations applied to images, such as image enhancement, rotation, resampling, registration, and so forth. However, unlike the processes which are normally applied to only one image or a single spectral band, the transformations mentioned here mainly refer to operations applied to multiple images or bands to create "new" images in order to highlight certain features or objects of interest.

Arithmetic operations such as subtraction, addition, multiplication, and division are commonly applied to perform transformation among images or spectral bands, and are also termed spectral or band math (Khorram et al., 2013, 2016). A representative of such kind of transformations is the calculation of vegetation index (Figure 4.6), such as the most frequently used NDVI (Rouse et al., 1974):

$$NDVI = \frac{NIR - Red}{NIR + Red} \tag{4.1}$$

(a)   (b)

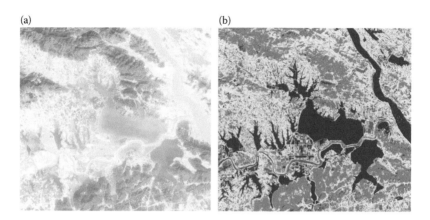

FIGURE 4.6   Normalized difference vegetation index generated from Landsat TM image on October 17, 2009. (a) Landsat TM image (RGB composite of bands 4, 3, 2) and (b) NDVI.

where *NIR* and *Red* denote reflectance collected from a near-infrared band and a red band, respectively. In real-world applications, various vegetation indices such as LAI have been developed to aid in the monitoring of vegetation; most of them rely on the absorption differences of vegetation in the red and near-infrared wavelengths, such as the soil-adjusted vegetation index (Huete, 1988) and enhanced vegetation index (Huete et al., 2002).

In addition to simple arithmetic operations, Principal Component Analysis (PCA) is another procedure that is frequently applied for image transformation, especially for information reduction on multispectral or particular hyperspectral images, as the multispectral imagery data are always correlated from one band to the other (Cheng and Hsia, 2003; Pandey et al., 2011). In image processing, the essence of PCA is to apply a linear transformation of multispectral band data to make a rotation and translation of the original coordinate system (Batchelor, 1978). Normally, PCA is performed on all bands of multispectral images without *a priori* information associated with image spectral characteristics. Derived PCAs represent the spectral information more efficiently than the original ones. The first principal component always accounts for the largest portion of variance, while other principal components subsequently account for the remaining variance. Due to its efficiency and information reduction characteristics, PCA has been frequently used for spectral pattern recognition and image enhancement (Cheng and Hsia, 2003; KwangIn et al., 2005; Pandey et al., 2011).

4.2.2.4   Image Segmentation

Image segmentation is the process of partitioning an image into multiple distinct segments each containing pixels with similar attributes. The goal of segmentation is to simplify and/or change the representation of an image into something that is more meaningful and easier to analyze with specific emphasis (Gunturk and Li, 2013). Hence, segmentation is typically used to identify objects or other relevant information in digital images. A wealth of methods and techniques has been developed to perform image segmentation, which can be either contextual or noncontextual. Based on working principles, image segmentation techniques can be further classified as thresholding methods, color-based segmentation, transform methods, and texture methods. The simplest non-contextual segmentation technique is thresholding, which takes no account of spatial relationships between features in an image, but simply groups pixels together on the basis of certain global attributes such as gray level or color. In contrast, contextual techniques consider both spectral and spatial relationships to advance feature extraction, for example, group together pixels with similar gray levels and close spatial locations.

The commonly used image classification can be considered as one special form of segmentation because it works mainly in the spectral domain without referring to spatial relationships among pixels. Classification is an important approach to distinguish different types of objects based on

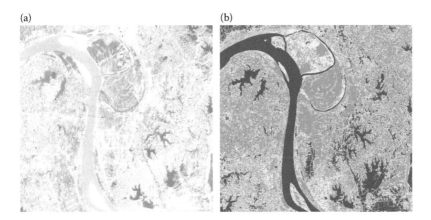

FIGURE 4.7 Unsupervised classification of Landsat–5 TM image on October 17, 2009, using the ISODATA method. (a) Landsat–5 TM image with RGB composite of bands 4, 3, 2 and (b) classified image.

their distinctive features (Acharya and Ray, 2005; Gonzalez and Woods, 2008; Giri, 2012). Land covers are identified and classified into different categories based on the differences of spectral features. In general, techniques developed for image classification in remote sensing can be divided into unsupervised classification and supervised classification.

- *Unsupervised classification*: Pixels in one image are automatically classified and then grouped into separate clusters, depending on the similarities of spectral features of each pixel, without human intervention (Lee et al., 1999; Fjortoft, 2003). These kinds of classifications are also termed as clustering, and the representative algorithms are K-means (Mac Queen, 1967) and Iterative Self-Organizing Data Analysis Technique (ISODATA) (Ball and Hall, 1964). Classification with unsupervised methods is simple and fast since it involves only statistical calculation of the input image. However, the final output highly depends on the number of clusters given by operators, and results in feature mixtures frequently, especially for those objects having similar spectral characteristics, such as water and shadows. In addition to ISODATA, a variety of algorithms have been developed for unsupervised classification, such as the K-means and based methods (Hara et al., 1994; Yuan et al., 2009), probabilistic methods (Fjortoft, 2003), and even hybrid methods (Lee et al., 1999) (Figure 4.7).
- *Supervised classification*: Compared to unsupervised approaches, supervised classifications require the user to select representative samples for each cluster as training sites (i.e., samples) beforehand, and the identified clusters thus highly depend on these predetermined training sites (Khorram et al., 2016). Therefore, the final output depends heavily on the cognition and skills of the image specialist for training site selection. Commonly used supervised classification algorithms include maximum likelihood (Ahmad and Quegan, 2012) and minimum-distance classification (Wacker and Landgrebe, 1972). Despite this, results from supervised classification are still much more accurate than those from unsupervised approaches.

4.3 COMMON SOFTWARE FOR IMAGE PROCESSING

With the advancement of computer sciences and remote sensing technologies, various tools, software packages, and programming languages have been developed and applied for image processing purposes. Here, some common software packages and programming languages frequently used for image processing in remote sensing are introduced.

4.3.1   ENVI

ENVI, an acronym for "the ENvironment for Visualizing Images," is a software application developed by the Exelis Visual Information Solutions (Exelis VIS) company, which specializes in remote sensing imagery processing and analysis. ENVI was first released in 1994, and written in IDL (Interactive Data Language). In contrast to the text-based IDL, ENVI has a suite of user-friendly Graphical User Interfaces (GUI) with a number of advanced scientific algorithms and wizard-based tools embedded for imagery visualization, analysis, and processing (Figure 4.8).

As shown in Figure 4.8, EVNI provides various algorithms and tools for image processing and analysis, including basic imagery reading modules to visualize images collected from different platforms in different formats, as well as pre-processing functions and further advanced spatial and spectral transformations. Compared to other image processing software, one of the advantages of ENVI lies in its distinct combination of spectral-based and file-based techniques through interactive manipulations which enables users to easily manipulate more than one image simultaneously for advanced processing steps. In addition, ENVI provides extension interfaces to external tools and functions, which enables users to create customized or application-oriented tools for different purposes. Due to its supereminent performance in image processing, ENVI has been used in a variety of industries, particularly in remote sensing.

4.3.2   ERDAS IMAGINE

ERDAS IMAGINE, a geospatial image processing application with raster graphics editor capabilities designed by ERDAS Inc., has also been widely applied to process and analyze remotely sensed imagery from different satellite platforms such as AVHRR, Landsat, SPOT, and LiDAR. Before the ERDAS IMAGINE suite, various products were developed by ERDAS Inc. under the name of ERDAS to assist in processing imagery collected from most optical and radar mapping sensors.

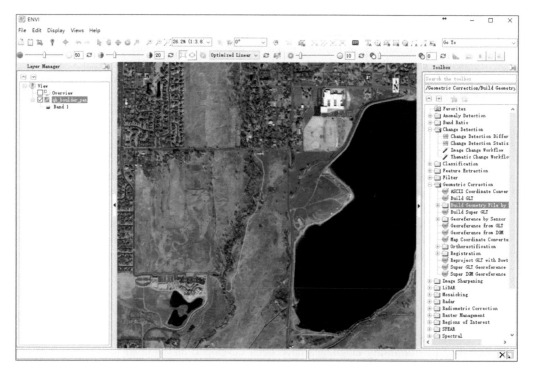

FIGURE 4.8   The GUI of ENVI version 5.0.

Similar to most image processing applications, ERDAS IMAGINE also provides a user-friendly GUI to support imagery visualization, mapping, and so forth.

The first version of ERDAS was released in 1978, whereas the ERDAS IMAGINE was provided in 1991. The latest version of ERDAS IMAGINE was released in 2015. Like all the previous products, ERDAS IMAGINE aims mainly at processing geospatial raster data by providing many solutions associated with image visualization, mapping, and data (e.g., raster, vector, LiDAR point) analysis in one, allowing users to perform numerous operations on imageries toward specific goals. It supports optical panchromatic, multispectral, and hyperspectral imagery, as well as radar and LiDAR data in a wide variety of formats.

By integrating multiple geospatial technologies, ERDAS IMAGINE can be used as a powerful package to process remotely sensed imagery supporting consolidated workflows. In addition, ERDAS IMAGINE is flexible, depending on users' needs. It provides three product tiers (i.e., Essentials, Advantage, and Professional) designed for all levels of users, which enables handling any geospatial analysis task. Due to the robust multicore and distributed batch processing, ERDAS IMAGINE is capable of handling tasks with a remarkable processing performance through dynamic modeling, even when dealing with massive datasets from any sensor.

4.3.3 PCI Geomatica

PCI Geomatica is a powerful geospatial image processing software suite commonly used by researchers and professionals to process and analyze remotely sensed data and imagery. It aims primarily at fast data processing and allowing users to accurately perform advanced analysis and operations on remotely sensed imagery. The latest version of the software is Geomatica 2015, which provides visualization tools that support a variety of satellite and aerial sensors, including the latest instruments. Meanwhile, hundreds of algorithms specifically optimized for performance and accuracy have been assembled and integrated in PCI to support fast and accurate geospatial imagery manipulation and analysis.

Due to its powerful workflows for orthorectification and mosaicking through OrthoEngine, and the automation capability to generate intelligent products, PCI allows for efficient handling of complex tasks and has been widely applied by many users around the world for image processing purposes. Compared to similar image processing packages, PCI features cutting-edge atmospheric correction workflows based on the automatic detection of cloud and haze within imagery, easier and faster extraction of digital elevation models, quick and direct access to data, and advanced Synthetic Aperture Radar (SAR) analysis as well.

4.3.4 ArcGIS

ArcGIS is a leading geographic information system (GIS) application that allows users to work with geospatial maps and perform geoprocessing on the input raw data resulting in the production of valuable information. The first ArcGIS suite was released in late 1999 by ESRI (Environmental Systems Research Institute). Prior to ArcGIS, ESRI had developed various products focusing mainly on the development of ArcInfo workstation and several GUI-based products such as the ArcView. However, these products did not integrate well with one another. Within this context, ESRI revamped its GIS software platform toward a single integrated software architecture, which finally resulted in the ArcGIS suite.

ArcGIS provides a comprehensive platform to manage, process, and analyze the input raster or vector data to extract valuable information. It is capable of managing geographic information in a database, creating and analyzing geospatial maps, discovering and sharing geographic information, and so forth. Key features of ArcGIS include: (1) a variety of powerful spatial analysis tools, (2) automated advanced workflows, (3) high-quality maps creation, (4) geocoding capabilities, and (5) advanced imagery support.

With the development of remote sensing, ArcGIS provides a suite of image processing and analysis tools enabling users to better understand the information locked in the imagery pixels. At present, ArcGIS is capable of efficient managing and processing of time-variant, multi-resolution imagery from multiple sources (e.g., satellite, aerial, LiDAR, and SAR), formats (e.g., GeoTIFF, HDF, [General Regularly-distributed Information in Binary form] [GRIB], and netCDF), and projections. In addition to the basic viewing and editing modules, ArcGIS provides a number of extensions that can be added to aid in complex tasks, including spatial analyst, geostatistical analyst, network analyst, 3D analyst, and so forth, which are capable of geoprocessing, data conversion, and analysis.

Due to its multiple functionalities, ArcGIS has been widely applied to process geospatial imagery in remote sensing. One of the significant features of ArcGIS is that it provides a model builder tool, which can be used to create, edit, and manage workflows for automatic sequential execution of geoprocessing tools. In other words, outputs of one tool are fed into another tool as input (Figure 4.9). The established model can be thought of as a new tool for batch processing, and it is of great help in handling large volumes of datasets (e.g., long-term satellite imagery) for multiple processing purposes (Figure 4.9).

4.3.5   MATLAB®

MATLAB is a high-level proprietary programming language developed by MathWorks Inc. that integrates computation, visualization, and programming in a user-friendly interactive environment. MATLAB has been widely used across disciplines for numeric computation, data analysis and visualization, programming and algorithm development, creation of user interfaces, and so forth. Since the basic data element of MATLAB is an array, it allows fast solution formulations for many numeric computing problems, in particular those involving matrix representations, such as images (i.e., two-dimensional numerical arrays). This means image processing operations can be easily

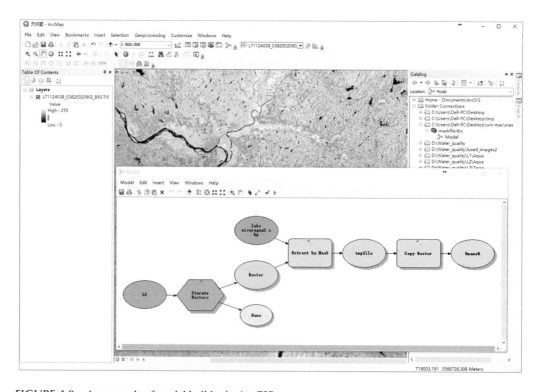

FIGURE 4.9   An example of model builder in ArcGIS.

expressed in a compact and clear manner toward a quick solution of image processing problems (Gonzalez et al., 2004).

With the development of an extensive set of algorithms and functions specializing in manipulating images, the capability of MATLAB is extended to the image processing domain. These comprehensive algorithms and functions are achieved through a toolbox termed the Image Processing Toolbox. With the aid of this toolbox, MATLAB can be easily applied to perform image analysis and processing including image segmentation, enhancement, registration, and transformations, as well as noise reduction and so forth. In addition, many algorithms and functions provided in the toolbox support multicore processors and even GPUs (i.e., graphics processing units), resulting in the acceleration of image processing, especially for computationally intensive workflows.

At present, MATLAB supports a diverse set of image types in different formats. Images achieved in standard data and image formats can be directly read into a matrix in MATLAB for visualization and even further manipulation purposes, as well as a number of specialized file formats, such as HDF and netCDF. Meanwhile, results or matrices acquired after processing can also be exported as raster datasets or images.

4.3.6 IDL

IDL, short for Interactive Data Language, is a scientific program with similar capabilities to MATLAB, also developed by Exelis VIS. It has been commonly used along with ENVI, an image processing software package built in IDL, for data analysis and image processing, particularly in remote sensing and medical imaging. Similar to other programming languages, IDL incorporates three essential capabilities including interactivity, graphics display, and array-oriented operation for data analysis. Its vectorized nature makes IDL capable of performing fast array computations, especially for numerically heavy computations, by taking advantage of the built-in vector operations.

With the capability of handling a large volume of data, IDL has been widely applied for image processing and analysis. In addition to the built-in iTools widgets for interactive image display, hundreds of algorithms and functions are provided for further advanced image manipulation and processing with capabilities including segmentation, enhancement, filtering, Fourier transform and wavelet transform, spectral analysis, and so forth. A distinctive feature of IDL is that it can be used to develop customized tools for use as extended modules in ENVI for specific purposes.

4.4 SUMMARY

In this chapter, a variety of commonly used image pre-processing techniques including atmospheric correction, radiometric correction, geometric correction, resampling, mosaicking, and gap filling are discussed, as well as advanced processing methods including image enhancement, image transformation, and image classification. In addition, image processing software and programming languages such as ENVI, ArcGIS, MATLAB, and IDL are also briefly introduced. In the next chapter, concepts of feature extraction in remote sensing will be formally introduced to expand the theoretical foundation of remote sensing.

REFERENCES

Abraham, R. and Simon, P., 2013. Review on mosaicing techniques in image processing. *International Journal of Software Engineering Research and Practices*, 3, 63–68.

Acharya, T. and Ray, A. K., 2005. *Image Processing: Principles and Applications.* Wiley InterScience, New Jersey.

Addink, E. A., 1999. A comparison of conventional and geostatistical methods to replace clouded pixels in NOAA-AVHRR images. *International Journal of Remote Sensing*, 20, 961–977.

Ahmad, A. and Quegan, S., 2012. Analysis of maximum likelihood classification on multispectral data. *Applied Mathematical Sciences*, 6, 6425–6436.

Baboo, D. S. S. and Devi, M. R., 2010. An analysis of different resampling methods in coimbatore, district. *Journal of Computer Science and Technology*, 10, 61–66.

Baboo, D. S. S. and Devi, M. R., 2011. Geometric correction in recent high resolution satellite imagery: A case study in coimbatore, Tamil Nadu. *International Journal of Computer Applications*, 14, 32–37.

Ball, G. H. and Hall, D. J., 1964. Some fundamental concepts and synthesis procedures for pattern recognition preprocessors. In: *International Conference on Microwaves, Circuit Theory, and Information Theory*, September, Tokyo, 113–114.

Batchelor, B. G., 1978. Digital image processing. *Electronics & Power*, 24, 863.

Bedi, S. S. and Khandelwal, R., 2013. Various image enhancement techniques—a critical review. *International Journal of Advanced Research in Computer Engineering*, 2, 1605–1609.

Berk, A., Bernstein, L. S., and Robertson, D. C., 1989. *MODTRAN: A Moderate Resolution Model for LOWTRAN 7*. Technical Report, May 12, 1986–May 11, 1987. Spectral Sciences, Inc., Burlington, MA.

Chander, G., Markham, B. L., and Helder, D. L., 2009. Summary of current radiometric calibration coefficients for landsat MSS, TM, ETM+, and EO-1 ALI sensors. *Remote Sensing of Environment*, 113, 893–903.

Chang, N.-B., Bai, K., and Chen, C.-F., 2015. Smart information reconstruction via time-space-spectrum continuum for cloud removal in satellite images. *IEEE Journal of Selected Topics in Applied Earth Observations and Remote Sensing*, 8, 1898–1912.

Chavez, P. S., 1988. An improved dark-object subtraction technique for atmospheric scattering correction of multispectral data. *Remote Sensing of Environment*, 24, 459–479.

Chen, H., Li, A., Kaufman, L., Hale, J., Haiguang, C., Li, A., Kaufman, L., and Hale, J., 1994. A fast filtering algorithm for image enhancement. *IEEE Transactions on Medical Imaging*, 13, 557–564.

Chen, J., Zhu, X., Vogelmann, J. E., Gao, F., and Jin, S., 2011. A simple and effective method for filling gaps in Landsat ETM+ SLC-off images. *Remote Sensing of Environment*, 115, 1053–1064.

Cheng, S.-C. and Hsia, S.-C., 2003. Fast algorithms for color image processing by principal component analysis. *Journal of Visual Communication and Image Representation*, 14, 184–203.

Deng, H., Wang, L., Liu, J., Li, D., Chen, Z., and Zhou, Q., 2013. Study on application of scale invariant feature transform algorithm on automated geometric correction of remote sensing images. In: *Computer and Computing Technologies in Agriculture VI*, 352–358. Editted by Li, D. and Chen, Y., Zhangjiajie, China.

Dodgson, N. A., 1997. Quadratic interpolation for image resampling. *IEEE Transactions on Image Processing*, 6, 1322–1326.

Dozier, J. and Frew, J., 1990. Rapid calculation of terrain parameters for radiation modeling from digital elevation data. *IEEE Transactions on Geoscience and Remote Sensing*, 28, 963–969.

Du, Y., Teillet, P. M., and Cihlar, J., 2002. Radiometric normalization of multitemporal high-resolution satellite images with quality control for land cover change detection. *Remote Sensing of Environment*, 82, 123–134.

Duggin, M. J. and Piwinski, D., 1984. Recorded radiance indices for vegetation monitoring using NOAA AVHRR data; atmospheric and other effects in multitemporal data sets. *Applied Optics*, 23, 2620.

Essery, R. and Marks, D., 2007. Scaling and parametrization of clear-sky solar radiation over complex topography. *Journal of Geophysical Research-Atmospheres*, 112, D10122.

Eugenio, F., Marques, F., and Marcello, J., 2002. A contour-based approach to automatic and accurate registration of multitemporal and multisensor satellite imagery. In: *IEEE International Geoscience and Remote Sensing Symposium*, 3390–3392. Toronto, Ontario, Canada.

Figueiredo, M. A. T. and Nowak, R. D., 2003. An EM algorithm for wavelet-based image restoration. *IEEE Transactions on Image Processing*, 12, 906–16.

Fjortoft, R., 2003. Unsupervised classification of radar images using hidden markov chains and hidden markov random fields. *IEEE Transactions on Geoscience and Remote Sensing*, 14, 675–686.

Gao, F., Morisette, J. T., Wolfe, R. E., Ederer, G., Pedelty, J., Masuoka, E., Myneni, R., Tan, B., and Nightingale, J., 2008. An algorithm to produce temporally and spatially continuous MODIS-LAI time series. *IEEE Geoscience and Remote Sensing Letters*, 5, 60–64.

Gianinetto, M. and Scaioni, M., 2008. Automated geometric correction of high-resolution pushbroom satellite data. *Photogrammetric Engineering & Remote Sensing*, 74, 107–116.

Giri, C. P., 2012. *Remote Sensing of Land Use and Land Cover: Principles and Applications*, CRC Press, Boca Raton, FL, pp. 1–469.

Gonzalez, R. C. and Woods, R. E., 2008. *Digital Image Processing*. 3rd edition, Pearson Prentice Hall, Upper Saddle River, NJ.

Gonzalez, R. C., Woods, R. E., and Eddins, S. L., 2004. *Digital Image Processing Using MATLAB*. Gatesmark Publishing, Knoxville, TN.

Gopalan, K., Jones, W. L., Biswas, S., Bilanow, S., Wilheit, T., and Kasparis, T., 2009. A time-varying radiometric bias correction for the TRMM microwave imager. *IEEE Transactions on Geoscience and Remote Sensing*, 47, 3722–3730.

Goshtasby, A., 1987. Geometric correction of satellite images using composite transformation functions. In: *The 21st International Symposium on Remote Sensing of Environment*, Ann Arbor, Michigan.

Gunturk, B. K., Li, X., 2013. *Image Restoration Fundamentals and Advances*. CRC Press, Boca Raton, FL.

Gurjar, S. B. and Padmanabhan, N., 2005. Study of various resampling techniques for high-resolution remote sensing imagery. *Journal of the Indian Society of Remote Sensing*, 33, 113–120.

Hadjimitsis, D. G. and Clayton, C., 2009. Darkest pixel atmospheric correction algorithm: A revised procedure for environmental applications of satellite remotely sensed imagery. *Environmental Monitoring and Assessment*, 159, 281–292.

Hadjimitsis, D. G., Papadavid, G., Agapiou, A., Themistocleous, K., Hadjimitsis, M. G., Retalis, A., Michaelides, S. et al. 2010. Atmospheric correction for satellite remotely sensed data intended for agricultural applications: impact on vegetation indices. *Natural Hazards and Earth System Sciences*, 10, 89–95.

Hall, F. G., Strebel, D. E., Nickeson, J. E., and Goetz, S. J., 1991. Radiometric rectification: Toward a common radiometric response among multidate, multisensor images. *Remote Sensing of Environment*, 35, 11–27.

Hara, Y., Atkins, R., Yueh, S., Shin, R., and Kong, J., 1994. Application of neural networks to radar image classification. *IEEE Transactions on Geoscience and Remote Sensing*, 32, 1994.

Herman, B. M., Browning, S. R., and Curran, R. J., 1971. The effect of atmospheric aerosols on scattered sunlight. *Journal of the Atmospheric Sciences*, 28, 419–428.

Huete, A., 1988. A soil-adjusted vegetation index (SAVI). *Remote Sensing of Environment*, 25, 295–309.

Huete, A., Didan, K., Miura, T., Rodriguez, E., Gao, X., and Ferreira, L., 2002. Overview of the radiometric and biophysical performance of the MODIS vegetation indices. *Remote Sensing of Environment*, 83, 195–213.

Hummel, R., 1977. Image enhancement by histogram transformation. *Computer Graphics and Image Processing*, 6, 184–195.

Inampudi, R. B., 1998. Image Mosaicing. In: *IGARSS '98. Sensing and Managing the Environment. 1998 IEEE International Geoscience and Remote Sensing. Symposium Proceedings*, 2363–2365.

Janzen, D. T., Fredeen, A. L., and Wheate, R. D., 2006. Radiometric correction techniques and accuracy assessment for Landsat TM data in remote forested regions. In: *Canadian Journal of Remote Sensing*, 330–340.

Kandasamy, S., Baret, F., Verger, A., Neveux, P., and Weiss, M., 2013. A comparison of methods for smoothing and gap filling time series of remote sensing observations—application to MODIS LAI products. *Biogeosciences*, 10, 4055–4071.

Kang, S., Running, S. W., Zhao, M., Kimball, J. S., and Glassy, J., 2005. Improving continuity of MODIS terrestrial photosynthesis products using an interpolation scheme for cloudy pixels. *International Journal of Remote Sensing*, 26, 1659–1676.

Kaufman, Y. J. and Sendra, C., 1988. Algorithm for automatic atmospheric corrections to visible and near-IR satellite imagery. *International Journal of Remote Sensing*, 9, 1357–1381.

Kaufman, Y. J. and Tanré, D., 1996. Strategy for direct and indirect methods for correcting the aerosol effect on remote sensing: From AVHRR to EOS-MODIS. *Remote Sensing of Environment*, 55, 65–79.

Kay, S., Hedley, J. D., and Lavender, S., 2009. Sun glint correction of high and low spatial resolution images of aquatic scenes: A review of methods for visible and near-infrared wavelengths. *Remote Sensing*, 1, 697–730.

Keys, R., 1981. Cubic convolution interpolation for digital image processing. *IEEE Transactions on Acoustics, Speech, and Signal Processing*, 29, 1153–1160.

Khorram, S., Nelson, S. A. C., Cakir, H., and van der Wiele, C. F., 2013. Digital image acquisition: Preprocessing and data reduction, in: Pelton, J. N., Madry, S., and Camacho-Lara, S. (Eds.) *Handbook of Satellite Applications*, 809–837.

Khorram, S., van der Wiele, C. F., Koch, F. H., Nelson, S. A. C., and Potts, M. D., 2016. *Principles of Applied Remote Sensing*. Springer, New York.

Kim, W., He, T., Wang, D., Cao, C., and Liang, S., 2014. Assessment of long-term sensor radiometric degradation using time series analysis. *IEEE Transactions on Geoscience and Remote Sensing*, 52, 2960–2976.

Kneizys, F. X., Shettle, E. P., and Gallery, W. O., 1981. Atmospheric transmittance and radiance: The LOWTRAN 5 code, in: Fan, R. W. (Ed.), *SPIE 0277, Atmospheric Transmission*. 116 (July 28, 1981), 116–124. SPIE, Washington D.C., United States.

KwangIn, K., Franz, M., and Scholkopf, B., 2005. Iterative kernel principal component analysis for image modeling. *Pattern Anal. Mach. Intell.* 27, 1351–1366.

Lagendijk, R. and Biemond, J., 1999. Basic methods for image restoration and identification, in: Bovik, A. (Ed.), *Handbook of Image and Video Processing*, 1–25. Academic Press, Massachusetts, USA.

Lee, J. S., Grünes, M. R., Ainsworth, T. L., Du, L. J., Schuler, D. L., and Cloude, S. R., 1999. Unsupervised classification using polarimetric decomposition and the complex wishart classifier. *IEEE Transactions on Geoscience and Remote Sensing*, 37, 2249–2258.

Li, Z., 2014. Fast Fourier transformation resampling algorithm and its application in satellite image processing. *Journal of Applied Remote Sensing*, 8, 83683.

Mac Queen, J., 1967. Some methods for classification and analysis of multivariate observations. In: *Proceedings of the Fifth Berkeley Symposium on Mathematical Statistics and Probability*, 231–297.

Maini, R. and Aggarwal, H., 2010. A Comprehensive review of image enhancement techniques. *Journal of Computing*, 2, 39–44.

Maxwell, S. K., Schmidt, G. L., and Storey, J. C., 2007. A multi-scale segmentation approach to filling gaps in Landsat ETM+ SLC-off images. *International Journal of Remote Sensing*, 28, 5339–5356.

Nielsen, A. A., Conradsen, K., and Simpson, J. J., 1998. Multivariate alteration detection (MAD) and MAF postprocessing in multispectral, bitemporal image data: New approaches to change detection studies. *Remote Sensing of Environment*, 64, 1–19.

Obata, K., Tsuchida, S., and Iwao, K., 2015. Inter-band radiometric comparison and calibration of ASTER visible and near-infrared bands. *Remote Sensing*, 7, 15140–15160.

Pandey, P. K., Singh, Y., and Tripathi, S., 2011. Image processing using principle component analysis. *International Journal of Computer Applications*, 15, 37–40.

Parker, J. A., Kenyon, R.V., and Troxel, D. E., 1983. Comparison of interpolating methods for image resampling. *IEEE Transactions on Medical Imaging*, 2, 31–39.

Pons, X. and Solé-Sugrañes, L., 1994. A simple radiometric correction model to improve automatic mapping of vegetation from multispectral satellite data. *Remote Sensing of Environment*, 48, 191–204.

Pons, X., Pesquer, L., Cristóbal, J., and González-Guerrero, O., 2014. Automatic and improved radiometric correction of Landsat imagery using reference values from MODIS surface reflectance images. *International Journal of Applied Earth Observation and Geoinformation*, 33, 243–254.

Reichenbach, S. E. and Geng, F., 2003. Two-dimensional cubic convolution. *IEEE Transactions on Image Processing*, 12, 857–865.

Richter, R., 1996a. A spatially adaptive fast atmospheric correction algorithm. *International Journal of Remote Sensing*, 17, 1201–1214.

Richter, R., 1996b. Atmospheric correction of satellite data with haze removal including a haze/clear transition region. *Computers & Geosciences*, 22, 675–681.

Roerink, G. J., Menenti, M., and Verhoef, W., 2000. Reconstructing cloudfree NDVI composites using fourier analysis of time series. *International Journal of Remote Sensing*, 21, 1911–1917.

Rouse, J. W., Hass, R. H., Schell, J. A., and Deering, D. W., 1974. Monitoring vegetation systems in the great plains with ERTS. In: *Third Earth Resources Technology Satellite (ERTS) Symposium*, pp. 309–317. Texas, United States.

Schott, J. R., Salvaggio, C., and Volchok, W. J., 1988. Radiometric scene normalization using pseudoinvariant features. *Remote Sensing of Environment*, 26, 1–16.

Silva Centeno, J. A. and Haertel, V., 1997. An adaptive image enhancement algorithm. *Pattern Recognition*, 30, 1183–1189.

Song, C., Woodcock, C. E., Seto, K. C., Lenney, M., and Macomber, S. A., 2001. Classification and change detection using landsat TM data: when and how to correct atmospheric effects. *Remote Sensing of Environment*, 75, 230–244.

Starck, J.-L., Murtagh, F., Candes, E. J., and Donoho, D. L., 2003. Gray and color image contrast enhancement by the curvelet transform. *IEEE Transactions on Image Processing*, 12, 706–717.

Teillet, P. M., Fedosejevs, G., Thome, K. J., and Barker, J. L., 2007. Impacts of spectral band difference effects on radiometric cross-calibration between satellite sensors in the solar-reflective spectral domain. *Remote Sensing of Environment*, 110, 393–409.

Thorne, K., Markharn, B., Slater, P., and Biggar, S., 1997. Radiometric calibration of landsat. *Photogrammetric Engineering & Remote Sensing*, 63, 853–858.

Toutin, T., 2004. Review article: Geometric processing of remote sensing images: Models, algorithms and methods. *International Journal of Remote Sensing*, 25, 1893–1924.

Verger, A., Baret, F., Weiss, M., Kandasamy, S., and Vermote, E., 2013. The CACAO method for smoothing, gap filling, and characterizing seasonal anomalies in satellite time series. *IEEE Transactions on Geoscience and Remote Sensing*, 51, 1963–1972.

Vermote, E. F., Tanré, D., Deuzé, J. L., Herman, M., and Morcrette, J. J., 1997. Second simulation of the satellite signal in the solar spectrum, 6s: an overview. *IEEE Transactions on Geoscience and Remote Sensing*, 35, 675–686.

Vicente-Serrano, S. M., Pérez-Cabello, F., and Lasanta, T., 2008. Assessment of radiometric correction techniques in analyzing vegetation variability and change using time series of Landsat images. *Remote Sensing of Environment*, 112, 3916–3934.

Wacker, A. G. and Landgrebe, D. A., 1972. *Minimum distance classification in remote sensing*. LARS Technichal Reports, paper 25.

Wang, D., Morton, D., Masek, J., Wu, A., Nagol, J., Xiong, X., Levy, R., Vermote, E., and Wolfe, R., 2012. Impact of sensor degradation on the MODIS NDVI time series. *Remote Sensing of Environment*, 119, 55–61.

Weiss, D. J., Atkinson, P. M., Bhatt, S., Mappin, B., Hay, S. I., and Gething, P. W., 2014. An effective approach for gap-filling continental scale remotely sensed time-series. *ISPRS Journal of Photogrammetry and Remote Sensing*, 98, 106–118.

Weisstein, E. W. Affine Transformation. From MathWorld—A Wolfram Web Resource. http://mathworld.wolfram.com/AffineTransformation.html. Accessed 2017.

Yang, C.-C., 2006. Image enhancement by modified contrast-stretching manipulation. *Optics & Laser Technology*, 38, 196–201.

Yang, X. and Lo, C. P., 2000. Relative radiometric normalization performance for change detection from multi-date satellite images. *Photogrammetric Engineering & Remote Sensing*, 66, 967–980.

Yuan, H., Van Der Wiele, C. F., and Khorram, S., 2009. An automated artificial neural network system for land use/land cover classification from landsat TM imagery. *Remote Sensing*, 1, 243–265.

Zhang, C., Li, W., and Travis, D., 2007. Gaps-fill of SLC-off Landsat ETM+ satellite image using a geostatistical approach. *International Journal of Remote Sensing*, 28, 5103–5122.

Zhu, X., Liu, D., and Chen, J., 2012. A new geostatistical approach for filling gaps in Landsat ETM+ SLC-off images. *Remote Sensing of Environment*, 124, 49–60.

Zitová, B. and Flusser, J., 2003. Image registration methods: a survey. *Image and Vision Computing*, 21, 977–1000.

Part II

Feature Extraction for Remote Sensing

5 Feature Extraction and Classification for Environmental Remote Sensing

5.1 INTRODUCTION

Human beings seeking to detect and extract information from imagery dates back to the time when the first photographic image was acquired, as early as the mid-nineteenth century. Motivated by the subsequent advances in photogrammetry, the invention of the airplane, improvements in the relevant instrumentations and techniques, the advent of digital imagery, and the capabilities of electronic processing, interest in efficiently extracting information from imagery to help with learning and decision-making has increased significantly (Wolf et al., 2000; Quackenbush, 2004).

With the advancement of remote sensing, a wealth of instruments has been deployed onboard various satellite and space-borne platforms dedicated to providing versatile remotely sensed data to monitor Earth's environment. As many remotely sensed imageries with high spatial, temporal, and spectral resolutions are available on a daily basis at the global scale, the data volume increases by many orders of magnitude, making it even harder to convert images into actionable information and knowledge through conventional manual interpretation approaches for further decision-making (Momm and Easso, 2011). Manual interpretation is time-consuming and labor-intensive; in addition, it is difficult to cope with large volume information embedded in remotely sensed data, particularly for remotely sensed images with fine resolutions in spectral (e.g., hyperspectral images) and spatial (e.g., panchromatic images) domains.

Along this line, many statistical and geophysical methods were developed to help retrieve information from different types of remote sensing imageries. Machine learning and/or data mining are relatively new methods for feature extraction. When performing feature extraction with machine learning or data mining in search of geospatial intelligence for a complex dataset, one of the major problems is the low efficiency issue stemming from the large number of variables involved. With more learning algorithms becoming available, feature extraction not only requires a huge amount of memory and computational power, but also results in a slow learning process with possible overfitting the training samples and poor generalization of the prediction to new samples (Zena and Gillies, 2015). Generally, the large amount of information retrieved from remotely sensed data makes it difficult to perform classification or pattern recognition for environmental decision-making, because the observed information can be miscellaneous and highly overlapped, although some are complementary with each other. Overall, these problems can be mainly attributed to the large amount of redundant or complementary information embedded in data in either spatial, temporal or spectral domain requiring tremendous efforts of data analyses and syntheses.

When the input data to an algorithm is prone to be redundant or too large to be managed, it is desirable to transform the raw data into a reduced form by keeping only the primary characteristics of the raw data. Key information embedded in such reduced forms can be well represented or preserved by a set of features to facilitate the subsequent learning process and improve generalization and interpretability toward efficient decision making. In image processing and pattern recognition, the techniques designed to construct a compact feature vector well representing the raw observations are referred to as feature extraction, which is largely related to dimensionality reduction (Sharma and Sarma, 2016). In some cases, the reduced form (i.e., feature vector) could even lead to better human interpretations, especially for hyperspectral data, which has more than several hundreds to one thousand total bands. Due to significant advantages in reducing the size and dimensionality of the raw data, feature extraction has been widely used to help with the problem of constructing and identifying certain types of features from the given input data to solve various problems via the use of machine learning, data mining, image compression, pattern recognition, and classification.

In fields of computational intelligence and information management such as machine learning and pattern recognition, feature extraction has become the most critical step prior to classification and decision-making, as the final performance of analysis is highly dependent on the quality of extracted features. A typical workflow of image processing and pattern recognition for environmental monitoring is presented in Figure 5.1, from which we can see that feature extraction is the first essential step of processing after the pre-processing. With the fast development of computer sciences and other relevant information technologies, having all the features of interest in an observed scene automatically identified at the push of a button, namely, a process of automatic feature extraction, is truly appealing and plausible. The ultimate goal is to develop automatic and intelligent techniques to cope with the problem of detecting and extracting informative features from the input data effectively and efficiently.

In this chapter, basic concepts and fundamentals associated with feature extraction, as well as a wealth of commonly applied feature extraction techniques that can be used to help with classification problems in remote sensing, will be introduced to aid in environmental decision-making. Different learning strategies summarized below will be thoroughly discussed:

- *Supervised Learning*: This involves a set of target values that may be fed into the learning model, allowing the model to adjust according to errors.
- *Unsupervised Learning*: This is required when there is not a set of target values for a model to learn, such as searching for a hidden pattern in a big dataset. Often, clustering analysis is conducted by dividing the big data set into groups according to some unknown pattern.
- *Semi-supervised Learning*: This is a class of supervised learning processes that make use of very small amounts of labeled data within a large amount of unlabeled data for training. In this way, we may guess the shape of the underlying data distribution and generalize better to new samples. These algorithms can perform well when we have a very small amount of labeled points and a large amount of unlabeled points.

In addition, metrics that can be used to evaluate the performance of feature extraction methods as well as perspectives of feature extraction will be presented.

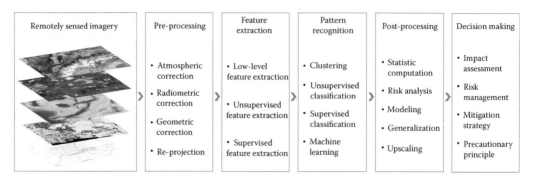

FIGURE 5.1   A typical framework of pattern recognition for environmental decision-making.

5.2 FEATURE EXTRACTION CONCEPTS AND FUNDAMENTALS

5.2.1 DEFINITION OF FEATURE EXTRACTION

Due to its efficacy in manipulating large amounts of information embedded in remotely sensed imagery, feature extraction has long been considered the Holy Grail of remote sensing. However, it is still difficult to develop a unified definition for feature extraction, and a wealth of variations can be found in the literature:

- Feature extraction is the process of transforming raw data into more informative signatures or characteristics of a system, which will most efficiently or meaningfully represent the information that is important for analysis and classification (Elnemr et al., 2016).
- Feature extraction is a process for extracting relevant information from an image. After detecting a face, some valuable information is extracted from the image which is used in the next step for identifying the image (Bhagabati and Sarma, 2016).
- Feature extraction is the process of transforming the input data into a set of features which can very well represent the input data. It is a special form of dimensionality reduction (Sharma and Sarma, 2016).
- Feature extraction is a process of deriving new features from the original features in order to reduce the cost of feature measurement, increase classifier efficiency, and allow higher classification accuracy (Akhtar and Hassan, 2015).
- Feature extraction is a process of extracting the important or relevant characteristics that are enclosed within the input data. Dimensionality or size of the input data will be subsequently reduced to preserve important information only (Ooi et al., 2015).
- Feature extraction is a special form of dimensionality reduction aiming at transforming the input data into a reduced representation set of features (Kumar and Bhatia, 2014).
- Feature extraction is one of the important steps in pattern recognition, aiming at extracting a set of descriptors, various characteristic attributes, and the relevant information associated to form a representation of input pattern (Ashoka et al., 2012; Jain et al., 2000).
- Feature extraction is the process of extracting and building features from raw data. Feature functions are utilized to extract and process informative features that are useful for prediction (Gopalakrishnan, 2009).
- Feature extraction is a dimensionality reduction method that finds a reduced set of features that are a combination of the original ones (Sánchez-Maroño and Alonso-Betanzos, 2009).
- Feature extraction refers to the extraction of linguistic items from the documents to provide a representative sample of their content. Distinctive vocabulary items found in a document are assigned to the different categories by measuring the importance of those items to the document content (Durfee, 2006).
- Feature extraction can be viewed as finding a set of vectors that represent an observation while reducing the dimensionality (Benediktsson et al., 2003).
- Feature extraction is a process that extracts a set of new features from the original features through some functional mapping (Wyse et al., 1980).

The above-listed definitions are all meaningful and informative, indicating that the key to feature extraction is to construct a compact feature vector to well represent the original data in a lower dimensionality space. However, it is clear that the definition varies among research domains and applications. By summarizing previous definitions, we may define feature extraction broadly as a general term referring to "the process of constructing a set of compact feature vectors by extracting the most relevant features from the input data to facilitate further decision-making by using the reduced representation (i.e., feature vector) instead of the original full-size data while still maintaining sufficient accuracy."

5.2.2 Feature and Feature Class

A compact feature vector is of critical importance to feature extraction, as the final performance depends largely on how well the extracted features can represent the original input data set. Nevertheless, it is difficult to obtain a universal definition of a feature, because each specific property of any object can be considered a possible feature. For example, an object can be discriminated by simply considering its intrinsic characteristics such as color, size, shape, edge, and other related properties (Lillesand et al., 1994; Tian, 2013). Therefore, any pattern associated with distinct properties of objects can be applied to construct a feature. For example, in speech recognition, features can be the length of sounds, relative power, and noise ratios. By contrast, in image processing, the resulting features could be several unique subsets of the image represented as Binary Large OBjects (BLOBs or blobs), isolated points, edges, continuous curves or connected patches. Therefore, the definition of a feature can vary substantially between disciplines. In machine learning and pattern recognition, a feature may refer to an individual measurable property of an observed phenomenon (Bishop, 2006). In computer vision and image processing, a feature can be a piece of information relevant for solving the computational task related to a certain application (Liu et al., 2012). Overall, the concept of a feature is very general, and a solid definition often depends largely on the domain of a specific problem or the type of application.

Given that, a feature can be referred to as a distinct pattern or structure associated with the input data, such as a set of unique points or small patches present in an image (Figure 5.2). Toward feature extraction, what constitutes a feature and what the feature actually represents are not issues; the key is to determine whether the resulting feature set is sufficient to facilitate further decision-making processes. There is no doubt that a good feature representation is essential to attain high accuracy in any image processing or pattern recognition tasks (Elnemr et al., 2016). However, no definite criterion is available to define a good feature set. Generally, a good feature should exhibit the following properties:

- *Informative*: The resulting feature should be expressive and perceptually meaningful, and be able to explain a certain level of information embedded in the input data.
- *Distinctive*: The neighborhood around the feature center varies enough to allow for a reliable discrimination between the features.
- *Nonredundant*: Features derived from different samples of the same class should be grouped in the same category, and each type of feature should represent a unique property of the input data.
- *Repeatable detections*: The resulting features should be the same in two different images of the same scene. In other words, the features should be resistant to changes in viewing conditions and noise, such as the presence of rotation and scaling effect.
- *Localizable*: The feature should have a unique location assigned to it, and changes in viewing conditions or directions should not affect its location.

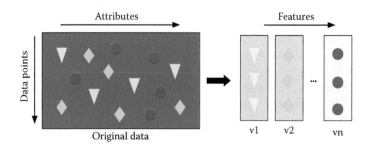

FIGURE 5.2   An illustrative example of selecting features from a given input data.

Nevertheless, the aforementioned properties are not the only criteria that can be used to evaluate a feature vector, and features should not be limited to these characteristics, because the resulting features are highly dependent on the specific problems at hand.

Features can be broadly categorized as low-level features and high-level features, although there is no distinct gap between them (Elnemr et al., 2016). In general, low-level features are fundamental features such as edges and lines as well as many other basic descriptors that can be easily detected without performing any complex manipulation, which can be further divided into general features and domain-specific features. The so-called "general features" mainly refer to those common features that can be directly detected from any given image. In other words, general features should be universal and application-independent. Three general features commonly used are:

- *Color features*: Color is one of the most important features of images because it is visually intuitive to human perception. Color features are often defined subject to a particular color space or model, and the most popular color spaces are RGB (red-green-blue) and HSV (hue-saturation-value). Based on these color spaces, a variety of color features including color histogram (Wang et al., 2009), color moment (Huang et al., 2010), color coherence vector (Pass et al., 1998), and color correlogram (Huang et al., 1997) can be then extracted.
- *Texture features*: In image processing, texture refers to a set of metrics designed to quantify the perceived information about the spatial arrangement of color or intensities in an image or selected region of an image (Haralick et al., 1973). As opposed to color, which is usually represented by the brightness of each individual pixel, texture is often measured based on a set of pixels by considering spatial or spectral similarities among pixels. Based on the domain from which features are extracted, textures can be divided into spatial texture features and spectral texture features.
- *Shape features*: Shape is an important geometrical cue used by human beings to discriminate real-world objects. A shape can be described by different parameters such as rectangularity, circularity ratio, eccentricity ratio, and center of gravity.

A list of object-based features that have been categorized into three classes, namely spectral, textural, and shape features is summarized in Table 5.1. Most of the object-based features in Table 5.1

TABLE 5.1

List of Object-based Features that have been Categorized into Three Classes: Spectral, Textural, and Shape Features (Chen et al., 2014)

Feature Category	Features
Spectral feature	Mean, Standard Deviation, Skewness, Ratio, Maximum, Minimum Mean of Inner Border, Mean of Outer Border, Mean Diff. to Darker Neighbors, Mean Diff. to Brighter Neighbors, Contrast to Neighbor Pixels, Edge Contrast of Neighbor Pixels, Std. Dev. To Neighbor Pixels, Circular Mean, Circular Std. Dev., Mean Diff. to Neighbors, Mean Diff. to Scene, Ratio to Scene
Texture GLCM feature	Angular Second Moment, Contrast, Correlation, Dissimilarity, Entropy, Homogeneity, Mean, Std. Dev.
Texture GLDV feature	Angular Second Moment, Contrast, Entropy, Mean
Shape feature	Area, Asymmetry, Border Index, Border Length, Compactness, Density, Elliptic Fit, Length, Main Direction, Radius of Largest Enclosed Ellipse, Radius of Smallest Enclosing Ellipse, Rectangular Fit, Roundness, Shape Index, Width

Source: Chen, X., Li, H., and Gu, Y., 2014. *2014 Fourth International Conference on Instrumentation and Measurement, Computer, Communication and Control*, Harbin, China, 539–543.

Note: The features and their abbreviations are shown in the right column.

are low-level features. In contrast, the so-called "domain-specific features" mainly refer to those application-dependent features, and thus are highly related to practical applications in the domain. For instance, fingerprints can be used as good features in human identity identification. Analogously, human faces can be detected for face recognition, while chlorophyll-a content can be considered a good feature for vegetation detection. In general, domain-specific features are confined to certain specific applications that are not universal across domains.

High-level features refer to those feature vectors further derived from the low-level features by using certain extra algorithms after basic feature extraction in the sense that hybrid algorithms are often employed. The primary difference between low- and high-level features lies in the complexity of the process used to extract the advanced features based on the low-level features. In Table 5.1, for example, shape features should be application-dependent, such as fingerprints, faces, and body gestures used for basic evidence for high-level human pattern recognitions, whereas spectral features usually serve as low-level features for subsequent high-level feature extraction. Although it is sometimes harder to obtain a high-level feature, it is of more help in understanding the designated target embedded in those low-level features.

5.2.3   Fundamentals of Feature Extraction

The concept of feature extraction refers to the construction of a set of feature vectors from the input data to facilitate the complex decision-making process. Generally, a process of feature extraction should consist of the following three essential steps: (1) feature detection, (2) feature construction, and (3) feature selection. Feature detection aims at computing abstractions of the input image by finding the inherent unique objects, such as edges or blobs. More specifically, it refers to a discriminative process involving making local decisions at every pixel (i.e., image point) to examine whether a given type of feature is presented at that point. The rationale is to find targets of interest that can be used for feature construction and to finally choose an appropriate feature detector. Based on the types of image features, common feature detectors can be grouped into three categories: (1) edge, (2) corner, and (3) blob. In terms of a feature detector, the desirable property is repeatability, which means that the detected features should be locally invariant, even in the presence of temporal variations or rotations. The same feature should be detectable in two or more different images of the same scene, regardless of changes in orientation or scale.

A good feature detector provides better chances of finding adequate unique objects that can be used for generating features from the given input image. Once objects of interest have been detected, a set of features can be constructed by transforming those candidate objects into different types of features, thus the process of generating or constructing new features from functions of the original detected objects is referred to as feature construction. This construction often involves a variety of image processing techniques, such as segmentation and clustering (Bouman and Shapiro, 1994; Agrawal et al., 2005), aimed at computing a set of descriptors by transforming fragmentary objects into feature vectors to well represent the original input image. Since the input remotely sensed image always contains a large amount of information, the resulting initial set of raw features can be diverse. Although a larger number of input features may increase the final estimation accuracy by providing more descriptions of the target to a certain extent, the increased number of features may in turn add more computational burden and result in new problems such as overfitting. The redundant or irrelevant features might make the feature vector too large to be efficiently managed by the implemented algorithms. Therefore, removing those redundant and irrelevant features from the initial set of raw features to reduce the number of input features is essential to facilitate advanced learning and subsequent classification.

The process of constructing a new and compact feature vector by selecting a subset of relevant features from the initial set of raw features based on certain principles or criteria is thus termed feature selection. In other words, feature selection works to select the most relevant attributes from the input data. This is achievable by removing redundant or irrelevant features and reducing the dimensionality of feature vectors, which facilitates the advanced learning process and improves the generalization (Zena and Gillies, 2015). Thus, feature selection techniques are frequently used in many domains to cope with problems with large input spaces, such as data mining, pattern recognition, and image processing, because too much information can reduce the effectiveness of further data manipulation due to complexities. Nevertheless, we should be aware that feature selection is different from feature construction, as the latter generates new feature vectors, whereas the former focuses on selecting a subset of features. In addition, applying a feature selection technique relies primarily on the input data, which contains a variety of either redundant or irrelevant features, as the dimensionality or space can be substantially reduced while still maintaining sufficient information to well represent the original target (Bermingham et al., 2015).

A feature selection method can be considered an integrated process of feature subset screening and quality assessment (Peng et al., 2005). By examining the manner of combining the screening algorithm and the model building, the feature selection methods can be divided into three primary categories including (1) wrappers, (2) filters, and (3) embedded methods (Das, 2001; Guyon and Elisseeff, 2003; Zena and Gillies, 2015). Some different applications for feature selection methods were summarized in Table 5.2. These methods were differentiated mostly by the preselected evaluation criterion (Guyon and Elisseeff, 2006).

The wrapper methods use a predictive model to score feature subsets, whereas the filter methods only use a proxy measure (e.g., correlation and mutual information) instead of the error rate to rank subsets of features (Guyon and Elisseeff, 2003). More specifically, the wrapper methods take advantage of the prediction performance of the given learning machine to evaluate the importance of each feature subset (Kohavi and Johnb, 1997). Thus, it has nothing to do with the chosen learning machine, as it is often used as a perfect black box (Phuong et al., 2005). Compared to the wrapper methods, filter-type methods are usually less computationally intensive, as the rationale of filter methods is mainly based on the filter metrics (e.g., mutual information) without incorporating learning to detect the similarity between a candidate feature subset and the desired output (Hall, 1999; Peng et al., 2005; Nguyen et al., 2009). Nevertheless, filter methods are vulnerable to redundant features, as the interrelationships between candidate features are not taken into account. Many experimental results show that although the wrapper methods have the disadvantage of being computationally inefficient, they often yield better performances (Zhuo et al., 2008). Embedded methods aim at reducing the computational complexity by incorporating feature selection as part of the training process, which is usually specific to the given learning machine (Guyon and Elisseeff, 2003; Duval et al., 2009; Zare et al., 2013). In general, extracting features from a given input data set is associated with combining various attributes into a reduced set of features, which is a combination of art and science, as the whole process involves the integration of advanced computational algorithms as well as the knowledge of the professional domain expert.

Overall, there are three primary aspects associated with feature extraction:

- *Feature detectors*: A good feature detector is of critical importance to the final extracted features, as the detected inherent objects in the original input data are fundamental elements for the construction of the initial set of raw features.
- *Feature construction*: The process of constructing features is the key to feature extraction, and how well the constructed features can represent the original target determines the final performance of the whole feature extraction process.
- *Dimensionality reduction*: Selecting a subset of features from the initial set of raw features by removing those redundant or irrelevant features may significantly improve the learning and generalization efficiency, which in turn advances the development and application of feature extraction techniques, particularly in domains dealing with large feature spaces, such as remote sensing applications.

TABLE 5.2

Different Feature Selection Methods and Their Characteristics

Methods	Types	Descriptions	References
Minimum-redundancy-maximum-relevance (mRMR) feature selection	Filter	The method aims to select good features according to the maximal statistical dependency criterion based on mutual information.	Peng et al. (2005)
Bayesian network	Filter	The method can be viewed as a search and optimization procedure where features are evaluated based on their likelihood.	Castro and Von Zuben (2009); Hruschka et al. (2004)
Correlation Feature Selection (CFS)	Filter	Features are evaluated on the basis of their correlation with the class.	Haindl et al. (2006); Hall (1999); Yu and Liu (2003)
Cascade Correlation Feature Selection (C2FS)	Wrapper	This new internal wrapper feature selection method selects features at the same time hidden units are being added to the growing C2 net architecture.	Backstrom and Caruana (2006)
Genetic algorithm	Wrapper	The method uses an evolutional way to optimize the feature subset.	Zhuo et al. (2008)
Sequential search	Wrapper	Candidate features are sequentially added to the subset until the further addition does not increase the classification performance.	Glass and Cooper (1965); Nakariyakul (2014)
Particle Swarm Optimization (PSO)	Wrapper	Features are selected according to the likelihood calculated by PSO.	Xue et al. (2013)
Support Vector Machine-Recursive Feature Elimination (SVM-RFE)	Embedded	The SVM-RFE method looks for the features that lead to the maximum margin separation between the classes as the features are ranked based on certain ranking criteria.	Guyon et al. (2002)
Kernel-Penalized SVM	Embedded	The method uses the scaling factors principle to penalize the use of features in the dual formulation of SVM by considering an additional term that penalizes the zero norm of the scaling factors.	Maldonado and Weber (2011)
Random Forests	Embedded	The method combines binary decision trees built based on several bootstrap samples, as each decision tree has maximal depth and is not pruned, and using different algorithms to attain generalization improvement.	Genuer et al. (2010)
Laplacian Score ranking + a modified Calinski–Harabasz index	Hybrid	The method sorts the features according to their relevance and evaluates the features considering them as a subset rather than individually based on a modified Calinski–Harabasz index.	Solorio-Fernández et al. (2016)
Information gain + wrapper subset evaluation + genetic algorithm	Hybrid	The method uses a combination of sample domain filtering and resampling to refine the sample domain and two feature subset evaluation methods to select reliable features.	Naseriparsa et al. (2013)

Although many high-end computational algorithms can substantially advance feature extraction, the knowledge of the domain expert is still critical because it is often difficult to quantitatively assess the accuracy or performance of each process so expert knowledge is thus of help. For instance, the selection of feature detectors and the number of features to be selected is often determined according to human intuition and interpretation.

5.3   FEATURE EXTRACTION TECHNIQUES

Feature extraction is motivated by the fact that data analysis tasks like environmental modeling often require mathematically and computationally convenient input. Real-world data, such as remotely sensed multispectral images, however, are usually complex, redundant, and highly variable. Thus, there is a need to extract useful features or representations from raw input in a compact manner. Although human interpretation and expert knowledge can substantially aid in the extraction of features, converting the original input into actionable information is often labor-intensive because the input data are often voluminous, whereas the human interpretation is usually less efficient and arbitrary. Hence, an advanced feature extraction technique is essential to automate detection in order to select the most relevant features from a set of candidate feature vectors, particularly in coping with problems with large inputs, for example, remotely sensed hyperspectral imagery.

To date, a wealth of methods and techniques have been developed for feature extraction purposes. Based on the domain on which the feature extraction is performed, feature extraction techniques can be broadly grouped into spectral- and spatial-based feature extraction. Considering the working principles used to extract features based on whether or not the labeled input data is applied, feature extraction techniques can also be divided into three categories: supervised, semi-supervised, and unsupervised methods.

5.3.1   SPECTRAL-BASED FEATURE EXTRACTION

In the spectral domain, the data value over each grid of remotely sensed imagery recorded at one particular channel represents the spectral information of one observed scene at a specific wavelength. Hence, the data size depends highly on the spatial resolution and the number of wavelengths of the sensors. With the advancement of remote sensing technologies, multispectral and hyperspectral imageries provide synergistic capabilities spatially and spectrally to depict the observed target with improved spectral characteristics. Although the increased spectral information in hyperspectral remote sensing enables the detection of targets in more details, the large amount of information may reduce the efficiency of data manipulation or data interpretation. Therefore, using several techniques to extract informative and salient features from remotely sensed imagery in the spectral domain is desirable; this has often been performed as pre-processing to remotely sensed imagery, in particular for hyperspectral data analysis. The extracted features will be more intuitive and efficient than the original data for further visualizations and calculations; in addition, the size of the data after feature extraction will be significantly reduced due to the dimensionality reduction. All feature extraction schemes performed in the spectral domain are collectively referred to as spectral-based feature extraction, regardless of the applied methods and approaches.

The process of spectral-based feature extraction involves statistical transformations or band math calculations among remotely sensed imagery at different spectral channels. The objective is to extract unique features by removing irrelevant information (i.e., noise reduction) to facilitate further classification or decision-making. Methods and approaches toward such a goal are numerous, and three commonly used techniques, including (1) thresholding method, (2) Principal Component Analysis (PCA), and (3) band math, are illustrated below for demonstration.

Thresholding is the simplest technique that has been commonly applied to extract features in imagery, and is actually a form of low-level feature extraction performed as a point operation on the input image (Nixon and Aguado, 2012). Theoretically, with a single threshold, it transforms a greyscale or color image into a binary image, with pixels having data values larger than the threshold labeled as one class (e.g., 0 or 1) and the remaining for the other class. Therefore, such a method works purely based on an arbitrary threshold to partition the data sets into two distinct classes, without referring to other criterion such as spatial relationships between features. Instead, global attributes like grey level or color are used. Since the method separates an image into multiple distinct segments with similar attributes, such a process is also termed as image segmentation.

(a)   (b)

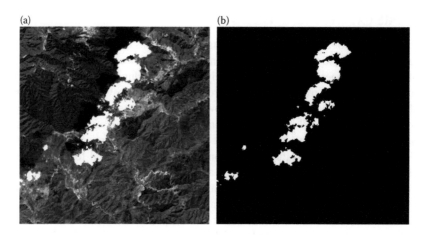

FIGURE 5.3 Clouds extracted from one observed Landsat-8 OLI scene by using the thresholding method. (a) Original image; (b) extracted clouds.

The process to extract features based on threshold is pretty simple and straightforward, but the key is to determine an optimal threshold value. For example, by setting a proper threshold value to the cloud-contaminated surface reflectance imagery observed by the Landsat 8 Operational Land Imager (OLI), clouds can be easily detected and extracted from the original imagery (Figure 5.3). The whole process can be modeled as

$$DN_i' = \begin{cases} 0, DN_i < \theta \\ 1, DN_i \geq \theta \end{cases} \tag{5.1}$$

where DN_i and DN_i' denote the value of a digit number at pixel i in the original image and the segmented binary image (e.g., Figure 5.3b), respectively. θ is the threshold value to be determined by an expert with *a priori* knowledge or other advanced methods.

Although the thresholding method is capable of extracting certain features effectively from a given input imagery, the determination of an optimal threshold is not easy without the aid of a series of experiments or *a priori* knowledge. Furthermore, one single threshold value may not suffice to handle all features in one image with various properties (Lv et al., 2017), for example, land use classification in complex urban regions. To cope with such a complex problem, two or more thresholds can be used to separate each type of feature sequentially.

PCA is a classic statistical technique which has been commonly used to decorrelate a set of possibly correlated variables by projecting the original space into different orthogonal spaces (Pearson, 1901). After the transformation, the resulting vectors are an uncorrelated orthogonal basis set, which is termed principal components (PCs). Theoretically, the total number of resulting PCs should not be greater than the dimension of the input data set, with the first PC accounting for the largest variance of the input data, and with each subsequent component in turn having the rest of highest variance while being orthogonal to the preceding components. Because of this advantage, PCA has been extensively used in processing remotely sensed imagery, especially for the purpose of dimensionality reduction.

Due to the fact that the resulting PCs are orthogonal to one another, PCA can be applied for feature extraction purposes by explaining the largest variance in one specific space because each PC can be considered a unique feature by integrating most of the relevant spectral information. Compared to other methods, PCA has great advantages due to its low complexity, the absence of pre-determined parameters and, last but not least, the fact that PCs are orthogonal to each other. An illustrative example of applying PCA to remotely sensed imagery is demonstrated in Figure 5.4. The first PC explained a total variance of 71.3%, which mainly represents the characteristics of

(a)   (b)   (c)

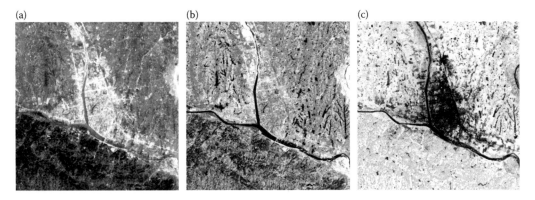

FIGURE 5.4   Principal component analysis performed on multispectral images. (a) Landsat-5 TM true color image; (b) the first principal component; (c) the second principal component.

the texture of the observed scene, whereas the second PC explained a total variance of 23.4%, by mainly emphasizing rivers and built-up areas.

Band math can be linked to the development of some bio-optical models that are functions in terms of certain bands. Remotely sensed imageries recorded at multiple wavelengths provide a synergistic opportunity to better monitor the changing Earth environment because of increasing spectral information content. Based on the absorption spectrum difference, objects viewed in one scene can be separated by generating composite images with specific features emphasized through certain mathematical calculations between images observed at different wavelengths. The process to generate a composite image by using spectral information at different wavelengths with mathematical tools is referred to as band math. Compared to the original imagery with multiple observations, the composite image should be more intuitive. One representative is the Normalized Difference Vegetation Index (NDVI) that is commonly used to assess whether the target being observed contains live green vegetation. NDVI is calculated based on the absorption difference at red and near-infrared wavelengths by green vegetation, which can be modeled as follows (Rouse et al., 1974):

$$NDVI = \frac{NIR - Red}{NIR + Red} \tag{5.2}$$

where *NIR* and *Red* stand for the surface spectral reflectance measurements acquired in the near-infrared and red wavelengths regions, respectively.

Statistically, the values of NDVI should vary between -1.0 and $+1.0$, with green vegetation often possessing positive values. Theoretically, negative values of NDVI (<-0.1) often correspond to non-vegetation objects, for example, water bodies, and values close to zero (-0.1 to $+0.1$) commonly correspond to barren areas of rock, sand or snow. By contrast, low, positive values of NDVI (0.2 to 0.4) should correspond to low-level vegetated areas like shrub and grassland, while high values (>0.4) indicate areas covered with dense green vegetation (e.g., temperate and tropical rainforests). Based on these characteristics, NDVI has considered a good graphic indicator for measuring green vegetation coverage on Earth's surface. Hence, the vegetation features can be easily extracted from a given remotely sensed imagery after calculating the NDVI values. The spatial distribution of green vegetation shown in Figure 5.4a can be easily detected and separated from other targets within the scene based on the calculated NDVI values (Figure 5.5). Similarly, by calculating different water feature-related indexes from a series of multitemporal Landsat 5-TM, 7-ETM+, and 8-OLI images, the spatiotemporal changes of the surface area of the water body in Lake Urmia in the Middle East during 2000–2013 were investigated (Rokni et al., 2014).

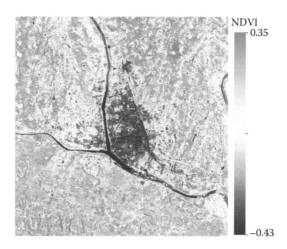

FIGURE 5.5 Calculated NDVI based on Landsat-5 TM surface reflectance at red and near-infrared wavelengths.

5.3.2 SPATIAL-BASED FEATURE EXTRACTION

Objects observed in one scene can be described in terms of not only spectral characteristics but also spatial arrangement. Differing from spectral-based methods relying simply on spectral differences among pixels to extract specific features from remotely sensed imagery, spatial-based feature extraction refers to extracting informative features by considering spatial relationships among pixels, aiming at describing shapes, structures, textures, and the size of a target on Earth's surface (Lv et al., 2017). Although a large variety of spatial feature extraction methods have been developed in the past decades, the methods taking advantage of spatial information for feature extraction are still relatively few compared to widely applied spectral-based methods.

In spatial feature extraction, region-growing methods are commonly used to segment images for feature extraction by considering spatial relationships among pixels, based on the assumption that pixels nearby (or neighboring pixels) often have similar data values or attributes. Therefore, the common practice is to find a data center and then compare this central pixel with its neighbors to examine their similarities. For example, spectrally similar and spatially closing pixels can be gathered to form one specific cluster. In spatial feature extraction methods, connectivity and similarity are two basic aspects commonly applied for feature extraction. Connectivity is defined in terms of pixel neighborhoods while similarity refers to texture property in either grey level or shape. A typical method is the statistical region merging, which begins by building the graph of pixels using 4-connectedness with edges weighted by the absolute value of the intensity difference and then sorting those edges in a priority queue and merging the current regions belonging to the edge pixels based on certain criteria (Boltz, 2004).

With the advent of high spatial resolution remotely sensed imagery, identifying small-scale features such as roads and rivers becomes feasible by using spatial-based feature extraction approaches (Han et al., 2012). Building a mathematical model to detect and extract contextual information is an effective means for spatial feature extraction. Among spatial-based feature extraction methods, Markov random field is one of the classic methods that is extensively used to extract spatial-contexture features; it works mainly based on the probabilistic theory to construct a segmentation model where regions are formed by spatial clusters of pixels with similar intensity (Moser and Serpico, 2013; Zhao et al., 2015). Morphological profiles are another powerful tool that is commonly used to automate spatial feature extraction. Relying on the morphology theory, a wealth of methods has been developed for extracting spatial features, such as gray level co-occurrence matrix (Haralick et al., 1973; Baraldi and Parmiggiani, 1995), pixel-based index (Huang et al.,

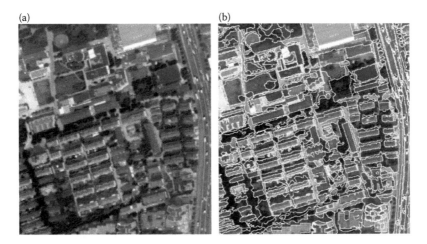

FIGURE 5.6   (a) One Worldview observed scene on August 13, 2010; (b) a segmented image from (a) based on shape, texture, and spectral characteristics.

2007), shape-size index (Han et al., 2012), and morphological profiles (Benediktsson et al., 2003; Huang et al., 2014; Chunsen et al., 2016). Although the fundamental theory of each method is different, the basic process is almost the same since these methods work mainly in the spatial domain to detect and extract features of the targets being observed.

In real-world applications, spatial and spectral attributes are always used in conjunction with one another to provide synergistic capability to aid in complex feature extraction tasks, such as the high-resolution imagery classification (Shahdoosti and Mirzapour, 2017). For instance, urban morphology is commonly characterized by a complex and variable coexistence of diverse and spatially and spectrally heterogeneous objects (Boltz, 2004), hence the classification of land use types in urban regions is by no means an easy task. To cope with such a complex problem, both spatial and spectral attributes should be considered simultaneously to extract multiple land use types effectively. Because features are detected and extracted for different objects, these methods are also referred to as object-based feature extraction (Taubenböck et al., 2010; Shruthi et al., 2011; Lv et al., 2017). For example, by considering both spatial and spectral attributes of objects observed in one Worldview image, different types of ground targets can be detected and extracted with high accuracy (Figure 5.6).

5.4   SUPERVISED FEATURE EXTRACTION

Feature extraction is considered an essential process in exploratory data analysis. To map the input data onto a feature space which reflects the inherent structure of the original data, various methods and approaches have been developed to automate the feature extraction process (Zhao et al., 2006). Considering to what extent the training set is required, the feature learning methods can be broadly categorized into three types: supervised, semi-supervised, and unsupervised (Figure 5.7). Conceptually, in supervised feature extraction, features are learned and extracted with labeled input data, and this learning process is usually referred to as "training" (Mohri, 2012). The labeled input data are termed training sets, which typically consist of a suite of paired training examples (i.e., samples). Commonly, each training set is a pairwise data vector containing an input data set and the corresponding output (i.e., target). For each input vector, the number of included samples is usually arbitrarily determined by the user, as representative samples are often determined typically based on the knowledge of experts. By analyzing the training sets, a supervised learning algorithm is capable of inferring a function between the input and output, which can then be applied for mapping

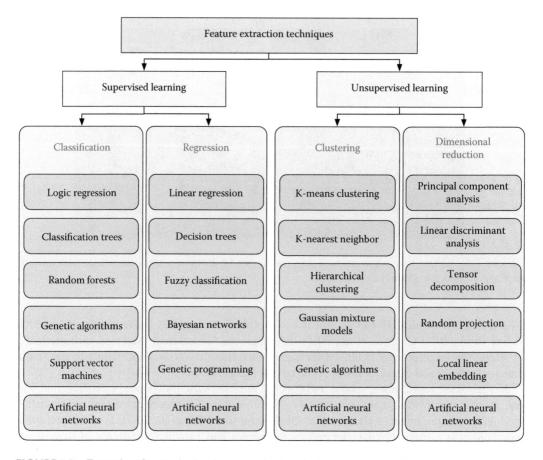

FIGURE 5.7 Examples of supervised and unsupervised methods to automate the feature extraction process.

new inputs. In general, a robust learning algorithm should work well for the unseen data, and this leads to the cross-validation in learning algorithms for performance evaluation.

In order to perform feature extraction in a supervised learning manner, one must go through the following primary steps to:

1. *Determine what data should be used*: This is essential for each exploratory data analysis; the user must first have a clear idea about what data are to be used as training inputs. In the case of land cover classification, for example, a possible training set could be constructed from a remotely sensed panchromatic imagery, a multispectral imagery, or hyperspectral imagery.

2. *Construct an optimal training set*: A supervised learning algorithm will generalize a function from the given training set (i.e., pairwise training samples and targets) and then apply this inferred function for mapping the unseen data. Thus, the training set should be comprehensive and representative, because the final accuracy of the generalized function depends largely on how well the input-output corresponding structure is modeled. In order to facilitate the advanced learning processes, the number of features is usually determined based on expert knowledge such that redundant information could be significantly reduced in the training sets. Otherwise, the learning burden could be very heavy due to a large amount of irrelevant information. Moreover, the learning algorithm might fail to generalize a proper function for the given input because of high dimensionality. On the other hand, the number of features should not be too small, as the training inputs should contain adequate

information to represent all possible cases of the target so as to accurately predict the unseen data.

3. *Select a suitable learning algorithm*: To date, a wide range of learning algorithms are available, and the user can select different algorithms toward a specific application by considering the strengths and weaknesses of each algorithm. In addition, the structure of the learning algorithm must be determined simultaneously; for example, the number of hidden layers and hidden neurons represents the structure of an Artificial Neural Network (ANN) model. Since there is no method satisfying all types of problems, the user should select the algorithm most suitable for the specified real-world application.

4. *Choose a stopping criterion for the learning process*: Once a training set and learning algorithm are determined, the learning process can be started by generating a set of models with the given learning algorithm to build relationships between dependent and independent variables in the training set. In supervised learning, however, certain control parameters (or stopping criteria) are required to stop the learning process, in particular machine learning algorithms, such as ANN and Genetic Programming (GP). These parameters can be tuned through an optimization algorithm, which can also be arbitrarily defined by the user by setting certain criterion via cross-validation.

5. *Examine the performance of the learned function*: After parameter optimization, the performance of the inferred function should be carefully evaluated. The accuracy is commonly assessed by mapping a separated subset that differs from the training set. Statistical comparisons will be performed among the predicted output and the desired value to check the overall accuracy of the inferred function. Once the accuracy meets the anticipated level, the whole learning process is finished and the inferred function can then be applied for mapping unseen data.

Supervised learning has been widely used in environmental remote sensing, and the most common applications of feature extraction in remote sensing are remotely sensed image classification and environmental modeling via machine learning tools. For example, Zhao and Du (2016) developed a spectral–spatial feature-based classification framework to advance hyperspectral image classification by using the trained multiple-feature-based classifier. In this framework, a balanced local discriminant-embedding algorithm was proposed for spectral feature extraction from high-dimensional hyperspectral data sets while a convolutional neural network was utilized to automatically find spatial-related features at high levels. With the aid of GP, Chang et al. (2014) successfully predicted the total organic carbon concentration in William H. Harsha Lake during 2008–2012 based on the *in situ* measurements and the fused satellite-based remote sensing reflectance imagery. The feature extraction performed in these two examples are realized in terms of supervised learning.

Despite the effectiveness of supervised learning algorithms, several major issues with respect to supervised learning should be noted as well. The first is the tradeoff between bias and variance, as the prediction error of a learned function is related to the sum of the bias and the variance of the learning algorithm (Geman et al., 1992; James, 2003). Generally, a learning algorithm with low bias should be flexible enough to fit data well. However, a large variance will be observed in the predicted output if the algorithm is too flexible (e.g., fits each training set differently). Thus, a good learning algorithm should be able to adjust this tradeoff automatically.

The second issue is related to the number of training data and the complexity of the function. A small amount of training data will suffice if the inherent function is simple; otherwise, a large volume of data is required if the true function is highly complex. The third issue is related to the dimensionality of the input space. If the dimensionality of the input features is high, the learning process can be very difficult because the high dimension inputs could confuse the learning algorithm, making it fail to generalize well or generalize at local optimization. Therefore, a large training set typically requires the learning algorithm to have low variance but high bias, and this motivates the development of dimensionality reduction algorithms.

The fourth issue is the noise level embedded in the desired output. In a situation where the predicted values for the desired output are often incorrect, the learning algorithm should stop fitting a function to the training set to avoid possible underfitting. In such a situation, early stopping as well as removing noisy samples from the training set prior to the learning process can be of help. In addition to the above-mentioned four major issues, other aspects such as the redundancy and heterogeneity of data should also be considered in performing supervised learning tasks.

5.5   UNSUPERVISED FEATURE EXTRACTION

In many real-world applications like environmental remote sensing, labeled data (e.g., sampled ground truthing) are usually limited due to the infeasibility or cost inefficiency of obtaining them. Hence, it is necessary to utilize unlabeled data. In contrast to supervised feature extraction, unsupervised feature extraction retrieves features from the given input using certain learning algorithms without referring to any labeled data. The ultimate goal is to detect and extract low-dimensional features, capturing the underlying structures embedded in high-dimensional data. Mathematically, a central case of unsupervised feature learning is the problem of density estimation in statistics (Jordan and Bishop, 2004). Feature learning in an unsupervised way does not utilize the labels of the data and only exploits the structure underlying the data. Hence, unsupervised learning algorithms work by relying mainly on the similarity among features.

The classical approach to unsupervised learning is clustering. Clustering refers to the task of grouping a set of objects into different categories based on similarities among them, that is, objects in the same group are more similar to each other than to those in other groups. Cluster analysis was first introduced by Driver and Kroeber in 1932 in anthropology analysis and then used by Zubin in 1938 and Robert Tryon in 1939 in psychology, while Cattell made it famously known in 1943 by using it for trait theory classification in personality psychology (Tryon, 1939; Cattell, 1943). To date, more than 100 clustering algorithms can be found in the literature. Based on the cluster model, clustering algorithms can be briefly categorized into four different groups: (1) density-based clustering (Ester et al., 1996; Kriegel et al., 2011); (2) distribution-based clustering (Moon, 1996; Carson et al., 2002); (3) centroid-based clustering (e.g., K-means) (Forgy, 1965); and (4) connectivity-based clustering (or hierarchical clustering) (Johnson, 1967; Everitt et al., 2011; Murtagh and Contreras, 2012). Despite a wide range of clustering algorithms, it should be noted that an algorithm that is designed for one kind of model will generally fail on a data set that contains a radically different kind of model (Murtagh and Contreras, 2012). Hence, to determine the most appropriate clustering algorithm for a particular problem, experimental tests are commonly conducted unless there is a mathematical reason to prefer one cluster model over another.

In addition to clustering algorithms, approaches such as PCA, independent principal analysis, local linear embedding, and unsupervised dictionary learning, as well as types of neural network-based methods can be also used as learning algorithms in unsupervised feature extraction toward different purposes (e.g., dimensionality reduction). In real-world applications, it is usually difficult to tell which method is the best one, and *a priori* knowledge of the input data can greatly aid in the selection of the most appropriate method. Otherwise, the performance of the method needs to be tested experimentally.

5.6   SEMI-SUPERVISED FEATURE EXTRACTION

Instead of performing feature extraction based on either completely labeled (supervised) or unlabeled (unsupervised) training inputs, features can be extracted more effectively by taking advantage of a large volume of unlabeled data together with a small amount of labeled data to provide synergistic capability in advancing the learning process. The process of learning features from both labeled and unlabeled data is thus referred to as semi-supervised feature extraction (Kuo et al., 2005; Belkin et al., 2006). Generally, a semi-supervised learning process falls between supervised learning and unsupervised learning. More accurately, it should be considered a class of supervised learning tasks to a certain extent because it makes use of labeled data in the learning process. In other words, the desired output values are provided for a subset of the training data whereas the remaining is unlabeled. Typically, the amount of labeled data used is relatively smaller than that of the unlabeled data. Despite this, the use of a small amount of labeled data in conjunction with unlabeled data may result in considerable improvement in learning accuracy (Zhu, 2008). Therefore, semi-supervised learning can be of great practical value in real-world applications.

By taking advantage of combined information from labeled and unlabeled data, semi-supervised learning attempts to surpass the performance that could be obtained from either supervised learning or unsupervised learning on each individual data set. In order to make use of unlabeled data, the structure of the input data should be limited to one of the following assumptions (Chapelle et al., 2006):

- *Continuity assumption*: Data close to each other are more likely to be labeled in the same class. This is generally assumed in supervised learning and should also be obeyed in the case of semi-supervised learning. This assumption yields a preference for geometrically simple decision boundaries even in low-density regions to guarantee that fewer points in different classes are close to each other.
- *Cluster assumption*: Discrete clusters can be formed and data in the same cluster tend to be labeled in the same class. This is a special case of the continuity assumption that gives rise to clustering-based feature learning.
- *Manifold assumption*: The data tend to lie on a manifold of much lower dimension than that of the input data. Learning can proceed using distances and densities defined on the manifold with both the labeled and unlabeled data to avoid dimensionality issues. The manifold assumption is practical, especially for high-dimensional data with a few degrees of freedom that are hard to model directly.

Due to the difficulty in acquiring a large volume of labeled data and the availability of vast amounts of unlabeled data, semi-supervised learning has recently gained more popularity. In the case of semi-supervised learning, unlabeled data are commonly used to either modify or reprioritize hypotheses obtained from labeled data alone to aid in feature extraction (Zhu, 2008). Toward semi-supervised learning, many different methods and algorithms have been developed in the past decades, such as generative models, which are considered the oldest semi-supervised learning method based on probabilistic theory (Zhu, 2008). Many other methods can be also applied, such as transductive support vector machine (Vapnick, 1998; Yu et al., 2012), information regularization (Corduneanu and Jaakkola, 2002; Szummer and Jaakkola, 2002), and graph-based methods (Camps-Valls et al., 2007). More details of these methods will be introduced in the following chapters.

5.7 IMAGE CLASSIFICATION TECHNIQUES WITH LEARNING ALGORITHMS

One of the feature extraction techniques is image classification which deserves further attention. Early studies of remote sensing image classification relied on a few statistical classifiers (i.e., either supervised or unsupervised learning) such as the Maximum Likelihood (ML) classifier (Ahmad, 2012.), the K-Nearest Neighbor (KNN) classifier (Blanzieri and Melgani, 2008), and the K-means clustering classifier (Phillips, 2002). Since the late 1990s, artificial intelligence-based classifiers such as machine learning (i.e., supervised learning) or data mining (i.e., unsupervised learning) classification techniques have been popular in this field. Various segmentation and classification approaches, such as neural computing (Canty 2009; Kavzoglu, 2009), fuzzy logic (Chen et al., 2009), evolutionary algorithms (Tso and Mather, 1999; Foody and Mathur, 2004a,b), and expert systems (Stefanov et al., 2001), have been widely applied to improve the quality of image classification.

Neural computing methods are data-driven methods that have high fault-tolerance in general (Giacinto and Roli, 2001; Canty, 2009; Kavzoglu, 2009). Many successful applications of remote sensing image classification use classifiers based on various ANNs (i.e., such as Backward Propagation (BP), Radial Basis Function (RBF), Self-Organized Mapping [SOM]) (Heermann and Khazenie, 1992; Hoi-Ming and Ersoy, 2005; Suresh et al. 2008), and global optimization techniques (such as Support Vector Machine [SVM]) (Foody and Mathur, 2004a,b). In ANNs, such as BP, RBF, and SOM, an input space is mapped onto a feature space through the hidden layer, resulting in a nonlinear classifier that outperforms most traditional statistical methods. However, these ANNs are all "black box" models, whose classification mechanisms are difficult to interpret. Problems such as overfitting, local minimum, and slow convergence speed are quite common for neural computing methods. SVM differs radically from ANNs because SVM training processes always give rise to a global minimum, and their simple geometric interpretation provides opportunities for advanced optimization. While ANNs are limited by multiple local minima, the solution to an SVM is global and unique; on the other hand, ANNs use empirical risk minimization whereas SVMs choose structural risk minimization.

Classifiers based on fuzzy logic are much easier to interpret because the classification is usually implemented according to rules summarized from the training dataset. Most fuzzy logic methods are hybrid methods; for example, the Fuzzy C-Means (FCM) algorithm (Fan et al., 2009) is a hybrid between fuzzy logic and the statistical algorithm (c-means). Classifiers based on Fuzzy Neural Networks (FNN) (Chen et al., 2009) and Fuzzy Artmap (FA) (Han et al., 2004), which are combinations of fuzzy logic and neural networks, were also reported. However, involving fuzzy logic in these hybrid algorithms (i.e., FCM, FNN, and FA) may enlarge the uncertainty in the final classification.

Evolutionary algorithms are another category of machine learning techniques that have been widely used in remote sensing image classification. Genetic Algorithm (GA), Evolutionary Programming (EP), and GP are several classical evolutionary algorithms with many successful applications (Agnelli et al., 2002; Ross et al., 2005; Makkeasorn et al., 2006; Awad et al., 2007; Chang et al., 2009; Makkeasorn and Chang, 2009). Classifiers based on Artificial Immune System (AIS) (Zhong et al., 2007) and swarm intelligence (Daamouche and Melgani, 2009) can also be included in this category. In addition, classifiers such as those based on expert system theory (Stefanov et al., 2001) and decision tree techniques (Friedl and Brodley, 1997) are also representative and important classification methods. The current progress in the literature can be summarized based on the above findings (Table 5.3). Hybrid learning algorithms integrating ML, FCM, FNN, FA, or KNN with

---

TABLE 5.3

Summary of Classification Methods in Image Processing

Statistical Methods		ML, KNN, K-means
Artificial	Neural Networks	BP, RBF, SOM
Intelligence	Global Optimization	SVM
	Fuzzy Logic	FCM, FA, FNN
	Evolutionary Algorithms	GA, EP, GP, AIS
Other Methods		Expert system, Decision tree

Source: Chang, N. B. (Ed.), 2012. *Environmental Remote Sensing and Systems Analysis.* CRC Press, Boca Raton, FL.

Note: Maximum likelihood (ML), K-Nearest Neighbor (KNN), Backward propagation (BP), radial basis function (RBF), self-organized mapping (SOM), support vector machine (SVM), Fuzzy C-means (FCM), fuzzy artmap (FA), fuzzy neural network (FNN), Genetic algorithm (GA), evolutionary programming (EP), and genetic programming (GP), artificial immune system (AIS).

ANN, SOM, RBF, BP SVM or GP to form unique learning systems for specific feature extraction can be anticipated.

5.8   PERFORMANCE EVALUATION METRIC

Developing a new algorithm to automate feature extraction is appealing and creative. For every proposed new algorithm or approach, the inventors normally claim several superiorities and advantages of their own algorithm relative to other methods. To support such claims, performance and accuracy of the proposed method should be rigorously evaluated. Regarding feature extraction, the performance of the methods can be assessed either qualitatively or quantitatively. Visual assessment is the most straightforward method that is commonly used to evaluate the performance of the resulting products. However, judgment of how to select or create these quality indexes or statistical indicators for performance evaluation depends highly on expert knowledge. A lack of quantitative statistics can make a comparison between methods difficult. Therefore, quantitative measures are essential to performance evaluation because the final accuracy of each method can be compared objectively.

To date, a variety of statistical indicators has been proposed and used to measure the quality of feature extraction results, especially for the problem of statistical classification. Among the proposed indexes, the Overall Accuracy (OA) and Kappa coefficient are two widely used measures. In practice, these measures are computed primarily based on a confusion matrix, which is also known as an error matrix (Stehman, 1997). A confusion matrix refers to a special kind of table that typically has two rows and two columns such that the number of false positives, false negatives, true positives, and true negatives is reported; an illustrative example is given in Figure 5.8. Each column of the matrix represents the instances in a predicted class while each row represents the instances in an actual class (Powers, 2011). With this matrix, accuracy metrics such as OA and Kappa coefficient can be easily computed (Fauvel et al., 2008; Chunsen et al., 2016; Imani and Ghassemian, 2016).

The OA is the fraction of correctly classified samples in relation to all samples for testing, which is often used to indicate the overall performance of the method. An overall accuracy can be calculated by summing the number of well-classified samples (diagonal elements) divided by the total number of all testing samples. Based on the confusion matrix shown in Figure 5.8, the overall accuracy can be computed as follows:

$$OA = \frac{TP + TN}{P + N} \tag{5.3}$$

Another measure is the Kappa coefficient, which measures the agreement between two raters each of which classifies N items into C mutually exclusive categories (Galton, 1892; Smeeton, 1985).

		Prediction condition	
	Total population	Prediction positive	Prediction negative
True condition	Condition positive (P)	True positive (TP)	False negative (FN)
	Condition negative (N)	False positive (FP)	True negative (TN)

FIGURE 5.8   A 2 × 2 confusion matrix with P positive instances and N negative instances. P is the total number of true positive instances (equivalent to $TP + FN$) and N is the total number of true negative instances (equivalent to $FP + TN$).

In other words, it is a measure of how the classification results compared to values assigned by chance. In contrast to the simple percent agreement calculation, the Kappa coefficient is generally thought to be more robust, because it considers the possibility of the agreement occurring by chance (Fauvel et al., 2008). Conceptually, it can be defined as:

$$KC = \frac{p_o - p_e}{1 - p_e} \tag{5.4}$$

where p_o denotes the observed proportionate agreement equivalent to the OA and p_e denotes the overall random agreement probability which can be calculated as:

$$p_e = \frac{TP + FN}{P + N} * \frac{TP + FP}{P + N} + \frac{FP + TN}{P + N} * \frac{FN + TN}{P + N} \tag{5.5}$$

Kappa values range from 0 to 1. A value of 0 means no agreement between the predicted condition and the actual condition, while a value of 1 indicates that the predicted condition and the actual condition are totally identical (i.e., perfect agreement). Hence, the larger the value of the Kappa coefficient the more accurate the result. However, some researchers expressed concerns that the Kappa coefficient is an overly conservative measure of agreement because it has a tendency to take the observed categories' frequencies as givens, making it unreliable for measuring agreement in situations with limited observations (Wu and Yang, 2005; Strijbos et al., 2006).

Apart from the overall accuracy and the Kappa coefficient to measure the general performance, the accuracy of class identification should also be assessed. Within such a context, statistics like errors of commission and/or errors of omission can be computed. Errors of commission are a measure of false positives, representing the fraction of values that were predicted to be in a class but do not belong to that class. In contrast, errors of omission are a measure of false negatives, representing the fraction of values that belong to a class but were predicted to be in a different class. Hence, the Commission Error (*CE*) of condition positive shown in Figure 5.8 can be calculated as:

$$CE = \frac{FN}{P} \tag{5.6}$$

Similarly, the Omission Error (*OE*) of positive can be computed as:

$$OE = \frac{FP}{TP + FP} \tag{5.7}$$

In addition, Producer Accuracy (*PA*) and User Accuracy (*UA*) are another two performance measures that are commonly computed and used for performance assessment of classification. The *PA* shows the probability that a value in a given class was classified correctly, which can be calculated by summing the number of pixels correctly classified in a particular category divided by the total number of pixels actually belonging to that category.

$$PA = \frac{TP}{TP + FP} \tag{5.8}$$

The *UA* shows the probability that a value predicted to be in a certain class really belongs to that class, which can be calculated as the fraction of correctly predicted values to the total number of values predicted to be in a class.

$$UA = \frac{TP}{P} \tag{5.9}$$

It is clear that the *PA* and *OE* complement each other, as a total value of them equals 1, so do the *UA* and *CE*.

Furthermore, statistics like Partition Coefficient (Bezdek, 1973), Fukuyama-Sugeno index (Fukuyama and Sugeno, 1989), Fuzzy Hyper Volume (Gath and Geva, 1989), β index (Pal et al., 2000), *Xie-Beni* index (Xie and Beni, 1991), and many others (Congalton, 1991; Wu and Yang, 2005), can be also applied. To assess the statistical significance of differences in the classification results, methods such as the McNemars test can be further applied, which is based upon the standardized normalized test statistic (Foody, 2004). The parameter Z_{12} in McNemars test is defined as:

$$Z_{12} = \frac{f_{12} - f_{21}}{\sqrt{f_{12} + f_{21}}}$$

(5.10)

where f_{12} denotes the number of samples classified correctly by classifier 1 and incorrectly by classifier 2. A positive Z_{12} indicates classifier 1 outperforms classifier 2 while a negative value shows vice versa. The difference in accuracy between two classifiers is considered to be statistically significant one $|Z_{12}| > 1.96$ (Fauvel et al., 2008; Imani and Ghassemian, 2016).

5.9 SUMMARY

In this chapter, the basic concepts and fundamentals of feature extraction were introduced, including definitions of feature, feature selection, and feature extraction. Based on the domain of interest, feature extraction methods can be grouped into spectral- and spatial-based feature extraction, yet by considering the working modes, feature extraction techniques can be divided into supervised, semi-supervised, and unsupervised methods. Illustrative examples were also provided for demonstration purposes. In addition, a set of different statistical indicators were introduced as performance measures for the evaluation of the resulting outcomes.

It is clear that developing a robust feature extraction workflow is by no means a simple task, since it requires us to gain a thorough understanding of the input data, to devote the requisite time for pre-processing, and to effectively apply the elements of image interpretation for decision analysis. In the next chapter, a wealth of traditional methods and approaches that were proposed for feature extraction will be introduced and discussed.

REFERENCES

Agnelli, D., Bollini, A., and Lombardi, L., 2002. Image classification: An evolutionary approach. *Pattern Recognition Letters*, 23, 303–309.

Agrawal, R., Gehrke, J., Gunopulos, D., and Raghavan, P., 2005. Automatic subspace clustering of high dimensional data. *Data Mining and Knowledge Discovery*, 11, 5–33.

Ahmad, A., 2012. Analysis of maximum likelihood classification on multispectral data. *Applied Mathematical Sciences*, 6, 6425–6436.

Akhtar, U. and Hassan, M., 2015. Big data mining based on computational intelligence and fuzzy clustering. In: Zaman, N., Seliaman, M. E., Hassan, M. F. and Marquez, F. P. G. (Eds.), *Handbook of Research on Trends and Future Directions in Big Data and Web Intelligence*, IGI Global, 130–148.

Ashoka, H. N., Manjaiah, D. H., and Rabindranath, B., 2012. Feature extraction technique for neural network based pattern recognition. *International Journal of Computer Science and Engineering*, 4, 331–340.

Awad, M., Chehdi, K., and Nasri, A., 2007. Multicomponent image segmentation using a genetic algorithm and artificial neural network. *IEEE Geoscience and Remote Sensing Letters*, 4, 571–575.

Backstrom, L. and Caruana, R., 2006. C2FS: An algorithm for feature selection in cascade neural networks. In: *The 2006 IEEE International Joint Conference on Neural Network Proceedings*, Vancouver, BC, Canada, 4748–4753.

Baraldi, A. and Parmiggiani, F., 1995. An investigation of the textural characteristics associated with gray level cooccurence matrix statistical parameters. *IEEE Transactions on Geoscience and Remote Sensing*, 33, 293–304.

Belkin, M., Niyogi, P., and Sindhwani, V., 2006. Manifold regularization: A geometric framework for learning from labeled and unlabeled examples. *Journal of Machine Learning Research*, 7, 2399–2434.

Benediktsson, J. A., Pesaresi, M., and Arnason, K., 2003. Classification and feature extraction for remote sensing images from urban areas based on morphological transformations. *IEEE Transactions on Geoscience and Remote Sensing*, 41, 1940–1949.

Bermingham, M. L., Pong-Wong, R., Spiliopoulou, A., Hayward, C., Rudan, I., Campbell, H., Wright, A. F., et al., 2015. Application of high-dimensional feature selection: Evaluation for genomic prediction in man. *Scientific Reports*, 5, 10312.

Bezdek, J. C., 1973. Cluster validity with fuzzy sets. *Cybernetics and Systems*, 3, 58–73.

Bhagabati, B. and Sarma, K. K., 2016. Application of face recognition techniques in video for biometric security. In: Gupta, B., Dharma, P., Agrawal, D. P., and Yamaguchi, S. (Eds.), *Handbook of Research on Modern Cryptographic Solutions for Computer and Cyber Security*, IGI Global, 460–478.

Bishop, C., 2006. *Pattern Recognition and Machine Learning, Technometrics*. Springer-Verlag, New York.

Blanzieri, E. and Melgani, F., 2008. Nearest neighbor classification of remote sensing images with the maximal margin principle. *IEEE Transactions on Geoscience and Remote Sensing*, 46, 1804–1811.

Boltz, S., 2004. Statistical region merging code. *IEEE Transactions on Pattern Analysis and Machine Intelligence*, 26, 1452–1458.

Bouman, C. A. and Shapiro, M., 1994. A multiscale random field model for Bayesian image segmentation. *IEEE Transactions on Image Processing*, 3, 162–177.

Camps-Valls, G., Bandos Marsheva, T. V., and Zhou, D., 2007. Semi-supervised graph-based hyperspectral image classification. *IEEE Transactions on Geoscience and Remote Sensing*, 45, 3044–3054.

Canty, M. J. 2009. Boosting a fast neural network for supervised land cover classification. *Computers & Geosciences*, 35, 1280–1295.

Carson, C., Belongie, S., Greenspan, H., and Malik, J., 2002. Blobworld: Image segmentation using expectation-maximization and its application to image querying. *IEEE Transactions on Pattern Analysis and Machine Intelligence*, 24, 1026–1038.

Castro, P. A. D. and Von Zuben, F. J., 2009. Learning Bayesian networks to perform feature selection. In: *2009 International Joint Conference on Neural Networks*, Atlanta, GA, USA, 467–473.

Cattell, R. B., 1943. The description of personality: Basic traits resolved into clusters. *Journal of Abnormal & Social Psychology*, 38, 476–506.

Chang, N. B., Daranpob, A., Yang, J. and Jin, K. R., 2009. Comparative data mining analysis for information retrieval of MODIS images: Monitoring lake turbidity changes at Lake Okeechobee, Florida. *Journal of Applied Remote Sensing*, 3, 033549.

Chang, N.-B., Vannah, B. W., Yang, Y. J., and Elovitz, M., 2014. Integrated data fusion and mining techniques for monitoring total organic carbon concentrations in a lake. *International Journal of Remote Sensing*, 35, 1064–1093.

Chapelle, O., Schölkopf, B., and Zien, A., 2006. *Semi-supervised Learning*. MIT Press, Cambridge, MA.

Chen, H. W., Chang, N. B., Yu, R. F., and Huang, Y. W., 2009. Urban land use and land cover classification using the neural-fuzzy inference approach with Formosat-2 data. *Journal of Applied Remote Sensing*, 3, 033558.

Chen, X., Li, H., and Gu, Y., 2014. Multiview Feature Selection for Very High Resolution Remote Sensing Images. In: *2014 Fourth International Conference on Instrumentation and Measurement, Computer, Communication and Control*, Harbin, China, 539–543.

Chunsen, Z., Yiwei, Z., and Chenyi, F., 2016. Spectral–spatial classification of hyperspectral images using probabilistic weighted strategy for multifeature fusion. *IEEE Geoscience and Remote Sensing Letters*, 13, 1562–1566.

Congalton, R. G., 1991. A review of assessing the accuracy of classifications of remotely sensed data. *Remote Sensing of Environment*, 37, 35–46.

Corduneanu, A. and Jaakkola, T., 2002. On information regularization. In: *Proceedings of the Nineteenth Conference on Uncertainty in Artificial Intelligence*, Acapulco, Mexico, 151–158.

Daamouche, A. and Melgani, F., 2009. Swarm intelligence approach to wavelet design for hyperspectral image classification. *IEEE Geoscience and Remote Sensing Letters*, 6(4), 825–829.

Das, S., 2001. Filters, wrappers and a boosting-based hybrid for feature selection. In: *Proceedings of the Eighteenth International Conference on Machine Learning (ICML'01)*, San Francisco, CA, USA, Morgan Kaufmann Publisher, 74–81.

Durfee, A., 2006. Text mining. In: Garson, G. D. and Khosrow-Pour (Eds.), *Handbook of Research on Public Information Technology*, IGI Global, 592–603.

Duval, B., Hao, J.-K., and Hernandez Hernandez, J. C., 2009. A memetic algorithm for gene selection and molecular classification of cancer. In: *Proceedings of the 11th Annual Conference on Genetic and Evolutionary Computation—GECCO '09*, New York, New York, USA, ACM Press.

Elnemr, H. A., Zayed, N. M., and Fakhreldein, M. A., 2016. Feature extraction techniques: Fundamental concepts and survey. In: Kamila, N. K. (Ed.), *Handbook of Research on Emerging Perspectives in Intelligent Pattern Recognition, Analysis, and Image Processing*, IGI Global, 264–294.

Ester, M., Kriegel, H. P., Sander, J., and Xu, X., 1996. A density-based algorithm for discovering clusters in large spatial databases with noise. In: *Proceedings of the 2nd International Conference on Knowledge Discovery and Data Mining*, Portland, Oregon, USA, 226–231.

Everitt, B. S., Landau, S., Leese, M., and Stahl, D., 2011. *Hierarchical Clustering, in Cluster Analysis*, 5th Edition. John Wiley & Sons, Ltd, Chichester, UK.

Fan, J., Han M., and Wang, J., 2009. Single point iterative weighted fuzzy C-means clustering algorithm for remote sensing image segmentation. *Pattern Recognition*, 42, 2527–2540.

Fauvel, M., Benediktsson, J. A., Chanussot, J., and Sveinsson, J. R., 2008. Spectral and spatial classification of hyperspectral data using SVMs and morphological profiles. *IEEE Transactions on Geoscience and Remote Sensing*, 46, 3804–3814.

Foody, G. M., 2004. Thematic map comparison: Evaluating the statistical significance of differences in classification accuracy. *Photogrammetric Engineering & Remote Sensing*, 70, 627–633.

Foody, G. M. and Mathur, A., 2004a. A relative evaluation of multiclass image classification by support vector machines. *IEEE Transactions on Geoscience and Remote Sensing*, 42, 1335–1343.

Foody, G. M. and Mathur, A., 2004b. Toward intelligent training of supervised image classifications: Directing training data acquisition for SVM classification. *Remote Sensing of Environment*, 93, 107–117.

Forgy, E. W., 1965. Cluster analysis of multivariate data: Efficiency versus interpretability of classifications. *Biometrics*, 21, 768–769.

Friedl, M. A. and Brodley, C. E., 1997. Decision tree classification of land cover from remotely sensed data. *Remote Sensing of Environment*, 61, 399–409.

Fukuyama, Y. and Sugeno, M., 1989. A new method of choosing the number of clusters for the fuzzy C-means method. In: *Procdings of 5th Fuzzy System Sympptom*, 247–250.

Galton, F., 1892. *Finger Prints*. Macmillan, London.

Gath, I. and Geva, A. B., 1989. Unsupervised optimal fuzzy clustering. *IEEE Transactions on Pattern Analysis and Machine Intelligence*, 11, 773–780.

Geman, S., Bienenstock, E., and Doursat, R., 1992. Neural Networks and the Bias/Variance Dilemma. *Neural Computation*, 4, 1–58.

Genuer, R., Poggi, J.-M., and Tuleau-Malot, C., 2010. Variable selection using random forests. *Pattern Recognition Letters*, 31, 2225–2236.

Giacinto, G. and Roli, F., 2001. Design of effective neural network ensembles for image classification purposes. *Image and Vision Computing*, 19, 699–707.

Glass, H. and Cooper, L., 1965. Sequential search: A method for solving constrained optimization problems. *Journal of the ACM*, 12, 71–82.

Gopalakrishnan, V., 2009. Computer aided knowledge discovery in biomedicine. In: Daskalaki, A. (Ed.), *Handbook of Research on Systems Biology Applications in Medicine*, IGI Global, 126–141.

Guyon, I. and Elisseeff, A., 2003. An introduction to variable and feature selection. *Journal of Machine Learning Research*, 3, 1157–1182.

Guyon, I. and Elisseeff, A., 2006. *An Introduction to Feature Extraction, in: Feature Extraction: Foundations and Applications*. Springers, Berlin, Heidelberg, 1–25.

Guyon, I., Weston, J., Barnhill, S., and Vapnik, V., 2002. Gene selection for cancer classification using support vector machines. *Machine Learning*, 46, 389–422.

Haindl, M., Somol, P., Ververidis, D., and Kotropoulos, C., 2006. Feature selection based on mutual correlation. In: *Progress in Pattern Recognition, Image Analysis and Applications*, Havana, Cuba, 569–577.

Hall, M. A., 1999. *Correlation-based Feature Selection for Machine Learning*. University of Waikato, Hamilton, NewZealand.

Han, Y., Kim, H., Choi, J., and Kim, Y., 2012. A shape–size index extraction for classification of high resolution multispectral satellite images. *International Journal of Remote Sensing*, 33, 1682–1700.

Han, M., Tang, X., and Cheng, L., 2004. An improved fuzzy ARTMAP network and its application in wetland classification. In: *Proceedings of 2004 IEEE International Geoscience and Remote Sensing Symposium*, Alaska, USA.

Haralick, R. M., Shanmugam, K., and Dinstein, I., 1973. Textural features for image classification. *IEEE Transactions on Systems, Man, and Cybernetics*, SMC-3, 610–621.

Heermann, P. D. and Khazenie, N., 1992. Classification of multispectral remote sensing data using a back-propagation neural network. *IEEE Transactions on Geoscience and Remote Sensing*, 30, 81–88.

Hoi-Ming, C. and Ersoy, O. K., 2005. A statistical self-organizing learning system for remote sensing classification. *IEEE Transactions on Geoscience and Remote Sensing*, 43, 1890–1900.

Hruschka, E. R., Hruschka, E. R., and Ebecken, N. F. F., 2004. Feature selection by Bayesian networks. In: Tawfik, A. Y. and Goodwin, S. D. (Eds.), *Advances in Artificial Intelligence*, Cairns, Australia, 370–379.

Huang, Z. C., Chan, P. P. K., Ng, W. W. Y., and Yeung, D. S., 2010. Content-based image retrieval using color moment and Gabor texture feature. In: *2010 International Conference on Machine Learning and Cybernetics (ICMLC)*, Qingdao, China.

Huang, X., Guan, X., Benediktsson, J. A., Zhang, L., Li, J., Plaza, A., and Dalla Mura, M., 2014. Multiple morphological profiles from multicomponent-base images for hyperspectral image classification. *IEEE Journal of Selected Topics in Applied Earth Observations and Remote Sensing*, 7, 4653–4669.

Huang, J., Kumar, S. R., Mitra, M., Zhu, W., and Zabih, R., 1997. Image indexing using color correlograms. In: *Conference on Computer Vision and Pattern Recognition (CVPR'97)*, Puerto Rico, USA, 762–768.

Huang, X., Zhang, L., and Li, P., 2007. Classification and extraction of spatial features in urban areas using high-resolution multispectral imagery. *IEEE Geoscience and Remote Sensing Letters*, 4, 260–264.

Imani, M. and Ghassemian, H., 2016. Binary coding based feature extraction in remote sensing high dimensional data. *Information Sciences*, 342, 191–208.

Jain, A. K., Duin, R. P. W., and Jianchang, M., 2000. Statistical pattern recognition: A review. *IEEE Transactions on Pattern Analysis and Machine Intelligence*, 22, 4–37.

James, G. M., 2003. Variance and bias for general loss functions. *Machine Learning*, 51, 115–135.

Johnson, S. C., 1967. Hierarchical clustering schemes. *Psychometrika*, 32, 241–254.

Jordan, M. and Bishop, C., 2004. Neural networks. In: Tucker, A. B. (Ed.), *Computer Science Handbook*, Second Edition, Chapman and Hall/CRC, Flordia, USA, 1–16.

Kavzoglu, T., 2009. Increasing the accuracy of neural network classification using refined training data. *Environmental Modelling & Software*, 24, 850–858.

Kohavi, R. and Johnb, G. H., 1997. Wrappers for feature subset selection. *Artificial Intelligence*, 97, 273–324.

Kriegel, H.-P., Kröger, P., Sander, J., and Zimek, A., 2011. Density-based clustering. *WIREs Data Mining and Knowledge Discovery*, 1, 231–240.

Kumar, G. and Bhatia, P. K., 2014. A detailed review of feature extraction in image processing systems. In: *2014 IEEE 4th International Conference on Advanced Computing & Communication Technologies*, Rohtak, India, 5–12.

Kuo, B. C., Chang, C. H., Sheu, T. W., and Hung, C. C., 2005. Feature extractions using labeled and unlabeled data. In: *International Geoscience and Remote Sensing Symposium (IGARSS)*, Seoul, South Korea, 1257–1260.

Lillesand, T. M., Kiefer, R. W., and Chipman, J. W., 1994. *Remote Sensing and Image Interpretation*. John Wiley and Sons, Inc., Toronto.

Liu, Z., Li, H., Zhou, W., and Tian, Q., 2012. Embedding spatial context information into inverted file for large-scale image retrieval. In: *Proceedings of the 20th ACM International Conference on Multimedia*, New York, USA, ACM Press, 199.

Lv, Z., Zhang, P., and Atli Benediktsson, J., 2017. Automatic object-oriented, spectral-spatial feature extraction driven by Tobler's first law of geography for very high resolution aerial imagery Classification. *Remote Sensing*, 9, 285.

Makkeasorn, A. and Chang, N. B., 2009. Seasonal change detection of riparian zones with remote sensing images and genetic programming in a semi-arid watershed. *Journal of Environmental Management*, 90, 1069–1080.

Makkeasorn, A., Chang, N. B., Beaman, M., Wyatt, C., and Slater, C., 2006. Soil moisture prediction in a semi-arid reservoir watershed using RADARSAT satellite images and genetic programming. *Water Resources Research*, 42, 1–15.

Maldonado, S. and Weber, R., 2011. Embedded feature selection for support vector machines: State-of-the-Art and future challenges. In: *Progress in Pattern Recognition, Image Analysis, Computer Vision, and Applications*, Pucón, Chile, 304–311.

Mohri, M., 2012. *Foundations of Machine Learning*. The MIT Press, Massachusetts, USA.

Momm, H. and Easso, G., 2011. Feature extraction from high-resolution remotely sensed imagery using evolutionary computation. In: Kita, E. (Ed.), *Evolutionary Algorithms*, InTech, 423–442.

Moon, T. K., 1996. The expectation-maximization algorithm. *IEEE Signal Processing Magazine*, 13, 47–60.

Moser, G. and Serpico, S. B., 2013. Combining support vector machines and Markov random fields in an integrated framework for contextual image classification. *IEEE Transactions on Geoscience and Remote Sensing*, 51, 2734–2752.

Murtagh, F. and Contreras, P., 2012. Algorithms for hierarchical clustering: An overview. *WIREs Data Mining and Knowledge Discovery*, 2, 86–97.

Nakariyakul, S., 2014. Improved sequential search algorithms for classification in hyperspectral remote sensing images. In: *Proceedings SPIE 9273, Optoelectronic Imaging and Multimedia Technology III*, 927328, Beijing, China.

Naseriparsa, M., Bidgoli, A.-M., and Varaee, T., 2013. A hybrid feature selection method to improve performance of a group of classification algorithms. *International Journal of Computer Applications*, 69, 28–35.

Nguyen, H., Franke, K., and Petrovic, S., 2009. Optimizing a class of feature selection measures. In: *NIPS 2009 Workshop on Discrete Optimization in Machine Learning: Submodularity, Sparsity & Polyhedra (DISCML)*, Vancouver, Canada.

Nixon, M. S. and Aguado, A. S., 2012. *Feature Extraction and Image Processing*, 2nd Edition. Academic Press, London, UK.

Ooi, C. S., Seng, K. P., and Ang, L.-M., 2015. Automated technology integrations for customer satisfaction assessment. In: Kaufmann, H.-R. (Ed.), *Handbook of Research on Managing and Influencing Consumer Behavior*, IGI Global, Hershey, Pennsylvania, USA, 606–620.

Pal, S. K., Ghosh, A., and Shankar, B. U., 2000. Segmentation of remotely sensed images with fuzzy thresholding, and quantitative evaluation. *International Journal of Remote Sensing*, 21, 2269–2300.

Pass, G., Zabih, R., and Miller, J., 1998. Comparing images using color coherence vectors. In: *Proceedings of the Fourth ACM International Conference on Multimedia*, Massachusetts, USA, 1–14.

Pearson, K., 1901. On lines and planes of closest fit to systems of points in space. *Philosophical Magazine*, 2, 559–572.

Peng, H., Long, F., and Ding, C., 2005. Feature selection based on mutual information criteria of max-dependency, max-relevance, and min-redundancy. *IEEE Transactions on Pattern Analysis and Machine Intelligence*, 27, 1226–1238.

Phillips, S. J., 2002. Acceleration of K-means and related clustering algorithms. In: Mount, D. M. and Stein, C. (Eds.), *Lecture Notes in Computer Science*. Springer Berlin Heidelberg, Berlin, Germany, 166–177.

Phuong, T. M., Lin, Z., and Altman, R. B., 2005. Choosing SNPs using feature selection. In: *Proceedings—2005 IEEE Computational Systems Bioinformatics Conference*, California, USA, 301–309.

Powers, D. M. W., 2011. Evaluation: From Precision, Recall and F-Measure to Roc, Informedness, Markedness & Correlation. *Journal of Machine Learning Technologies*, 2, 37–63.

Quackenbush, L. J., 2004. A review of techniques for extracting linear features from imagery. *Photogrammetric Engineering & Remote Sensing*, 70, 1383–1392.

Rokni, K., Ahmad, A., Selamat, A., and Hazini, S., 2014. Water feature extraction and change detection using multitemporal landsat imagery. *Remote Sensing*, 6, 4173–4189.

Ross, B. J., Gualtieri, A. G., and Budkewitsch, P., 2005. Hyperspectral image analysis using genetic programming. *Applied Soft Computing*, 5, 147–156.

Rouse, J. W., Haas, R. H., Schell, J. A., and Deering, D. W., 1974. Monitoring vegetation systems in the great Okains with ERTS. In: *Third Earth Resources Technology Satellite-1 Symposium*, Texas, USA, 325–333.

Sánchez-Maroño, N. and Alonso-Betanzos, A., 2009. Feature selection. In: Shapiro, S. C. (Ed.), *Encyclopedia of Artificial Intelligence*, IGI Global, Hershey, Pennsylvania, USA, 632–638.

Shahdoosti, H. R. and Mirzapour, F., 2017. Spectral–spatial feature extraction using orthogonal linear discriminant analysis for classification of hyperspectral data. *European Journal of Remote Sensing*, 50, 111–124.

Sharma, M. and Sarma, K. K., 2016. Soft-computational techniques and Spectro-temporal features for telephonic speech recognition. In: Bhattacharyya, S., Banerjee, P., Majumdar, D., and Dutta, P. (Eds.), *Handbook of Research on Advanced Hybrid Intelligent Techniques and Applications*, IGI Global, Hershey, Pennsylvania, USA, 161–189.

Shruthi, R. B. V., Kerle, N., and Jetten, V., 2011. Object-based gully feature extraction using high spatial resolution imagery. *Geomorphology*, 134, 260–268.

Smeeton, N. C., 1985. Early history of the Kappa statistic (response letter). *Biometrics*, 41, 795.

Solorio-Fernández, S., Carrasco-Ochoa, J. A., and Martínez-Trinidad, J. F., 2016. A new hybrid filter–wrapper feature selection method for clustering based on ranking. *Neurocomputing*, 214, 866–880.

Stefanov, W. L., Ramsey, M. S., and Christensen, P. R., 2001. Monitoring urban land cover change: An expert system approach to land cover classification of semiarid to arid urban centers. *Remote Sensing of Environment*, 77, 173–185.

Stehman, S. V., 1997. Selecting and interpreting measures of thematic classification accuracy. *Remote Sensing of Environment*, 62, 77–89.

Strijbos, J.-W., Martens, R. L., Prins, F. J., and Jochems, W. M. G., 2006. Content analysis: What are they talking about? *Computers & Education*, 46, 29–48.

Suresh, S., Sundararajan, N., and Saratchandran, P., 2008. A sequential multi-category classifier using radial basis function networks. *Neurocomputing*, 71, 1345–1358.

Szummer, M. and Jaakkola, T., 2002. Information regularization with partially labeled data. In: *Proceedings of Advances in Neural Information Processing Systems*, 15, 1025–1032.

Taubenböck, H., Esch, T., Wurm, M., Roth, A., and Dech, S., 2010. Object-based feature extraction using high spatial resolution satellite data of urban areas. *Journal of Spatial Science*, 55, 117–132.

Tian, D. P., 2013. A review on image feature extraction and representation techniques. *International Journal of Multimedia and Ubiquitous Engineering*, 8, 385–395.

Tryon, R. C., 1939. *Cluster Analysis: Correlation Profile and Orthometric (factor) Analysis for the Isolation of Unities in Mind and Personality*. Edwards Brother. Inc. lithoprinters Publ.

Tso, B. C. K. and Mather, P. M., 1999. Classification of multisource remote sensing imagery using a genetic algorithm and Markov random fields. *IEEE Transactions on Geoscience and Remote Sensing*, 37, 1255–1260.

Vapnick, V. N., 1998. *Statistical Learning Theory*. Wiley, New York.

Wang, X.-Y., Wu, J.-F., and Yang, H.-Y., 2009. Robust image retrieval based on color histogram of local feature regions. *Multimedia Tools and Applications*, 49, 323–345.

Wolf, P., Dewitt, B., and Mikhail, E., 2000. *Elements of Photogrammetry with Applications in GIS*. McGraw-Hill Education, New York, Chicago, San Francisco, Athens, London, Madrid, Mexico City, Milan, New Delhi, Singapore, Sydney, Toronto.

Wu, K.-L. and Yang, M.-S., 2005. A cluster validity index for fuzzy clustering. *Pattern Recognition Letters*, 26, 1275–1291.

Wyse, N., Dubes, R., and Jain, A. K., 1980. A critical evaluation of intrinsic dimensionality algorithms. In: Gelsema, E. S. and Kanal, L. N. (Eds.), *Pattern Recognition in Practice*, North-Holland Publishing Company, Amsterdam, Netherlands, 415–425.

Xie, X. L. and Beni, G., 1991. A validity measure for fuzzy clustering. *IEEE Transactions on Pattern Analysis and Machine Intelligence*, 13, 841–847.

Xue, B., Zhang, M., and Browne, W. N., 2013. Particle swarm optimization for feature selection in classification: A multi-objective approach. *IEEE Transactions on Cybernetics*, 43, 1656–1671.

Yu, L. and Liu, H., 2003. Feature selection for high-dimensional data: A fast correlation-based filter solution. In: *Proceedings of the Twentieth International Conference on Machine Learning (ICML-2003)*, Washington, DC, USA, 1–8.

Yu, X., Yang, J., and Zhang, J., 2012. A transductive support vector machine algorithm based on spectral clustering. *AASRI Procedia*, 1, 384–388.

Zare, H., Haffari, G., Gupta, A., and Brinkman, R. R., 2013. Scoring relevancy of features based on combinatorial analysis of Lasso with application to lymphoma diagnosis. *BMC Genomics*, 14, S14.

Zena, M. H. and Gillies, D. F., 2015. A review of feature selection and feature extraction methods applied on microarray data. *Advances in Bioinformatics*, 2015, Article ID 198363, 1–13.

Zhao, W. and Du, S., 2016. Spectral-spatial feature extraction for hyperspectral image classification: A dimension reduction and deep learning approach. *IEEE Transactions on Geoscience and Remote Sensing*, 54, 4544–4554.

Zhao, H., Sun, S., Jing, Z., and Yang, J., 2006. Local structure based supervised feature extraction. *Pattern Recognition*, 39, 1546–1550.

Zhao, J., Zhong, Y., and Zhang, L., 2015. Detail-preserving smoothing classifier based on conditional random fields for high spatial resolution remote sensing imagery. *IEEE Transactions on Geoscience and Remote Sensing*, 53, 2440–2452.

Zhong, Y., Zhang, L., Gong, J., and Li, P., 2007. A supervised artificial immune classifier for remote-sensing imagery. *IEEE Transactions on Geoscience and Remote Sensing*, 45, 3957–3966.

Zhu, X., 2008. Semi-Supervised Learning Literature Survey. Computer Sciences TR 1530. University of Wisconsin—Madison.

Zhuo, L., Zheng, J., Li, X., Wang, F., Ai, B., and Qian, J., 2008. A genetic algorithm based wrapper feature selection method for classification of hyperspectral images using support vector machine. In: *In Proceedings of SPIE 7147, The International Society for Optical Engineering*, Guangzhou, China, 71471J–71471J–9.

6 Feature Extraction with Statistics and Decision Science Algorithms

6.1 INTRODUCTION

With the fast development of air-borne and space-borne remote sensing technologies, large volumes of remotely sensed multispectral, hyperspectral, and microwave images have become available to the public. Such massive data sources often require a large amount of memory for storage and computational power for processing. With proper feature extraction techniques, these images may provide a huge amount of information to help better understand Earth's environment. Traditional feature extraction methods involve using regression, filtering, clustering, transformation, and probabilistic theory as opposed to modern feature extraction methods that heavily count on machine learning and data mining. Nevertheless, these massive and various data sources are prone to be redundant, which in turn may complicate traditional feature extraction processes and even result in overfitting issues in machine learning or data mining (Liao et al., 2013; Huang et al., 2014; Romero et al., 2016). Hence, the extraction of specific features of interest from complex and redundant data inputs is of great importance to the exploitation of these data sources on a large scale.

Due to its efficacy in transforming the original redundant and complex inputs into a set of informative and nonredundant features, feature extraction has long been considered a crucial step in image processing and pattern recognition, as well as remote sensing and environmental modeling, because it facilitates the subsequent data manipulation and/or decision making (Elnemr et al., 2016). In the remote sensing community, feature extraction techniques have been widely used for image processing, typically for pattern recognition and image classification. In image classification and pattern recognition, feature extraction is often considered a special form of dimensionality reduction which aims to construct a compact and informative feature space by removing the irrelevant and redundant information from the original data space (Elnemr et al., 2016). Practical applications typically involve road extraction (Gamba et al., 2006), urban and building detection (Bastarrika et al., 2011), oil spill detection (Brekke and Solberg, 2005), change detection (Celik, 2009), burned areas mapping (Bastarrika et al., 2011), surface water body mapping (Feyisa et al., 2014), hyperspectral image classification (Chen et al., 2013; Qian et al., 2013), and so forth.

As elaborated in Chapter 5, feature extraction can be performed either on the spatial or spectral domain; hence, methods of feature extraction can be developed by making use of various theories such as simple filtering, mathematical morphology, clustering, regression, spatial/spectral transformation, classification, and so on. Traditional feature extraction approaches work in practice based on conventional mathematical theories like filtering, regression, spatial/spectral transformation, and others requiring less computational resources. Yet those advanced methods taking advantage of artificial intelligence (e.g., artificial neural network, genetic algorithm, support vector machine, genetic programming), as well as other advanced optimization theories (e.g., particle swarm optimization), demand a lot more computational resources. Therefore, those advanced methods perceptually involve high-performance computing issues (i.e., compression, storage, and performance-driven load distribution for heterogeneous computational grids) in various real-world applications.

In this chapter, a suite of traditional feature extraction approaches that rely on statistics and decision science principles, such as filtering, morphology, decision trees, transformation, regression, and probability theory, will be introduced with specific focus on the mathematical foundations of each kind of method. Since numerous methods and approaches found in the literature share similar principles, these analogous approaches will be grouped into the same category in a logical order. Chapter 7, which focuses on machine learning and data mining for advanced feature extraction, will follow these traditional feature extraction techniques.

6.2   STATISTICS AND DECISION SCIENCE-BASED FEATURE EXTRACTION TECHNIQUES

In this section, a suite of feature extraction algorithms that are commonly seen in the literature will be introduced in light of statistics and decision science to aid in remote sensing domain-specific practices. In addition to a holistic description of the principles and basics of each individual method, the foundational mathematical theories will be emphasized. As found in the literature, traditional feature extraction techniques typically include those methods based on theories in association with filtering, clustering, regression, transformation, and other statistical theories, and each of them will be delineated sequentially in the following subsections.

6.2.1   Filtering Operation

Remotely sensed data are collected in forms of either panchromatic, multispectral or hyperspectral images at various spatiotemporal scales. The embedded color, shape, and textual characteristics are three typical features that can be extracted to represent the property of an image (Tian, 2013). These retrieved features can be further referred to as the "fingerprint" or "signature" associated with a given image. In general, color features are often defined and are subject to a specific color space or model, such as RGB (red-blue-green), HSV (hue, saturation, value) also known as HSB (hue, saturation, brightness), and LUV. Note that LUV stands for non-RGB color space that decouples the "color" (chromaticity, the UV part) and "lightness" (luminance, the L part) of color to improve object detection. These fixed color spaces will in turn limit further explorations of spectral information at other wavelengths (e.g., multispectral and hyperspectral) because only three spectral components are required with respect to these color spaces. To overcome this constraint, other kinds of techniques are often used to better detect and extract the relevant features embedded in each image.

In practice, filtering is generally considered the simplest method for feature extraction, which has also been routinely used in image processing to detect and extract the targets of interest from a given remotely sensed image. Thresholding, which can be considered a representative of such a technique, is actually a form of low-level feature extraction method performed as a point operation on the input image by applying a single threshold to transform any greyscale (or color image) into a binary map (Nixon and Aguado, 2012). An illustrative example of thresholding-based feature extraction was given in the previous chapter, as shown in Figure 5.3. For instance, clouds can be easily detected with the aid of human vision, since clouds are brighter relative to other terrestrial objects in RGB color space. Evidenced by this property, an empirical threshold value can then be applied to one spectral band (e.g., blue band) to extract the observed clouds by simply using a Boolean operator (see Equation 5.2 for details).

Technically, thresholding can also be treated as a decision-based feature extraction method. In most cases, one or two empirical threshold values will fulfill the need to detect and extract the desired features with good accuracy. However, this does not hold for some extreme conditions such as the extraction of water bodies from remotely sensed images in mountainous areas, where shadows are often an obstacle. This is mainly due to the fact that both targets always show similar spectral properties optically. In other words, it is difficult to obtain a satisfying result by simply applying threshold values to one or two fixed spectral bands. Thus, more complex thresholding networks should be developed by making use of external information such as elevation data and microwave satellite images. This often leads to the creation of a multilayer stacked decision tree framework. More details regarding decision tree classifiers will be introduced in the following subsections.

Thresholding techniques work mainly by relying on spectral differences between various targets to extract the desired features; hence the threshold values for the same target could even vary between images due to radiometric distortions caused by various factors such as illumination conditions. To account for such drawbacks, more flexible approaches should be applied. In image interpretation, the shapes of targets are often considered good features for further pattern recognition, since the perimeter of an object can be easily perceived by human vision. Hence, detecting the shape features from a given imagery is critical to the subsequent feature extraction. Essentially, the shape of an object is commonly treated as a step change in the intensity levels (Nixon and Aguado, 2012).

In the remote sensing community, filtering approaches have been widely used in image processing to detect and extract shape features, for example, linear features such as roads and rivers. In order to extract the perimeter of an object or linear features like roads and rivers from a remotely sensed imagery, a suite of convolutional filters has been proposed for the extraction of edge features. Among them, "Roberts cross operator" and "Sobel operator" are the two most well-known filters, and they have been extensively used in practical applications. As one of the first edge detectors, the Roberts cross operator was initiated by Lawrence Roberts in 1963 (Davis, 1975), with two convolutional kernels (or operators) formulated as:

$$t_1 = \begin{bmatrix} 1 & 0 \\ 0 & -1 \end{bmatrix} \text{ and } t_2 = \begin{bmatrix} 0 & 1 \\ -1 & 0 \end{bmatrix} \tag{6.1}$$

According to Roberts (1963), the produced edges from an edge detector should be well defined, while the intensity of edges should correspond closely to what a human would perceive with little noise introduced by the background (Davis, 1975).

The Roberts cross operator works as a differential filter aiming to approximate to the gradient of an image through discrete differentiation by computing the summation of the squared differences between diagonally adjacent pixels. Let $I(i,j)$ be a pixel (at the location (i,j)) in the original image X, while G_x is the convoluted pixel value with the first kernel (e.g., t_1) and G_y is the convoluted pixel value with the second kernel (e.g., t_2); the gradient can be then defined as:

$$\nabla I(i,j) = G(x,y) \cong \sqrt{G_x^2 + G_y^2} \tag{6.2}$$

which can be further written as

$$\nabla I(i,j) \cong \left| I_{i,j} - I_{i+1,j+1} \right| + \left| I_{i+1,j} - I_{i,j+1} \right| \tag{6.3}$$

It is clear that this operation will highlight changes in intensity in a diagonal direction, hence it enables the detection of changes between targets (i.e., edges). An example of enhanced edge features in an observed scene through the application of Roberts filters as well as two other edge detectors can be used to sharpen our understanding (Figure 6.1). The results indicate that the edge features and linear features have been better characterized compared to the original image without applying Roberts cross operation (Figure 6.1a). Despite the simplicity and capability of this filter, it is observed that the Roberts cross suffers greatly from sensitivity to noise due to its convolutional nature (Davis, 1975).

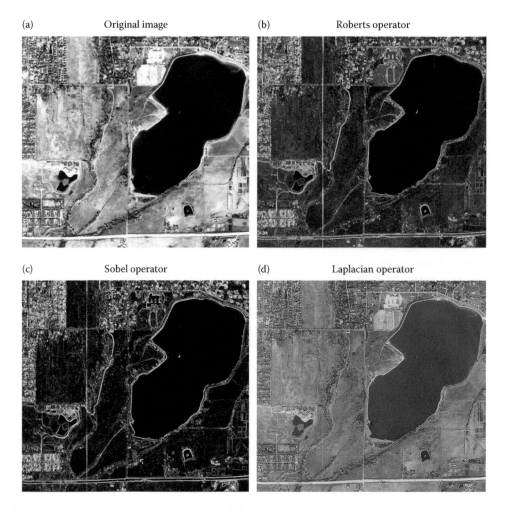

FIGURE 6.1 Edge features extracted by three different edge detectors. (a) RGB composite image; (b) edge features detected by performing Roberts cross operation; (c) features detected by Sobel operation; and (d) features detected by Laplacian operation.

In order to better detect edge features, an enhanced discrete differentiation operator, the Sobel filter (also known as Sobel–Feldman operator), was developed by Irwin Sobel and Gary Feldman in 1968 with the aim of computing an approximation of the gradient of the image intensity function. Differing from the Roberts cross operator, the Sobel filter is an isotropic image gradient operator that uses two separable and integer-valued 3×3 kernels to calculate approximations of the derivatives by convolving with the input image in the horizontal and vertical directions, respectively. Two kernels are formulated as follows:

$$t_1 = \begin{bmatrix} 1 & 2 & 1 \\ 0 & 0 & 0 \\ -1 & -2 & -1 \end{bmatrix} \text{ and } t_2 = \begin{bmatrix} -1 & 0 & 1 \\ -2 & 0 & 2 \\ -1 & 0 & 1 \end{bmatrix} \tag{6.4}$$

Since the Sobel kernels have a larger window size than the Roberts cross operator, the Sobel operator will yield larger degrees of accuracy in detecting and extracting edge features. More precisely, the derived edge features from the Sobel filters will be much clearer and brighter (i.e., with larger contrast) to human vision (Figure 6.1c). Aside from the Roberts and Sobel operators, there also exist other analogue filters, such as the Laplacian filters. A commonly used convolutional kernel is

$$t = \begin{bmatrix} 0 & 1 & 0 \\ 1 & -4 & 1 \\ 0 & 1 & 0 \end{bmatrix} \tag{6.5}$$

Theoretically, the Laplacian filters approximate a second-order derivative on the original image, which in turn highlights regions of rapid intensity change in particular (Figure 6.1d). Because the second derivatives are very sensitive to noise, a Gaussian smoothing is often performed on the original image before applying the Laplacian filter to counter this constraint (Reuter et al., 2009). Since all these filtering-based techniques are commonly used to enhance certain specific features, they also fall into the image enhancement category in the image processing community to some extent.

Due to their simplicity and relatively less expensive computation cost properties, filtering techniques have been widely exploited for practical applications. For instance, Gabor filters were also implemented for edge detection (Mehrotra et al., 1992) and texture classification (Clausi and Ed Jernigan, 2000). In addition to the direct application of filter approaches for feature extraction, various filters were also used in conjunction with other techniques to aid in feature extraction practices. For instance, Gamba et al. (2006) proposed an adaptive directional filtering procedure to enhance urban road extraction from high-resolution optical and Synthetic Aperture Radar (SAR) images, in which the filtering scheme was used to capture the predominant directions of roads. Similarly, Kang et al. (2014) used an edge-preserving filter to aid in hyperspectral image classification, and the results indicated that the incorporation of the edge-preserving filtering in the classification scheme resulted in higher classification accuracy.

6.2.2 Mathematical Morphology

Similar to filtering operators, morphological operators are another kind of promising filter that have been widely used in computer vision for geometric structure information analysis. Essentially, the foundation of morphological processing is in the mathematically rigorous field of describing shapes using set theory, geometry, and topology, hence such processing procedures are generally termed mathematical morphology (Serra, 1992; Soille and Pesaresi, 2002; Soille, 2004). In image processing, morphological operators refer to a variety of image filters that process images based on morphological information (e.g., size and shape). As opposed to many methods based on the spectral property of pixels, mathematical morphology concentrates on the spatial relationships between groups of pixels and treats the objects present in an image as sets (Soille and Pesaresi, 2002).

In image processing, mathematical morphology is commonly used to examine interactions between an image and a set of structuring elements using certain operations, while the structuring element acts as a probe for extracting or suppressing specific structures of the image objects (Plaza, 2007). More specifically, morphological operations apply a structuring element to filter an image, while the value of each pixel in the output image is based on a comparison of the corresponding pixel in the input image with its neighbors. By choosing a proper size and shape of the neighborhood, a morphological operation that is sensitive to specific shapes in the input image can be constructed. The output of filtering process depends fully on the matches between the input image and the structuring element and the operation being performed (Quackenbush, 2004).

A structuring element is a small binary image which is actually a small matrix of pixels with values of ones or zeros. Technically, in morphological operations, structuring elements play the same role as convolutional kernels in traditional linear image filtering, yet the basic operations of

(a)

1	1	1	1	1
1	1	1	1	1
1	1	1	1	1
1	1	1	1	1
1	1	1	1	1

(b)

0	0	1	0	0
0	0	1	0	0
1	1	1	1	1
0	0	1	0	0
0	0	1	0	0

(c)

0	0	1	0	0
0	1	1	1	0
1	1	1	1	1
0	1	1	1	0
0	0	1	0	0

FIGURE 6.2 An example of three simple structuring elements with different shapes. The blue square denotes the origin of the structuring elements. (a) Square-shaped 5 × 5 element, (b) Cross-shaped 5 × 5 element, and (c) Diamond-shaped 5 × 5 element.

morphology are nonlinear in nature (Davies, 2012). An example of three simple structuring elements is demonstrated for more understanding in Figure 6.2. There are two basic aspects associated with the structuring element. One is related to the size and the other to the shape. As indicated, the size of an element is determined by the dimension of a matrix in general. A common practice is to have a structuring matrix with an odd dimension, since the origin of the element is commonly defined as the center of the matrix (Pratt, 2007). The shape of an element depends fully on the pattern of ones and zeros distributed over the matrix grid. In practical usages, the structuring element may be applied over the input image, acting as a filter to compare with the input image block by block based on certain operations for the detection and extraction of specified geometrical features similar to the given structuring element (Tuia et al., 2009).

Aside from the structuring element, another critical factor in morphological image analysis is the morphological operation. The two most fundamental morphological operations are dilation and erosion (Soille, 2004). Conceptually, both operations rely on translating the structuring element to various points over the input image and then examining the intersection between the translated element coordinates and the input image coordinates. If g is a binary image to analyze and B is a structuring element, dilation ($\delta_{B(g)}$) and erosion ($\epsilon_{B(g)}$) can be mathematically represented as (Tuia et al., 2009):

$$\delta_{B(g)} = g \oplus B = \bigcup_{b \in B} g_{-b} \tag{6.6}$$

$$\epsilon_{B(g)} = g \ominus B = \bigcap_{b \in B} g_{-b} \tag{6.7}$$

As indicated, dilation expands the image by adding pixels in the structuring element, that is, a union between g and B. On the contrary, erosion is used to perform an intersection between them. This kind of analysis (based on binary images) is often called binary morphology, which can also be extended to grayscale images by considering them as a topographic relief. However, in grayscale morphology, the pointwise minimum and maximum operators will be used instead of the intersection and union, respectively (Tuia et al., 2009). More specifically, dilation adds pixels to the boundaries of objects in an image (i.e., grows boundary regions), while erosion is used to remove pixels on object boundaries (i.e., shrinks boundary regions). According to this principle, the number of pixels added or removed from the objects depends totally on the size and shape of the given structuring element.

The graphs in Figure 6.3 show a schematic illustration of dilation and erosion, comparatively. A practical application of these two operations to an image is shown in Figure 6.4. It is clear that

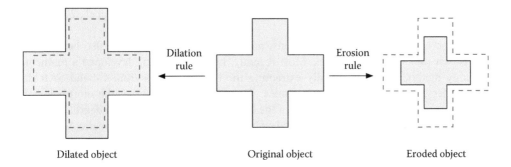

FIGURE 6.3 A schematic illustration of (left) dilation and (right) erosion.

the dilation operation has a unique effect; gaps between different regions are reduced and small intrusions into boundaries of a region are filled in (Figure 6.4a). For example, the road shown at the bottom of the Figure 6.4a confirms this observation. In contrast, the erosion operation shrinks the objects' boundaries, resulting in holes and gaps between different regions which become larger when small details are eliminated (Figure 6.4b). In addition to the two basic operations of dilation and erosion, many morphological operations in practical use are represented as compound operations based on dilation and erosion, such as opening, closing, hit and miss transform, thickening, thinning, and so forth (Tuia et al., 2009).

In the remote sensing community, morphological operators have been widely used in various practical applications; common usages include edge detection, noise removal, image enhancement, and image segmentation. In most cases, mathematical morphology is used to advance automatic pattern recognition, in particular the detection of targets in urban areas since an accurate extraction of shape and size features is essential to an automatic extraction process (Benediktsson et al., 2005; Chaudhuri et al., 2016). Due to its pronounced efficacy, mathematical morphology has been extensively used in detection and extraction of various terrestrial targets from remotely sensed high-resolution optical/SAR imageries, such as roads (Mohammadzadeh et al., 2006; Valero et al., 2010),

(a)   (b)

FIGURE 6.4 Morphologically transformed images by performing (a) dilation and (b) erosion operations based on a square-shaped 5 × 5 structure element.

bridges (Chen et al., 2014b), rivers (Sghaier et al., 2017), buildings (Chaudhuri et al., 2016), dwelling structures (Kemper et al., 2011), and so on.

In addition to these ordinary applications, morphological concepts have also been applied to aid in disaster management practices. Most recently, Lee et al. (2016) developed a mathematical morphology method for automatically extracting the hurricane eyes from C-band SAR data to advance understanding of hurricane dynamics. The results indicated that the morphology-based analyses of the subsequent reconstructions of the hurricane eyes showed a high degree of agreement with results derived from reference data based on National Oceanic and Atmospheric Administration (NOAA) manual work. Similarly, Chen et al. (2017) developed an object-oriented framework for landslide mapping based on Random Forests (RF) and mathematical morphology. The RF was used as a dimensionality reduction tool to extract landslides' relevant features, while a set of four closing and opening morphology operations were subsequently applied to optimize the RF classification results to map the landslides with higher accuracy. Moreover, morphological operators have also been applied in astronomy. For instance, Aragón-Calvo et al. (2007) developed a multiscale morphology filter to automatically segment cosmic structure into a set of basic components. Due to the distinct advantage of scale independence in segmentation, anisotropic features such as filaments and walls were well identified in this cosmic structure study.

6.2.3 DECISION TREE LEARNING

Decision tree learning refers to creating a tree-like structure model to perform a set of decision rules on original observations and internal results step-wise to retrieve some specific targets or conclusions. Generally, a decision tree is a layer-stacked technique framework that enables the division of a given data set into many smaller subsets following a tree structure. Considering whether the dependent variable is categorical (discrete) or numeric (continuous), decision trees can be further divided into classification trees and regression trees, respectively. A general term for both procedures is widely known as ClAssification and Regression Tree (CART) analysis (Breiman et al., 1984).

Compared with other methods, the most appealing property of decision trees is their capability for breaking down a complex decision-making process into a collection of simpler decisions, thus facilitating the complexity of modeling. This unique property renders decision trees a popular predictive modeling approach for use as a feature extractor in support of data mining and machine learning for classification in the next stage (Rokach and Maimon, 2008). In remote sensing, a decision tree is often used as an excellent feature extractor and/or classifier since it is computationally efficient and requires no prior distribution assumptions for the input data. Moreover, a decision tree sometimes performs even better in practice than a maximum likelihood classifier and a support vector machine (Otukei and Blaschke, 2010). The distinction between RF and decision tree can be described below.

6.2.3.1 Decision Tree Classifier

Decision tree classifiers are routinely used as a popular classification method in various pattern recognition problems such as image classification and character recognition (Safavian and Landgrebe, 1991). Decision tree classifiers are observed to work more effectively, in particular for complex classification problems, due to their flexible and computationally efficient features. Furthermore, as claimed by Friedl and Brodley (1997), decision tree classifiers excel over many conventional supervised classification methods, like maximum likelihood estimator, through several advantages. Specifically, no distribution assumption is required by decision tree classifiers with respect to the input data. This unique nature renders decision tree classifiers a higher flexibility to handle various data sets, regardless whether numeric or categorical, even with missing values. In addition, decision tree classifiers are essentially nonparametric. Moreover, decision trees are capable of handling nonlinear relations between features and classes. Finally, the classification process through a tree-like structure is always intuitive and interpretable.

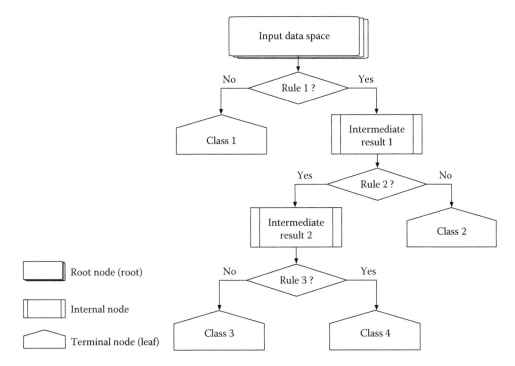

FIGURE 6.5   A schematic structure of a decision tree (binary tree).

In general, a decision tree consists of three essential components: a root node, several internal nodes, and a set of terminal nodes (also called leaves). An illustrative example of a decision tree structure is shown in Figure 6.5. As indicated, for each internal and terminal node (child node), there should exist a parent node showing the data source. Meanwhile, regarding the root node and each internal node (parent node), two or more children nodes will be generated from these parent nodes based on various decision rules. If each parent node is split into two descendants, the decision tree is often known as a binary tree (e.g., Figure 6.5), and the inherent decision rule can be expressed as a dyadic Boolean operator such that the data points are split simply based on whether the condition rule is satisfied or not. Among these three types of nodes, the root node involves the input data space, while the other two kinds of nodes correspond to partitioned subspaces. As opposed to root and internal nodes, the terminal nodes (i.e., leaves) refer to the final determined outputs of the whole decision-making process, which cannot be further partitioned; corresponding class labels (the majority class) will then be assigned. When developing a decision tree, the most critical process is to split each internal node and the root node with various decision rules or learning algorithms. In practice, there exist various learning algorithms of which the most well known is the CART algorithm, which is a binary recursive partitioning procedure (Breiman et al., 1984).

In the CART algorithm, a splitting rule is inherently defined as a determination function used to maximize the purity (or homogeneity) of the training data as represented by the resulting descendant nodes. Typically, an impurity function is defined to examine the goodness of split for each node, and the Gini diversity index is commonly used as a popular measure for the impurity function. Mathematically, the impurity measurement of the node t is usually defined as the follows:

$$i(t) = 1 - \sum_{j=1}^{K} P_j(t)^2 \tag{6.8}$$

where $P_j(t)$ denotes the posterior probability of class j presenting in node t. This probability is often defined as the proportion between the number of training samples that go to node t labeled as class j and the total number of training samples within node t:

$$P_j(t) = \frac{N_j(t)}{N(t)}, j = 1,2,\ldots,K \tag{6.9}$$

Taking a binary node for example, the goodness of the split s for the node t can be calculated as:

$$\Delta i(s,t) = i(t) - P_R \cdot i(t_R) - P_L \cdot i(t_L) \tag{6.10}$$

where P_R and P_L are the proportions of the samples in node t that go to the right descendant t_R and the left descendant t_L, respectively. Essentially, the goodness of the split s should be maximized to eventually achieve the lowest impurity in each step toward the largest purity in the terminal nodes.

Analogous to a machine learning process, the stopping criterion is also required by decision trees to stop the split process. In the CART algorithm, the stopping criterion is commonly defined as:

$$\max_s \Delta i(s,t) < \beta \tag{6.11}$$

where β is a predetermined threshold. The split process will continue until it meets such a stopping criterion. In other words, the decision tree will stop growing, which implies that the training process of the decision tree classifier is complete. In general, a decision tree classifier-based process starts from the root node, and the unclassified data points in the root node are partitioned into different internal nodes following a set of splitting rules before they finally arrive at terminal nodes (leaves) where a class label will be assigned to each of them.

Decision tree has been extensively used in support of data mining and machine learning, aiming to extract a target variable based on several input variables and a set of decision rules (Friedl et al., 1999; Liu et al., 2008). Due to their nonparametric and top-down framework, decision trees have been widely used in many practical remote sensing applications for feature extraction such as hyperspectral image classification (Chen and Wang, 2007), snow cover extraction (Liu et al., 2008), invasive plant species detection (Ghulam et al., 2014), and so on.

6.2.3.2   RF Classifier

RF, or random decision forests, is a suite of decision trees created by drawing a subset of training data through a bagging approach (Breinman, 2001). More specifically, RF consists of a combination of decision trees where each tree is constructed using an independently sampled random vector from the input set, while all trees in the forest maintain a consistent distribution (Pal, 2005; Tian et al., 2016). In practice, about two-thirds of the original input will be randomly selected to train these trees through the bagging process, while the remaining one-third will not be used for training. Instead, that portion of the data is used for internal cross-validation in order to check the performance of the trained trees (Belgiu and Drăguţ, 2016). In other words, there is no need to perform cross-validation to get an unbiased estimate of the test set error since it has already been done in the process of constructing RF. In general, RF tries to construct multiple CART models by making use of different samples and different initial variables, which in turn renders RF, accounting for the inherent drawback of overfitting associated with conventional decision trees (Hastie et al., 2009).

In general, two parameters are required to perform an RF-based classification, namely, the number of trees and the number of variables randomly chosen at each split (Winham et al., 2013). Each node in a tree will be split with a given number of randomly sampled variables from the input feature space. In RF, the Shannon entropy (or Gini index) is routinely used as the splitting function (or attribute selection measure) to measure the impurity of an attribute with respect to the classes

(Pal, 2005). In prediction, each tree votes for a class membership for each test sample, and the class with maximum votes will be considered the final class (Ni et al., 2017).

Unlike many other classification methods typically based on one classifier, hundreds of classifiers can be constructed in RF and a final prediction is always obtained by combining all these decisions with an optimal function (e.g., plurality vote). In traditional and advanced feature extraction practices, ensemble learning methods use multiple learning algorithms to obtain better predictive performance than could be obtained from any single learning algorithm. In fact, RF is regarded as an ensemble classifier of decision tree in which decision tree plays the role of a meta model. This ensemble learning nature renders RF many desirable advantages, for example, high accuracy, robustness against overfitting the training data, and integrated measures of variable importance (Chan and Paelinckx, 2008; Guo et al., 2011; Stumpf and Kerle, 2011). In addition, no distribution assumption is required for the input data, hence it can be used to process various data sets. Nevertheless, like many other statistical learning techniques, RF is also observed to be prone to bias once the number of instances is distributed unequally among the classes of interest (Winham et al., 2013). However, because of its outstanding advantages, RF has been widely used for remote sensing classification in terms of various applications, for example, laser data point clouds classification (Ni et al., 2017), LiDAR and multispectral image-based urban scene classification (Guo et al., 2011), land cover classification and mapping (Fan, 2013; Tian et al., 2016), hyperspectral image classification (Ham et al., 2005), and landslide mapping (Stumpf and Kerle, 2011; Chen et al., 2014a).

6.2.4 CLUSTER ANALYSIS

Cluster analysis, or clustering, was initiated in anthropology in 1932 and then introduced to psychology in the late 1930s; its usage for trait theory classification in personality psychology in the 1940s caused it to be more widely known to the public (Tryon, 1939; Cattell, 1943; Bailey, 1994). Essentially, clustering itself is not a specific algorithm, but is instead a general term referring to various tasks or processes of grouping a set of targets/objects with similar characteristics into the same class while isolating those with different properties into other classes. The core of clustering is related to various algorithms that have the capability of detecting and isolating distinct features into different groups. Therefore, the difference between various clustering algorithms primarily lies in their notion of what constitutes a cluster and how to detect clusters efficiently. Clustering can thus be considered a knowledge discovery or multi-objective optimization problem.

To date, more than one hundred clustering algorithms have been developed for various applications found in the literature. The reason for such numerous clustering algorithms can be ascribed to the fact that the notion of a cluster is difficult to define because it varies significantly in properties between algorithms (Estivill-Castro, 2002). In other words, different clustering algorithms are produced by employing different cluster models. Thus, understanding cluster models is critical to the realization of the differences between the various clustering algorithms as cluster models act as the core of each algorithm. As found in the literature, popular notions of clusters include groups with small distances among the cluster members, dense regions in the data space, intervals or particular statistical distributions. Typical cluster models associated with these notions include connectivity models, centroid models, density models, distribution models, and many others (Estivill-Castro, 2002). They are described below.

6.2.4.1 Connectivity-Based Clustering

Also known as hierarchical clustering, connectivity-based clustering aims to divide the input data set into an extensive hierarchy of clusters that could merge with each other at certain distances rather than a single partitioning (Rokach and Maimon, 2005). The relevant principle for such methods is that objects are more related to those nearby rather than to those farther away. In other words, clusters are formed based simply on distances between objects, and objects of the same kind are more likely close to each other in the space domain. Thus, a cluster is determined largely by the maximum distance needed to connect objects within this cluster, and hence different clusters will be formed under different distances (Everitt, 2011). Therefore, connectivity-based clustering methods will differ largely by the distance functions used in each method. In addition to the selection of distance functions, the linkage criteria also need to be decided. Popular choices include single linkage clustering (Sibson, 1973) and complete linkage clustering (Defays, 1977). Despite the efficacy of clustering objects into different groups, it has been observed that connectivity-based clustering methods are prone to outliers (e.g., resulting in additional clusters or causing other clusters to merge) in practical applications. Moreover, the computational burden of manipulating large data sets will be huge since it is difficult to compute an optimal distance due to the high dimensionality (Estivill-Castro, 2002; Everitt, 2011).

6.2.4.2 Centroid-Based Clustering

This kind of method assumes that clusters can be represented by various central vectors, as the objects in the data set can then be assigned to the nearest cluster center. Therefore, it is essential to determine optimal cluster centers, hence the clustering process can be in turn treated as an optimization problem as the goal is to find certain optimal cluster centers to minimize the squared distance from each cluster. The most well-known centroid-based clustering method is k-means clustering, which aims to partition a set of observations into k clusters in which each observation belongs to the cluster with the least variance. Given a data set $X = \{x_1, x_2, \ldots, x_n\}$ with n observations, it can be partitioned into k clusters $S = \{S_1, S_2, \ldots, S_k; k \leq n\}$ by making use of the k-means clustering method. Mathematically, the clustering is used to solve the following optimization problem by minimizing the squared distance within the cluster,

$$\arg_S \min \sum_{i=1}^{k} \sum_{x \in S_i} \left\| x - u_i \right\|^2 \tag{6.12}$$

where u_i is the mean value of S_i. Compared to other clustering, finding an optimal solution to k-means clustering is often computationally complex. Commonly, an iterative refinement technique is used to solve the problem. More details related to the modeling process can be found in MacKay (2003). Despite the computational complexity, k-means clustering is still featured in several distinct applications, including the Voronoi structure-based data partitioning scheme, the nearest neighbor classification concept, and model-based clustering basis.

An experimental example of k-means clustering for land use and land cover classification can be seen in Figure 6.6. The results show an adequate accuracy in classifying different land cover types. Compared to the true color image, the classified map exhibits more contrast between different features, in particular the water bodies, which are largely highlighted in the classified map. In recent years, a set of analogs has been developed based on the foundation of k-means clustering, including X-means clustering (Ishioka, 2000), G-means clustering (Hamerly and Elkan, 2004), and the most widely used fuzzy clustering (Dell'Acqua and Gamba, 2001; Modava and Akbarizadeh, 2017).

6.2.4.3 Density-Based Clustering

As the name suggests, in density-based clustering, clusters are constructed based on the density of grouped data points. More specifically, clusters are defined as areas covering dense data points (Kriegel et al., 2011). As opposed to many other clustering methods (e.g., k-means clustering) in which every data point will be assigned to one cluster, in density-based clustering query data in sparse regions are often considered to be outliers or noises. Among various density-based clustering methods, the most popular one is Density-Based Spatial Clustering of Applications with Noise (DBSCAN), as featured by a well-defined cluster model called "cluster-reachability" (Ester et al., 1996). The working principle of density-based clustering is similar to connectivity-based clustering, as data points within certain distance thresholds will be grouped to form clusters. However, a

(a) (b)

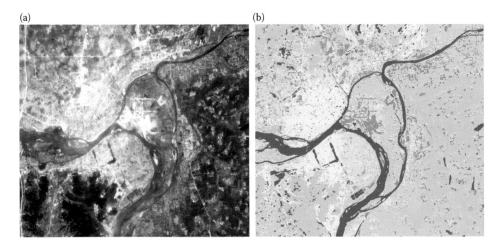

FIGURE 6.6 One observed Landsat TM scene and the corresponding classified result from *k*-means clustering. (a) true color image by given bands 3, 2, and 1 to the RGB space, respectively; (b) the classified result (6 classes) from the *k*-means method.

density criterion is also required by the density-based clustering. In other words, only connecting points satisfying the density criterion will be retained.

In addition to DBSCAN, some other types of density-based clustering methods are detailed in the literature as well, such as the Ordering Points to idenTify the Clustering Structure (OPTICS) (Ankerst et al., 1999), a generalized form of DBSCAN, which works regardless of an appropriate value for the range parameter. Nevertheless, such methods are always associated with one key drawback: they expect some kind of density drop to detect cluster borders. In contrast to many other clustering methods, density-based clustering methods only consider density-connected objects to form a cluster. Thus, the shape of the cluster is often arbitrary. Nevertheless, such methods may perform poorly in dealing with data sets with Gaussian mixtures since it is hard for them to model such data sets precisely.

6.2.4.4 Distribution-Based Clustering

Distribution models are given as prototypes and constraints of clusters, as only objects most likely belonging to the same distribution will be gathered to form the clusters. The principle of this approach resembles the generation of artificial data sets, as objects are randomly sampled based on *a priori* distribution. In practical usages, the most widely used methods are Gaussian Mixture Models (GMM) (Roberts et al., 1998; Jian and Vemuri, 2011), which are often performed by making use of the expectation-maximization algorithm (Dempster and Laird, 1977). Essentially, given a fitted GMM, the clustering algorithm will assign each data point to the relevant cluster yielding the highest posterior probability. In other words, each query data can be assigned to more than one cluster, and hence this method can be considered a type of fuzzy/soft clustering, even probabilistic clustering, as each query data is weighted by a set of posterior probability. Like most clustering methods, the number of desired clusters should also be defined before initiating model fitting, and the complexity of the model will inevitably increase as the number of desired clusters increases. Since each data point will be assigned a score to each cluster, GMM clustering enables the accommodation of clusters with different sizes and correlation structures. This flexible property renders GMM clustering more appropriate to use than fixed methods like *k*-means clustering.

With their distinct capabilities in dimensionality reduction as well as feature extraction, clustering methods have been widely used in the field of computer vision for various practical applications involving data mining, machine learning, pattern recognition, data compression, image analysis, and so on. In general, the selection of an appropriate clustering algorithm and its parameter settings (e.g., the distance function, a density threshold or the number of expected clusters) depends largely on the input data set (i.e., an algorithm specifically for one kind of model would generally fail on a data set involving different kinds of models) as well as the further usage or objective of the derived results. Overall, the clustering method with respect to a particular problem often needs to be selected experimentally or with *a priori* knowledge about the data set as well as the intended use of the results.

Clustering methods have long been widely used for feature learning and feature extraction to aid in remote sensing applications; they include building extraction from panchromatic images (Wei and Zhao, 2004), aerial laser cloud data (Tokunaga and Thuy Vu, 2007), SAR image segmentation by making use of spectral clustering (Zhang et al., 2008), fuzzy c-means clustering (Tian et al., 2013), and many other practices such as street tracking (fuzzy clustering) (Dell'Acqua and Gamba, 2001), coastline extraction (fuzzy clustering) (Modava and Akbarizadeh, 2017), geometrical structure retrieval (density-distance-based clustering) (Wu et al., 2017), and so on. In recent years, with the advances of big remote sensing data such as high-resolution hyperspectral imageries, many of the existing methods could fail in handling these data sets due to the curse of high dimensionality, which in turn stimulates the development of new clustering algorithms that focus on subspace clustering (Kriegel et al., 2012). An example of such a clustering algorithm is Clustering in QUEst (CLIQUE) (Agrawal et al., 2005). In order to advance hyperspectral imagery classification, Sun et al. (2015) proposed an improved sparse subspace clustering method to advance the band subset selection based on the assumption that band vectors can be sampled from the integrated low-dimensional orthogonal subspaces and each band can be sparsely represented as a linear or affine combination of other bands within its subspace. The experimental results indicated that such a subspace clustering method could significantly reduce the computational burden while improving the classification accuracy.

6.2.5   REGRESSION AND STATISTICAL MODELING

In remote sensing, statistical modeling approaches like regression have been widely used for feature extraction. Toward feature extraction, the commonly used modeling technique is regression-based, such as linear regression (Cho and Skidmore, 2006; Feyisa et al., 2014), least squares regression/ partial least squares regression (Pu, 2012; Li et al., 2014), logistic regression (Cheng et al., 2006; Qian et al., 2012; Khurshid and Khan, 2015), ridge regression (Imani and Ghassemian, 2015a; Yuan and Tang, 2017), and so on. In general, these methods are performed on low-level radiometric products like spectral reflectance at certain wavelength ranges by making use of a set of physically meaningful mathematic models. In this section, two popular regression-based techniques, linear regression and logistic regression, are illustrated below as representative methods with practical remote sensing applications.

6.2.5.1   Linear Extrapolation and Multivariate Regression

This method is commonly used in remote sensing for certain domain specific index derivations based on several spectral bands' information in order to emphasize and better extract certain terrestrial features from multispectral imageries. The most well-known example is the Normalized Difference Vegetation Index (NDVI), which is routinely calculated as the ratio of the subtraction of the near-infrared (*NIR*) and red (*Red*) bands to their summation (Rouse et al., 1974),

$$\text{NDVI} = \frac{NIR - Red}{NIR + Red} \tag{6.13}$$

In most cases, this kind of process is also considered to be image enhancement or data mining, since the vegetation information is emphasized through a data mining scheme. Nevertheless, we prefer to consider such a framework to be a feature extraction process, since informative features (e.g., vegetation) are extracted from multiple spectral bands toward a dimensionality reduction.

Such a method has been widely used in many remote sensing applications. For instance, McFeeters (1996) proposed the Normalized Difference Water Index (NDWI) based on the spectral differences of water bodies in the green and near-infrared wavelength ranges for surface water extraction purposes. The NDWI is formulated as follows (McFeeters, 1996):

$$NDWI = \frac{Green - NIR}{Green + NIR} \tag{6.14}$$

where pixels with positive NDWI values (NDWI > 0) are considered to be covered by water and negative values are nonwater. In recent years, to account for drawbacks associated with NDWI, a set of enhanced water indexes has been introduced seeking possible accuracy improvements, such as Modified Normalized Difference Water Index (MNDWI) (Xu, 2006) and Automated Water Extraction Index (AWEI) (Feyisa et al., 2014). Despite the usage of different spectral bands, these indexes are still derived in a manner of regression and extrapolation based on several empirical models. In the literature, there are many applications using such derivatives for specific feature extraction purposes, for example, red edge position extraction (Cho and Skidmore, 2006), flower coverage estimation (Chen et al., 2009), and burned areas mapping (Bastarrika et al., 2011).

Multivariate regression is another popular technique commonly used for detection and extraction of certain specific features. An example is the empirical algorithm operationally used for deriving chlorophyll-*a* concentrations in aquatic environments. Based on the *in situ* chlorophyll-*a* concentration measurements and the relevant spectral reflectance observations, a fourth-order polynomial empirical relationship was established for chlorophyll-*a* concentration estimation from a suite of optical remotely sensed images. The algorithm can be formulated as (O'Reilly et al., 1998):

$$\log_{10}(chl\text{-}a) = a_0 + \sum_{i=1}^{4} a_i * R^i \tag{6.15}$$

where a_i is the sensor-specific coefficient derived empirically and R is a local discriminative ratio of blue-to-green bands remote sensing reflectance. Taking Moderate-Resolution Imaging Spectroradiometer (MODIS) for example, where the coefficients a_0–a_4 are 0.2424, –2.7423, 1.8017, 0.0015, and –1.2280, respectively, the R can be calculated as

$$R = \log_{10}\left(\frac{R_{rs}443 > R_{rs}488}{R_{rs}547}\right) \tag{6.16}$$

where the operator > means to find the largest reflectance value of $R_{rs}443$ and $R_{rs}488$. The chl-*a* maps over the dry and wet seasons of Lake Nicaragua and Lake Managua are presented in Figure 6.7. These four subdiagrams in Figure 6.7 exhibit a seasonality effect in a comparative way. In general, the water quality in the dry season is worse than that of the wet season. However, based on the Probability Density Function (PDF) of all the band values, the input band values of Equation 6.15 do not follow the normality assumption as a linear regression equation in (6.15) implies (Figures 6.8 and 6.9). Hence, the predicted values of chl-*a* concentrations do not follow the normality assumption closely either (Figures 6.10 and 6.11). This finding gives rise to some insights about the inadequacy of using a linear regression model to infer the water quality conditions of the two tropical shallow lakes.

6.2.5.2 Logistic Regression

In statistics, logistic regression commonly refers to a regression model mainly used to measure the relationship between categorical dependent variables and one or more independent variables by estimating probabilities using a logistic function (Walker and Duncan, 1967; Freedman,

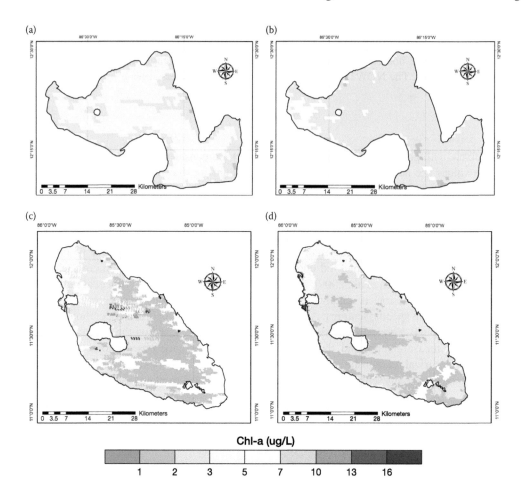

Chl-a (ug/L)

1   2   3   5   7   10   13   16

FIGURE 6.7 Chl-*a* concentration maps of Lake Managua and Lake Nicaragua for dry season and wet season, respectively. (a) Lake Managua (Dry Season/March 04, 2016); (b) Lake Managua (Wet Season/September 08, 2016); (c) Lake Nicaragua (Dry Season/March 01, 2016); and (d) Lake Nicaragua (Wet Season/September 03, 2016).

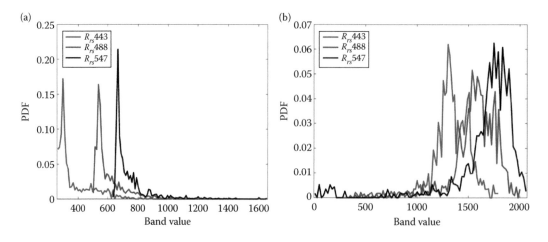

FIGURE 6.8 Band PDF for Lake Managua, (a) dry season; (b) wet season. (Note that the *X* axes do not stand for the original reflectance value, and they have been multiplied by a scale factor for convenience of expression.)

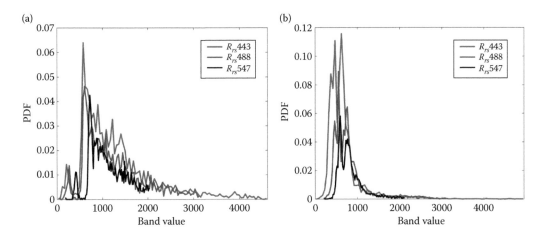

FIGURE 6.9 Band PDF for Lake Nicaragua, (a) dry season; (b) wet season. (Note that the X axes do not stand for the original reflectance value, and they have been multiplied by a scale factor for convenience of expression.)

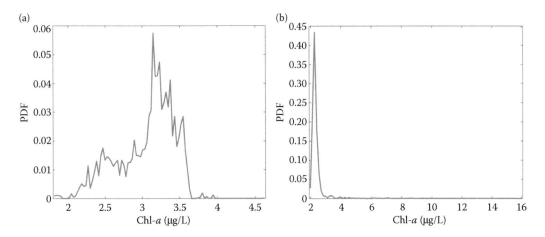

FIGURE 6.10 PDF of Chl-a concentrations in Lake Managua, (a) dry season; (b) wet season.

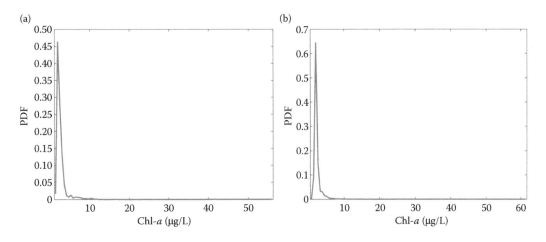

FIGURE 6.11 PDF of Chl-a concentrations in Lake Nicaragua, (a) dry season; (b) wet season.

2009). Conceptually, a simplified form of a logistic regression problem can be written as follows (binomial):

$$y = \begin{cases} 1, & \beta_0 + \beta_1 x_1 + \beta_2 x_2 + \cdots + \beta_k x_k > 0 \\ 0 & \text{else} \end{cases} \tag{6.17}$$

where the logistic regression is used to solve the regression $y' = \beta_0 + \beta_1 x_1 + \beta_2 x_2 + \cdots + \beta_k x_k$ to find the parameters β that best fit the Equation 6.17. Therefore, the underlying principle of logistic regression is analogous to linear regression. However, the assumptions of logistic regression are totally different. First, the conditional distribution should be Bernoulli distribution rather than Gaussian distribution since the dependent variables are dichotomous (binary). Second, the predicted values of logistic regression are probabilities of a particular outcome and hence vary from 0 to 1.

In practical uses, logistic regression can be binomial, multinomial or ordinal. As shown in Equation 6.17, binomial logistic regression is used to solve the problem in which the expected outcomes of the dependent variables are dyadic (e.g., 0 and 1, representing dyadic problems like win/loss, pass/fail, alive/dead). As opposed to binomial logistic regression, the outcomes of dependent variables for multinomial logistic regression can be three or even more (Hosmer and Lemeshow, 2000). Ordinal logistic regression thus means that the dependent variables are ordered.

Due to the categorical outcomes' nature, logistic regression has been widely used as a probabilistic, linear classifier, by projecting the input vector onto a set of hyperplanes as each of them corresponds to one specific class. Unlike the distance- and density-based clustering algorithms, in logistic regression the "distance" from the input vector to a hyperplane is represented as the probability that the input belongs to the corresponding class. Mathematically, given an input feature space $X = \{x_1, x_2, \ldots, x_k\}$ and a stochastic variable Y, logistical regression models the conditional probability of the vector X belonging to the class $y_j, j = 1, 2, \ldots, J$ as (Hosmer and Lemeshow, 2000; Cheng et al., 2006):

$$\log\left(\frac{p_j}{1 - p_j}\right) = X * W = \beta_{0j} + \beta_{1j} x_1 + \beta_{2j} x_2 + \cdots + \beta_{kj} x_k \tag{6.18}$$

and in turn we obtain

$$p_j = \frac{e^{(\beta_{0j} + \beta_{1j} x_1 + \beta_{2j} x_2 + \cdots + \beta_{kj} x_k)}}{1 + \sum_{l=1}^{J-1} e^{(\beta_{0l} + \beta_{1l} x_1 + \beta_{2l} x_2 + \cdots + \beta_{kl} x_k)}} \tag{6.19}$$

where β is the weight matrix to be optimized. If the Jth class is the baseline, the logistic regression model can be written in terms of $J - 1$ logit transformations as:

$$\log\left(\frac{p_1}{p_J}\right) = \beta_{01} + \beta_{11} x_1 + \beta_{21} x_2 + \cdots + \beta_{k1} x_k$$

$$\log\left(\frac{p_2}{p_J}\right) = \beta_{02} + \beta_{12} x_1 + \beta_{22} x_2 + \cdots + \beta_{k2} x_k \tag{6.20}$$

$$\vdots$$

$$\log\left(\frac{p_{J-1}}{p_J}\right) = \beta_{0(J-1)} + \beta_{1(J-1)} x_1 + \beta_{2(J-1)} x_2 + \cdots + \beta_{k(J-1)} x_k$$

and hence

$$p_J = \cfrac{1}{1 + \sum_{l=1}^{J-1} e^{(\beta_{0l} + \beta_{1l}x_1 + \beta_{2l}x_2 + \cdots + \beta_{kl}x_k)}} \tag{6.21}$$

The model's prediction is thus the class with maximal probability:

$$\arg\max_\beta \left(\cfrac{e^{(\beta_{0j} + \beta_{1j}x_1 + \beta_{2j}x_2 + \cdots + \beta_{kj}x_k)}}{1 + \sum_{l=1}^{J-1} e^{(\beta_{0l} + \beta_{1l}x_1 + \beta_{2l}x_2 + \cdots + \beta_{kl}x_k)}} \right) \tag{6.22}$$

and the optimal weight matrix β^* can in turn be estimated using the maximum likelihood method (Hosmer and Lemeshow, 2000).

In feature extraction, logistic regression is commonly used to reduce dimensionality of the input feature space by extracting the most relevant features based on the predicted probability of each feature class. Cheng et al. (2006) developed a systematic approach based on logistic regression for the feature selection and classification of remotely sensed images. The experimental results performed on both multispectral (Landsat ETM+) and hyperspectral (Airborne Visible/Infrared Imaging Spectrometer) images showed that the logistic regression enabled the reduction of the number of features substantially without any significant decrease in the classification accuracy. Similar work can be also found in Khurshid and Khan (2015). In addition, logistic regression can be further extended to structured sparse logistic regression by adding a structured sparse constraint (Qian et al., 2012). On the other hand, more advanced regression analyses can be conducted through either data mining or machine learning although the implementation of these methods are not straightforward but are always physically meaningful.

6.2.6 LINEAR TRANSFORMATION

One of the essential tasks of feature extraction is to detect and extract a set of the most relevant features from the original data set to reduce the dimensionality. Toward such a goal, many techniques have been developed. The most popular methods are those that attempt to project or decompose the original data inputs into a set of components, and then only the most relevant components will be extracted and used for dimensionality reduction purposes. In this section, three popular methods working with such a principle will be introduced below, including principal component analysis, linear discriminate analysis, and wavelet transform.

6.2.6.1 Principal Component Analysis (PCA)

PCA is a statistical procedure that is commonly used to transform a set of possibly correlated variables into a set of linearly uncorrelated subsets based on an orthogonal linear transformation. Mathematically, it is defined as an orthogonal linear transformation that projects the original data set to a new coordinate space such that the largest variance is projected onto the first coordinate (also known as first principal component) while the subsequent largest variance lies on the second coordinate under the constraint that it is orthogonal to the proceeding component, and so on (Jolliffe, 2002). Essentially, PCA aims to find a linear transformation $z = W^T x$, where $x \in \mathfrak{R}^d, z \in \mathfrak{R}^r$, and $r < d$, to maximize the variance of the data in the projected space (Prasad and Bruce, 2008).

Given a data matrix $X = \{x_1, x_2, \ldots, x_i\}, x_i \in \mathfrak{R}^d$, the transformation can be defined by a set of p-dimensional vectors of weights $W = \{w_1, w_2, \ldots, w_p\}, w_p \in \mathfrak{R}^k$ that map each vector x_i of X to a new space, that is

$$t_{k(i)} = W_k^T x_i \tag{6.23}$$

In order to maximize the variance, the first weight W_1 thus has to satisfy the following condition:

$$W_1 = \arg\max_{\|W\|=1}\left\{\sum_i (x_i \cdot W)^2\right\} \tag{6.24}$$

which can be further expanded as

$$W_1 = \arg\max_{\|W\|=1}\left\{\|X \cdot W\|^2\right\} = \arg\max_{\|W\|=1}\{W^T X^T X W\} \tag{6.25}$$

A symmetric matrix like $X^T X$ can be easily solved by finding the largest eigenvalue of the matrix, as W is the corresponding eigenvector. Once W_1 is obtained, the first principal component can be derived by projecting the original data matrix X onto the W_1 in the transformed space. The further components can be acquired in a similar manner after subtracting the previously derived components.

Since the number of principal components is usually determined by the number of significant eigenvalues with respect to the global covariance matrix, the derived components always have a lower dimension than the original data set (Prasad and Bruce, 2008). These components often retain as much of the variance in the original dataset as possible. A set of six principal components derived from a PCA analysis based on Landsat TM multispectral imageries is shown in Figure 6.12. The first two principal components (Figure 6.12a,b) have explained more than 95% of the variances of these multispectral images. The remaining four components are thus considered to be noise, which can be discarded for dimensionality reduction. Compared to each individual band of the original multispectral image, the information content of the first principal component is more abundant, which makes it a good data source for further data analysis such as classification. Because of this unique feature, PCA has been extensively used in various data analyses for dimensionality reduction, especially in manipulating high-dimensional data sets (Farrell and

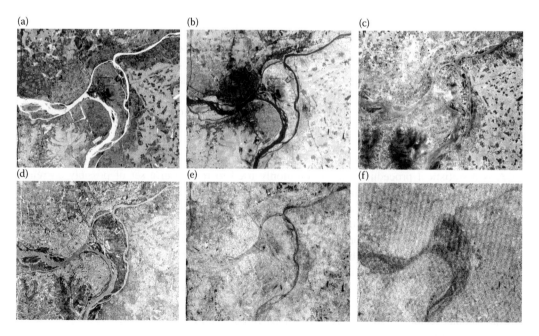

FIGURE 6.12 Principal components derived from the Landsat TM multispectral image shown in Figure 6.6a. A total of 6 components are shown in (a)–(f) sequentially, with the explained variance of 68.5%, 27.2%, 3.3%, 0.6%, 0.3%, and 0.1%, respectively.

Mersereau, 2005; Celik, 2009; Lian, 2012). However, several drawbacks and constraints have been observed associated with PCA, for example, the scaling effects (principal components are not scale invariant) (Rencher, 2003; Prasad and Bruce, 2008). In recent years, many enhanced PCA methods have been proposed toward various applications such as kernel PCA (Schölkopf et al., 1997), scale-invariant PCA (Han and Liu, 2012, 2014), and even more advanced techniques like independent component analysis (Stone, 2004; Wang and Chang, 2006) and projection pursuit (Friedman and Tukey, 1974; Chiang et al., 2001).

6.2.6.2 Linear Discriminant Analysis (LDA)

Analogous to logistic regression and PCA, LDA is a generalization of Fisher's linear discriminant, which also attempts to represent dependent variables as a linear combination of a set of independent features (Fisher, 1936; McLachlan, 2004). It is commonly used as a feature reduction tool to reduce the high-dimensional feature space by projecting the original data into a small dimension set. Unlike PCA, which ultimately intends to minimize mean square error between the original and reduced spaces, LDA seeks to maximize the interclass (i.e., between class) variance and minimize the intraclass (i.e., within class) variance (Dundar and Fung, 2009; Wurm et al., 2016). In LDA, the feature contribution is quantified in terms of the degree of difference between the features as well as the individual contribution to the discrimination of classes (Wurm et al., 2016). Therefore, the number of features can be significantly reduced to the one having the highest impact on the discrimination.

Given a high-dimensional data matrix $X = \{x_1, x_2, \ldots, x_i\}, x_i \in \mathfrak{R}^d$, there should exist a linear mapping function f enabling the transformation of every original data point x_i in X into a low-dimensional vector $Z = \{z_1, z_2, \ldots, z_j\}, z_j \in \mathfrak{R}^r$ $(r \leq d)$ based on a transformation matrix W. That is,

$$z_i = f(x_i) = W^T x_i \tag{6.26}$$

Analogous to many other feature extraction methods, the transformation matrix W can be computed by solving an optimization problem in terms of fulfilling a given maximization criterion of separability among classes. In LDA, this equals finding the best discrimination of the investigated groups by maximizing the ratio of the interclass variance to intraclass variance to measure the disparity of the groups (Wurm et al., 2016). The transformation matrix W can be optimized as:

$$W^* = \arg\max_W \left(\frac{W^T \overline{S} W}{W^T \underline{S} W} \right) \tag{6.27}$$

where $\overline{S}$ denotes the interclass variance and $\underline{S}$ is the intraclass variance, and can be modeled as follows:

$$\overline{S} = \sum_{k=1}^{C} n_k (u_k - u)(u_k - u)^T \tag{6.28}$$

$$\underline{S} = \sum_{k=1}^{C} \left(\sum_{i=1}^{n_k} (x_{ik} - u_k)(x_{ik} - u_k)^T \right) \tag{6.29}$$

where n_k is the number of samples in the kth class, u is the mean of the entire training set, u_k is the mean of the kth class, and x_{ik} is the ith sample in the kth class.

It is clear that the interclass variance is calculated as the square sum of the dispersion of the mean discriminant variables of each class (u_k) from the mean of all discriminant variable elements, and the intraclass variance is defined as the square sum of the dispersion of the discriminant variables of single objects from their class means (Wurm et al., 2016). The solution to Equation 6.27 can be obtained by solving the following eigenvalue problem:

$$\bar{S}W = \Lambda \underline{S}W \tag{6.30}$$

which can be further written as:

$$\underline{S}^{-1}\bar{S}W = \Lambda W \tag{6.31}$$

In remote sensing, LDA has been widely used for land cover classification from various remotely sensed imageries, in particular hyperspectral images because of their high-dimensional space (Liao et al., 2013; Yuan et al., 2014; Shahdoosti and Mirzapour, 2017). Despite its good performance in many applications, conventional LDA has the inherent limitation of becoming intractable when the number of input features exceeds the training samples' size (Bandos et al., 2009; Shahdoosti and Mirzapour, 2017). In order to extend the application of LDA to many practical cases, a number of adaptations have been implemented to conventional LDA, which in turn yields many enhanced LDA, such as regularized LDA (Bandos et al., 2009), orthogonal LDA (Duchene and Leclercq, 1988), uncorrelated LDA (Bandos et al., 2009), stepwise LDA (Siddiqi et al., 2015), two-dimensional LDA (Imani and Ghassemian, 2015b), and so on.

6.2.6.3 Wavelet Transform

The wavelet transform was developed as an alternative to the conventional short-time Fourier transform to overcome the associated frequency and time resolution problems. Essentially, wavelet transform decomposes an input signal into a series of components with localized frequency and time resolution based on a set of basic functions, and the derived components are in turn termed wavelets. The fundamental idea behind wavelet transform is to analyze signal at various scales. To some extent, the process of wavelet transform can be considered a type of multi-resolution analysis.

To perform wavelet analysis, a basic wavelet, also known as the mother wavelet, is routinely required, as other wavelet basis functions are created by shifting and scaling the mother wavelet. Given a mother wavelet $\varphi(t)$, the wavelet basis functions can be generated through the following function:

$$\varphi_{a,b}(t) = \frac{1}{\sqrt{a}}\varphi\left(\frac{t-b}{a}\right) \tag{6.32}$$

where a $(a > 0)$ and b are scaling and shifting factors, respectively. It is clear that the wavelet functions will be dilated when $a > 1$ and contracted when $a < 1$ relative to the mother wavelet. The $1/\sqrt{a}$ is used as a modulation coefficient to normalize the energy of the wavelets (Bruce et al., 2001). The most popular mother wavelet is the Morlet wavelet (also called Gabor wavelet), because it is closely related to human perception in both hearing and vision (Bernardino and Santos-Victor, 2005). The Morlet wavelet is modeled as a wavelet composed of a complex exponential multiplied by a Gaussian window, which can be expressed as:

$$\varphi_\sigma(t) = c_\sigma \pi^{-(1/2)} e^{-(1/2)t^2}\left(e^{i\sigma\lambda} - e^{-(1/2)t^2}\right) \tag{6.33}$$

where c_σ is a normalization constant:

$$c_\sigma = \left(1 + e^{-\sigma^2} - 2e^{-(3/4)\sigma^2}\right)^{-(1/2)} \tag{6.34}$$

Based on the admissibility criterion, all wavelet functions must oscillate with an average value of zero and finite support. Given an input signal $x(t)$, the projection onto the subspace of one wavelet function yields:

$$x_a(t) = \int_R W\{x,a,b\} \cdot \varphi_{a,b}(t) d_b \tag{6.35}$$

and the wavelet coefficient W can be obtained as:

$$W\{x,a,b\} = \langle x, \varphi_{a,b} \rangle = \int_R x(t) \cdot \varphi_{a,b}(t) d_t \tag{6.36}$$

Because of its multi-resolution capability, wavelet transform has been widely used for remotely sensed data analysis; however, many applications were confined to image compression (e.g., DeVore et al., 1992; Walker and Nguyen, 2001) and image fusion (e.g., Zhou et al., 1998; Nunez et al., 1999). Later, wavelet transform was introduced into the feature extraction domain, and was then extensively used for various practices. A fundamental reason why the wavelet transform is an excellent tool for feature extraction is its inherent multi-resolution properties, which enable it to project a signal onto a basis of wavelet functions to separate features at different scales by changing the scaling and shifting parameters with respect to the features to be extracted (Mallat, 1989; Bruce et al., 2002). The challenge here is related to how the wavelet coefficients can be interpreted to represent various features, and a common approach is to compute coefficient distribution over the selected wavelet functions (Ghazali et al., 2007). In feature extraction practices, wavelet transform has been widely used for various applications, for example, target detection (Bruce et al., 2001), dimensionality reduction of hyperspectral image (Bruce et al., 2002), forest mapping (Pu and Gong, 2004), vegetation phenology feature extraction (Martínez and Gilabert, 2009), hyperspectral image classification (Qian et al., 2013), and so on.

6.2.7 PROBABILISTIC TECHNIQUES

In remote sensing, many probabilistic techniques taking advantage of probability theory such as Bayes theorem have been used in feature extraction, particularly for image classification. The probability theory framework renders the classification process physically meaningful as the partitioning of one pixel into one specific class is fully determined by the posterior probability. In other words, the decision is made fully based on mathematical evidence. In this section, two popular probabilistic classifiers, maximum likelihood classifier and Naive Bayes classifier, will be introduced.

6.2.7.1 Maximum Likelihood Classifier (MLC)

Due to its probabilistic property, MLC is one of the most popular classification methods in remote sensing, in which the partition of a pixel into one corresponding class will be simply based on its likelihood (or probability). Essentially, MLC is a supervised classification method derived from the Bayes theorem, as the classification of each cell to one specific class represented in the class samples (or training set) is determined based on both the variance and covariance of the class signatures. With the assumption that the cells in each class sample should be normally distributed, a class can then be characterized by the mean vector and the covariance matrix. Given these two characteristics for each cell value, the membership of the cells to the class can be determined based on their statistical probabilities (likelihoods). More specifically, probability density functions will be built for each class based on the training data set, and then all unlabeled pixels will be classified based on the relative likelihood (probability) of that pixel occurring within each class's probability density function.

Mathematically, the likelihood that a pixel with feature vector X belongs to class k can be defined as a posterior probability (Ahmad and Quegan, 2012):

$$P(k|X) = \frac{P(k)P(X|k)}{P(X)} \tag{6.37}$$

where $P(X|k)$ is the conditional probability to observe X from class k (or probability density function). $P(k)$ is the prior probability of class k, the values of which are usually assumed to be equal to each other due to the lack of sufficient reference data. $P(X)$ is the probability that the X is observed, which can be further written as follows:

$$P(X) = \sum_{k=1}^{N} P(k)P(X|k) \tag{6.38}$$

where N is the total number of classes. Commonly, $P(X)$ is assumed to be a normalization constant in order to ensure $\sum_{k=1}^{N} P(k|X)$ sums to 1 (Ahmad and Quegan, 2012). A pixel x will be assigned into the class k once it satisfies the following criterion:

$$x \in k, \quad \text{if} \quad P(k|X) > P(j|X), \quad \text{for all } j \neq k \tag{6.39}$$

For mathematical reasons, ML often assumes the distribution (or probability density function) of the data in a given class to be a multivariate Gaussian distribution; the likelihood can then be expressed as follows:

$$P(k|X) = \frac{1}{(2\pi)^{N/2} |\Sigma_k|^{1/2}} \exp\left(-\frac{1}{2}(X - u_k)\Sigma_k^{-1}(X - u_k)^T\right) \tag{6.40}$$

where N is the number of data sets (e.g., bands for multispectral image), X is the whole data set of N bands, u_k is the mean vector of class k, and Σ_k is the variance-covariance matrix of class k. $|\Sigma_k|$ is thus the determinant of Σ_k.

Due to its probability theory principle, MLC has been widely used in remote sensing for classification. Applications include, but are not limited to: forest encroachment mapping (Tiwari et al., 2016), rice crop mapping (Chen et al., 2011), land cover change detection (Otukei and Blaschke, 2010), salt farm mapping (Hagner and Reese, 2007), and water quality mapping (Jay and Guillaume, 2014). The performance of MLC has been thoroughly compared with many other classification methods in the literature, such as decision tree, logistic regression, artificial neural network, and support vector machine. More details can be found in Frizzelle and Moody (2001), Hagner and Reese (2007), Kavzoglu and Reis (2008), and Hogland et al. (2013). Further investigations showed that MLC may be ineffective in some cases, for example, classifying spectrally similar categories and classes having subclasses (Kavzoglu and Reis, 2008). To account for these problems, methods like PCA could be used to aid in the classification process. In addition, many extended MLC methods have been developed, such as hierarchical MLC (Ediriwickrema and Khorram, 1997) and calibrated MLC (Hagner and Reese, 2007).

6.2.7.2 Naive Bayes Classifier

Analogous to MLC, the Naive Bayes classifier is also a probabilistic estimator frequently used for classification problems. Essentially, the Naive Bayes classifier works by following the Bayes' theorem principle. An unclassified feature vector $X = (x_1, x_2, \ldots, x_p)$ can be classified into the class C_k once it yields the largest posterior probability $P(C_k|X)$. Based on the Bayes' theorem, the conditional probability $P(C_k|X)$ can be calculated as:

$$P(C_k|X) = \frac{P(X|C_k)P(C_k)}{P(X)} \tag{6.41}$$

where $P(C_k)$ is the priori probability of class C_k. $P(X|C_k)$ is the likelihood (or conditional probability) of the feature vector X falling into the class C_k. $P(x)$ is the priori probability of predictor X.

Since $P(X)$ is independent of the class vector C and the feature values, it is thus equivalent to a constant. Therefore, the critical component to the calculation of the conditional probability lies in the estimation of the priori probability $P(C_k)$ and the class-conditional probability $P(X|C_k)$. In practice, the priori probability $P(C_k)$ can be estimated from the training dataset as the portion of samples within the training dataset taking the class label C_k:

$$P(C_k) = \frac{N_{C_k}}{N} \tag{6.42}$$

where N is the number of the training dataset and N_{C_k} is the number of training samples with class label C_k.

The estimation of the conditional probability $P(X|C_k)$ can be defined as a joint probability as follows:

$$
\begin{aligned}
P(C_k, X) &= P(C_k, x_1, x_2, \ldots, x_p) \\
&= P(x_1|x_2, \ldots, x_p, C_k)P(x_2, \ldots, x_p, C_k) \\
&= P(x_1|x_2, \ldots, x_p, C_k)P(x_2|x_3 \ldots, x_p, C_k)P(x_3, \ldots, x_p, C_k) \\
&= P(x_1|x_2, \ldots, x_p, C_k)P(x_2|x_3 \ldots, x_p, C_k) \cdots P(x_{p-1}|x_p, C_k)P(x_p|C_k)P(C_k)
\end{aligned}
\tag{6.43}
$$

To simplify the work, the Naive Bayes assumes that the presence of a feature in one class is conditionally independent of other features, which means that they have the same conditional probability:

$$P(x_i|x_{i+1}, \ldots, x_p, C_k) = P(x_i|C_k) \tag{6.44}$$

Then the joint probability model can be expressed as:

$$
\begin{aligned}
P(C_k|x_1, x_2, \ldots, x_p) &\propto P(C_k, x_1, x_2, \ldots, x_p) \\
&\propto P(C_k)P(x_1|C_k)P(x_2|C_k) \cdots P(x_p|C_k) \\
&\propto P(C_k)\prod_{i=1}^{p} P(x_p|C_k)
\end{aligned}
\tag{6.45}
$$

Based on the independent assumptions, the conditional probability $P(C_k|X)$ can be further written as:

$$P(C_k|X) = \frac{P(C_k)}{P(X)}\prod_{i=1}^{p} P(x_p|C_k) \tag{6.46}$$

In practice, the Naive Bayes classifier can handle both discrete and continuous variables (Chang et al., 2012). If X contains a finite number of discrete features x_i, then the estimation of $P(x_i|C_k)$ is equal to the calculation of the proportion of training samples x_i within class C_k, which can be expressed as:

$$P(x_i|C_k) = \frac{N_{C_k}(x_i)}{N_{C_k}} \tag{6.47}$$

However, if x_i is a continuous variable, then discretization should be performed. A common approach is to assume a normal distribution for the decomposed conditional probabilities:

$$P(x_i|C_k) = N(x_i, \mu_{C_k}, \sigma^2_{C_k}) \tag{6.48}$$

The parameters μ and σ can be directly estimated from the training dataset.

In remote sensing, the Naive Bayes classifier is often used as another popular probabilistic method for classification problems. As opposed to other classifiers, the popularity of the Naive Bayes classifier is enhanced by the following properties (Wu et al., 2008). First, the Naive Bayes classifier does not require a complicated iterative parameter estimation scheme, thus it is easy to construct, making it able to handle huge datasets. Second, the Bayesian scheme makes the classification process easy to understand and interpret, even for users unskilled in classifier technology. Finally, the classification result has much higher accuracy. In practice, applications of Naive Bayes classifier are relatively fewer compared to methods such as MLC. In the past years, the Naive Bayes classifier has been successfully used for multi-label learning (Zhang et al., 2009), image classification (Liu et al., 2011), text classification (Feng et al., 2015), and so forth.

6.3   SUMMARY

In this chapter, a suite of feature extraction methods based on statistic and decision science principles were introduced, focusing primarily on their theoretical foundations with some illustrative examples for practical applications. More specifically, methods discussed in this chapter include filtering operation, morphology, decision trees, clustering algorithms, linear regression, PCA, wavelet transform, MLC, and Naive Bayes classifier. All these techniques have been extensively used in remote sensing for feature extraction, mainly for dimensionality reduction and feature selection. In the next chapter, a set of artificial intelligence-based methods that are widely applied for feature extraction will be described in detail.

REFERENCES

Agrawal, R., Gehrke, J., Gunopulos, D., and Raghavan, P., 2005. Automatic subspace clustering of high dimensional data. *Data Mining and Knowledge Discovery*, 11, 5–33.

Ahmad, A. and Quegan, S., 2012. Analysis of maximum likelihood classification on multispectral data. *Applied Mathematical Sciences*, 6, 6425–6436.

Ankerst, M., Breunig, M. M., Kriegel, H., and Sander, J., 1999. OPTICS: Ordering points to identify the clustering structure. In: *Proceedings of the 1999 ACM SIGMOD International Conference on Management of Data*, 49–60, Pennsylvania, USA.

Aragón-Calvo, M. A., Jones, B. J. T., van de Weygaert, R., and van der Hulst, J. M., 2007. The multiscale morphology filter: identifying and extracting spatial patterns in the galaxy distribution. *Astronomy & Astrophysics*, 474, 315–338.

Bailey, K., 1994. Numerical taxonomy and cluster analysis. In: K. D. Bailey (Ed.) *Typologies and Taxonomies: An Introduction to Classification Techniques*, SAGE Publications Ltd., Thousand Oaks, California, USA, 34, 24.

Bandos, T., Bruzzone, L., and Camps-Valls, G., 2009. Classification of hyperspectral images with regularized linear discriminant analysis. *IEEE Transactions on Geoscience and Remote Sensing*, 47, 862–873.

Bastarrika, A., Chuvieco, E., and Martín, M. P., 2011. Mapping burned areas from Landsat TM/ETM+ data with a two-phase algorithm: Balancing omission and commission errors. *Remote Sensing of Environment*, 115, 1003–1012.

Belgiu, M. and Drăguţ, L., 2016. Random forest in remote sensing: A review of applications and future directions. *ISPRS Journal of Photogrammetry and Remote Sensing*, 114, 24–31.

Benediktsson, J. A., Palmason, J. A., and Sveinsson, J. R., 2005. Classification of hyperspectral data from urban areas based on extended morphological profiles. *IEEE Transactions on Geoscience and Remote Sensing*, 43, 480–491.

Bernardino, A. and Santos-Victor, J., 2005. A real-time Gabor primal sketch for visual attention. In: *2nd Iberian Conference on Pattern Recognition and Image Analysis*, 335–342, Estoril, Portugal.

Breinman, L., 2001. Random forests. *Machine Learning*, 45, 5–32.

Breiman, L., Friedman, J. H., Olshen, R. A., and Stone, C. J., 1984. *Classification and Regression Trees*. Wadsworth & Brooks/Cole Advanced Books & Software.

Brekke, C. and Solberg, A. H. S., 2005. Oil spill detection by satellite remote sensing. *Remote Sensing of Environment*, 95, 1–13.

Bruce, L. M., Koger, C. H., and Li, J., 2002. Dimensionality reduction of hyperspectral data using discrete wavelet transform feature extraction. *IEEE Transactions on Geoscience and Remote Sensing*, 40, 2331–2338.

Bruce, L. M., Morgan, C., and Larsen, S., 2001. Automated detection of subpixel hyperspectral targets with continuous and discrete wavelet transforms. *IEEE Transactions on Geoscience and Remote Sensing*, 39, 2217–2226.

Cattell, R. B., 1943. The description of personality: Basic traits resolved into clusters. *Journal of Abnormal Psychology*, 38, 476–506.

Celik, T. 2009. Unsupervised change detection in satellite images using principal component analysis and k-Means clustering. *IEEE Geoscience and Remote Sensing Letters*, 6, 772–776.

Chan, J. C. W. and Paelinckx, D., 2008. Evaluation of Random Forest and Adaboost tree-based ensemble classification and spectral band selection for ecotope mapping using airborne hyperspectral imagery. *Remote Sensing of Environment*, 112, 2999–3011.

Chang, N.-B., Han, M., Yao, W., and Chen, L.-C., 2012. Remote sensing assessment of coastal land reclamation impact in Dalian, China, using high-resolution SPOT images and support vector machine. In: *Environmental Remote Sensing and Systems Analysis*. CRC Press, Boca Raton, FL, USA, 249–276.

Chaudhuri, D., Kushwaha, N. K., Samal, A., and Agarwal, R. C., 2016. Automatic building detection from high-resolution satellite images based on morphology and internal gray variance. *IEEE Journal of Selected Topics in Applied Earth Observations and Remote Sensing*, 9, 1767–1779.

Chen, W., Li, X., Wang, Y., Chen, G., and Liu, S., 2014a. Forested landslide detection using LiDAR data and the random forest algorithm: A case study of the Three Gorges, China. *Remote Sensing of Environment*, 152, 291–301.

Chen, Y., Nasrabadi, N. M., and Tran, T. D., 2013. Hyperspectral image classification via kernel sparse representation. *IEEE Transactions on Geoscience and Remote Sensing*, 51, 217–231.

Chen, C., Qin, Q., Zhang, N., Li, J., Chen, L., Wang, J., Qin, X., and Yang, X., 2014b. Extraction of bridges over water from high-resolution optical remote-sensing images based on mathematical morphology. *International Journal of Remote Sensing*, 35, 3664–3682.

Chen, J., Shen, M., Zhu, X., and Tang, Y., 2009. Indicator of flower status derived from in situ hyperspectral measurement in an alpine meadow on the Tibetan Plateau. *Ecological Indicators*, 9, 818–823.

Chen, C.-F., Son, N.-T., Chen, C.-R., and Chang, L.-Y., 2011. Wavelet filtering of time-series moderate resolution imaging spectroradiometer data for rice crop mapping using support vector machines and maximum likelihood classifier. *Journal of Applied Remote Sensing*, 5, 53525.

Chen, T., Trinder, J., and Niu, R., 2017. Object-oriented landslide mapping using ZY-3 satellite imagery, random forest and mathematical morphology, for the Three-Gorges Reservoir, China. *Remote Sensing*, 9, 333.

Chen, J. and Wang, R., 2007. A pairwise decision tree framework for hyperspectral classification. *International Journal of Remote Sensing*, 28, 2821–2830.

Cheng, Q., Varshney, P. K., and Arora, M. K., 2006. Logistic regression for feature selection and soft classification of remote sensing data. *IEEE Geoscience and Remote Sensing Letters*, 3, 491–494.

Chiang, S.-S., Chang, C.-I., and Ginsberg, I. W., 2001. Unsupervised target detection in hyperspectral images using projection pursuit. *IEEE Transactions on Geoscience and Remote Sensing*, 39, 1380–1391.

Cho, M. A. and Skidmore, A. K., 2006. A new technique for extracting the red edge position from hyperspectral data: The linear extrapolation method. *Remote Sensing of Environment*, 101, 181–193.

Clausi, D. A. and Ed Jernigan, M., 2000. Designing Gabor filters for optimal texture separability. *Pattern Recognition*, 33, 1835–1849.

Davies, E. R. (Ed.), 2012. Mathematical morphology. In: *Computer and Machine Vision: Theory, Algorithms, Practicalities,* Academic Press, Massachusetts, USA, 185–208.

Davis, L., 1975. A survey of edge detection techniques. *Computer Graphics and Image Processing*, 4, 248–260.

Defays, D., 1977. An efficient algorithm for a complete link method. *The Computer Journal*, 20, 364–366.

Dell'Acqua, F. and Gamba, P., 2001. Detection of urban structures in SAR images by robust fuzzy clustering algorithms: the example of street tracking. *IEEE Transactions on Geoscience and Remote Sensing*, 39, 2287–2297.

Dempster, A. and Laird, N., 1977. Maximum likelihood from incomplete data via the EM algorithm. *Journal of the Royal Statistical Society. Series B*, 39, 1–38.

DeVore, R. A., Jawerth, B., and Lucier, B. J., 1992. Image compression through wavelet transform coding. *IEEE Transactions on Information Theory*, 38, 719–746.

Duchene, J. and Leclercq, S., 1988. An optimal transformation for discriminant and principal component analysis. *IEEE Transactions on Pattern Analysis and Machine Intelligence*, 10, 978–983.

Dundar, M. M. and Fung, G., 2009. Kernel fisher's discriminant with heterogeneous kernels. In: G. Camps-Valls and L. Bruzzone (Eds.), *Kernel Methods for Remote Sensing Data Analysis*, John Wiley & Sons, Inc, Chichester, West Sussex, UK, 111–124.

Ediriwickrema, J. and Khorram, S., 1997. Hierarchical maximum-likelihood classification for improved accuracies. *IEEE Transactions on Geoscience and Remote Sensing*, 35, 810–816.

Elnemr, H. A., Zayed, N. M., and Fakhreldein, M. A., 2016. Feature extraction techniques: Fundamental concepts and survey. In: *Handbook of Research on Emerging Perspectives in Intelligent Pattern Recognition, Analysis, and Image Processing*, IGI Global.

Ester, M., Kriegel, H. P., Sander, J., and Xu, X., 1996. A density-based algorithm for discovering clusters in large spatial databases with noise. In: *Procedings of 2nd International Conference Knowledge Discovery and Data Mining*, 226–231, Oregon, USA.

Estivill-Castro, V., 2002. Why so many clustering algorithms. *ACM SIGKDD Explorations Newsletter*, 4, 65–75.

Everitt, B., 2011. *Cluster Analysis, Wiley Series in Probability and Statistics*. SAGE Publications, Inc., 2455 Teller Road, Thousand Oaks California 91320 United States of America.

Fan, H., 2013. Land-cover mapping in the Nujiang Grand Canyon: Integrating spectral, textural, and topographic data in a random forest classifier. *International Journal of Remote Sensing*, 34, 7545–7567.

Farrell, M. D. and Mersereau, R. M., 2005. On the impact of PCA dimension reduction for hyperspectral detection of difficult targets. *IEEE Geoscience and Remote Sensing Letters*, 2, 192–195.

Feng, G., Guo, J., Jing, B.-Y., and Sun, T., 2015. Feature subset selection using naive Bayes for text classification. *Pattern Recognition Letters*, 65, 109–115.

Feyisa, G. L., Meilby, H., Fensholt, R., and Proud, S. R., 2014. Automated water extraction index: A new technique for surface water mapping using Landsat imagery. *Remote Sensing of Environment*, 140, 23–35.

Fisher, R. A., 1936. The use of multiple measurements in taxonomic problems. *Annals of Eugenics*, 7, 179–188.

Freedman, D. A., 2009. *Statistical Models: Theory and Practice, Technometrics*. Cambridge University Press, New York.

Friedl, M. A. and Brodley, C. E., 1997. Decision tree classification of land cover from remotely sensed data. *Remote Sensing of Environment*, 61, 399–409.

Friedl, M. A., Brodley, C. E., and Strahler, A. H., 1999. Maximizing land cover classification accuracies produced by decision trees at continental to global scales. *IEEE Transactions on Geoscience and Remote Sensing*, 37, 969–977.

Friedman, J. H. and Tukey, J. W., 1974. A projection pursuit algorithm for exploratory data analysis. *IEEE Transactions on Computers*, C-23, 881–890.

Frizzelle, B. G. and Moody, A., 2001. Mapping continuous distributions of land cover: A comparison of maximum-likelihood estimation and artificial neural networks. *Photogrammetric Engineering & Remote Sensing*, 67, 693–705.

Gamba, P., Dell'Acqua, F., and Lisini, G., 2006. Improving urban road extraction in high-resolution images exploiting directional filtering, perceptual grouping, and simple topological concepts. *IEEE Geoscience and Remote Sensing Letters*, 3, 387–391.

Ghazali, K. H., Mansor, M. F., Mustafa, M. M., and Hussain, A., 2007. Feature extraction technique using discrete wavelet transform for image classification. In: *The 5th Student Conference on Research and Development-SCORed*, 5–8, Selangor, Malaysia.

Ghulam, A., Porton, I., and Freeman, K., 2014. Detecting subcanopy invasive plant species in tropical rainforest by integrating optical and microwave (InSAR/PolInSAR) remote sensing data, and a decision tree algorithm. *ISPRS Journal of Photogrammetry and Remote Sensing*, 88, 174–192.

Guo, L., Chehata, N., Mallet, C., and Boukir, S., 2011. Relevance of airborne lidar and multispectral image data for urban scene classification using Random forests. *ISPRS Journal of Photogrammetry and Remote Sensing*, 66, 56–66.

Hagner, O. and Reese, H., 2007. A method for calibrated maximum likelihood classification of forest types. *Remote Sensing of Environment*, 110, 438–444.

Ham, J., Chen, Y., Crawford, M. M., and Ghosh, J., 2005. Investigation of the random forest framework for classification of hyperspectral data. *IEEE Transactions on Geoscience and Remote Sensing*, 43, 492–501.

Hamerly, G. and Elkan, C., 2004. Learning the k in kmeans. *Advances in Neural Information Processing Systems*, The MIT Press, Cambridge, USA, 16, 1–8.

Han, F. and Liu, H., 2012. Transelliptical component analysis. In: *Advances in Neural Information Processing Systems*, 1–9, Nevada, USA.

Han, F. and Liu, H., 2014. High dimensional semiparametric scale-invariant principal component analysis. *IEEE Transactions on Pattern Analysis and Machine Intelligence*, 36, 2016–2032.

Hastie, T., Tibshirani, R., and Friedman, J., 2009. *The Elements of Statistical Learning, Methods, Springer Series in Statistics*. Springer, New York.

Hogland, J., Billor, N., and Anderson, N., 2013. Comparison of standard maximum likelihood classification and polytomous logistic regression used in remote sensing. *European Journal of Remote Sensing*, 46, 623–640.

Hosmer, D. and Lemeshow, S., 2000. *Applied Logistic Regression*. John Wiley & Sons, Inc.

Huang, X., Guan, X., Benediktsson, J. A., Zhang, L., Li, J., Plaza, A., and Dalla Mura, M., 2014. Multiple morphological profiles from multicomponent-base images for hyperspectral image classification. *IEEE Journal of Selected Topics in Applied Earth Observations and Remote Sensing*, 7, 4653–4669.

Imani, M. and Ghassemian, H., 2015a. Ridge regression-based feature extraction for hyperspectral data. *International Journal of Remote Sensing*, 36, 1728–1742.

Imani, M. and Ghassemian, H., 2015b. Two dimensional linear discriminant analyses for hyperspectral data. *Photogrammetric Engineering & Remote Sensing*, 81, 777–786.

Ishioka, T., 2000. Extended K-means with an efficient estimation of the number of clusters. In: *Proceedings of the Seventeenth International Conference on Machine Learning Table of Contents*, 17–22, Hong Kong, China.

Jay, S. and Guillaume, M., 2014. A novel maximum likelihood based method for mapping depth and water quality from hyperspectral remote-sensing data. *Remote Sensing of Environment*, 147, 121–132.

Jian, B. and Vemuri, B. C., 2011. Robust point set registration using Gaussian mixture models. *IEEE Transactions on Pattern Analysis and Machine Intelligence*, 33, 1633–1645.

Jolliffe, I. T., 2002. *Principal Component Analysis, Springer Series in Statistics*. Springer-Verlag, New York.

Kang, X., Li, S., and Benediktsson, J. A., 2014. Spectral-spatial hyperspectral image classification with edge-preserving filtering. *IEEE Transactions on Geoscience and Remote Sensing*, 52, 2666–2677.

Kavzoglu, T. and Reis, S., 2008. Performance analysis of maximum likelihood and artificial neural network classifiers for training sets with mixed pixels. *GIScience & Remote Sensing*, 45, 330–342.

Kemper, T., Jenerowicz, M., Pesaresi, M., and Soille, P., 2011. Enumeration of dwellings in Darfur Camps from GeoEye-1 satellite images using mathematical morphology. *IEEE Journal of Selected Topics in Applied Earth Observations and Remote Sensing*, 4, 8–15.

Khurshid, H. and Khan, M. F., 2015. Segmentation and classification using logistic regression in remote sensing imagery. *IEEE Journal of Selected Topics in Applied Earth Observations and Remote Sensing*, 8, 224–232.

Kriegel, H.-P., Kröger, P., Sander, J., and Zimek, A., 2011. Density-based clustering. *WIREs Data Mining and Knowledge Discovery*, 1, 231–240.

Kriegel, H.-P., Kröger, P., and Zimek, A., 2012. Subspace clustering. *WIREs. Data Mining and Knowledge Discovery*, 2, 351–364.

Lee, I. K., Shamsoddini, A., Li, X., Trinder, J. C., and Li, Z., 2016. Extracting hurricane eye morphology from spaceborne SAR images using morphological analysis. *ISPRS Journal of Photogrammetry and Remote Sensing*, 117, 115–125.

Li, X., Zhang, Y., Bao, Y., Luo, J., Jin, X., Xu, X., Song, X., and Yang, G., 2014. Exploring the best hyperspectral features for LAI estimation using partial least squares regression. *Remote Sensing*, 6, 6221–6241.

Lian, H., 2012. On feature selection with principal component analysis for one-class SVM. *Pattern Recognition Letters*, 33, 1027–1031.

Liao, W., Pizurica, A., Scheunders, P., Philips, W., and Pi, Y. 2013. Semisupervised local discriminant analysis for feature extraction in hyperspectral images. *IEEE Transactions on Geoscience and Remote Sensing*, 51, 184–198.

Liu, Y.-F., Guo, J.-M., and Lee, J.-D., 2011. Halftone image classification using LMS algorithm and naive bayes. *IEEE Transactions on Image Processing*, 20, 2837–2847.

Liu, Y., Zhang, D., and Lu, G., 2008. Region-based image retrieval with high-level semantics using decision tree learning. *Pattern Recognition*, 41, 2554–2570.

MacKay, D. J. C. (Ed.), 2003. An example inference task: Clustering. In: *Information Theory, Inference and Learning Algorithms*, Cambridge University Press, Cambridge, UK, 284–292.

Mallat, S. G., 1989. A theory for multiresolution signal decomposition: The wavelet representation. *IEEE Transactions on Pattern Analysis and Machine Intelligence*, 11, 674–693.

Martínez, B. and Gilabert, M. A., 2009. Vegetation dynamics from NDVI time series analysis using the wavelet transform. *Remote Sensing of Environment*, 113, 1823–1842.

McFeeters, S. K., 1996. The use of the normalized difference water index (NDWI) in the delineation of open water features. *International Journal of Remote Sensing*, 17, 1425–1432.

McLachlan, G., 2004. *Discriminant Analysis and Statistical Pattern Recognition*. Wiley-Interscience.

Mehrotra, R., Namuduri, K. R., and Ranganathan, N., 1992. Gabor filter-based edge detection. *Pattern Recognition*, 25, 1479–1494.

Modava, M. and Akbarizadeh, G., 2017. Coastline extraction from SAR images using spatial fuzzy clustering and the active contour method. *International Journal of Remote Sensing*, 38, 355–370.

Mohammadzadeh, A., Tavakoli, A., and Valadan Zoej, M. J., 2006. Road extraction based on fuzzy logic and mathematical morphology from pan-sharpened ikonos images. *The Photogrammetric Record*, 21, 44–60.

Ni, H., Lin, X., and Zhang, J., 2017. Classification of ALS point cloud with improved point cloud segmentation and random forests. *Remote Sensing*, 9, 288.

Nixon, M. S. and Aguado, A. S. (Eds.), 2012. Low-level feature extraction (including edge detection). In: *Feature Extraction & Image Processing for Computer Vision*, Academic Press, London, UK, 137–216.

Nunez, J., Otazu, X., Fors, O., Prades, A., Pala, V., Arbiol, R., Núnez, J., Otazu, X., Fors, O., Prades, A., Palà, V., and Arbiol, R., 1999. Multiresolution-based image fusion with additive wavelet decomposition. *IEEE Transactions on Geoscience and Remote Sensing*, 37, 1204–1211.

O'Reilly, J. E., Maritorena, S., Mitchell, B. G., Siegel, D. A., Carder, K. L., Garver, S. A., Kahru, M., and McClain, C., 1998. Ocean color chlorophyll algorithms for SeaWiFS. *Journal of Geophysical Research: Oceans*, 103, 24937–24953.

Otukei, J. R. and Blaschke, T., 2010. Land cover change assessment using decision trees, support vector machines and maximum likelihood classification algorithms. *International Journal of Applied Earth Observation and Geoinformation*, 12, S27–S31.

Pal, M., 2005. Random forest classifier for remote sensing classification. *International Journal of Remote Sensing*, 26, 217–222.

Plaza, A. J., 2007. Morphological hyperspectral image classification: A parallel processing perspective. In: Chein-I Chang (Ed.), *Hyperspectral Data Exploitation: Theory and Applications*, John Wiley & Sons, Inc, Chichester, West Sussex, UK, 353–378.

Prasad, S. and Bruce, L. M., 2008. Limitations of principal components analysis for hyperspectral target recognition. *IEEE Geoscience and Remote Sensing Letters*, 5, 625–629.

Pratt, W. K., 2007. Morphological image processing. In: Nick Efford (Ed.), *Digital Image Processing*, Pearson Education, London, UK, 421–463.

Pu, R., 2012. Comparing canonical correlation analysis with partial least squares regression in estimating forest leaf area index with multitemporal landsat TM imagery. *GIScience & Remote Sensing*, 49, 92–116.

Pu, R. L. and Gong, P., 2004. Wavelet transform applied to EO-1 hyperspectral data for forest LAI and crown closure mapping. *Remote Sensing of Environment*, 91, 212–224.

Qian, Y., Ye, M., and Zhou, J., 2012. Hyperspectral image classification based on structured sparse logistic regression and three-dimensional wavelet texture features. *IEEE Transactions on Geoscience and Remote Sensing*, 51, 2276–2291.

Qian, Y., Ye, M., and Zhou, J., 2013. Hyperspectral image classification based on structured sparse logistic regression and three-dimensional wavelet texture features. *IEEE Transactions on Geoscience and Remote Sensing*, 51, 2276–2291.

Quackenbush, L. J., 2004. A review of techniques for extracting linear features from imagery. *Photogrammetric Engineering & Remote Sensing*, 70, 1383–1392.

Rencher, A. C., 2003. Principal component analysis. In: *Methods of Multivariate Analysis*. John Wiley & Sons, Inc., New York, USA, 380–407.

Reuter, M., Biasotti, S., Giorgi, D., Patanè, G., and Spagnuolo, M., 2009. Discrete Laplace–Beltrami operators for shape analysis and segmentation. *Computers & Graphics*, 33, 381–390.

Roberts, L. G., 1963. *Machine Perception of Three-Dimensional Solids*, Massachusetts Institute of Technology, Massachusetts, USA.

Roberts, S. J., Husmeier, D., Rezek, I., and Penny, W., 1998. Bayesian approaches to Gaussian mixture modeling. *IEEE Transactions on Pattern Analysis and Machine Intelligence*, 20, 1133–1142.

Rokach, L. and Maimon, O., 2005. Clustering methods. In: Oded Maimon and Lior Rokach (Eds.), *Data Mining and Knowledge Discovery Handbook*, Springer, New York, USA, 321–352.

Rokach, L. and Maimon, O., 2008. *Data Mining With Decision Trees: Theory and Applications*. World Scientific Publishing Co., Inc. River Edge, NJ, USA.

Romero, A., Gatta, C., and Camps-Valls, G., 2016. Unsupervised deep feature extraction for remote sensing image classification. *IEEE Transactions on Geoscience and Remote Sensing*, 54, 1349–1362.

Rouse, J. W., Hass, R. H., Schell, J. A., and Deering, D. W., 1974. Monitoring vegetation systems in the great plains with ERTS. In: *Third Earth Resources Technology Satellite (ERTS) Symposium*, 309–317, Texas, USA.

Safavian, S. R. and Landgrebe, D., 1991. A survey of decision tree classifier methodology. *IEEE Transactions on Systems, Man, and Cybernetics: Systems*, 21, 660–674.

Schölkopf, B., Smola, A., and Müller, K. R., 1997. Kernel principal component analysis. In: *International Conference on Artificial Neural Networks*, 583–588, Lausanne, Switzerland.

Serra, J., 1992. *Image Analysis and Mathematical Morphology*. Academic Press, New York.

Sghaier, M. O., Foucher, S., and Lepage, R., 2017. River extraction from high-resolution SAR images combining a structural feature set and mathematical morphology. *IEEE Journal of Selected Topics in Applied Earth Observations and Remote Sensing*, 10, 1025–1038.

Shahdoosti, H. R. and Mirzapour, F., 2017. Spectral–spatial feature extraction using orthogonal linear discriminant analysis for classification of hyperspectral data. *European Journal of Remote Sensing*, 50, 111–124.

Sibson, R., 1973. SLINK: An optimally efficient algorithm for the single-link cluster method. *The Computer Journal*, 16, 30–34.

Siddiqi, M. H., Ali, R., Khan, A. M., Park, Y. T., and Lee, S., 2015. Human facial expression recognition using stepwise linear discriminant analysis and hidden conditional random fields. *IEEE Transactions on Image Processing*, 24, 1386–1398.

Soille, P., 2004. *Morphological Image Analysis*. Springer-Verlag, Berlin, Germany.

Soille, P. and Pesaresi, M., 2002. Advances in mathematical morphology applied to geoscience and remote sensing. *IEEE Transactions on Geoscience and Remote Sensing*, 40, 2042–2055.

Stone, J. V., 2004. *Independent Component Analysis : A Tutorial Introduction*. MIT Press, Cambridge.

Stumpf, A. and Kerle, N., 2011. Object-oriented mapping of landslides using random forests. *Remote Sensing of Environment*, 115, 2564–2577.

Sun, W., Zhang, L., Du, B., Li, W., Lai, M. Y., 2015. Band selection using improved sparse subspace clustering for hyperspectral imagery classification. *IEEE Journal of Selected Topics in Applied Earth Observations and Remote Sensing*, 8, 2784–2797.

Tian, D. P., 2013. A review on image feature extraction and representation techniques. *International Journal of Multimedia and Ubiquitous Engineering*, 8, 385–395.

Tian, X., Jiao, L., and Zhang, X., 2013. A clustering algorithm with optimized multiscale spatial texture information: application to SAR image segmentation. *International Journal of Remote Sensing*, 34, 1111–1126.

Tian, S., Zhang, X., Tian, J., and Sun, Q., 2016. Random forest classification of wetland landcovers from multi-sensor data in the arid region of Xinjiang, China. *Remote Sensing*, 8, 954.

Tiwari, L. K., Sinha, S. K., Saran, S., Tolpekin, V. A., and Raju, P. L. N., 2016. Forest encroachment mapping in Baratang Island, India, using maximum likelihood and support vector machine classifiers. *Journal of Applied Remote Sensing*, 10, 16016.

Tokunaga, M. and Thuy Vu, T., 2007. Clustering method to extrct buildings from airborne laser data. In: *IEEE 2007 International Geoscience and Remote Sensing Symposium*, 2018–2021, Barcelona, Spain.

Tryon, R. C., 1939. *Cluster Analysis: Correlation Profile and Orthometric (factor) Analysis for the Isolation of Unities in Mind and Personality*. Edwards Brother.

Tuia, D., Pacifici, F., Kanevski, M., and Emery, W. J., 2009. Classification of very high spatial resolution imagery using mathematical morphology and support vector machines. *IEEE Transactions on Geoscience and Remote Sensing*, 47, 3866–3879.

Valero, S., Chanussot, J., Benediktsson, J. A., Talbot, H., and Waske, B., 2010. Advanced directional mathematical morphology for the detection of the road network in very high resolution remote sensing images. *Pattern Recognition Letters*, 31, 1120–1127.

Walker, S. H. and Duncan, D. B., 1967. Estimation of the probability of an event as a function of several independent variables. *Biometrika*, 54, 167.

Walker, J. S. and Nguyen, T. Q., 2001. Wavelet-based image compression. In: *The Transform and Data Compression Handbook*. CRC Press, Boca Raton, FL, USA, 1–10.

Wang, J. and Chang, C.-I., 2006. Applications of independent component analysis in endmember extraction and abundance quantification for hyperspectral imagery. *IEEE Transactions on Geoscience and Remote Sensing*, 44, 2601–2616.

Wei, Y. and Zhao, Z., 2004. Urban building extraction from high-resolution satellite panchromatic image using clustering and edge detection. In: *IEEE 2004 International Geoscience and Remote Sensing Symposium*, 2008–2010, Anchorage, AK, USA.

Winham, S. J., Freimuth, R. R., and Biernacka, J. M., 2013. A weighted random forests approach to improve predictive performance. *Statistical Analysis and Data Mining*, 6, 496–505.

Wu, J., Chen, Y., Dai, D., Chen, S., and Wang, X., 2017. Clustering-based geometrical structure retrieval of man-made target in SAR images. *IEEE Geoscience and Remote Sensing Letters*, 14, 279–283.

Wu, X., Kumar, V., Ross Quinlan, J., Ghosh, J., Yang, Q., Motoda, H., McLachlan, G. J., Ng, A., Liu, B., Yu, P. S., Zhou, Z.-H., Steinbach, M., Hand, D. J., and Steinberg, D., 2008. Top 10 algorithms in data mining. *Knowledge and Information Systems*, 14, 1–37.

Wurm, M., Schmitt, A., and Taubenbock, H., 2016. Building types' classification using shape-based features and linear discriminant functions. *IEEE Journal of Selected Topics in Applied Earth Observations and Remote Sensing*, 9, 1901–1912.

Xu, H., 2006. Modification of normalised difference water index (NDWI) to enhance open water features in remotely sensed imagery. *International Journal of Remote Sensing*, 27, 3025–3033.

Yuan, H. and Tang, Y. Y., 2017. Spectral–spatial shared linear regression for hyperspectral image classification. *IEEE Transactions on Cybernetics*, 47, 934–945.

Yuan, H., Tang, Y. Y., Lu, Y., Yang, L., and Luo, H., 2014. Spectral-spatial classification of hyperspectral image based on discriminant analysis. *IEEE Journal of Selected Topics in Applied Earth Observations and Remote Sensing*, 7, 2035–2043.

Zhang, X., Jiao, L., Liu, F., Bo, L., and Gong, M., 2008. Spectral clustering ensemble applied to SAR image segmentation. *IEEE Transactions on Geoscience and Remote Sensing*, 46, 2126–2136.

Zhang, M.-L., Peña, J. M., and Robles, V., 2009. Feature selection for multi-label naive Bayes classification. *Information Sciences*, 179, 3218–3229.

Zhou, J., Civco, D. L., and Silander, J. A., 1998. A wavelet transform method to merge Landsat TM and SPOT panchromatic data. *International Journal of Remote Sensing*, 19, 743–757.

7 Feature Extraction with Machine Learning and Data Mining Algorithms

7.1 INTRODUCTION

Some feature extraction algorithms with statistics and decision science principles for basic feature extraction were introduced in Chapter 6. Yet many different machine learning and/or data mining techniques can also be applied to identify patterns, retrieve rules, and perform classification or regression within a data set for feature extraction. To some extent, the feature extraction algorithms designs based on statistics and decision science principles may be limited by their learning capacity and inference mechanism. However, they can be used as a feature extractor tool to retrieve some basic features in support of high-end, subsequent feature extraction through machine learning and/or data mining algorithms. With deepened representation power based on better learning capability and inference mechanism, artificial intelligence (AI)-based classifiers may be more helpful in establishing the complex relations between various independent variables such as surface reflectance and environmental constituents. In Figure 7.1, the flowchart illustrates such an integrated feature extraction philosophy in which a feature extractor may perform some basic feature extraction activities with respect to the object-based search as *a priori* knowledge in support of more sophisticated feature extraction using machine learning and/or data mining techniques. Nevertheless, it is not always necessary to include the traditional feature extraction techniques as a feature extractor for advanced feature extraction.

Machine learning or data mining algorithms can be used to classify features based on supervised, unsupervised, semi-supervised, and even hybrid (e.g., reinforcement) learning processes. Notable machine learning and data mining techniques with regression and/or classification capabilities in the AI regime include genetic algorithm (GA)/Genetic Programming (GP) (Chen et al., 2008; Chang et al., 2012a, 2013; Song et al., 2012), Artificial Neural Networks (ANN) (Doerffer and Schiller, 2007; Ioannou et al., 2011), Support Vector Machine (SVM) (Filippi and Archibald, 2009; Huang et al., 2002), and Particle Swarm Optimization (PSO), which will be highlighted in this chapter. Some hybrid models may be applicable for ensemble learning (Nieto et al., 2013).

Most of the machine learning algorithms count on supervised learning, in which the labeled independent data pairs are first given to establish a possible function to link the input data (independent variables) to the target data (dependent data). Data mining techniques perform unsupervised learning that is most closely related to information retrieval from the remotely sensed images without labeled data sets. In this case, the learning algorithm groups variables into distinct clusters based on the inherent similarities between them. In contrast, semi-supervised learning takes advantage of both supervised and unsupervised methods to better solve problems. In practice, it uses unprocessed input data to first mine the rules and then it improves the accuracy by using limited labeled data to generate a reliable function explaining the relationship between the input and target values.

In the realm of AI, ANN models are inspired by the natural neural network of the human nervous system. The first ANN, called "Perceptron," was invented by psychologist Frank Rosenblatt (Rosenblatt, 1958). It was originally designed to model how the human brain processed visual data, retrieved patterns, and learned to recognize objects. ANN is a neurocomputing model made up of

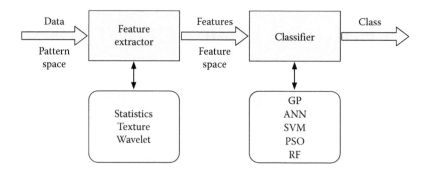

FIGURE 7.1   A typical feature extraction procedure. GP: Genetic programming; ANN: Artificial neural network; SVM: Support vector machine; PSO: Particle swarm optimization; RF: Random forests.

many simple processing nodes in a highly interconnected network, which may learn the input data and retrieve the needed information systematically by the dynamic state of the nodes and the status of the network in response to external inputs. Whereas information is contained in the overall activation "state" of the network rather than a stand-alone memory, knowledge is represented by the network itself instead of a set of instructions set aside in a rule base. As a consequence, ANNs are neither sequential nor linear and deterministic processes (Caudill, 1989). Because of this model architecture, ANNs are universal approximators with a high tolerance to error, which may be suitable for discovering regularities, identifying rules or capturing associations within a set of patterns when dealing with big data based on either supervised or unsupervised learning. This type of "system thinking" in a high-dimensional environment is particularly useful when there is no given information about the existing relationships between variables or when the relationships are not possible to describe correctly with traditional linear, sequential, and deterministic approaches. Extended concepts associated with ANNs have since been used to study human cognition, leading to the formalization of the fast and deep learning processes (Matusugu et al., 2003; Huang et al., 2006a,b).

To overcome the heavy computational time required for solving ANN models, one of the well-known fast learning algorithms is the Extreme Learning Machine (ELM). ELM is designed as a feedforward neural network model for fast classification or regression with a single layer of hidden nodes (Huang et al., 2006a,b). It was developed to calculate network weights between output and hidden layers in a single iteration via a random assignment and thus can largely minimize learning time while still generating accurate predictions with a minimal amount of training data (Huang et al., 2006a,b). ELM is unique because the weights between hidden nodes and outputs are randomly determined and assigned in a single step for the final prediction of the target values, which essentially ends up running a linear model. With such a neat structure, ELM is able to produce good results thousands of times faster than the same type of ANN trained using the backpropagation algorithm (Huang et al., 2006a,b). Nevertheless, ELM might not have deep thinking power when outliers are abundant.

On the other hand, Fukushima (1980) discussed the idea of a self-organizing neural network model for an embedded mechanism of pattern recognition in machine learning with the implication of deep learning. Only very recently did deep learning receive wide attention in AI, right after Google DeepMind's AlphaGo program defeated South Korean Master Se-dol Lee in the board game Go in early 2016. Among several deep learning architectures, Convolutional Neural Networks (CNNs), used by the AlphaGo program, demonstrate superior performance when compared to other deep machine learning methods in pattern recognition. CNNs are a class of deep, multilayer feedforward ANNs designed to reduce the focused features over initial layers of CNNs and require minimal pre-processing time; they have been successfully applied to analyze visual imagery. Later applications showed the effectiveness of using CNNs for handwriting recognition and face detection (Matusugu et al., 2003). One more expansion of the deep learning algorithms in machine learning is the Deep Belief Network (DBN). As a generative graphical model or alternatively a class of deep neural network, DBN is composed of one visible layer on the front to "see" the environment and multiple layers of latent variables ("hidden units") on the back to "think about" the inputs and judge the environment with a minimal level of discrepancies. With such a cognitive structure to "learn" the data and "infer" the environment, DBN is closer to a real-world neural system biologically.

GP is a popular machine learning method; more specifically, it is an evolutionary computing algorithm, essentially working based on the Darwin principles, as problems are solved in a manner of mimicking the natural evolutionary processes (Goldberg, 1989; Davis, 1991; Koza, 1992). The GP is a subset of evolutionary algorithms (EA). EAs comprise the GP, evolution strategies, evolutionary programming, and GAs (Ryan, 2000). Although each of the above evolves differently, the basic algorithm is still designed based on the same evolutionary concept. GP is particularly powerful for digging up the inherent relationships between variables when the convoluted relationships are very difficult to describe adequately with linear, sequential, and deterministic approaches. The GP algorithms are able to sort out a few highly nonlinear and nonstationary relationships and generate a "functional" rather than a "function" at the endpoint of their search.

Overall, the terms AI, machine learning, and deep learning are distinguished in Figure 7.2. We may visualize them as concentric squares; AI is the idea that came first and is the largest, and then machine learning such as ANN, GP, and ELM blossomed later, and finally and formally deep learning appeared, which became popular in 2016 and later. Of course, more details may be cast into these concentric squares in later stages.

In this chapter, basic concepts and fundamental theories associated with machine learning- or data mining-based feature extraction practices are used to help with classification and regression problems in remote sensing. Illustrative examples are introduced to aid in environmental monitoring and earth observation. Evaluation of the performance of feature extraction methods are presented as well for demonstrating the learning strategies in applications.

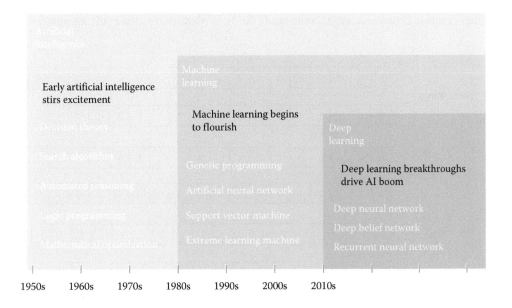

FIGURE 7.2 The evolutionary pathway of artificial intelligence.

7.2 GENETIC PROGRAMMING

7.2.1 MODELING PRINCIPLES AND STRUCTURES

Holland (1975) first developed GA, which has long been considered the basis of evolutionary computing (Figures 7.3 and 7.4). The GA selects parent 1 and parent 2 in the population pool, both of which mimic the gene structure in a chromosome for evolution numerically. Then both parents may experience crossover or mutation occasionally for improvement of fitness. A goodness of fit solution is one which can successfully deal with all the examples. When choosing crossover, GA will experience a stepwise procedure to choose a random point on the two parents, split parents at this crossover point, and create children by exchanging tails. Mutation may occur at any time to alter each gene in the chromosome independently with a probability p_m, in which p_m is called the mutation rate. The mutation rate varies typically between 1/(population size) and 1/(chromosome length).

Sharing the same idea of natural selection and genetic recombination, GP is a type of GA evolving populations over many generations to solve problems with its flexible binary tree structured representation (Figure 7.5). Note that variables A, B, and C in Figure 7.5 need to be expressed as binary variables (i.e., chromosomes) to follow the evolutionary computation rules of GA, as shown in Figure 7.3. Koza (1992) advanced evolutionary computing by developing the GP techniques that are still widely used today (Figure 7.6). Given the ideas associated with what Figures 7.3 through 7.5 may combine, GP is a specialization of GAs that employs machine learning techniques to identify system behaviors based on empirical data. As shown in Figure 7.5, all variables are defined as binary variables, and each node in a binary tree contains either an operator or an input variable in the beginning stage of evolution that may be regarded as a computer program. In the computational process, GP mimics natural evolutionary pathways to optimize computer programs in order to converge toward a target solution. Specifically, the first step is to create a set of chromosomes (i.e., parents) to randomly represent variables, namely, the population pool initialization. In general, a larger population will yield better performance in modeling the problem because of the higher chance of generating an optimal function (Francone, 1998). In each generation, the performance of programs associated with the population will be evaluated to examine their fitness. The program will continue to evolve toward an improved generation of programs until it meets the user-defined stopping criterion (Francone, 1998; Nieto et al., 2013). With respect to the evolving process, a new

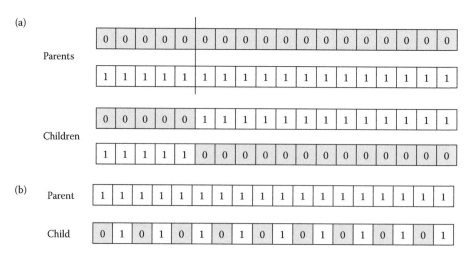

FIGURE 7.3 Definition of binary variables in the pools of parents and children for a conventional genetic algorithm. (a) The two child chromosome structures after one step of crossover from parents and (b) the single child chromosome structure after several steps of selection, crossover, and mutation.

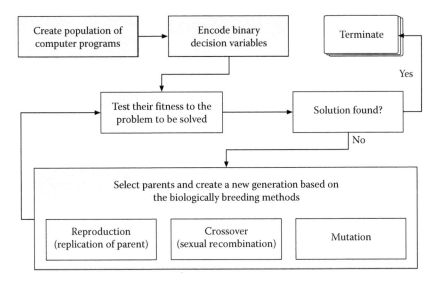

FIGURE 7.4 Flowchart for the conventional genetic algorithm.

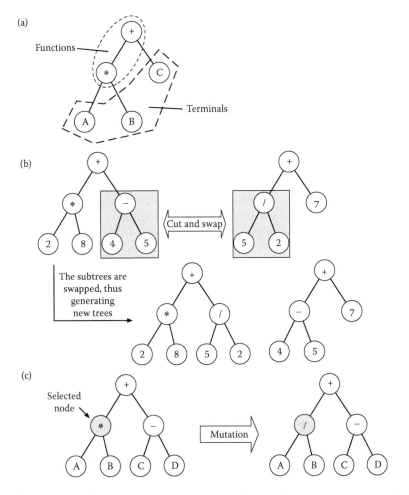

FIGURE 7.5 The tree recombination and mutation processes for evolutionary computation in GP. (a) General structure of a GP, (b) crossover between programs, and (c) program mutation.

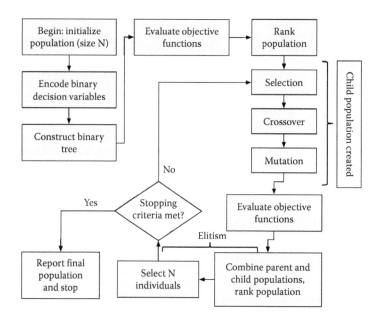

FIGURE 7.6 Flowchart for the GP.

generation is often formed by replacing programs with poor performance and then applying three principal search or genetic operators to the better-fitting programs. The three principal genetic operators include (Figure 7.6):

- *Reproduction*: Good fitted programs are directly copied into the new generation without making any change.
- *Crossover*: Instructions or nodes within best-fit programs will be exchanged to create a new program. Specifically, three types of crossovers exist that can be used for the creation of new programs (Engelbercht, 2007):
  - *Asexual*: Only one program is used when generating the new program.
  - *Sexual*: Two programs are used to create the new program.
  - *Multirecombination*: More than two programs are used to create the new program.
- *Mutation*: Abrupt changes are randomly made to the best-fit programs to create a new program, which in turn results in genetic diversity within a population. Since the mutation is randomly applied, it can occur in all programs, even those selected for crossover.

Such an iteration will continue until the stopping criterion can be met. Then the final stage of binary tree may be used to retrieve the information into an algebraic equation that is normally a highly nonlinear equation (i.e., a white-box).

GP can decode system behaviors based on empirical data for symbolic regression by examining observation data sets using association, path analysis, classification, clustering, and forecasting (Seifert, 2004). With supervised learning strategies, GP is very instrumental as no prior knowledge is required with respect to the relationships between the inputs and the target values. The Discipulus® software was used in this chapter for demonstration; it has a similar methodological flow as shown in Figures 7.5 and 7.6 (Francone, 1998).

7.2.2 Illustrative Example

The waterbody in Tampa Bay, Florida has long been an issue of eutrophication. In practice, space-borne sensors such as Moderate Resolution Imaging Spectroradiometer (MODIS) are able to provide remotely sensed ocean color images to monitor the aquatic dynamics. However, changes in nitrogen concentrations are not tied to changes in ocean color, making the assessment of eutrophication difficult. The use of other relevant factors to infer the changes in nitrogen concentrations through a nonlinear relationship based on a representative GP model becomes a challenge under a supervised learning procedure (Chang et al., 2013).

Tampa Bay Water Atlas provides historical *in situ* data via online databases through University of South Florida. The data collections go back more than 20 years. More than 222,000 nitrogen data points were obtained from 711 sampling sites. The modeling process mainly includes the following steps: (1) extract the *in situ* data from the Tampa Bay Water Atlas to be used as ground truth, (2) clean up the *in situ* data points based on the relevant locations and dates of sampling, (3) download remote sensing data that are temporally and spatially synchronous to the *in situ* data, (4) pre-process the remote sensing data (atmospheric, geometric, radiometric corrections, etc.), (5) extract the remote sensing data based on the given locations and dates of the *in situ* data; (6) export the paired data into the GP modeling platform, and (7) perform symbolic regression analyses via Discipulus®.

Figure 7.7 shows a schematic flowchart of the inverse modeling analysis for this numerical example. As indicated, the daily level 1/2 ocean color products collected by the MODIS sensor with a sinusoidal projection in the Hierarchical Data Format (HDF) format had to be first reprojected into the Universal Transverse Mercator geographic coordinate system via the MODIS Reprojection Tool and saved in GeoTIFF format with a spatial resolution of 1 km × 1 km for each grid. All the raw data were processed by the SeaWIFS Data Analysis System (SeaDAS) software package. From the SeaDAS main menu, we selected several products related to our case study. These include: (1) the MODIS data with bandwidth between 405 and 683 nm, and (2) two other products of information concerning Colored Dissolved Organic Material (CDOM) and chlorophyll-a (chl-*a*) because both are regarded as relevant factors in association with the Total Nitrogen (TN) concentration. After reprojection, data pairs between raster images and field measurements were created by extracting the corresponding grids from the raster images based on the given location and dates of the *in situ* data in a Geographical Information System (GIS) (i.e., ArcGIS). After calibration, the TN estimates from the GP model can then be compared with the associated unseen *in situ* data for validation purposes. The date of MODIS images and corresponding MODIS bands selected for use in the study are summarized in Tables 7.1 and 7.2.

Overall, space-borne sensors mainly detect optical properties of aquatic targets, despite their relatively low spatial resolution compared to objects such as macroplankton, microplankton, phytoplankton, and ultraplankton in bays and estuaries. Nevertheless, there are some constraints in

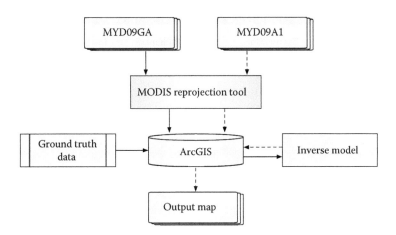

FIGURE 7.7   Workflow for the GP regression—inverse modeling analysis.

TABLE 7.1

The Date of MODIS Images Selected from Clear Days for Use in the Study

	Jan.	Feb.	March	April	May	June	July	Aug.	Sept.	Oct.	Nov.	Dec.
2007			6, 13	24		26		21				4
2008										30		
2009						9		25	29	6, 13		

ocean color remote sensing. For instance, in turbid water, CDOM can be a big obstacle in retrieving chl-*a* concentration due to the overlap in light absorbance, such that the electromagnetic spectral responses of CDOM strongly interfere with those in retrieving chl-*a* concentration (Chang et al., 2013). In addition, space-borne sensors cannot detect the nutrients in water bodies directly since nutrients such as dissolved nitrogen and phosphorus do not directly change the optical property of water. In other words, indirect approaches like inverse modeling should be applied to estimate the nutrient concentrations by characterizing relations with some external factors (Chang et al., 2013). The GP modeling framework in Figure 7.8 can be tied to those measurable water body optical properties with possible overlapped reflectance bands to reach quantifiable estimation inferred within various GP models. Discipulus® looks through the best 30 programs among the millions of GP-based models for calibration and validation.

After training, the established GP models must be carefully validated before applying to predict unseen data. In practice, unseen data independent of the calibration data are routinely used to avoid autocorrelation between validation data and calibration data. Commonly, two-thirds of paired data points are used for model calibration (i.e., model training and building) while the remaining one-third are independently used for model validation. The purpose of such an arrangement is to guarantee the general utility of the calibrated GP model for inferring TN concentrations over the entire Tampa Bay, or even other coastal water bodies. Comparisons between measured and estimated TN concentrations showed good accuracy of the calibrated GP model, with R-squared values of 0.75 and 0.63 based on the calibration data set and the validation data set, respectively (Figure 7.8). To prioritize the best-fitted TN models, single and multiple input parameters were

TABLE 7.2

The Selected MODIS Bands for Use in the Study

Band	Bandwidth[a]	Spectral Radiance[b]	Required SNR[c]
8	405–420	44.9	880
9	438–448	41.9	838
10	483–493	32.1	802
11	526–536	27.9	754
12	546–556	21	750
13	662–672	9.5	910
14	673–683	8.7	1087
15	743–753	10.2	586
16	862–877	6.2	516

[a] Bands 1 to 19 are in nm; Bands 20 to 36 are in μm.
[b] Spectral radiance values are in W/m^2 μm sr.
[c] SNR = Signal-to-noise ratio.

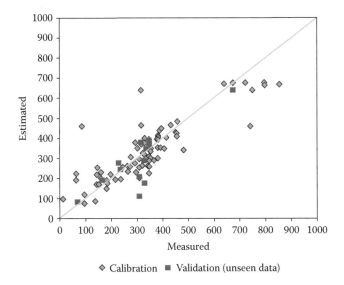

FIGURE 7.8   Comparisons between measured and estimated TN values.

automatically compared by the GP modeling process. Further investigations indicated that the development of GP models depends largely on bands 8, 10, 11, and 16, which can be deduced from the statistics in Table 7.3, in which the frequency of use of selected MODIS bands in the study is summarized by the Discipulus®. In Figure 7.9, seasonal TN maps in 2007 were produced based on the validated GP model expressed by Equation 7.1. Note that $V0$ and $V1$ represent the inputs of CDOM and chl-a in relation to the TN prediction, which are part of the convoluted functions of $X2$ and $X5$.

$$\text{Total Nitrogen}\left(\frac{\text{mg}}{\text{L}}\right) = \frac{X1 - V0}{X5 + X3 + X1 - V0} + 1.505081653594971$$

$$X1 = (X2)^2 + X3 + X5 + V1$$

$$X2 = \left[\frac{(0.4281637668609619)(X3)}{(0.7233922481536865)(V0)}\right] + 1.77857518196106$$

$$X3 = \frac{(X4)\left[(2)^{\text{Integer}(X5)}\right]}{X5} \tag{7.1}$$

$$X4 = \left[(-16)(X5)^2 + 1.907608032226563\right]^2$$

$$X5 = \left[\frac{(1.549970149993897)(V0)^2}{V1}\right] - 1.907608032226563$$

$$V0 = \text{Extracted Chlorophyll} - a\left(\frac{\text{mg}}{\text{m}^3}\right)$$

$$V1 = \text{CDOM(ppb)}$$

TABLE 7.3

The Frequency of Use of Selected MODIS Bands by Discipulus® in the Study

Band	Bandwidth/nm	Frequency	Rank
8	405–420	1.00	1
9	438–448	0.83	6
10	483–493	1.00	1
11	526–536	1.00	1
15	743–753	0.90	5
16	862–877	1.00	1

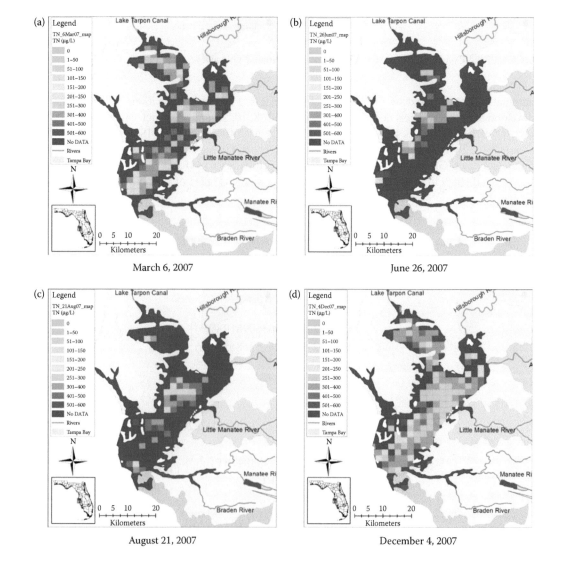

FIGURE 7.9   The measured and estimated TN values.

7.3 ARTIFICIAL NEURAL NETWORKS

The human brain is composed of over one hundred billion nerve cells called neurons that are interconnected by axons (Figure 7.10a). Stimuli from the external environment or inputs from sensory organs are accepted by dendrites for information transfer driven by induced electric impulses. The message can be sent from one neuron to another neuron, which enables the message to quickly travel through the neural network to handle the issue when it is acute or remain when it is not critical (Figure 7.10a). In general, ANN models refer to the MultiLayer Perceptron (MLP) to mimic this natural nerve system for neural computing. An ANN model with one input layer, one hidden layer, and one output layer (the simplest modeling structure of MLP) has an interconnected group of nodes, akin to the vast network of neurons in a brain. In Figure 7.10b, each circular node of the ANN model represents an artificial neuron and arrows represent the possible connections between neurons, indicating the pathway for the flow of information which imitates biological neurons of the human brain. In such an ANN model, all the neurons are connected by links and they interact with each other; each node can take input data and perform simple mathematical operations on the data. The result or output of these operations is passed to other neurons via a link in the neural network, and each link is associated with weight when the signal (information flow) passes through it. Note that the information flow is unidirectional and the link does not receive any information back. Memory in ANNs is distributed, internalized, short term, and content addressable. The result or output at each node is called its node value or activation. Hence, ANNs are capable of learning and generalization, which takes place by tuning weight values to improve the overall prediction accuracy step wise (Figure 7.10c).

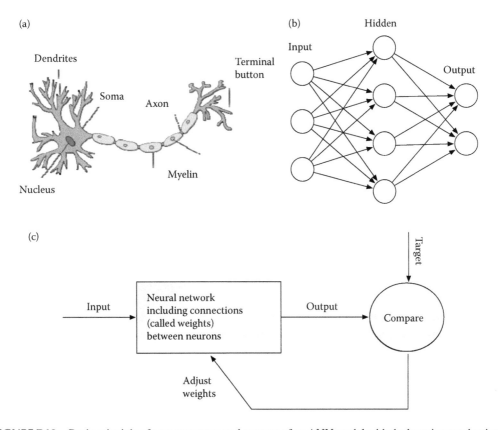

FIGURE 7.10 Basic principle of a nerve system and concept of an ANN model with the learning mechanism. (a) A natural nerve system (Adapted from Ibrić et al., 2012. *Pharmaceutics*, 4, 531–550.), (b) an ANN modeling structure, and (c) the learning and error correction mechanism.

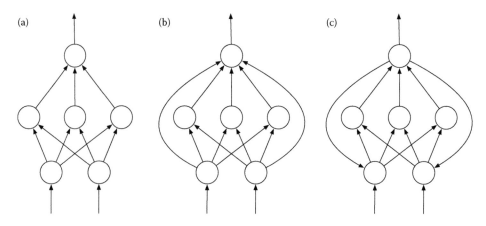

FIGURE 7.11 Types of ANN models based on the simple neural network. (a) A basic ANN structure, (b) feedforward ANN, and (c) feedback ANN.

There are two ANN topologies, including feedforward and feedback loops with fixed inputs and outputs (Figure 7.11). Both topologies can be used in: (1) pattern generation, pattern recognition, and classification, (2) prediction/forecasting, (3) clustering and categorization, (4) function approximation, (5) optimization, (6) content-addressable memory, and (7) control after carrying out the learning and generalization by tuning the weight of each link holistically (Jain et al., 1996). To achieve this goal, we normally count on supervised learning such that a neural network learns by existing examples that users submit to the algorithm. The learning process triggers the change of the network's weights so that it can produce the desired output for an input once the training step is finished. In the topology diagrams shown in Figure 7.11, the feedforward and feedback neural network models may generate a "good or desired" output after adjusting the weights a couple of times for improvement during the iterative training, testing, and validation procedures sequentially at the expense of computational time.

Among ANNs, the "backpropagational neural networks" (BPNs) are the most popular, with backpropagation serving as an abbreviation for the backwards propagation of error associated with the feedback neural network models. As shown in Figure 7.11, no matter how complex the ANN models, they are still grossly oversimplified compared to the structure of real neurons. Nevertheless, these models are created to develop understanding of what networks of simple units can do. The learning in BPNs is a supervised learning process that occurs with each cycle. When an ANN model is initially presented with an input pattern it makes a random "guess" of weights as to what it might be. It then sees how far its predicted value was from the target value and makes an appropriate adjustment to its connection weights based on a gradient descent learning rule for updating the weights of the inputs to artificial neurons toward final convergence in the BPN algorithm. Before the final convergence, many cycles are usually needed during which the neural network is presented with a new input pattern through a forward activation flow of outputs and the backwards error for weight adjustment toward final convergence. In machine learning, this improvement is called the delta rule in a single-layer neural network, which is a special case of the more general backpropagation algorithm.

7.3.1 SINGLE-LAYER FEEDFORWARD NEURAL NETWORKS AND EXTREME LEARNING MACHINE

The Single-Layer Feedforward Neural Network (SLFN) algorithm is flexible, comprehensive, and effective when large data (input/output) are available. SLFN is able to address complex problems and features easy-to-create behavior examples, even with changing problem conditions. This algorithm, which can learn by example and solve nonlinear problems, does not change the structure of the network or activation functions. Instead, it adjusts the weights between nodes. Let

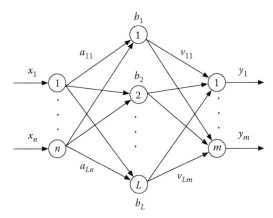

FIGURE 7.12 The generic model architecture of SLFN and ELM with n input units, L hidden units and m output units. (Chang, N. B., Bai, K. X., and Chen, C. F., 2015. *IEEE Journal of Selected Topics in Applied Earth Observations*, 99, 1–19.)

us look at the mathematical architecture of SLFN in Figure 7.12. Assume that there is a training set $\{(x_i, y_i) \mid x_i \in \mathbf{R}^n, y_i \in \mathbf{R}^m, i = 1, 2, 3,\ldots, N\}$ with N training samples which will be used to establish n-input and m-output for function generalization. If there are L numbers of hidden nodes and the SLFN is standard, then the activation function $g(x)$ can be modeled as:

$$\sum_{i=1}^{L} v_i g_i(x_j) = \sum_{i=1}^{L} v_i g(a_i \cdot x_j + b_i) \quad (j = 1, 2, 3,\ldots, N) \tag{7.2}$$

where $a_i = [a_{i1}, a_{i2},\ldots, a_{in}]^T$ $(i = 1, 2, 3,\ldots, L)$ is the weight vector connecting the ith hidden node with the input nodes, and $v_i = [v_{i1}, v_{i2},\ldots, v_{in}]^T$ $(i = 1, 2, 3,\ldots, L)$ is the weight vector connecting the ith hidden node with the output nodes. Here, b_i is the bias of the ith hidden node and $g(x)$ is the activation function.

This activation function may be sine, cosine, radial basis, threshold (i.e., symmetric hard limit transfer function), piece-wise linear, and other nonlinear functions such as sigmoidal function and Radial Basis Function (RBF) (Figure 7.13). These activation functions may be deployed in an MLP to enhance the learning capacity if a single layer with limited hidden neurons cannot achieve the learning goal. Two layers of sigmoidal units (i.e., $f(x) = 1/(1 + e^{-x})$) in both hidden layers represent such enhancement (Figure 7.14).

With these equations, it is possible to depict the forward calculation process of SLFN to generate the predictions Y for N number of training inputs. The general expression of this is

$$HV = Y \tag{7.3}$$

Here H denotes the hidden layer output matrix of the neural network, which can be expressed as

$$H = \begin{bmatrix} g(a_1 \cdot x_1 + b_1) & \cdots & g(a_L \cdot x_1 + b_L) \\ \vdots & \ddots & \vdots \\ g(a_1 \cdot x_N + b_1) & \cdots & g(a_L \cdot x_N + b_L) \end{bmatrix}_{N \times L} \tag{7.4}$$

$$V = \begin{bmatrix} v_1^T \\ \vdots \\ v_L^T \end{bmatrix}_{L \times m}, \quad Y = \begin{bmatrix} y_1^T \\ \vdots \\ y_N^T \end{bmatrix}_{N \times m} \tag{7.5}$$

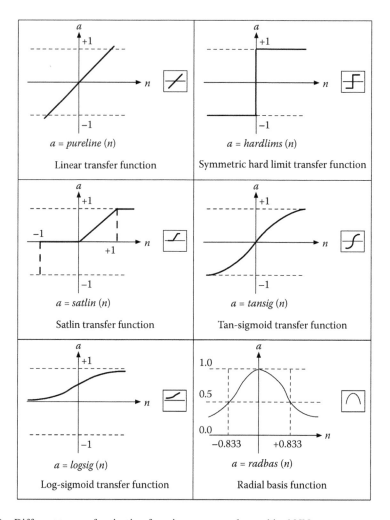

FIGURE 7.13 Different types of activation functions commonly used in ANNs.

where g is the activation function. After this process is completed, the SLFN is trained, and it then tries to find the specific sets of $\{V^*, a_i^*, b_i^*, i = 1, 2, 3,..., L\}$ to minimize the difference between approximations and targets (Chang et al., 2009, 2012a)

$$\left\| H(a_1^*,..., a_L^*, b_1^*,..., b_L^*)V^* - Y \right\| = \min(\left\| H(a_1,..., a_L, b_1,..., b_L)V - Y \right\|)_{a_i, b_i, V} \qquad (7.6)$$

The process of minimizations seeks the gradient descent-based algorithms that adjust the weights and biases through iteration of backward propagation in which the symbol "∂" corresponds to the delta rule. This is expressed as

$$W_k = W_{k-1} - \eta \frac{\partial E(W)}{\partial W} \qquad (7.7)$$

where η is the learning rate and E is the error left in each predictive iteration. These gradient descent-based algorithms are good for a majority of problems and applications. The only drawback associated with such algorithms is that the learning process needs to be run iteratively, which in turn makes them slower than what is required. Moreover, they also have the problems associated with local minima fitting instead of global minima, and even overfitting issues.

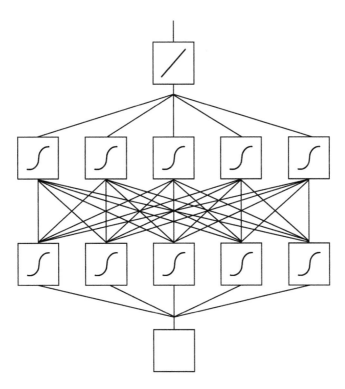

FIGURE 7.14   An ANN model with two hidden layers and sigmoidal activation function.

ELM is a feedforward neural network fast learning machine that inclines to global optimum and is very effective. Empirical results reveal that ELM and its variants are more effective and accurate than other state-of-the-art learning algorithms. ELM was originally proposed by Huang et al. (2006a,b), with the aim of improving learning speed and prediction accuracy via SLFNs (Huang et al., 2006a,b). As a fast learning algorithm, ELM is different from the traditional learning algorithm in feedforward networks whose parameters normally need to be tuned back and forth with iterative methods from nonlinear optimization. In ELM, the input weights and hidden layer biases can be randomly assigned with the hidden output layer remaining unaltered if the initiation functions are vastly differentiable (Huang et al., 2006a,b). Subsequently, the preparing procedure is comparable to finding a base standard arrangement for the straight framework $HV = Y$ since the yield weights, V, are really the main variable of SLFN to be tuned. So, for ELM, Equation 7.6 now becomes:

$$\| \boldsymbol{H}(a_1,..., a_L, b_1,..., b_L)\boldsymbol{V}^* - \boldsymbol{Y} \| = \min(\| \boldsymbol{H}(a_1,..., a_L, b_1,..., b_L)\boldsymbol{V} - \boldsymbol{Y} \|)_V \qquad (7.8)$$

The linear framework above can be solved efficiently in a least square manner, and the output weights can be calculated as

$$\boldsymbol{V}^* = \boldsymbol{H}^{\dagger} \qquad (7.9)$$

where $\boldsymbol{H}^{\dagger}$ is the Moore–Penrose Generalized inverse of $\boldsymbol{H}$ (Huang et al., 2006a,b).

Because ELM does not need to adjust the input weights and the bias of its hidden layer, ELM can significantly reduce the training time. As an effective solution to the SLFNs, ELM has been applied to various domains such as computer vision (Tang et al., 2016), control (Yu et al., 2012), system modeling and prediction (Zhao et al., 2013), image/video understanding and processing (Decherchi et al., 2013; Bai et al., 2014), time series analysis (Butcher et al., 2013), and remote sensing (Chang et al., 2015).

Due to the randomness in allocating the input weights and hidden layer bias, ELM is capable of making a direct response to the inputs and finding a solution with extraordinarily fast speed. However, when dealing with big data with larger variations, such as environmental monitoring or earth observations, the presence of a large number of outliers could seriously affect the prediction accuracies of ELM because the one-time and direct assignment of random weights cannot capture the acute need of bias correction during prediction, even with the inclusion of the normalization of the raw data set.

There are two ways to improve the performance of ELM when tackling the issues of outliers in the training data set via reinforcement learning strategies. On one hand, weights can be determined directly by the training errors from ELM. Along this line, Huang et al. (2012) proposed the Regularized ELM (RELM), based on the original ELM. RELM uses a regularization term to balance the weight between the norm of the training error and that of the output, so that the final performance can be improved. For the same purpose, based on ELM's objective function, Huynh and Won (2008) proposed the weighted ELM (WELM) by imposing penalty weights on training errors. On the other hand, based on RELM's objective function, Deng et al. (2009) further proposed the weighted regularized ELM (WRELM) by imposing additional special weighting factors on training errors. Those factors are estimated by the statistics of the training error from RELM. RELM, WELM, and WRELM are all promising for dealing with outlier problems using similar computational time as ELM. Horata et al. (2013) also proposed ELM-based algorithms to reduce outlier impacts, but those algorithms require much more computational time than traditional ELM.

7.3.2 Radial Basis Function Neural Network

RBF neural network model is an extended case from the basic ANN modeling structure. Like the general structure in Figure 7.15a, an RBF model typically has three layers: (1) an input layer, (2) a hidden layer with a nonlinear RBF activation function, and (3) a linear output layer. The reason for having a nonlinear RBF activation function is that the nervous system contains many examples of neurons with "local" or "tuned" receptive fields, and hence this local tuning is due to network properties (Figure 7.15b).

Examples in neuroscience include orientation-selective cells in visual cortex and somatosensory cells responsive to specific body regions. In this situation, the sigmoidal function that is normally defined as the linear activation function in an ANN model is no longer appropriate for representation

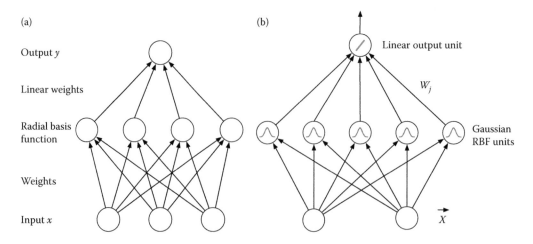

FIGURE 7.15   The modeling structure of the RBF neural network model. (a) A general architectural of RBF and (b) an RBF with Gaussian activation function.

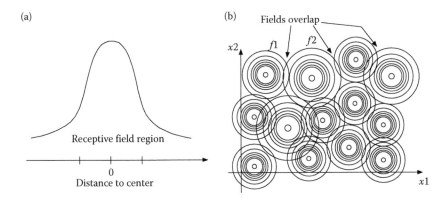

FIGURE 7.16 The Gaussian activation function in an RBF. (a) Gaussian local response function and (b) receptive field region.

of the network properties, and the Gaussian function (Figure 7.16a) using the exponential function of distance squared instead of dot product, as defined in the RBF kernel function (Equation 7.10), is more appropriate with local receptive fields (Figure 7.16b). Note that the overlapped local receptive fields provide each cluster some interactions with neighboring clusters locally. Thus, it is imperative for us to use distance instead of dot product to compute the degree of "match" reflecting local response. The RBF kernel function is shown below

$$y_j = \exp\left(\frac{-\left\|\vec{X} - \overrightarrow{\mu_j}\right\|^2}{\sigma_j^2}\right) \tag{7.10}$$

where y_j is the jth linear output unit from the neuron, $\overrightarrow{\mu_j}$ and σ_j^2 are the mean and variance of the jth cluster, and $\vec{X}$ is the input vector.

Typically, RBF networks are trained via a three-step process. The first step is with respect to the selection of center vectors of the RBF functions in the hidden layer. In general, centers can be randomly sampled from a set of examples in a supervised learning process, or they can be determined using k-means clustering in an unsupervised learning process. The second step simply fits a linear model with weights in association with layer's outputs via a dot product formulation with respect to some objective function to predict the target value and produce the error terms based on the difference between the predicted and observed values. The third step can be performed to adjust all the parameters associated with an RBF network through a backpropagation loop (Schwenker et al., 2001). This is an example of reinforcement learning in fact.

RBFs really enlighten for low-dimensional manifolds embedded in high-dimensional spaces associated with the local receptive fields. In low-dimensional spaces (i.e., limited inputs in the local receptive field), we can simply use nonparametric approaches to locally estimate density function by using a small number of neighboring samples to characterize the local receptive field (i.e., a Parzen window for classification) or a table-lookup interpolation scheme. But in high-dimensional spaces, we cannot afford to process the entire inputs due to the curse of the dimensionality issue. However, we can place RBF units only where they are needed by either supervised or unsupervised learning methods. For unsupervised learning, for instance, the goal is to mitigate the curse of the dimensionality issue by using the K-means clustering method that is initialized from randomly chosen points from the training set or a Kohonen Self-Organizing Map (SOM) that maps the space toward choosing centers in the input space for training the RBF via an unsupervised learning process (Kohonen, 1990; Tan et al., 2005; Wehrens and Buydens, 2007).

7.4 DEEP LEARNING ALGORITHMS

7.4.1 DEEP LEARNING MACHINE

From the genesis of neural networks to the invention of fast and deep learning models, a suite of machine learning tools appears in the technological hub that dominates the last decade of research for feature extraction. These major deep learning algorithms generally include: (1) deep autoencoder (DAE), (2) CNNs, (3) DBNs, and (4) recurrent neural networks (RNN), all of which can be applied for unsupervised, semi-supervised, and supervised learning processes.

DAE is a very nice way to perform nonlinear dimensionality reduction with respect to a hierarchical structure (Hinton and Salakhutdinov, 2006). Whereas CNN is designed based on prior knowledge of the biological vision system with the aid of deep learning through multiple hidden layers, DBN, inspired by physics, is evolved from the Hopfield Network regarding leaping out of the local optimum or oscillation of states through a weight-energy-probability algorithm. It is a physics-based method to transfer state; regardless, current energy is introduced from the visible layer to the hidden layer ending up with a probability to change the state (Hinton et al., 2006, 2010). The RNN is a class of neural network whose connections of units form a directed cycle to enhance the learning process (Grossberg, 2013). This type of model architecture warrants its capability to work with time series data (Grossberg, 2013; Lipton et al., 2015).

The four major deep learning methods described above have different model architectures while sharing the similar concept of neural networks when considering trade-offs among dimensionality reduction, prediction effectiveness, and computational efficiency (Figure 7.17). Overall, representation power increases at the expense of model sophistication and computational complexity, and model selection depends on trade-offs between representation power and computation complexity (Figure 7.18) (Wang and Raj, 2017). A more detailed discussion of CNN and DBN can be seen in the following subsections.

Natural images often have the property of being stationary, such as objects at the ground level. When this phenomenon is salient, the statistics of one part of the image are the same as any other part. Such local 2-D structure helps image processing for classification based on object characteristics such as edges, corners, endpoints, and so on. Following this philosophy, while nodes embedded in

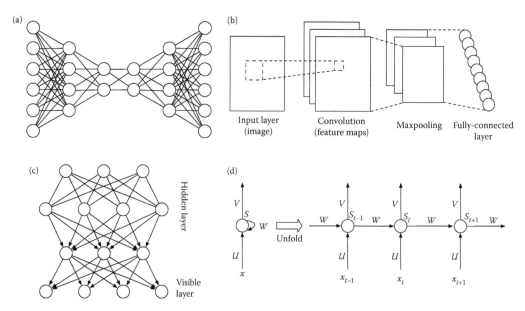

FIGURE 7.17 Block diagrams of four major deep learning model architectures. (a) AE, (b) CNN, (c) DBN, and (d) RNN.

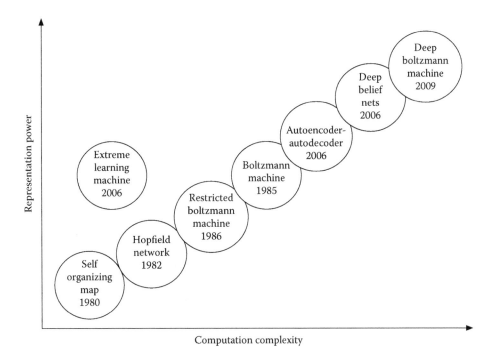

FIGURE 7.18 Trade-off between the representation power and computation complexity of several models, which guides the development of better models.

standard ANN models take input from all nodes in the previous layer, CNNs enforce that a node receives only a small set of features which are spatially or temporally close to each other. This kind of "translation invariance" is called "receptive fields" from one layer to the next, leading to handling only low dimensional local structure gradually in the first few layers of CNNs. Such a design can be achieved by selecting neighboring pixels that will have high correlations and local features (edges, corners, etc.), while distant pixels (features) are uncorrelated. The 2-D planes of nodes (or their outputs) at subsequent layers after dimensionality reduction in a CNN are called feature maps, on which each node connects to a different overlapping receptive field of the previous layer. At the same time, each node in a feature map has the same weights associated with the target feature, and each feature map searches the full previous layer to see if and how often its feature occurs.

A typical CNN structure is presented in Figure 7.19, which is comprised of 5 layers, excluding the input layer, all of which contain weights that are trainable in a supervised learning process when carrying out a series of convolution and pooling to explore major features and narrow down the focal points stepwise for final classification. The proposed CNN is further connected with a fully connected layer and an RBF layer for classification in the end. Note that the output layer is designed as an RBF with Gaussian connection whereas the full connection layers before RFB are MLPs. Before the MLPs, two pairs of "convolutional layer" and "subsampling or pooling layer" are designed to extract local receptive fields and carry out spatial or temporal subsampling, respectively, while tuning the shared weights driven by the RBF kernel functions in this feedforward loop. Feature maps are extracted throughout the convolutional and subsampling layers. Each plane in these two types of layers is a feature map representing a set of units whose weights are constrained to be identical.

Training a pure MLP requires using a huge amount of data to learn independent weights at each spatial or temporal location. CNNs are a special structure of MLPs used to exploit spatial or temporal invariance through learning high-level features corresponding to a high level of abstraction in a feedforward loop. A CNN is implemented effectively by reducing the number of parameters in the network, as the input in CNN is 2-D image data or 3-D volume data nowadays.

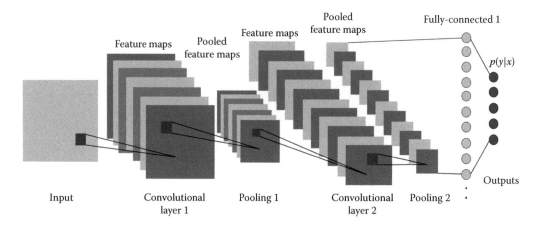

FIGURE 7.19 The structure of a CNN, consisting of convolutional, pooling, and fully-connected layers (Adapted from Albelwi, S. and Mahmood, A., 2017. A Framework for Designing the Architectures of Deep Convolutional Neural Networks. *Entropy*, 19, 242. DOI:10.3390/e19060242).

7.4.2 BAYESIAN NETWORKS

Bayesian networks, also known as belief networks or Bayes nets (BN), often use graphical structures to depict the probabilistic relationships among a set of random variables (Friedman et al., 1997, 2000). In these BNs, each individual node represents a random variable with specific propositions reasoning about an uncertain domain and the arcs connecting the nodes represent probabilistic dependencies among those random variables. The dependence between random variables is defined by the conditional probability associated with each node. Commonly, there is only one constraint on the arcs in a BN as one cannot return to a node simply by following directed arcs. BNs are capable of handling multivalued random variables in nature because their random variables are composed of two dimensions, including a range of prepositions and probability assigned to each of the prepositions. With this design, the architecture of BNs is an ideal model for integrating prior knowledge and observed data for various applications. BNs are, therefore, capable of learning the causal-effect relationships and predicting future events in various problem domains, even with missing data.

The concept of probability assigned to each of the prepositions in BNs requires further illustration with the aid of a Boltzmann machine in physics, creating a unique learning process with either Feed-Forward or Feed-Backward process. Boltzmann distribution, invented by Ludwig Boltzmann, is the probability of particles in a system over various possible states:

$$F(s) \propto e^{-\frac{E_s}{kT}} \tag{7.11}$$

where E_s denotes the corresponding energy of the state s, k is the Boltzmann's constant, and T is the thermodynamic temperature. Boltzmann factor is, therefore, defined as the ratio of two distributions characterized by the difference of energies, as below:

$$r = \frac{F(s_1)}{F(s_2)} = e^{\frac{E_{s2} - E_{s1}}{kT}} \tag{7.12}$$

With Equation 7.12, the probability can be defined as the term of each state divided by a normalizer:

$$P_{s_i} = \frac{e^{-(E_{si}/kT)}}{\sum_j e^{-(E_{sj}/kT)}} \tag{7.13}$$

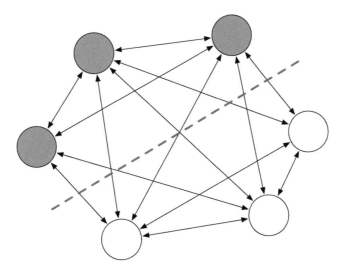

FIGURE 7.20 The architecture of a Boltzmann machine with hidden units (blue) and visible units (white). The red dashed line is highlighted for conceptual separation.

Boltzmann Machine is a stochastic Hopfield Network with a visible layer and a hidden layer (Figure 7.20) in which only visible nodes relate to inputs whereas hidden nodes are mainly used as complementary to visible units in describing the distribution of data (Ackley et al., 1985).

In this context, the model can be considered conceptually as consisting of two parts, that is, the visible part and the hidden part. We can set a state to State 1 regardless of the current state with the following probability:

$$p = \frac{1}{1 + e^{-(\Delta E / kT)}} \tag{7.14}$$

where ΔE is the difference of energies when the state is on and off. Ideally, the higher the temperature T, the more likely the state will change. The probability of a higher energy state changing to a lower energy state will be always larger than that of the reverse process, which is consistent with the thermodynamic principle.

Restricted Boltzmann Machine (RBM), invented by Smolensky (1986), is an extended version of Boltzmann Machine restricted by one rule: there are no connections either between visible nodes or between hidden nodes. On the other hand, Deep Believe Networks (DBNs) is a probabilistic generative model consisting of multiple layers of stochastic hidden nodes being stacked, whose building blocks are RBMs. Each RBM has two layers (see Figure 7.21). The first layer consists of visible nodes (i.e., observable data variables) and the second layer consists of hidden nodes (i.e., latent variables). DBNs are not just stacking RBMs together. Only the top layer has the bi-directional connections, whereas the bottom and middle layers do not. When taking an image processing issue into account, the visible layer can read out the image at first and pass the images to hidden layers for learning and retrieving the low-level features in the first hidden layer and the high-level features in the second hidden layer in the stacked RBM, respectively (Figure 7.22). Note that the bi-directional architecture in the RBM would allow the reconstruction of the weights back and forth between the two hidden layers. However, this is not possible between the visible layer and the first hidden layer given the unidirectional architecture.

The energy of the joint configuration of the visible and hidden nodes in a DBN is given by

$$E(v,h) = -\sum_{i=1}^{V_N}\sum_{j=1}^{H_M} w_{i,j} v_i h_j - \sum_{i=1}^{V_N} a_i v_i - \sum_{j=1}^{H_M} b_j h_j \tag{7.15}$$

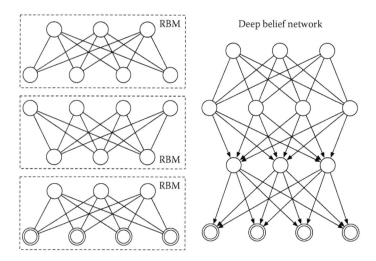

FIGURE 7.21   The architecture of a DBN with one visible unit of RBM (bottom) and two hidden units of RBM (middle and top).

where model parameters are $w_{i,j}$, a_i, and b_j. $w_{i,j}$ is the symmetric weight between visible node v_i and hidden node h_j. a_i and b_j are the bias terms for visible nodes and hidden nodes, respectively. V_N and H_M are the numbers of visible nodes and hidden nodes, respectively. The optimal set of parameters ($w_{i,j}$, a_i, and b_j) can be determined by maximizing the probability of visible nodes, which is described as the following:

The joint probability distribution of the visible-hidden node pair can be defined as (Bishop, 2006):

$$P(v,h) = \frac{e^{-E(v,h)}}{\sum_{v'}\sum_{h'} e^{-E(v',h')}} \tag{7.16}$$

where the denominator is formulated for summing over all possible pairs of visible-hidden nodes. By marginalizing over the space of hidden nodes, Equation 7.16 becomes the probability assigned to a visible node.

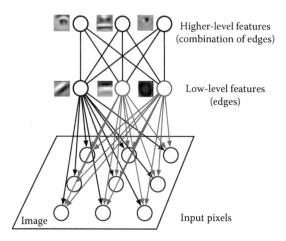

FIGURE 7.22   Stacking up for image analysis in a DBN.

$$P(v) = \sum_h P(v,h) = \frac{\sum_h e^{-E(v,h)}}{\sum_{v'}\sum_{h'} e^{-E(v',h')}} \tag{7.17}$$

The best parameter of $\{w_{i,j}, a_i, b_j\}$ of an RBM with N training data samples can be found by the objective function

$$\max_{w_{i,j},a_i,b_j}\left\{\frac{1}{N}\sum_{k=1}^{N}\log P(v^k)\right\} \tag{7.18}$$

whose aim is to increase the model probability for these training samples (Hinton, 2010).

Given the training samples and the structure of an RBM, the optimal solution to Equation 7.18 completes the RBM. The DBN is stacked by multiples of RMB. This type of deep learning architecture has three variants, however, as shown in Figure 7.23. Whereas a sigmoid belief net and a deep belief net have been modularized for different formations flexibly, a Deep Boltzmann Machine (DBM), as shown in Figure 7.23c, is built with a multilayer architecture in which each unit of RBM captures complicated, higher-order correlations between the activities of hidden features in the layer below with a bi-directional connection. Similar to DBN, DBMs have the potential of learning internal representations that become increasingly complex, and have high-level representations built from a large supply of unlabeled sensory inputs. Unlike DBN, the approximate inference procedure in DBM can incorporate not only an initial bottom-up pass embedded in DBN, but also top-down feedback, allowing DBMs to better propagate uncertainties and minimize biases (Salakhutdinov and Hinton, 2009). On the contrary, Sigmoid Believe Network (SBN) is a simplified version of DBN. If we connect stochastic binary neurons in a directed acyclic graph, we get an SBN that does not have an equilibrium sample from the top-level RBM by performing alternating Gibbs sampling while performing a top down pass to get states for all the other layers (Neal, 1992).

For example, the DBM in Figure 7.24 consists of three RBMs with a stacked multilayer architecture from which the ensemble learning capacity may be signified. The first RBM is 3×180 in size (visible nodes $\times$ hidden nodes) in which the RBM's image input via visible nodes is passed to the middle level RBM unit (hidden nodes). The second RBM unit is 180×60 in size. After the middle level RBM unit has learned the low-level features, the RBM's output is passed

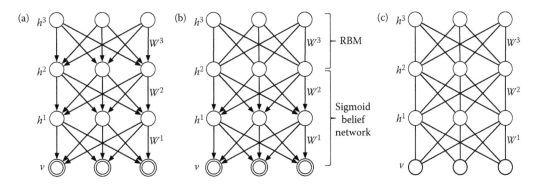

FIGURE 7.23 Comparison of architectures of three variants of deep learning model. (a) Sigmoid belief net, (b) deep belief net, and (c) deep Boltzmann machine.

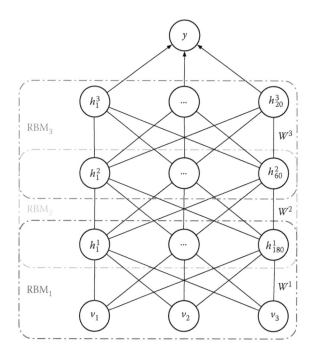

FIGURE 7.24 The architecture of DBM for model formulation. There are three inputs (v) and one output (y).

to the upper level RBM unit for learning high-level features. The upper level RBM unit (i.e., third one from the bottom) is 60×20 in size. To simplify the illustration, the weight (w) is depicted, while the biases (a and b) are omitted. After the layer-by-layer pretraining and final retrieval in the DBM, all the parameters (w, a, and b) in the DBM are fine-tuned via backpropagation techniques, forming a hybrid learning process.

7.4.3 ILLUSTRATIVE EXAMPLE

Monitoring water quality dynamics to address ecosystem status and public health concerns over a vast lake area is a challenge in environmental sustainability assessment. Satellite remote sensing offers various niches to provide inclusive and timely water quality mapping for different aquatic environments. The quality of image processing, fusion, reconstruction, and feature extraction depends on our ability to retrieve multifaceted features embedded in the satellite imagery. This illustrated example aims to evaluate and compare the effectiveness of fast and deep learning models on feature extraction of water quality from remote sensing images with 30 m spatial resolution. Practical implementation was assessed by the reconstruction of the spatial water quality patterns of a key constituent, chl-a, for one day in both wet and dry seasons in Lake Nicaragua in Central America for eutrophication assessment. In this supervised classification, the data set is fed into the three feature extraction algorithms with fast and deep learning capacity, including ELM, SLFN, and DBN to carry out a comparative study of water quality patterns with respect to chl-a concentrations. Note that the DBN herein is formulated as the same one depicted in Figure 7.24. Customized MATLAB computer codes were developed to process these image data with respect to the predetermined three neural-network-based feature extraction approaches.

When multiple models for feature extraction are developed, there must be a set of statistical indexes chosen for comparison to evaluate and rank these inversion models (i.e., classifiers). In general, a model is deemed accurate if the predicted values closely resemble the observed values in the validation stage. In this study, model performance for different algorithms/inversion methods was analyzed by using five statistical indexes, including root-mean-squared error (*rmse*), ratio of the standard deviations (CO), mean percentage of error (PE), Pearson product moment correlation coefficient (r-value), and coefficient of determination (R^2). Those four statistical indexes are defined in the following Equations 7.19 through 7.23. Theoretically, a model performs well if $rmse$ is close to zero, CO is close to one, PE is close to zero, or r (or R^2) is close to one. The equations are given below:

- Root-mean-squared error ($rmse$):

$$rmse = \sqrt{\frac{\sum_{i=1}^{N}(y_{pi} - y_{oi})^2}{N}} \tag{7.19}$$

where y_{pi} is the ith predicted value and y_{oi} is the ith observed value. N is the total number of data pairs involved. $rmse$ closer to 0 indicates that the model is getting better.

- Ratio of standard deviation of predicted to observed values (CO):

$$CO = \sqrt{\frac{\sum_{i=1}^{N}(y_{pi} - \bar{y}_p)^2}{\sum_{i=1}^{N}(y_{oi} - \bar{y}_o)^2}} \tag{7.20}$$

where $\bar{y}_{pi}$ is the mean of predicted values and $\bar{y}_{oi}$ is the mean of the observed values. CO closer to 1 indicates that the model is getting better.

- Mean of percent error (PE):

$$PE = \frac{\sum_{i=1}^{N}((y_{pi} - y_{oi})/y_{oi} \times 100\%)}{N} \tag{7.21}$$

The PE indicates a better model when its value approaches zero.

- Correlation coefficient (r):

$$r = \frac{\sum_{i=1}^{N}(y_{oi} - \bar{y}_o)(y_{pi} - \bar{y}_p)}{\sqrt{\sum_{i=1}^{N}(y_{oi} - \bar{y}_o)^2 \sum_{i=1}^{N}(y_{pi} - \bar{y}_p)^2}} \tag{7.22}$$

r indicates a better model as it approaches 1.

- Coefficient of determination (R^2):

$$R^2 = \left[\frac{\sum_{i=1}^{N}(y_{oi} - \bar{y}_o)(y_{pi} - \bar{y}_p)}{\sqrt{\sum_{i=1}^{N}(y_{oi} - \bar{y}_o)^2 \sum_{i=1}^{N}(y_{pi} - \bar{y}_p)^2}} \right]^2 \tag{7.23}$$

where x indicates independent variable and y indicates dependent variable.

For comparison, SLFN, DBM, and ELM are simulated over scenarios with the same numbers of hidden nodes in the hidden layer that has 14 nodes to seek for final competition based on such a common basis. Finally, three feature extraction models are put together for comparison in Table 7.4.

The ground truth data set needs to be grouped randomly into three subsets of 75%, 15%, and 15% for model training, testing, and validation, respectively (Table 7.4). This random grouping scheme would allow the feature extraction models to reach their best supervised learning outcome. In Table 7.4, findings suggest that ELM has relatively better prediction accuracy in the training stage while its explanatory power also partially exceeds that of others during the testing and validation

TABLE 7.4

Statistical Indices of Different Algorithms Based on the Random Grouping Scheme

	Training			Testing			Validation		
	SLFN	ELM	DBN	SLFN	ELM	DBN	SLFN	ELM	DBN
RMSE	0.30	0.17	0.38	0.31	0.50	0.40	0.67	0.85	0.69
CO	0.57	0.90	0.39	0.93	2.11	0.91	0.12	0.22	0.28
PE	−24.67	−1.92	−15.61	−19.03	−33.33	−8.94	−34.45	11.66	−53.93
RSQ	0.73	0.91	0.41	0.42	0.47	0.26	0.97	0.92	0.54
R²	0.53	0.82	0.17	0.18	0.22	0.07	0.93	0.84	0.29

Note: The values marked in red represent the best value in each category.

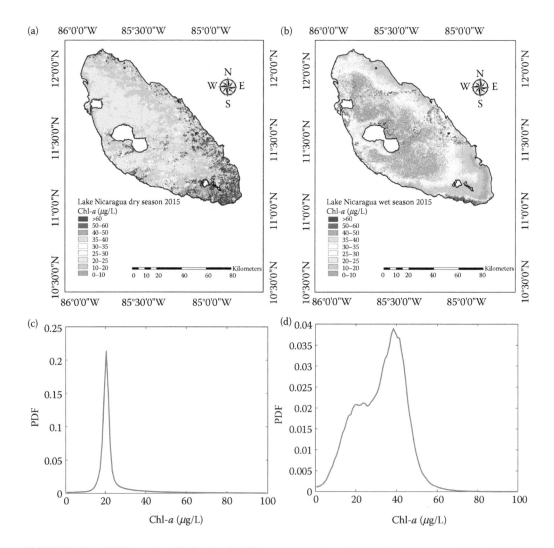

FIGURE 7.25 ELM-based prediction for the Chl-*a* concentrations in lake Nicaragua on (a) March 16, 2015 (dry season) and (b) October 5, 2015 (wet season); their estimated probability density function on (c) March 16, 2015 (dry season) and (d) October 5, 2015 (wet season).

stage. Such an outcome also reveals the potential of the fast learning over the deep learning methods. Based on the ELM feature extraction model, the content-based mapping in Figure 7.25 presents the spatial distribution of chl-*a* concentrations over both wet and dry seasons for seasonal eutrophication assessment. Overall, the water quality in the wet season is generally worse than that of the dry season in terms of chl-*a* concentrations, probably due to the rainfall-runoff impact, which washes out nutrient-laden particles from the lake watershed and enriches the nutrient contents in the waterbody while receiving enough sun light to trigger more growth of algal blooms.

7.5 SUPPORT VECTOR MACHINE

SVM is an advanced regression and classification approach proposed by Vapnik and his group at AT&T Bell Laboratories in the 1990s (Vapnik, 2000). In recent years, kernel methods, such as SVM and RBN, have become more popular image processing methods for feature extraction and classification (Melgani and Bruzzone, 2004; Filippi and Archibald, 2009; Bovolo et al., 2010; Giacco et al., 2010; Mūnoz-Marí et al., 2010; Ratle et al., 2010; Tarabalka et al., 2010; Salah, 2014). Similar to RBN, SVM works by relying on the calculation of a distance-based measure between data points in a proper Hilbert data space to split data points into different clusters (Courant and Hilbert, 1954). With the aid of the concept of structural risk minimization, this property of a distance-based measure improves the robustness of SVM via filtering more noises during the training processes, which in turn renders higher accuracy in classification over many other widely used classifiers such as decision tree, maximum likelihood, and neural network-based approaches (Huang et al., 2002; Bazi and Melgani, 2006). As opposed to the aforementioned popular classifiers, an SVM classifier is essentially binary by nature. Given that multiple classes are usually considered in remote sensing image processing, the manner of constructing a multi-class SVM classifier is an issue of concern.

7.5.1 CLASSIFICATION BASED ON SVM

To illustrate the binary nature of SVM, we may place some training sets in a two-dimensional data space to demonstrate a linearly separable case (Figure 7.26a). The training dataset can be expressed as $\{x_i, y_i\}$, where $i = 1,2,\dots, N$ denotes the number of training samples in each feature, $x_i = (x_i^1, x_i^2)$ are the two features to be separated, while $y_i \in \{-1,1\}$ represent the corresponding class labels. Toward a linearly separable case, theoretically, the training samples can be easily isolated into two distinct sections by a line. Such a plane for separation is formally known as a separating hyper-plane, of which the dimensions can be even larger than two. If we cannot separate the datasets, they are named linearly inseparable data spaces.

Typically, a separating hyper-plane can be defined as follows:

$$w \cdot x + b = 0 \tag{7.24}$$

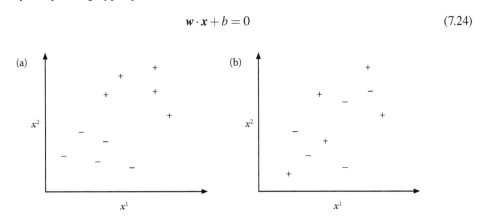

FIGURE 7.26 Two-dimensional (a) linearly separable and (b) linearly inseparable data spaces.

where $\boldsymbol{w}$ represents the direction of the plane and b is a constant scalar denoting the bias. It is clear that there exist various options for defining a separating hyper-plane once the data points are linearly separable.

Essentially, an SVM classifier is designed to search for an Optimal Separating Hyper-plane (OSH), by which the various data points can be effectively split into different classes. As a consequence, the training process of an SVM classifier is equivalent to solving the following constraint optimization problem:

$$\min \frac{\|\boldsymbol{w}\|^2}{2} \tag{7.25}$$

$$s.t. \ y_i(\boldsymbol{w} \cdot \boldsymbol{x}_i + b) \geq 1, \ \forall i = 1, 2, ..., N \tag{7.26}$$

The above equations represent the conception of structural risk minimization, where to minimize $\|\boldsymbol{w}\|$ is equivalent to maximizing the separating margin. This optimization nature renders SVM a better generalization performance than many other classifiers.

In practice, the Lagrange function is introduced to solve the above constraint optimization problem, which can be modeled as:

$$\min J(\boldsymbol{w}, b, \boldsymbol{a}) = \frac{1}{2} \boldsymbol{w}^T \boldsymbol{w} - \sum_{i=1}^{N} a_i [y_i(w \cdot x_i + b) - 1] \tag{7.27}$$

where $\boldsymbol{a} = (a_1, a_2, ..., a_N)$ and the non-negative constant a_i is the Lagrange multiplier. Such an optimization problem can be solved from its dual problem (Boyd and Vandenberghe, 2004). Let the optimal solution for the problem as w_0 and b_0, given a new data input $\mathbf{x}'$, it's class label $\boldsymbol{y}'$ can be predicted based on the well trained SVM classifier as:

$$\boldsymbol{y}' = \text{sgn}(w_0 \cdot \mathbf{x}' + b_0) \tag{7.28}$$

In practice, the training data points, in most cases, are commonly not linearly separable, however. To make them separable, these data points always need to be first transformed into a higher dimension data space. Toward such a goal, various kernel functions are often employed, and the mapping process can be defined as:

$$k(x, x_i) = \phi(x) \cdot \phi(x_i) \tag{7.29}$$

where $\phi(x)$ denotes the transformation from the original data space to a linearly separable data space. In fact, such a kernel function $k(x,x)$ does exist, as is guaranteed by the Mercer's theorem (Vapnik, 2000), as does the transformation $\phi(x)$. In practice, there exists a variety of kernels, and some popular ones are summarized in Table 7.5.

Through such a transformation, the original optimization problem behind the training process of the SVM classifier has changed to:

$$\min \left[\frac{\|\boldsymbol{w}\|^2}{2} + C \sum_{i=1}^{N} \xi_i \right] \tag{7.30}$$

$$s.t. \ y_i(\boldsymbol{w} \cdot \phi(\boldsymbol{x}_i) + b) > 1 - \xi_i, \ \forall i = 1, 2, ..., N \tag{7.31}$$

where ξ_i denotes a relaxation term that is used to allow for some misclassification of data points, however, such a relaxation is not unlimited; instead, it is confined by the penalty term $C\Sigma_{i=1}^{N}\xi_i$. With such a projection, the above optimization problem can be easily solved following the method of Lagrange multiplier as the linearly separable case.

Since the mathematical foundation of SVM is based on statistic theory, by nature it should be considered an advanced statistic learning method. Nevertheless, the classification process of an SVM classifier can still be represented as a three-layer network (Figure 7.27). Analogous to many classic networks, the first layer here is also the input layer and the last layer is the output layer. The middle layer is mainly constituted by a set of nodes, and the number of nodes (e.g., N in Figure 7.27) represents the size of the training dataset. Mathematically, the output of an SVM classifier can be expressed as:

$$y = \text{sgn}\left(\sum_{i=1}^{N} a_i^* y_i k(x_i, x) + b_0\right) \tag{7.32}$$

in which a_i^* denotes the optimal solution of the Lagrange multiplier.

TABLE 7.5

Some Types of Kernel Functions Commonly Used in SVM

Linear kernel	$k(x, x_i) = x \cdot x_i$
Polynomial kernel	$k(x, x_i) = (x \cdot x_i + 1)^d,\ d \in N$
Radial basis function	$k(x, x_i) = e^{-\gamma\|x - x_i\|^2},\ \gamma > 0$
Sigmoid function	$k(x, x_i) = \tanh(\gamma(x \cdot x_i) + c)$

Source: Adapted from Chang, N. B., Han, M., Yao, W., and Chen, L. C., 2012b. *Environmental Remote Sensing and Systems Analysis*, Boca Raton, FL, USA, CRC Press.

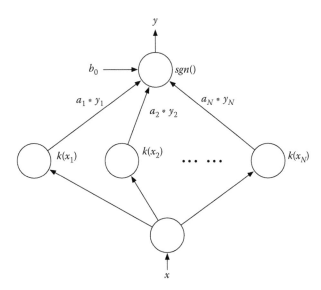

FIGURE 7.27 The use of equivalent RBN to describe an SVM classification process. (Adapted from Chang, N. B., Han, M., Yao, W., and Chen, L. C., 2012b. *Environmental Remote Sensing and Systems Analysis*, Boca Raton, FL, USA, CRC Press.)

7.5.2 Multi-Class Problem

The fundamental theory described above can only entail the usage of SVM in terms of a binary classifier. Remote sensing image classification, however, requires addressing problems when there are more than two classes. To deal with the complexity of these multi-class problems, many complicated schemes and methods have been developed to combine a set of binary classifiers to form a multi-class classifier (Mathur and Foody, 2008). Typically, there exist two important approaches, namely, the "one against all" approach (Bottou et al., 1994) and the "one against one" approach (Knerr et al., 1990), commonly used to combine multiple binary classifiers.

With respect to the "one against all" approach, the underlying principle is to train a set of K binary SVM classifiers and then separate each class from the other K-1 classes. Given an unlabeled data point x, K predictions will be made based on K sets of optimized parameters $\{w_i, b_i\}$ ($i = 1,2,...,K$) via K different binary classifiers:

$$y_i = w_i \cdot x + b_i, \ i = 1,2,...,K. \tag{7.33}$$

Theoretically, the data point x is prone to be partitioned into the ith class once the y_i is positive; however, the final determination is based on the global maximum value of y_i, and the larger the y_i, the higher the confidence it belongs to the ith class. The whole process can be modeled as:

$$y = \arg\max_i\{w_i \cdot x + b_i, \ i = 1,2,...,K\} \tag{7.34}$$

in which y is the final class label given to the data point x by the multi-class SVM classifier.

As opposed to the "one against all" approach, the "one-against-one" approach trains a set of binary classifiers to split each pair of classes. Therefore, a total number of $K(K-1)/2$ binary classifiers are required. Given an unlabeled data point x, predictions will be yielded from the $K(K-1)/2$ binary classifiers. Different from the "one against all" approach in which the final decision is based on the prediction with the largest positive value, in the "one-against-one" approach, the final classification is determined in an ensemble manner as the class label will be given to the one with the maximum number of votes because each binary classifier will give a "vote" to the class it chooses.

Compared to many traditional statistical methods and even neural networks models, SVM has been demonstrated to have better performance in dealing with many classification problems, although such good performances depend largely on the optimization of parameters associated with SVM (Bazi and Melgani, 2006). In practice, cross validation is applied to optimize parameters. Typically, the training dataset will be partitioned into different parts for training and validation simultaneously. The optimization is performed to achieve higher accuracy in the validation process. In SVM, to search for the optimal parameters, the training and validating processes are often required to be run iteratively, which in turn makes the learning process time-consuming. To advance the optimization process in SVM, more advanced optimization approaches like GA and PSO can be applied to create hybrid feature extraction algorithms (Garšva and Danenas, 2014).

7.5.3 Illustrative Example

The proposed classification approach for the change detection of land use and land cover (LULC) in a coastal region (Dalian Development Area, China) follows the following four major steps (Figure 7.28). First, normalized difference vegetation index (NDVI) and texture features were also extracted from the original SPOT-4 images as additional features to construct multi-feature image sets. Second, training datasets and testing datasets were constructed based on the ground truth data. Third, the classifier was trained using the training datasets and calibrated with the corresponding testing datasets. Finally, the classification map over the study area was created by classifying the full image sets with the well-trained SVM classifier. In this illustrative example, the "one against

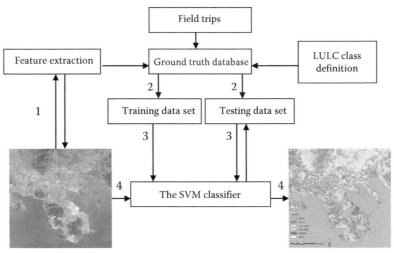

Major experimental steps:
1. Extract multiple features from the original remote sensing images
2. Construct the training dataset and testing dataset based on the ground truth database
3. Train the classifier with the training dataset, then test its performance with the testing dataset

FIGURE 7.28 Flowchart of the proposed multi-class SVM classification approach. (Adapted from Chang, N. B., Han, M., Yao, W., and Chen, L. C., 2012b. *Environmental Remote Sensing and Systems Analysis*, Boca Raton, FL, USA, CRC Press.)

all" approach was applied to meet our demand of LULC classification for the six prescribed classes, leading to balancing prediction accuracy and processing speed.

In this study, six major LULC types, including water bodies, forests, grasslands, bare fields, buildings, and roads, were established to train the SVM classifier. Given the fact that the LULC types in urban regions can be affected by many factors, for example, multitemporal variations of grasslands due to urban sprawl, water coverage due to coastal land reclamation, and buildings due to population increase and migration would reflect the main pathway of the urbanization process, defining a robust training set is by no means an easy task. To build a well-defined training set and testing set for each class, we selected a total of 1,000 data points from the ground truth data set to ensure adequate feature samples, with half of them used for the training data set and the rest for testing dataset. To better assess the performance of the SVM classifier, other external classifiers such as RBF, decision tree (DT), and Naive Bayes (NB) classifier were also used for the inter-comparison purpose. Finally, the well-trained SVM classifier with the best accuracy was used to classify the remotely sensed images over the study area.

With respect to the default SVM classifier (no optimization was applied), all parameters only used the default values (γ of 1/8 and C of 1). The confusion matrix shown in Table 7.6 fully revealed the performance of the SVM classifier. It is clear that the SVM classifier performed pretty well, with an overall accuracy of 95.6%. Further comparisons with other external classifiers such as RBF, NB, and DT are summarized in Table 7.7. Among these classifiers, RBF, as a classic neural computing technique, is another popular kernel method in addition to SVM. In this study, the RBF classifier was implemented based on the MATLAB toolbox (Nabney, 2002). In contrast, NB is a popular probabilistic method based on the Bayesian theorem (Domingos and Pazzani, 1997). The DT classifier is established using the CART algorithm. Both the NB classifier and the DT classifier were implemented based on the statistic toolbox with MATLAB. As shown in Table 7.7, the SVM classifier outperformed all the other classifiers since it yielded best overall accuracy and Kappa coefficient. Finally, the LULC map was generated with the SVM classifier based on the 2003 SPOT image for the land use assessment (Figure 7.29).

TABLE 7.6

Confusion Matrix of the SVM Classifier Based on NDVI, Texture, and Spectral Features

		Predictions						
		Water Body	Forest	Grass Land	Bare Field	Building	Road	PA (%)
Actual	Water body	995	2	0	0	3	0	99.5
conditions	Forest	0	1445	46	0	9	0	96.3
	Grass land	0	63	936	0	1	0	93.6
	Bare field	0	0.0	1	456	35	8	91.2
	Building	1	16	3	26	2398	56	95.9
	Road	0	0.0	0	3	44	453	90.6
	UA(%)	99.9	94.7	94.9	94.0	96.3	87.6	
OA(%)					95.6			
Kappa coefficient					0.9416			

Source: Adapted from Chang, N. B., Han, M., Yao, W., and Chen, L. C., 2012b. *Environmental Remote Sensing and Systems Analysis*, Boca Raton, FL, USA, CRC Press.

TABLE 7.7

Comparisons of Classification Accuracies with Respect to Different Classifiers

	SVM	RBF	NB	DT
Water body	99.5	99.4	98.6	99.2
Forest	96.3	89.9	85.3	91.5
Grass land	93.6	76.1	87.6	87.0
Bare field	91.2	75.6	70.4	81.2
Building	95.9	82.3	83.9	89.8
Road	90.6	61.0	67.8	74.2
OA (%)	95.6	83.5	84.7	89.4
Kappa	0.9416	0.7877	0.8045	0.8632

Source: Adapted from Chang, N. B., Han, M., Yao, W., and Chen, L. C., 2012b. *Environmental Remote Sensing and Systems Analysis*, Boca Raton, FL, USA, CRC Press.

Note: Radial basis function neural network (RBF), Naïve Bayes (NB), and decision tree (DT).

7.6   PARTICLE SWARM OPTIMIZATION MODELS

The particle swarm optimization (PSO) algorithm was invented by James Kennedy and Russell C. Eberhart in 1995 (Kennedy and Eberhart, 1995) and formalized by Kennedy et al. (2001). It is a behavioral model of a flock of birds which searches for food randomly group wise in some habitat areas, given the distance from the food. In PSO, a bird is called a particle, and a flock of birds is called a swarm. From a computational point of view, its location relative to the food is defined by the fitness value. Each particle has one velocity to determine its flying direction and distance depending on its inertia coefficient, social component, and cognitive component. Thus, the swarm (i.e., all the particles) has one fitness value defined by a function, as the optimization process is to solve this function. All the particles perform space searching by following the most optimal particle at present.

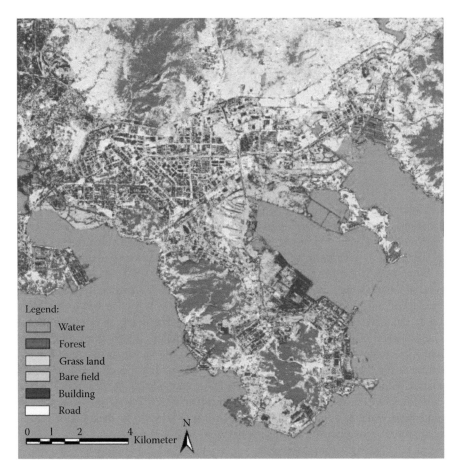

Legend:
- Water
- Forest
- Grass land
- Bare field
- Building
- Road

0 1 2 4 Kilometer   N

FIGURE 7.29 LULC map of Dalian development area in China produced by using the SVM classifier based on the 2007 SPOT 4 image. (Adapted from Chang, N. B., Han, M., Yao, W., and Chen, L. C., 2012b. *Environmental Remote Sensing and Systems Analysis*, Boca Raton, FL, USA, CRC Press.)

In a mathematical context, PSO is initialized to be a group of random particles that evolve automatically in a searching space toward the optimal solution, while each particle is updated stepwise as it is confined by two extremes. The first one is called personal best (pbest), which is defined as a cognitive component, referring to the optimal solution found by the particle itself. Likewise, the other one is global best (gbest), which is defined as a social component, referring to the optimal solution found by the whole swarm. As an evolutionary computation algorithm, each particle in the PSO evolves automatically, which is mainly reflected by the update of the position of each particle via a velocity vector. Therefore, there are two aspects associated with such a velocity vector, namely, the speed and direction. Essentially, for each particle in the swarm, its current position in the searching space is collectively determined by its previous velocity state as well as the location relative to that of the pbest and the gbest. Due to the nature of global (gbest) and local (pbest) searching principles, particles in the swarm can be stochastically accelerated toward their best position at each step, thereby pushing all particles to continuously search for the optimal solution with the most promising way found at each step in the solution space.

It is clear that the update of velocity and position of each particle in the swarm is the most critical step in the PSO algorithm, as the optimization process is related to these updated steps. Mathematically, the velocity update of each particle in the swarm can be expressed as follows.

Mathematically, let $x_i = (x_i^1, x_i^2, \ldots, x_i^D), i = 1, 2, \ldots, N$ be the position of the ith particle (a swarm of N particles) in a D-dimension search space, and $v_i = (v_i^1, v_i^2, \ldots, v_i^D)$ is the velocity of this particle. Given $x_i = (x_i^1, x_i^2, \ldots, x_i^D)$ is the personal best position of the ith particle and $p_g = (p_g^1, p_g^2, \ldots, p_g^D)$ is the global best position in the swarm, the evolving process to search for the optimum solution can be expressed as (Chen and Leou, 2012; Wang et al., 2015):

$$v_i^d(t+1) = \omega \cdot v_i^d(t) + c_1 \cdot r_1(t) \cdot (p_i^d(t) - x_i^d(t)) + c_2 \cdot r_2(t) \cdot (p_g^d(t) - x_i^d(t)) \tag{7.35}$$

$$x_i^d(t+1) = x_i^d(t) + v_i^d(t+1) \tag{7.36}$$

where $d = 1, 2, \ldots, D$ and t is the iteration counter ω denotes the inertia weight controlling the impact of previous velocities on the current velocities. c_1 and c_2 are two learning constraints that control the contributions of cognitive information to the local and global solutions, respectively. r_1 and r_2 are two random scaling factors ranging between 0 and 1. A common value range is [0, 1.2] for w and [0, 2] for c_1 and c_2. Generally, the three right terms in the velocity update equation represent different components in the PSO algorithm (Kennedy et al., 2001). The first term $\omega x_i^d(t)$ refers to the inertia component, which represents the basic solution as it was originally heading. Typically, a value between 0.8 and 1.2 is given to the inertial coefficient w. It is clear that a smaller value of w will speed up the convergence of the swarm to optima, whereas a larger value will encourage the particles' exploration of the entire search space, thereby making the optimization process take a longer time. The second term $r_1(t) \cdot (p_i^d(t) - x_i^d(t))$ is commonly known as the cognitive component, which represents the particle's memory of the historical searching space where the high individual fitness exists. Commonly, a value close to 2 is given to c_1 to modulate the searching steps in finding its individual best candidate solution $x_i^d(t)$ The third term $c_2 \cdot r_2(t) \cdot (p_g^d(t) - x_i^d(t))$ is termed the social component, which is applied to guide the particle to move to the best region that the swarm has found at present. Analogously, a value approximate to 2 is given to c_2.

In general, the PSO algorithm involves only three steps toward an optimized solution, including: (1) evaluate the fitness of each particle stepwise, (2) update individual and global best fitness values and particle positions, and (3) update the velocity and position of each particle at the end of each step. PSO can be used to enhance the feature extraction algorithms developed based not only on the statistics and decision science principles, such as the linear discriminate analysis (Lin and Chen, 2009), but also the machine learning and data mining approaches, such as support vector machine (Garšva and Danenas, 2014), and developed advanced optimization approaches, such as GA and PSO, for parameter optimization in SVM, creating a hybrid feature extraction algorithm.

7.7 SUMMARY

The disciplines of feature extraction we discussed in this chapter stem from a traditional computational science point of view and use this perspective to explore, innovate, and finally design better tools for feature extraction in remote sensing. Among these innovative algorithms, a relatively new branch of image processing techniques in connection to neural science and Darwin principle emerges with big advancements at an unprecedented pace, reflecting an era when associationism modeling of the brain became popular. Many of the different AI-based feature extraction techniques in this regard are found to be useful in real world applications.

In the system of systems engineering context, different branches of AI serve as bases for feature extraction from remotely sensed images as compared to different branches of statistics and decision science, which served as bases for feature extraction in the previous chapter. Varying levels of hybrid feature extraction methods can be innovated and designed based on proper integration of these algorithms over the two different regimes for better feature extraction. Such a quantitative engineering paradigm for hybrid feature extraction outlines the typical steps involved in the engineering of a suite of powerful remote sensing products for dealing with more challenging issues. In the future, the performance of ensemble learning should be evaluated and improvements in this regard should be regarded as necessary steps for exploring more complex feature extraction issues.

REFERENCES

Ackley, D. H., Hinton, G. E., and Sejnowski, T. J., 1985. A learning algorithm for Boltzmann machine. *Cognitive Science*, 9, 147–169.

Albelwi, S. and Mahmood, A., 2017. A Framework for Designing the Architectures of Deep Convolutional Neural Networks. *Entropy*, 19, 242. DOI:10.3390/e19060242

Bai, Z., Huang, G. B., Wang, D. W., Wang, H., and Westover, M. B., 2014. Sparse extreme learning machine for classification. *IEEE Transactions on Cybernetics*, 44, 1858–1870.

Bazi, Y. and Melgani, F., 2006. Toward an optimal SVM classification system for hyperspectral remote sensing images. *IEEE Transactions on Geoscience and Remote Sensing*, 44, 3374–3385.

Bishop, C. M., 2006. *Pattern Recognition and Machine Learning*. Springer, Berlin.

Bottou, L., Cortes, C., Denker, J. S., Drucker, H., Guyon, I., Jackel, L. D., LeCun, Y. et al., 1994. Comparison of classifier methods: A case study in handwritten digit recognition. In: *Pattern Recognition, 1994. Vol. 2—Proceedings of the 12th IAPR International Conference on Computer Vision & Image Processing*, Jerusalem, Israel, 77–82.

Bovolo, F., Bruzzone, L., and Carlin, L., 2010. A novel technique for subpixel image classification based on support vector machine. *IEEE Transactions on Image Processing*, 19, 2983–2999.

Boyd, S. P. and Vandenberghe, L., 2004. *Convex Optimization*. Cambridge University Press, Cambridge, UK.

Butcher, J. B., Verstraeten, D., Schrauwen, B., Day, C. R., and Haycock, P. W., 2013. Reservoir computing and extreme learning machines for non-linear time-series data analysis. *Neural Networks*, 38, 76–89.

Caudill, M., 1989. Neural Network Primer: Part I, AI Expert.

Chang, N. B., Bai, K. X., and Chen, C. F., 2015. Smart information reconstruction via time-space-spectrum continuum for cloud removal in satellite images. *IEEE Journal of Selected Topics in Applied Earth Observations*, 99, 1–19.

Chang, N. B., Daranpob, A., Yang, J., and Jin, K. R., 2009. A comparative data mining analysis for information retrieval of MODIS images: Monitoring lake turbidity changes at Lake Okeechobee, Florida. *Journal of Applied Remote Sensing*, 3, 033549.

Chang, N. B., Han, M., Yao, W., and Chen, L. C., 2012b. Remote sensing assessment of coastal land reclamation impact in Dalian, China, using high resolution SPOT images and support vector machine. In: *Environmental Remote Sensing and Systems Analysis*, Boca Raton, FL, USA, CRC Press.

Chang, N., Xuan, Z., Yang, Y., 2013. Exploring spatiotemporal patterns of phosphorus concentrations in a coastal bay with MODIS images and machine learning models. *Remote Sensing of Environment*, 134, 100–110.

Chang, N. B., Yang, J., Daranpob, A., Jin, K. R., and James, T., 2012a. Spatiotemporal pattern validation of Chlorophyll-a concentrations in Lake Okeechobee, Florida using a comparative MODIS image mining approach. *International Journal of Remote Sensing*, 33, 2233–2260.

Chen, H. Y. and Leou, J. J., 2012. Multispectral and multiresolution image fusion using particle swarm optimization. *Multimedia Tools and Applications*, 60, 495–518.

Chen, L., Tan, C., Kao, S., and Wang, T., 2008. Improvement of remote monitoring on water quality in a subtropical reservoir by incorporating grammatical evolution with parallel genetic algorithms into satellite imagery. *Water Research*, 42, 296–306.

Courant, R. and Hilbert, D., 1954. Methods of mathematical physics: Vol. I. *Physics Today*, 7, 17.

Davis, L., 1991. *Handbook of Genetic Algorithms*. Van Nostrand Reinhold, New York, United States.

Decherchi, S., Gastaldo, P., Zunino, R., Cambria, E., and Redi, J., 2013. Circular-ELM for the reduced-reference assessment of perceived image quality. *Neurocomputing*, 102, 78–89.

Deng, W., Zheng, Q., and Chen, L., 2009. Regularized extreme learning machine. In: *IEEE Symposium on Computational Intelligence and Data Mining*, 389–395.

Doerffer, R. and Schiller, H., 2007. The MERIS Case 2 water algorithm. *International Journal of Remote Sensing*, 28, 517–535.

Domingos, P. and Pazzani, M., 1997. On the optimality of the simple Bayesian classifier under zero-one loss. *Machine Learning*, 29, 103–130.

Engelbercht, A., 2007. *Computational Intelligence: An Introduction*. Wiley, New York.

Filippi, A. M. and Archibald, R., 2009. Support vector machine-based endmember extraction. *IEEE Transactions on Geoscience and Remote Sensing*, 47, 771–791.

Francone, D. F., 1998. *Discipulus Software Owner's Manual, version 3.0 DRAFT.* Machine Learning Technologies, Inc., Colorado.

Friedman, N., Geiger, D., and Goldszmidt, M., 1997. Bayesian network classifiers. *Machine Learning*, 29, 131–163.

Friedman, N., Linial, M., Nachman, I., and Pe'er, D., 2000. Using bayesian networks to analyze expression data. *Journal of Computational Biology*, 7, 601–620.

Fukushima, K. 1980. Neocognitron: A self-organizing neural network model for a mechanism of pattern recognition unaffected by shift in position. *Biological Cybernetics*, 36, 93–202.

Garšva, G. and Danenas, P., 2014. Particle swarm optimization for linear support vector machines based classifier selection. *Nonlinear Analysis: Modelling and Control*, 19, 26–42.

Giacco, F., Thiel, C., Pugliese, L., Scarpetta, S., and Marinaro, M., 2010. Uncertainty analysis for the classification of multispectral satellite images using SVMs and SOMs. *IEEE Transactions on Geoscience and Remote Sensing*, 48, 3769–3779.

Goldberg, D., 1989. *Genetic Algorithms in Search, Optimization, and Machine Learning.* Addison-Wesley, New York.

Grossberg, S., 2013. Recurrent neural networks. *Scholarpedia*, 8, 1888.

Hinton, G. E., 2010. A practical guide to training restricted boltzmann machines. *Momentum*, 9, 926.

Hinton, G. E., Osindero, S., and The, Y. W., 2006. A fast learning algorithm for deep belief nets. *Neural computation*, 18, 1527–1554.

Hinton, G. E. and Salakhutdinov, R. R., 2006. Reducing the dimensionality of data with neural networks. *Science*, 313, 504–507.

Holland, J. M., 1975. *Adaptation in Natural and Artificial Systems.* University of Michigan Press, Ann Arbor, MI.

Horata, P., Chiewchanwattana, S., and Sunat, K., 2013. Robust extreme learning machine. *Neurocomputing*, 102, 31–44.

Huang, C., Davis, L. S., and Townshend, J. R. G., 2002. An assessment of support vector machines for land cover classification. *International Journal of Remote Sensing*, 23, 725–749.

Huang, G. B., Zhou, H., Ding, X., and Zhang, R., 2012. Extreme learning machine for regression and multiclass classification. *IEEE Transactions on Systems, Man, and Cybernetics, Part B (Cybernetics)*, 42, 513–529.

Huang, G. B., Zhu, Q. Y., and Siew, C. K., 2006a. Real-time learning capability of neural networks. *IEEE Transactions on Neural Networks*, 17, 863–878.

Huang, G. B., Zhu, Q. Y., and Siew, C. K., 2006b. Extreme learning machine: theory and applications. *Neurocomputing*, 70, 489–501.

Huynh, H. T. and Won, Y., 2008. Weighted least squares scheme for reducing effects of outliers in regression based on extreme learning machine. *International Journal of Digital Content Technology and its Applications*, 2, 40–46.

Ibrić, S., Djuriš, J., Parojčić, J., and Djurić, Z., 2012. Artificial neural networks in evaluation and optimization of modified release solid dosage forms. *Pharmaceutics*, 4, 531–550.

Ioannou, I., Gilerson, A., Gross, G., Moshary, F., and Ahmed, S., 2011. Neutral network approach to retrieve the inherent optical properties of the ocean from observations of MODIS. *Applied Optics*, 50, 3168–3186.

Jain, A. K., Mao, J., and Mohiuddin, K., 1996. Artificial Neural Networks: A Tutorial. IEEE Computer Special Issue on Neural Computing. https://pdfs.semanticscholar.org/f6d5/92af572ebf89dabb0a06aa22804905 5923fb.pdf. Accessed on July 2017.

Kennedy, J. and Eberhart, R., 1995. Particle swarm optimization. In: *Proceedings of the IEEE International Conference on Neural Networks*, Piscataway, NJ, IEEE Press, volume IV, 1942–1948.

Kennedy, J., Eberhart, R. C., and Shi, Y., 2001. *Swarm Intelligence. Morgan Kaufmann Publishers*, San Francisco, CA, USA.

Knerr, S., Personnaz, L., and Dreyfus, G., 1990. Single-layer learning revisited: A stepwise procedure for building and training a neural network. *Optimization Methods and Software*, 1, 23–34.

Kohonen, T., 1990. The self-organizing map. *Proceedings of the IEEE*, 78, 1464–1480.

Koza, J. R., 1992. *Genetic Programming: On the Programming of Computers by Means of Natural Selection.* MIT Press, Cambridge, MA, USA.

Lin, S. W. and Chen, S. C., 2009. PSOLDA: A particle swarm optimization approach for enhancing classification accuracy rate of linear discriminant analysis. *Applied Soft Computing*, 9, 1008–1015.

Lipton, Z. C., Berkowitz, J., and Elkan, C., 2015. A critical review of recurrent neural networks for sequence learning. arXiv preprint arXiv:1506.00019.

Mathur, A. and Foody, G. M., 2008. Multiclass and binary SVM classification: Implications for training and classification users. *IEEE Geoscience and Remote Sensing Letters*, 5, 241–245.

Matusugu, M., Mori, K., Mitari, Y., and Kaneda, Y., 2003. Subject independent facial expression recognition with robust face detection using a convolutional neural network. *Neural Networks*, 16, 555–559.

Melgani, F. and Bruzzone, L., 2004. Classification of hyperspectral remote sensing images with support vector machines. *IEEE Transactions on Geoscience and Remote Sensing*, 42, 1778–1790.

Mūnoz-Marí, J., Bovolo, F., Gómez-Chova, L., Bruzzone, L., and Camp-Valls, G., 2010. Semisupervised one-class support vector machines for classification of remote sensing data. *IEEE Transactions on Geoscience and Remote Sensing*, 48, 3188–3197.

Nabney, I., 2002. *NETLAB: Algorithms for Pattern Recognition*. Springer Verlag, Germany.

Neal, R., 1992. Connectionist learning of belief networks. *Artificial Intelligence*, 56, 71–113.

Nieto, P., Fernandez, J., Juez, F., Lasheras, F., and Muniz, C., 2013. Hybrid modeling based on support vector regression with genetic algorithms in forecasting cyanotoxins presence in the Trasona reservoir Northern Spain. *Environmental Research*, 122, 1–10.

Ratle, F., Camps-Valls, G., and Weston, J., 2010. Semisupervised neural networks for efficient hyperspectral image classification. *IEEE Transactions on Geoscience and Remote Sensing*, 48, 2271–2282.

Rosenblatt, F., 1958. The perceptron: A probabilistic model for information storage and organization in the brain. *Psychological Review*, 65, 386.

Ryan, C., 2000. *Automatic re-engineering of software using genetic programming*. Kluwer Academic Publisher, Massachusetts, USA.

Salah, M., 2014. Combining Pixel-based and Object-based Support Vector machines using Bayesian Probability Theory. In: *ISPRS Annals of the Photogrammetry, Remote Sensing and Spatial Information Sciences, Volume II-7, 2014, ISPRS Technical Commission VII Symposium*, September 29 – October 2, 2014, Istanbul, Turkey.

Salakhutdinov, R. and Hinton, G., 2009. Deep Boltzmann machines. In: *Proceedings of the 12th International Conference on Artificial Intelligence and Statistics (AISTATS)*, Clearwater Beach, Florida, USA.

Schwenker, F., Kestler, H. A., and Palm, G., 2001. Three learning phases for radial-basis-function networks. *Neural Networks*, 14, 439–458.

Seifert, J., 2004, *Data Mining: An Overview*. CRS Report for Congress.

Smolensky, P., 1986. *Information Processing in Dynamical Systems: Foundations of Harmony Theory*. Technical report, DTIC Document.

Song, K., Li, L., Tedesco, L., Li, S., Clercin, N., Hall, B., Li, Z., and Shi, K., 2012. Hyperspectral determination of eutrophication for a water supply source via genetic algorithm-partial least squares (GA-PLS) modeling. *Science of the Total Environment*, 426, 220–232.

Tan, P. N., Steinbach, M., and Kumar, V., 2005. *Introduction to Data Mining*. Pearson, London, United Kingdom.

Tang, J., Deng, C., and Huang, G. B., 2016. Extreme learning machine for multilayer perceptron. *IEEE Transactions on Neural Networks and Learning Systems*, 27, 809–821.

Tarabalka, Y., Fauvel, M., Chanussot, J., and Benediktsson, J. A., 2010. SVM- and MRF-based method for accurate classification of hyperspectral images. *IEEE Geoscience and Remote Sensing Letters*, 7, 736–740.

Vapnik, V., 2000. *The Nature of Statistical Learning Theory*. Springer-Verlag, Germany.

Wang, H. and Raj, B., 2017. On the Origin of Deep Learning. arXiv:1702.07800v4, https://arxiv.org/pdf/1702.07800.pdf. Accessed on July 2017.

Wang, C., Liu, Y., Chen, Y., and Wei, Y., 2015. Self-adapting hybrid strategy particle swarm optimization algorithm. *Soft Computing*, 20, 4933–4963.

Wehrens, R. and Buydens, L. M. C., 2007. Self- and super-organizing maps in R: The Kohonen package. *Journal of Statistical Software*, 21, 1–19.

Yu, Y., Choi, T. M., and Hui, C. L., 2012. An intelligent quick prediction algorithm with applications in industrial control and loading problems. *IEEE Transactions on Automation Science and Engineering*, 9, 276–287.

Zhao, Z. P., Li, P., and Xu, X. Z., 2013. Forecasting model of coal mine water inrush based on extreme learning machine. *Applied Mathematics & Information Sciences*, 7, 1243–1250.

Part III

Image and Data Fusion for Remote Sensing

8 Principles and Practices of Data Fusion in Multisensor Remote Sensing for Environmental Monitoring

8.1 INTRODUCTION

The monitoring of Earth's environment has entered a totally new era due to the unparalleled advantage of remote sensing for mapping Earth's surface. Compared to data sparsely collected via laborious and costly manual sampling, remotely sensed data enable us to have vast spatial coverage along with high spectral and temporal resolution. Since the 1970s, a wealth of satellite remote sensing instruments with versatile capabilities have been routinely and consecutively sent into space to monitor the varying status of Earth's environment and the changing states of the ecosystem.

However, due to the inherent constraints of platform and payload design, orbital features, and signal-to-noise ratio in remote sensing, Environmental monitoring and earth observations taken from one individual satellite sensor are commonly restricted to limited spatial, temporal, or spectral coverage and resolution. For instance, remotely sensed imagery from the Landsat Thematic Mapper (TM) has a spatial resolution of 30-m, which enables us to monitor the Earth's environment in more detail. In contrast, it has a temporal resolution of 16 days, making it difficult to be routinely used for the monitoring of parameters such as air temperature that demand daily or even hourly records. On the other hand, the Moderate resolution Imaging Spectroradiometer (MODIS) flying onboard Terra and Aqua satellites is capable of providing multispectral images on a daily basis with a spatial resolution ranging from 250-m to 1000-m, depending on the product. Due to its high temporal resolution and moderate spatial resolution, MODIS imagery has been extensively used for daily monitoring of the Earth. Nevertheless, most of the practices were performed on a large scale that is hardly feasible for urban-scale studies. When identifying different types of plant species or mineral resources using satellite remote sensing, hyperspectral imagery is always used due to the high spectral resolution. In general, remotely sensed imagery recorded by one individual sensor cannot suffice for all types of environmental monitoring applications.

With the advent of high spatial, high temporal, and high spectral resolution imagery from multiple remote sensing instruments, blending and fusing information from multiple sensors into a single dataset by making use of certain image processing techniques and algorithms is considered a valuable tool in remote sensing science. The primary objective of multisensor data fusion is to retrieve distinct features from different image/data sources to attain more information than can be obtained from each single data set alone. This chapter will describe the following three image and data fusion technology hubs although some cases are intertwined with each other. They include: (1) multispectral remote sensing-based fusion techniques, (2) hyperspectral remote sensing-based fusion techniques, and (3) microwave remote sensing-based fusion techniques. By taking advantage of fused data, we can gain more insight regarding targets than would be possible by making use of each individual data set separately.

8.2 CONCEPTS AND BASICS OF IMAGE AND DATA FUSION

For those unfamiliar with image and data fusion, we will first distinguish a family of remote sensing fusion terminologies. The following terms in the literature, including data fusion, image fusion, and sensor fusion, deserve our attention. Although all these terms refer to a fusion process, there still exist certain differences among them in the domain specific category due to the different data sources applied for fusion purposes. As shown in Figure 8.1, data fusion encompasses all kinds of fusion processes. In other words, all processes involving blending information from multiple data sources into one integrated data set with more features than each individual data set is known as data fusion. The data sources used for data fusion can be all types of information, such as space- or air-borne remotely sensed imageries, discrete data points, vector files, camera images, and so on. Thus, data fusion is a generalized term for all fusion processes in data science. The concept of data fusion was initialized in the late 1950s. Data fusion was formally defined by Hall and Llinas (1997), and refers to combining data from multiple sensors and related information from associated databases to achieve improved accuracy and more specific inferences than could be achieved by the use of a single sensor alone. Similarly, Wald (1999) emphasized that data fusion should be regarded as a synergistic framework instead of simply a collection of tools and methods for integration. This is because the data fusion process involves not only fusion algorithms, but also many data processing steps and thus should be viewed in a systematic manner.

In contrast, sensor fusion refers to blending data from disparate sensors to attain more information than could be achieved by each individual sensor (Elmenreich, 2002). Specifically, data sources used for sensor fusion are derived from disparate sensors. Once the data sources are provided in the form of imagery, the fusion process is termed image fusion. Several famous definitions of image fusion can be found in the literature, including:

- Image fusion is the integration of two or more different images to form a new image by using a certain algorithm (Van Genderen and Pohl, 1994).
- Image fusion is the process of combining information from two or more images of a scene into a single fused image that is more informative and more suitable for visual perception or computer processing (Ardeshir Goshtasby and Nikolov, 2007).
- Image fusion is a process of combining images obtained by sensors of different wavelengths simultaneously in a view of the same scene to form a composite image. The fused image is produced to improve image content and to make it easier for the user to detect, analyze, recognize, and discover targets and increase his or her situational awareness (multisensor fusion) (Chang et al., 2014a).

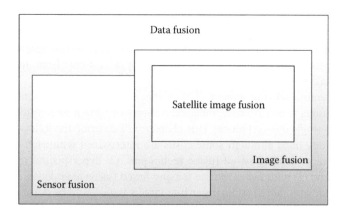

FIGURE 8.1   Schematic illustration of the relationships among data fusion, sensor fusion, and image fusion.

- Image fusion is the process of integrating relevant information from two or more images into a single one, where the fused image is more informative than any of the input images (Haghighat et al., 2011).

Image fusion has been widely used across disciplines. For example, fusion can be performed on medical images like Computed Tomography (CT), Magnetic Resonance Imaging (MRI), and/or Positron Emission Tomography (PET) images to aid in the detection of complex tumor issues. PET/CT scanning as well as MRI techniques can be collectively used for medical diagnosis at the molecular level. They are modern noninvasive functional imaging methods with a possibility of image fusion for tumor treatment. In environmental remote sensing, sensor fusion, image fusion, data fusion, and information fusion can be collectively used for holistic environmental investigation at the global level. However, some situations in image processing require both high spectral and high spatial resolution in a single image for earth observations. Hence, fusing images from multiple sensors in which one has high spectral resolution and the other one high spatial resolution is viable for achieving the purpose of image or data fusion, resulting in both high spatial and high spectral resolution. If the images are acquired from satellite sensors, the fusion process is referred to as satellite image fusion. In the remote sensing domain, satellite image fusion can be broadly summarized as the integration of data derived from satellite-based instruments in the form of images with multispatial, multitemporal, or multispectral characteristics to generate a more informative fused image and advance earth observation and environmental monitoring. In the past decades, a variety of methods and techniques were developed for image or data fusion. According to Pohl and Van Genderen (1998), these methods can be typically classified into the following three categories (Figure 8.2).

8.2.1 PIXEL-LEVEL FUSION

Pixel-level fusion is a kind of data fusion performed at the lowest processing level. Namely, it refers to the fusion of pre-processed images on a pixel-by-pixel basis after accurate image registration. In other words, a new data value is calculated for each separate pixel by making use

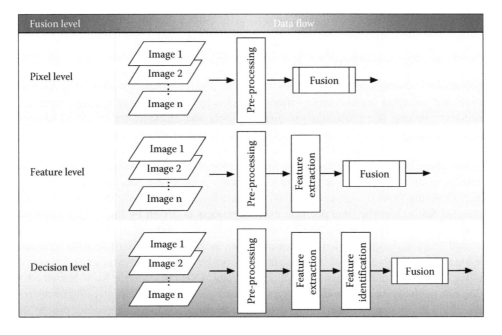

FIGURE 8.2 Three classic levels of image fusion.

of several mathematical operators to blend information from other images at this pixel. A typical example of pixel-level fusion is producing the Normalized Difference Vegetation Index (NDVI) by blending information from two images measured at the near-infrared and red wavelengths to create a new image in which the features of vegetation cover are largely enhanced. NDVI can be produced by:

$$NDVI = \frac{NIR - Red}{NIR + Red} \tag{8.1}$$

where *NIR* is the observed reflectance at the near-infrared band, and *Red* is the relevant reflectance at the red band. As indicated, the fusion process is a mathematic computation between the observed reflectance images at two different wavelengths on a pixel-by-pixel basis.

8.2.2 FEATURE-LEVEL FUSION

Feature-level fusion is conducted at the intermediate level, unlike pixel-level fusion which performs at the lowest processing level. Feature-level fusion refers to the combination of the extracted target features from each individual sensor or spectral domain to produce a more informative fused image (Ross and Govindarajan, 2005). The fusion process at the feature level requires detecting, segmenting, and extracting distinctive features of objects at different domains prior to the fusion scheme. Therefore, distinct features are firstly extracted from the input images and then blended together. In other words, the fusion targets are features extracted from each image rather than the original data value at the pixel level. As described in Chapter 5, features involved in image fusion can be radiometric (e.g., intensities) and geometric (e.g., shapes, sizes, height). By fusing these features, a more informative common feature space is created to aid in monitoring and identifying targets of interest from a heterogeneous environment. The core of image fusion at the feature level is the accurate selection and extraction of features from multiple images collected from multispectral, hyperspectral, or microwave remote sensing sensors. Diversified features could complicate the key characteristics of the target of interest, and inaccurate feature extraction may result in an inaccurate fusion outcome (Chang et al., 2016). Feature-level fusion has been widely applied in land use/land cover mapping, such as forestry mapping (Reiche et al., 2013).

8.2.3 DECISION-LEVEL FUSION

Decision-level fusion refers to the fusion process at a more advanced processing level, which combines the extracted features from different images by making use of external decision rules to reinforce common interpretations via different tools and elucidate the observed targets with multiple aspects (Pohl and Van Genderen, 1998). The fusion processes commonly involve feature extraction and decision-making. Decision rules are established for the features extracted from each individual image before directly fusing the final decisions by blending different feature extractions and interpretations through one or more mathematical operators (i.e., decision rules) for decision-making. Differing from feature-level fusion, decision-level fusion highly depends on the quality of extracted features as the final decision-making process is driven by the type of features used for final data fusion (Chang et al., 2016). In regard to decision-making, decision rules may include either hard decision or soft decision rules, or even both for comparison. Hard decision methods refer to the weighted sum score method (i.e., weighted sensors by inverse of covariance and sum to derive score function), Boolean operators (e.g., applying logical AND/OR), and the M-of-N method, while soft decision methods include the use of Bayesian, fuzzy variable, and Dempster–Shafer approaches (Waltz, 2001). An example of decision-level fusion is creating a land use/land cover map by using a simple binary tree with proper thresholds for reflectance values at different wavelengths. Different land use/land cover classes are thus detected and extracted from multispectral remote sensing images, which can be considered distinctive decision features over the area of interest toward the final classifier fusion via a decision tree (DT) for decision-level fusion.

In addition to the aforementioned three classical image or data fusion categories, data fusion techniques can also be grouped into different categories based on the relations among the input data sources. According to Durrant-Whyte (1988), data fusion can be classified into (1) competitive, (2) complementary, and (3) cooperative. Competitive fusion occurs when two or more sensors provide observations at the same location and degree of freedom; by contrast, complementary fusion occurs when a set of information sources offer different information complementary to the same geometric feature in different degrees of freedom. Finally, cooperative fusion happens when one sensor is dependent on the information of another sensor's prior observation (Chang et al., 2016).

8.3 IMAGE AND DATA FUSION TECHNOLOGY HUBS

Numerous image and data fusion technologies have been developed in the past few decades. Basically, these technology hubs can be categorized into the following three broad areas in terms of multispectral, hyperspectral, and microwave-based fusion although some of them are intertwined with each other. Each of the above three types of remote sensing sensors has its strengths and weaknesses, providing synergistic opportunities to work together. By fusing these separate data sources, we are cyber-enabled to improve the discriminatory capability over the use of either sensor or data source alone. Some of the multisensor data fusion and multi-class feature extraction applications involved in a hybrid approach are even more advantageous for dealing with highly challenging missions than two sensors simply being fused together with or without auxiliary ground-based datasets to aid in feature extraction.

8.3.1 MULTISPECTRAL REMOTE SENSING-BASED FUSION TECHNIQUES

High-Resolution Multispectral (HRM) images have been widely used in many remote sensing applications. For remote sensing instruments, the wavelength of the panchromatic band often covers a wide spectrum range that typically spans a large portion of the visible part of the spectrum. This connection between the two types of images reveals high application potential for image and data fusion. The earliest image fusion technology is the panchromatic-sharpening (or pan-sharpening for short) technique in which a high-resolution panchromatic image and a low-resolution multispectral image are fused together to produce an HRM image (Figure 8.3). Well-known examples of image fusion of such a data set may include, but are not limited to, the SPOT-XS (3 bands, 20-m) and SPOT-P (panchromatic, 10-m) images as well as the Landsat 7 Enhanced Thematic Mapper Plus (ETM+) multispectral bands (6 bands, 30-m) and the panchromatic band (15-m). In other words, image fusion is mainly designed to extract the high frequency details from a panchromatic image and inject them into a multispectral image for image enhancement with some unsupervised learning methods. In addition, image and data fusion can be made possible across different or the same optical sensors onboard differing satellite platforms. To clarify, these image and data fusion techniques can be categorized into the following five broad groups: pre-processor, feature-based classification, uncertainty-based approaches, matrix factorization approaches, and hybrid approaches, as described below.

8.3.1.1 Image and Data Fusion with the Aid of Pre-Processors

Unsupervised learning methods are often used to produce pre-processors and generate compact representations with respect to independent or uncorrelated components for image and data fusion. In other words, most unsupervised learning methods can produce compact representations that either preserve the attributes or reduce the dimensionality of the input for the panchromatic band to achieve better data fusion between panchromatic and multispectral images. Popular methods of its kind include Principal Component Analysis (PCA) (Chavez et al., 1991), wavelet

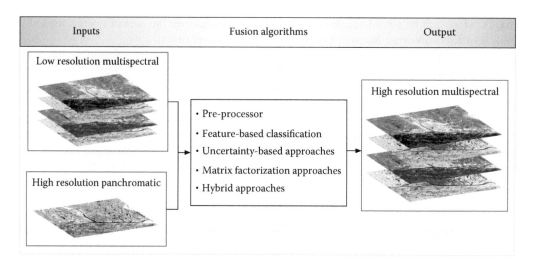

FIGURE 8.3   Schematic illustration of using pan-sharpening to fuse a high-resolution panchromatic image with low-resolution multispectral images to generate high-resolution multispectral images.

decomposition method (Li et al., 1995), Independent Component Analysis (ICA) (Aiazzi et al., 2002), compressed sensing-based panchromatic sharpening method with a dictionary from Low-Resolution Multispectral (LRM) images, and High-Resolution Panchromatic (HRP) images in which the HRP is sparse (Jiang et al., 2012; Zhu and Bamler, 2013). The chronological review below may provide a more comprehensive understanding.

In the early 1990s, some of the most commonly used pan-sharpening techniques for image fusion included Brovey transform, Intensity-Hue-Saturation (IHS), PCA, High Pass Filters (HPF), spherical coordinates, and color normalized transform, in which IHS is perhaps the most common technique. For instance, the use of SPOT panchromatic information to complement the six spectral bands of Landsat TM images was conducted via three methods, including IHS, PCA, and HPF, and the results revealed that IHS distorted the spectral characteristics the least (Chavez et al., 1991). With the capabilities of IHS, fusing panchromatic information into the multispectral bands of SPOT images with respect to the radiometric properties of the sensor was conducted by transforming the original multispectral image into a new 3-dimensional system in which one of the axes stands for intensity. After replacing this intensity with the panchromatic band followed by the inverse transformation, the data fusion process is completed and the multispectral band is enhanced (Pellemans et al., 1993). In addition, multisensor image fusion using wavelet transform was developed and applied to elevate the performance of image and data fusion in the mid-1990s (Li et al., 1995). Nunez et al. (1999) developed the multi-resolution fusion technique in which the multispectral and panchromatic images were decomposed into approximation based on the wavelet decomposition principles. These principles include (Nunez et al., 1999): (1) the multispectral image is interpolated to be consistent with the size of the panchromatic image, (2) the detail sub-band coefficients of the panchromatic image are injected into the corresponding sub-band of the multispectral image by a predefined rule, and (3) the inverse transformation is conducted and the multispectral band is enhanced. This concept was known as "à trous" wavelet transform-based fusion in which the image is convolved with the cubic spline filter and decomposed into wavelet planes for image enhancement before the inverse transformation.

In the 2000s and thereafter, there were two different directions of advancement in image and data fusion. On one hand, more methods using mathematical transforms were employed for enhancement and image fusion; on the other hand, spatiotemporal fusion with multitemporal information was developed to include more information content over the spatial, temporal, and even spectral domains. With the extension of the concept of mathematical transforms, data fusion is generalized by the ARSIS concept of high spatial and spectral resolution images (Ranchin and Wald, 2000; Ranchin et al., 2003). The ARSIS concept (from its French name "Amélioration de la Résolution Spatiale par Injection de Structures") uses different mathematical transforms between the panchromatic band and multispectral images to better produce a description of the information content of both images. With such a multiscale method (i.e., the ARSIS concept), the missing spatial information (high frequency) between the two images represented by the wavelet coefficients is then extracted and used for the construction of the synthetic image (Shi et al., 2005). Following this, the Discrete Wavelet Transform (DWT) (Aiazzi et al., 2002), ICA (Aiazzi et al., 2002), the generalized Laplacian pyramid (Aiazzi et al., 2002), curvelets (Choi et al., 2005), Additive Wavelet Luminance Proportional (AWLP) (Otazu et al., 2005), combined adaptive PCA approach and contourlets (Shah et al., 2008), contourlet transform-based edge learning (Upla et al., 2011), multiscale dual bilateral filter (Hu and Li, 2011), and wavelet transform and Taguchi methods (Pradhan et al., 2016) were used to extract the high frequency information from the HRP image for image enhancement of the LRM image.

8.3.1.2 Image and Data Fusion with the Aid of Feature Extractors

Unsupervised learning methods are sometimes used to produce feature extractors with respect to texture and spectral features, which aim to build the detection capability in satellite images by combining multispectral classification and texture filtering. With this capability, multiple spectral and spatial features may be retrieved for final classification. For example, the Gray Level Co-occurrence Matrix (GLCM) method can provide spatial information to support a spectral classification in which the gray value of an image is transformed into the co-occurrence matrix to help generate various output images produced through adopting different spatial measures (Zhang, 1999). In addition, Pesaresi and Benediktsson (2001) and Benediktsson et al. (2003) proposed a series of spatial features extracted by morphological and differential morphological profiles for enhancement. Along these lines, the Pixel Shape Index (PSI) was designed to integrate shape and spectral features into a Support Vector Machine (SVM) that warrants self-adaptability with the aid of high-dimensional property and a swift learning pace in feature space simultaneously (Zhang et al., 2006). This machine learning approach of a single classifier (e.g., SVM) with inputs from multiple features is favored because of its effective and fast processing, which can also reduce the dominance effects of the spectral information if the input vector covers both spatial and spectral features (Zhang et al., 2006). Therefore, multiple classifiers can be applied whenever necessary to improve the overall performance of data fusion at the decision level (Benediktsson and Kanellopoulos, 1999; Petrakos et al., 2001; Briem et al., 2002).

8.3.1.3 Uncertainty-Based Approaches for Multi-Resolution Fusion

Multi-resolution fusion of satellite images such as pan-sharpening can be performed with the aid of uncertainty-based methods to improve the quality of the fused products. Uncertainty-based methods for data fusion include fuzzy theory (Carrara et al., 2008), Bayesian data fusion (Fasbender et al., 2008), probability theory (Li et al., 2013), and Dempster-Shafer evidence theory (Lu et al., 2015). There are many extensions within the context of probability theory in this regard. For example, the restoration-based method (Li and Leung, 2009) first resamples the LRM images to the scale of the high-resolution panchromatic image. The relationship between these two kinds of images is then used to restore the resampled multispectral images in which the HRP image is modeled as a linear combination of the ideal HRM images plus the observation noise, and the resampled LRM images are regarded as the noisy blurred versions of the ideal multispectral images. The ideal HRM images are then predicted based on the HRP and the resampled LRM images with a closed-form solution of the derived fused images. Additionally, Zhang et al. (2012) developed an adjustable uncertainty-based image fusion method to describe the inverse problem of image fusion via relationships between the desired HRM images, the observed LRM images, and the HRP image with the aid of the maximum *a posteriori* framework.

Use of correlation analyses to identify the relationship between pixel values spatially and temporally to help predict the fused images can be regarded as a delicate extension of the uncertainty-based image fusion method. For instance, Landsat satellite imageries have a high spatial resolution of 30-m, but a long temporal resolution of 16 days. Free coarse resolution sensors with a daily repeat cycle, such as MODIS, are incapable of providing detailed information at the spatial scale because of low spatial resolution. Thus, fusing Landsat and MODIS imageries may produce promising products that feature both high spatial and temporal resolution. One of the popular methods of its kind is the Spatial and Temporal Adaptive Reflectance Fusion Model (STARFM) proposed by Gao et al. (2006). STARFM was designed to build relationships between the predicted pixel and the candidate pixels spatially and temporally in a matrix form. On the basis of STARFM, the data fusion model has been developed for predicting high spatial resolution surface reflectance through data blending with low spatial resolution scenes of MODIS while preserving its high temporal resolution. Many methods with similar capabilities were developed for enhancement along this line for different applications. They include (1) the Spatial Temporal Adaptive Algorithm for mapping Reflectance Change (STAARCH) (Hilker et al., 2009), (2) the extended STARFM (ESTARFM) (Zhu et al., 2010; Xie et al., 2016), (3) the SParse-representation-based SpatioTemporal reflectance Fusion Model (SPSTFM) (Huang and Song, 2012), (4) the Spatial Temporal Data Fusion Approach (STDFA) or Modified Spatial Temporal Data Fusion Approach (MSTDFA) (Wu et al., 2012, 2015), (5) the modified ESTARFM (mESTARFM) (Fu et al., 2013), (6) the Spatiotemporal Image Fusion Model (STI-FM) (Hazaymeh and Hassan, 2015), and (7) the unmixing-based STARFM (USTARFM) (Yang et al., 2016). These various forms derived from STARFM (i.e., STAARCH, ESTARFM, SPSTFM, STDFA, MSTDFA, mESTARFM, STI-FM, and USTARFM) have not yet been fully tested with respect to a common basis, therefore, more future work is required in this regard.

8.3.1.4 Matrix Factorization Approaches across Spatiotemporal and Spectral Domains

This data fusion technology can handle any data set expressed in a matrix form from attribute-based representations to ontologies, spatiotemporal associations, and inherent networks over temporal, spectral, and spatial domains. With an even more sophisticated mathematical construct, the matrix factorization fusion methods focus on exploring a specific target relation and exploiting associated data constrained by the context (Lee and Seung, 1999; Zitnik and Zupan, 2015). For example, Zitnik and Zupan (2015) formalized the theory of matrix factorization fusion in which the penalized matrix tri-factorization was formulated for simultaneously factorizing data matrices to reveal hidden associations. Such a method is favored over several state-of-the-art multiple kernel learning methods due to its higher accuracy. On top of using linear algebra as a tool, optimization theory was also proposed to help. Wei et al. (2016) proposed a spatiotemporal fusion method using sparse representation instead of the traditional linear mixture method when handling spontaneously changing content in the terrestrial system. Within this context, the cluster and joint structural sparsity of the sparse coefficients may be employed as *a priori* to increase prediction accuracy. This can be achieved by using a new optimization model that is formulated by the semi-coupled dictionary learning and structural sparsity method leading to the prediction of the unknown high-resolution image from known images (Wei et al., 2016).

8.3.1.5 Hybrid Approaches

The integrated Inhomogeneous Gaussian Markov Random Fields (IGMRFs) and the maximum *a posteriori* estimation method proposed by Joshi and Jalobeanu (2010) belong to the category of hybrid method. Within this hybrid method, an LRM image was modeled as an aliased and noisy version of the corresponding fused image with high spatial and spectral resolution during the optimization process based on two types of uncertain inputs. In essence, the decimation matrix was estimated from the high resolution panchromatic image and the low resolution multispectral images, known as the data term (likelihood term). Then the IGMRF parameters estimated from the panchromatic image were used as the prior term. An optimization technique was finally used to minimize the sum of the data and the prior terms, leading to the estimation of the unknown high resolution multispectral image for each of the multispectral bands. On the other hand, as proposed by Shackelford and Davis (2003), the length–width extraction algorithm that is akin to GLCM was able to extract the length and width of connected groups of pixels that are spectrally similar to one another based on the fuzzy relationships. This is complementary to the measures of the spatial features in a certain direction within a spectrally similar neighborhood.

8.3.1.6   Environmental Applications

Data fusion methods based on multispectral remote sensing can be used to advance the monitoring capacity of the atmospheric environment. By using a Bayesian data fusion framework, Fasbender et al. (2009) successfully proposed a space–time prediction scheme (i.e., a data-model fusion scheme) to account for secondary information sources contributing to the variation of nitrogen dioxide concentration. Similarly, Berrocal et al. (2012) proposed two separate space-time data fusion schemes to combine ground-level ozone concentration monitored at a point level and numerical modeling output at a grid cell. These two schemes included a Gaussian Markov Random Fields (GMRF) smoothed downscaler and a modified smoothed downscaler featured by spatially varying random weights. Such spatially varying random weights may be defined through a latent Gaussian process and an exponential kernel function, respectively. The results yielded a range of predictive gain from 5% to 15% in overall predictive mean square error over prior downscaler models without data fusion.

Differing from many previous studies of data fusion for improving either spatial, temporal, or spectral resolution toward better prediction accuracy, data fusion schemes can also be applied to optimize data quality and spatial coverage simultaneously. In this case, the terminology of data merging might be favored since the goal is to perform gap filling over space to increase spatial coverage instead of spatial resolution. This is especially true for remotely sensed products, such as Aerosol Optical Depth (AOD) products in which data gaps are often severe due to cloud contamination. For instance, with the aid of a traditional universal Kriging approach and a spatial statistical data fusion scheme, Puttaswamy et al. (2013) successfully merged two satellite-based daily AOD measurements with ground-based observations over the continental United States to improve spatial coverage and overall accuracy. Similarly, multiple AOD products were merged using the maximum likelihood estimation method to create a consistent dataset of AOD products with improved spatial coverage (Xu et al., 2015).

Data fusion methods are also used to aid in better feature extraction for water quality monitoring. For instance, the Integrated Data Fusion and Machine learning (IDFM) algorithm was developed with MODIS and Landsat data sources, leading to the generation of fused images with better temporal and spatial resolution for feature extraction via machine learning. These fused images were retrieved by machine learning or data mining algorithms for monitoring challenging water quality constituents of concern such as total organic carbon, total nitrogen, total phosphorus, and microcystin in fresh water lakes or coastal bays (Chang et al., 2014a,b, 2015). With the aid of Genetic Programing (GP) methods, for instance, Chang et al. (2014a) successfully mapped the distribution of microcystin toxins in Lake Erie by using remotely sensed imagery fused from the MODIS and the Medium Resolution Imaging Spectrometer (MERIS). The results revealed that GP models performed better than traditional two-band models in quantifying the concentrations of microcystin in water bodies due to the inclusion of the additional spectral reflectance data embedded in hyperspectral MERIS images. The unique hyperspectral band helped improve the prediction accuracy, especially in the range of low microcystin concentrations. Similarly, Imen et al. (2015) performed the IDFM with the ANN-based machine learning tool for featuring the total suspended sediment concentrations in Lake Mead.

Fused data can also be used for environmental and ecosystem health assessment. On one hand, data fusion was applied to enhance the temporal resolution of satellite imagery for environmental health assessment associated with the West Nile Virus outbreak in Los Angeles (Liu and Weng,

2012); on the other hand, for automated benthic habitat mapping in a portion of the Florida Keys by combining multiple data sources (hyperspectral, aerial photography, and bathymetry data), the fused dataset associated with both the pixel- and feature-level fusion schemes was preclassified by two machine learning algorithms and one classical algorithm (i.e., random forest, SVM, and k-Nearest Neighbor) to generate object-based habitat maps through an ensemble analysis of outcomes from the three aforementioned machine learning classifiers (Zhang, 2015).

Finally, image and data fusion can also be applied for multitemporal change detection of land use and land cover. These recent spatiotemporal adaptive data fusion algorithms may include, but are not limited to: (1) developing the length–width extraction algorithm for land use classification (Shackelford and Davis, 2003), (2) innovating a spatial unmixing algorithm to promote the synergistic use of MERIS full resolution and Landsat TM data for vegetation dynamics studies in which a linear mixing model was employed to preserve the spectral signature of the low resolution image as much as possible for all the thematic classes (Zurita-Milla et al., 2008), (3) evaluating forest disturbance based on STARFM (Hilker et al., 2009), (4) predicting changes in land use and phenology change (i.e., seasonal change of vegetation) (Huang and Song, 2012), (5) generating advanced daily land surface temperature using the extreme learning machine at Landsat resolution (Huang et al., 2013; Weng et al., 2014), (6) fusing RapidEye and MODIS imageries to explore vegetation dynamics of semi-arid rangelands (Tewes et al., 2015), leading to examine forest disturbance and loss with a 12-year NDVI time series of an 8-day interval and a 30-m spatial resolution dataset feasible to detect even tiny changes on a daily basis (Schmidt et al., 2015), and (7) integrating multiple temporal, angular, and spectral features to improve land cover classification accuracy (Chen et al., 2017).

Recent major studies taking advantage of various data fusion, statistical, and other inference methods toward different environmental applications were summarized in Table 8.1. The table shows that many methods originally proposed to fuse land band products were also used to fuse sea band products, for example, STARFM, ESTARFM, and USTARFM. A potential barrier is that those STARFM-like methods might not work well with input images that have strong disturbances in atmospheric and/or oceanic environments that could be highly dynamic, driven by various bio-geophysical and bio-geochemical factors. More research to advance this area is an acute need for the future.

8.3.2 Hyperspectral Remote Sensing-Based Fusion Techniques

Hyperspectral imaging, which is a relatively new remote sensing technology, may deploy hundreds of spectral bands over the electromagnetic spectrum to collect image data for the same study area. With this advantage, hyperspectral imaging can provide powerful, detailed classification for earth observation, environmental monitoring, and resource exploitation (e.g., mineral exploitation) because intrinsic spectral properties of objects or species at the ground level can be discerned properly by suitable hyperspectral bands. However, many regions in the world have a high spatial and spectral heterogeneity simultaneously, given that some unique landscape environments are present in the form of linear/narrow shapes or small patches requiring the use of images with high spatial resolution for classification. In land use and land cover classification, intraclass variations are defined as samples in the same class that may have different spectral signatures. The problem of intraclass variations becomes more serious if a hyperspectral image also has high spatial resolution. This is because intraclass variations exist in both the spectral and spatial domains with intertwined influences. Therefore, fusion between hyperspectral and multispectral (or Light Detection and Ranging [LiDAR]) remote sensing imageries becomes necessary to minimize the intraclass variations, although band conversion issues require further attention. These data fusion techniques can be categorized into the following three broad groups (Figure 8.4).

TABLE 8.1

Recent Advances in Applying Multispectral Data Fusion Methods for Environmental Applications

Reference	Fusion Method	Data Source	Objective	Enhanced Properties
Land Environment and Change Detection				
Huang and Song (2012)	SPSTFM	MODIS and Landsat surface reflectance	To produce better land use and phenology change	Spatial Temporal
Tewes et al. (2015)	ESPSTFM	RapidEye and MODIS	To detect daily vegetation dynamics	Spatial Temporal
Wu et al. (2012); Wu et al. (2015)	STDFA, MSTDFA	MODIS and Landsat NDVI products	To detect daily NDVI changes	Spatial Temporal
Huang et al. (2013); Weng et al. (2014)	Bilateral filtering	MODIS and Landsat surface reflectance	To detect daily land surface temperature and urban heat island effects	Spatial Temporal
Water Environment				
Wang et al. (2010)	Fuzzy Integral	Chlorophyll-a data from multiple neural network predictions	To improve the monitoring capacity of chlorophyll-a concentration	Overall accuracy
Mattern et al. (2012)	Bayesian Data Fusion	Kriging and regression tree based nitrate concentration map	To create a unified nitrate contamination map for groundwater quality mapping	Overall accuracy
Chang et al. (2014a,b) Doña et al. (2015) Imen et al. (2015)	STARFM	MODIS and Landsat surface reflectance	To generate a fused imagery for water quality mapping	Spatial Temporal
Fok (2015)	Spatiotemporal combination	TOPEX/Poseidon Geosat-Follow-on Envisat	To generate a gridded altimetry data set for ocean tides modeling	Spatial Temporal
Shi et al. (2015)	BME	Chlorophyll-a data from SeaWiFS, MERIS and MODIS	To create a unified chlorophyll-a dataset with improved spatial coverage	Spatial Temporal
Huang et al. (2016)	ESTARFM	VIIRS and Landsat surface reflectance	To generate a fused imagery for surface water mapping	Spatial Temporal
Du et al. (2016)	Pan-sharpening	Sentinel-2 Imagery	To produce high spatial resolution water index	Spatial
Atmospheric Environment				
Puttaswamy et al. (2013)	UK and SSDF	MODIS, GOES, AERONET	To merge daily AOD from satellite and ground-based data to achieve optimal data quality and spatial coverage	Spatial

(Continued)

TABLE 8.1 (*Continued*)

Recent Advances in Applying Multispectral Data Fusion Methods for Environmental Applications

Reference	Fusion Method	Data Source	Objective	Enhanced Properties
Xu et al. (2015)	UK, MODIS Dark Target and Deep Blue algorithms	MODIS, MISR and SeaWiFS, AERONET, and the China Aerosol Remote Sensing NETwork	To merge daily AOD dataset over mainland China by integration of several AOD products	Spatial
Environmental and Ecosystem Health				
Liu and Weng (2012)	STARFM	ASTER and MODIS	To assess the West Nile Virus outbreak in Los Angeles, USA	Temporal
Zhang (2015)	Pixel/feature-level data fusion, Object-based Image Analysis, machine learning, ensemble analysis	Hyperspectral, aerial photography, and bathymetry data	To generate a fused dataset for object-based benthic habitat mapping in a portion of the Florida Keys, USA	Spatial Temporal

Note: BME: Bayesian maximum entropy; STARFM: Spatial and Temporal Adaptive Reflectance Fusion Model; ESTARFM: Enhanced Spatial and Temporal Adaptive Reflectance Fusion Model; SPSTFM: SParse-representation-based SpatioTemporal reflectance Fusion Model; SADFAT: Spatiotemporal Adaptive Data Fusion Algorithm for Temperature mapping; MSTDFA: Modified Spatial and Temporal Data Fusion Approach; UK: Universal Kriging; SSDF: Spatial Statistical Data Fusion; DEM: Digital Elevation Model.

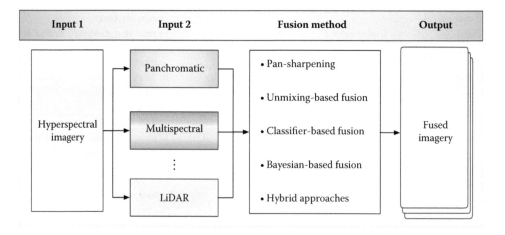

FIGURE 8.4   The process of fusing hyperspectral imagery with other data sources.

8.3.2.1   Data Fusion between Hyperspectral and Multispectral Images

To overcome limitations due to intraclass variations, data fusion at the pixel level with the aid of HRM images may provide improved earth observation and environmental monitoring spatially and spectrally. Like pan-sharpening for multispectral imagery, the use of high-resolution panchromatic data to enhance hyperspectral imagery after co-registration is entirely applicable. For instance,

Hardie et al. (2004) developed a maximum *a posteriori* estimator for enhancing the spatial resolution of a hyperspectral imagery. In addition, most of the available hyperspectral sensors such as Earth Observing-1 (EO-1)/Hyperon and Airborne Visible/InfraRed Imaging Spectrometer (AVIRIS) collect data with a relatively coarse spatial resolution (i.e., 20–30-m), and this can be improved by data fusion using the satellite images collected from WorldView, QuickBird, GeoEye-1, IKONOS, SPOT, and/or others. Along these lines, Zurita-Milla et al. (2008) developed an unmixing-based data fusion technique to generate images that exhibit the merit of both the spatial resolution of Landsat TM and the spectral resolution of the MERIS sensor simultaneously. Their results showed that it is possible to successfully preserve the MERIS spectral resolution while downscaling MERIS full resolution data to a Landsat-like spatial resolution. Some other studies even used a single source of primary information, which is then fused with textural features derived from the initial image (Fauvel et al., 2008), segmented features obtained through object-based image analysis (Huang and Zhang, 2010, 2012), and panchromatic images (Xu et al., 2017) for enhancement. On the other hand, even without the data fusion at the pixel level, integrating supervised and unsupervised feature recognition and performing data fusion at the decision level with classifier fusion techniques may still improve classification accuracy by alleviating the impact from trivial intraclass variations (Du, 2009).

8.3.2.2 Data Fusion between Hyperspectral Images and LiDAR Data

The complexity of regional landscapes, such as urban, coastal, or forest areas, triggered the integration of active remote sensing technologies with hyperspectral datasets to provide reliable, delicate, fused, and automatic regional mapping and change detection. Fusion of hyperspectral and LiDAR remote sensing data for the classification of complex areas appeared in the late 1990s. Jimenez et al. (1999) conducted both feature- and decision-level fusion using this type of data fusion method. In feature-level fusion, the inherent characteristics of hyperspectral feature space were retrieved by feature extraction algorithms where relevant features for classification are chosen to reduce the dimensionality. Then more effort is delivered to the final classification. Yet, in decision-level fusion, a local discrimination is performed at each sensor independently and the final decision can be made by including the results of multiple local discriminations with the aid of different classifier fusion techniques such as majority voting, max rule, min rule, average rule, and neural network. Many applications of its kind have been undertaken for solving challenging classification issues such as coastal mapping (Elasksher, 2008), forest monitoring (Dalponte et al., 2008), and tree species classification (Dalponte et al., 2012). In the case of coastal mapping, different sensors were designed to cover different ranges of the electromagnetic spectrum, providing synergistic opportunities for data fusion (Elasksher, 2008). For example, LiDAR data were prepared for geometric range measurements, operating with a wavelength of about 1000 nm; on the other hand, AVIRIS images collected the spectral reflectance of the ground, operating with a wavelength band from 380 nm to 2,500 nm. Bigdeli et al. (2014) conducted a decision-level fusion of hyperspectral and LiDAR data from which features are retrieved through texture and topographical information. After the feature extraction step, SVM classifiers were applied to both data sets separately based on each feature space, and the results of the two classifiers were fused through a Naive Bayes method. The Naive Bayes method is thus deemed a classifier fusion method that combines the results of several SVM classifiers. Man et al. (2015) presented pixel- and feature-level fusion of hyperspectral and LiDAR data for urban land use classification to evaluate the efficiency of combined pixel- and object-based classifiers for urban land use classification.

8.3.2.3 Hybrid Approach

Yokoya et al. (2016) emphasized the use of a hybrid method, including one matrix factorization method and two pan-sharpening methods, to conduct spatial data fusion for multispectral images while performing spectral unmixing to preserve the spectral properties for hyperspectral images, leading to the improvement of the geological mining survey.

8.3.2.4 Environmental Applications

Recent major advancements taking advantage of various multisensory or multisource data fusion, classifier fusion, and machine learning toward different environmental applications are summarized in Table 8.2. For instance, Zhang and Xie (2013) presented pixel-/feature-level fusion strategies followed by a classification scheme with the aid of the SVM to improve vegetation mapping in the coastal Everglades in Florida. In one of the most sophisticated studies in this emerging area of interest, Stavrakoudis et al. (2014) investigated the effectiveness of combining multispectral and hyperspectral satellite imagery for accurate forest species mapping using a decision fusion approach. This type of classification can be extended to simultaneously make use of the spectral information of various families of textural features (Dalla Mura et al., 2011), morphological profiles (Huang et al., 2009; Dalla Mura et al., 2011), and image segmentation derived features (Huang and Zhang, 2010, 2012) toward decision-level fusion. Following this philosophy, one similar decision-level fusion used two fuzzy classifications, namely two fuzzy output SVM, in support of the independent classification of the hyperspectral image, which is then resampled to the spatial resolution of the multispectral image (Huang and Zhang, 2010, 2012). The two sources of decision were combined using a prespecified fusion operator in the final stage to produce fusion results at the decision level toward better land cover classification.

8.3.3 MICROWAVE REMOTE SENSING-BASED FUSION TECHNIQUES

A synthetic aperture radar (SAR) may have longer wavelengths that are mostly unaffected by weather and clouds and may be deployed onboard a space- or air-borne platform. SAR is an active remote sensing system with a wavelength of C band or L band (see Chapter 3). The phase value of the SAR image is calculated by the apparent distance between the ground target and the radar antenna, providing more precious information. With a large virtual antenna and unique signal processing techniques, SAR products provide much higher spatial resolution than that of a regular aperture radar. By combining the phase components of two SAR co-registered images of the same area, Interferometric synthetic aperture radar (InSAR) can involve the information fusion of these two SAR images to extract the subtle patterns of Earth's surface changes through the phase change identification. In general, the environmental applications of InSAR include three categories for: (1) detecting ground deformation caused by natural hazards or anthropogenic activities, (2) mapping the extent and progression of natural hazards such as earthquakes, volcanic eruptions, landslides, mudflows, floods, and so on, and (3) monitoring tiny landscape changes by analyzing SAR intensity and coherence images.

Direct interpretation of SAR or InSAR data for environmental applications is by no means an easy task because the geometry and spectral range of SAR is very different from optical remote sensing imagery. However, optical remote sensing imagery and SAR or InSAR data are complementary, often meaning that it is possible to fuse SAR and optical remote sensing imagery for advanced analyses. Since SAR is generated by a computer, the computer screen can be viewed as an equivalent sensor to mimic the counterpart of optical remote sensing.

The goal of fusing SAR and optical remote sensing imageries is to preserve the accurate spectral information from the optical sensors and the textural information from the microwave sensors for advanced earth observations and environmental monitoring (Figure 8.5). For example, in the image stacking or vector stacking approach (Leckei, 1990), each pixel is represented by a vector that contains visible components from different optical sensors, texture features obtained by different SAR sensors, and ancillary data collected from field campaigns. Classification is usually performed by using a parametric statistical or machine learning technique. This type of fusion can be made possible at the pixel-, feature-, or decision-level. In other words, SAR-based data fusion methods may create a new composite feature space that is structured by a suite of input vectors from multiple sources. This process leads to either a parametric or non-parametric classification within a new feature space.

TABLE 8.2

Recent Advances in Applying Hyperspectral Data Fusion Methods for Environmental Applications

Reference	Fusion Method	Data Source	Objective	Enhanced Properties
Water Environment				
Guo et al. (2016)	Inversion-based Fusion	Hyperion dataset, HJ-1-CCD and MERIS image	To fuse different resolution images for inland water quality mapping	Spatial Spectral
Pan et al. (2016)	Statistical models	LIDAR data and hyperspectral imagery	To generate orthowaveforms for estimating shallow water bathymetry and turbidity	Texture
Land Environment and Change Detection				
Huang and Zhang (2012)	Multi-level decision fusion	Multi/hyperspectral imagery	To improve the land cover classification	Spatial Spectral
Zhang and Xie (2013)	Pixel- / feature-level fusion strategy with RF and SVR for classification	Merge 20-m hyperspectral imagery from AVIRIS with 1-m Digital Orthophoto Quarter Quads	To monitor vegetation mapping in the coastal Everglades	Spatial Spectral
Stavrakoudis et al. (2014)	Fuzzy SVM	QuickBird and EO-1 Hyperion hyperspectral image	To improve the forest species identification and mapping	Spatial Spectral
Bigdeli et al. (2014)	Naive Bayes method for classifier fusion at decision level	Hyperspectral and LIDAR data	To improve the land cover classification	Spatial Spectral
Man et al. (2015)	Pixel-based (SVM) and object-based	Air-borne laser system and CASI	To improve the land cover classification	Spatial Spectral
Yokoya et al. (2016)	One matrix factorization method and two pan-sharpening methods	EnMAP and Sentinel-2	To preserve constituent spectra using spectral unmixing and improve the geological mining survey	Spatial Spectral
Liu and Li (2016)	Data field modeling	ROSIS-03 University of Pavia dataset and AVIRIS Indian Pines dataset	To improve the land cover classification	Spatial Spectral
Xu et al. (2017)	VM, IGS, MABR, NMF	EO-1 and TG-1	To improve the land cover classification	Spatial Spectral
Chen et al. (2017)	STARFM, Spatiospectral fusion, Spatial hyperspectral fusion, and Spatial angular fusion	Landsat-8 (OLI), MODIS, China Environment 1A series (HJ-1A), and ASTER DEM data	To improve land cover classification accuracy	Temporal, Angular, Multispectral, Hyperspectral, Topographic

Note: VM: variational method; IGS: improved Gram-Schmidt; NMF: Non-negative matrix factorization; MABR: multi-resolution analysis and nonlinear PCA band reduction; RF: random forest; ROSIS-03: the Reflective Optics Systems Imaging Spectrometer; SVR: support vector machine; AVIRIS: Airborne Visible Infrared Imaging Spectrometer.

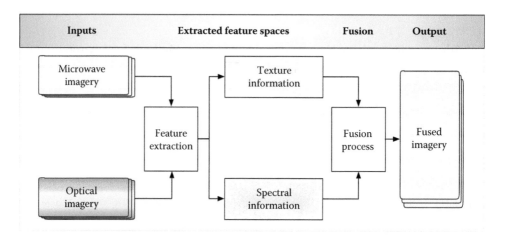

FIGURE 8.5   A schematic illustration of fusing microwave imagery with optical images.

A few important extensions of this approach are data fusion by blending texture statistics derived from the conditional joint probability density functions of the GLCM with radar (SAR) (Barber and LeDrew, 1991), multispectral or hyperspectral data with radar (SAR) (Bruzzone et al., 1999; Amarsaikhan and Douglas, 2004), or LiDAR information with radar (SAR) (Huang et al., 2011; Pedergnana et al., 2012; Ghosh et al., 2014). Different types of data fusion in relation to SAR or InSAR can be categorized into the following three broad groups:

8.3.3.1   Data Fusion between SAR and the Optical Imageries

SAR imaging techniques can be complementary to multispectral or hyperspectral remote sensing techniques because they are independent of solar illumination, cloud coverage, and weather conditions. This synergy gives rise to a unique insight and provides a precious opportunity to combine different data sources, for example, SAR and optical imageries, to improve the overall classification accuracy relative to the quality of any single-source imagery. For instance, urban mapping using coarse resolution SAR and optical imageries for land use and land cover classification has long been a popular topic. Solberg et al. (1994) applied a Bayesian method to fuse the ERS-1 SAR and Landsat TM images for land cover classification toward multitemporal change detection. This data fusion method led to a significant improvement in classification accuracy. Serpico and Roli (1995) and Bruzzone et al. (1999) fused optical and microwave remote sensing images with an ANN-based classifier to support multitemporal change detection of land use and land cover. The fusion operator at the decision level was designed based on the application of the Bayes rule for the "compound" classification of pairs of multisource images acquired at different dates (Duda and Hart, 1973). The temporal correlation between the multisource images acquired at two dates was depicted by using the joint probabilities related to possible changes of land use and land cover with respect to the expectation-maximization principle (Granville and Rasson, 1995).

In the 2000s, more extended work explored the potential value of SAR imagery for data fusion. Haack et al. (2002) fused Radarsat-1 and Landsat TM images by means of an PCA for urban land cover classification. Alparone et al. (2004) presented a novel multisensor image fusion algorithm, which extends pan-sharpening of multispectral data in which SAR texture is used to modulate the generalized intensity of the multispectral image, after the generalized intensity is enhanced by injection of high-pass details extracted from the available panchromatic image via the à trous wavelet decomposition. Chibani (2006) fused SAR and SPOT images by means of the à trous wavelet decomposition for land observations. Waske and Benediktsson (2007) presented a well-known data fusion in which the multitemporal SAR data set and multitemporal optical data set were fused for land use and land cover change detection. The former contains ENVISAT ASAR alternating polarization and ERS-2 precision images, whereas the latter contains Landsat-5 TM images and a SPOT 5 scene. The prediction accuracies associated with certain classifiers including a maximum likelihood classifier, a Decision Tree (DT), a DT-based classifier ensemble, and SVM were compared at the decision level. Hong et al. (2009) took LRM images and successfully fused them with high-resolution SAR by integrating wavelet and IHS transformations to not only maintain the spatial features of SAR images but also preserve the spectral property of the multispectral images simultaneously.

Not until the 2010s were the polarization effects thoroughly investigated in SAR-based data fusion. Zhang et al. (2010) established a pixel-level fusion for air-borne optical and SAR images based on the block-based synthetic variable ratio (Block-SVR) and compared it against IHS, PCA, and wavelet-based methods to confirm the advantage. Amarsaikhan et al. (2010) fused TerraSAR-X and QuickBird multispectral data and compared the performance of different data fusion techniques. This study led to the enhancement of urban features using a refined Bayesian classification. Lu et al. (2011) fused both dual polarized HH-HV (H-Horizontal, V-Vertical) ALOS PALSAR, and RADARSAT-2 data with Landsat using different image fusion methods including PCA, DWT, HPF, and Normalized Multiplication (NMM). In a comparable way, they applied the most commonly used maximum likelihood classification method for land cover mapping in the Brazilian Amazon. Reiche et al. (2013) presented a feature-level fusion study based on multitemporal ALOS PALSAR and Landsat data for monitoring and mapping of tropical forest degradation and deforestation. The extracted SAR and optical sub-pixel fraction features were fused after using a DT classifier for processing the separate SAR and optical input data streams.

8.3.3.2 Data Fusion between SAR and LiDAR or Laser Altimeter

LiDAR is a representative laser scanning system. LiDAR instruments use the time-of-flight measurement principle to estimate the distance to an object, such as a building, a tree, or a moving vehicle, at the ground level. With accurate range measurements between the sensor and a reflecting object, the 3-D position of an object at the ground level is defined by means of the known position of the sensor. LiDAR is currently a well-known method for the measurement of topography, generating the database of the high-resolution Digital Elevation Model (DEM). In principle, SAR data are processed as 2-dimensional image data containing the following information per resolution cell: (1) intensity of the backscattered signal, (2) the range between the antenna and the resolution cell, and (3) the phase of the received radiation affected by the range between the SAR antennae. When compared to LiDAR and optical imagery, the main advantage of space-borne SAR is the temporal frequency of imaging combined with the capability of the global coverage. Using LiDAR data for improved interpretation via information retrieval together with better feature extraction of SAR data is a topic of increasing interest in data fusion. With LiDAR-SAR synergy, phase information embedded in InSAR data may be used to extract coherence maps and 3-dimensional information for landscape monitoring. Currently, simultaneous InSAR data are available from the TanDEM-X system to create unprecedented opportunities to generate Polarimetric Interferometric SAR (PolInSAR) in support of some advanced analyses. Emphasis is placed on the sensitivity of fused products to shape, orientation, and dielectric properties of the scattering target, such as forest structure (Holopainen et al., 2010).

On the other hand, satellite radar altimetry was often used for monitoring absolute water-level changes over large inland water bodies and wetlands at coarse spatial resolution (a few hundred meters) (Lee et al., 2009). Within this arena, an important space-borne LiDAR is the Geoscience Laser Altimetry System (GLAS) onboard NASA's ICESat (Ice, Cloud, and land Elevation Satellite), which has produced data for a large number of studies in recent years. Slatton et al. (2001) fused InSAR and laser altimeter data to estimate surface topography and vegetation heights. Gamba and Houshmand (2002) further combined SAR, LiDAR, and aerial images to extract land cover and DEM for simultaneous extraction of land cover, Digital Terrain Model (DTM), and 3-dimensional shapes of buildings.

8.3.3.3　Data Fusion between SAR, Polarimetric, and Interferometric SAR or Others

This type of data fusion began in the mid-1990s. Bloch (1996) first applied the theory of information and data fusion with respect to some operators. Crawford (1999) presented the data fusion of airborne polarimetric and InSAR for classification of coastal environments. Solaiman et al. (1999) developed a fuzzy-based multisensor data fusion classifier and applied it to land cover classification using ERS-1/JERS-1 SAR composites. In the 2000s, advanced space-borne SAR systems became available to offer better spatial resolution across polarimetric parameters. These include ALOS PALSAR, TerraSAR-X, and RADARSAT-2. The polarimetric SAR data can be used for measuring landscape changes, forest scattering characteristics, fire disturbances, species differentiation, and canopy structures, producing classifications for different forest areas with pattern discrimination. Jouan and Allard (2004) combined features extracted from PolSAR and hyperspectral data for land use mapping in a system prototype called Intelligent Data Fusion System as a part of the Earth Observation Application Development Program sponsored by the Canada Space Agency. Abdikan et al. (2008) presented the fusion of two types of SAR images (PALSAR and RADARSAT-1) with a multispectral SPOT image. Chanussot et al. (1999) proposed fuzzy data fusion techniques for linear features detection in multitemporal SAR images to improve the detection of road systems. The most sophisticated combination of its kind in the literature is probably the case of Goodenough et al. (2008), which fused ALOS PALSAR L-band quad-pol data of polarimetric SAR, hyperspectral, and LiDAR data for useful forest information extraction. Due to the heterogeneity across these image data sources, radiometric correction and image orthorectification were performed on the hyperspectral data acquired by the NASA air-borne AVIRIS sensor. The LiDAR data from the Terra Surveys/UVic air-borne system were corrected and used to create a high-resolution DEM (2 m) which was transformed further into the PALSAR domain by applying a radar simulation algorithm (Niemann et al., 2007). The simulated radar DEM image was then used to help select ground control points for orthorectification of the PALSAR HSV-coding image. The orthorectified HSV-coding image, hyperspectral AVIRIS data, and DEM were finally fused for a non-parametric classification using LOGIT (Goodenough et al., 2005). Shimoni et al. (2009) investigated the complementarity and alternative fusion of different bands (such as L- and P-bands), polarimetric SAR (PolSAR), and polarimetric interferometric (PolInSAR) data for land cover classification. A large feature set was derived from each of these four modalities and a two-level fusion method associated with logistic regression as feature-level fusion and the ANN method for decision level fusion. Kim et al. (2009) and Lu et al. (2009) performed an integrated analysis of PALSAR/RADARSAT-1, InSAR, and the ENVISAT altimeter for mapping the absolute water level changes in the Louisiana wetland. Lu et al. (2011) further processed PALSAR L-band HH and RADARSAT C-band HH as well as Landsat TM multispectral images separately at first, followed by data fusion with the wavelet-merging method for land surface classification. QuickBird image sampling plots were used to generate some ground truth data for final verification of the mapping accuracy of impervious surfaces.

8.3.3.4　Environmental Applications

Measuring water-level changes in a water body is required for the comprehensive assessment of flood hazards (Coe, 1998). Fusion between SAR and optical images has been a popular tool for flood inundation mapping since the mid-1990s (e.g., Hess et al., 1995; Tholey et al., 1997; Townsend and Walsh, 1998; Bourgeau-Chavez et al., 2001; Townsend, 2002; Ahtonen et al., 2004; Dey et al., 2008; Ramsey et al., 2009; Schumann et al., 2009; Bwangoy et al., 2010; Mason et al., 2010). Multitemporal SAR images can be merged to map the progression of flooding over time and indicate the maximum flood extent. Other than monitoring flooding issues, fusion of RADARSAT and ENVISAT SAR images were also used for oil spill detection (Solberg et al., 2007). However, vegetation structure could affect the flood mapping. Two types of methods used to infer vegetation structure may be classified based on SAR images. The first type of method places the emphasis on SAR backscattering returns from single-polarization or multiple-polarization SAR images in order to properly infer vegetation structure and attributes (Saatchi et al., 2007). In this context, the relationship between radar backscattering and vegetation structure may be determined by using ground truth and optical images. The second type of method depends on the InSAR products associated with both phase and coherence images to delicately estimate vegetation structure (Crawford 1999; Slatton et al., 2001; Simard et al., 2006; Walker et al., 2007). Lu et al. (2010) further discussed fusion ideas across different radar images and data fusion technologies for natural hazards characterization such as flood mapping. As Lu et al. (2010) pointed out, during-event SAR images can be fused with pre-event optical images to produce a flood inundation map even within the vegetated area based on the "double bounce" effect driven by the canopy structure, resulting in very high backscattering coefficients. D'Addabbo et al. (2014) exploited a multitemporal stack of SAR intensity images for flood detection when taking into account different spectral signatures in association with inundation conditions. They showed that in the discrimination process, the use of interferometric coherence information can further help to improve flood detection. Hence, based on Bayesian networks, a new data fusion approach was developed to analyze an inundation event with respect to the time series of COSMO-SkyMed stripmap SAR images. Xiao et al. (2016) explored the object-oriented fusion of RADARSAT-2 polarimetric SAR and HJ-1A multispectral data for land cover classification.

The fusion of SAR and LiDAR data has been well evaluated through a wide variety of environmental applications such as sea ice monitoring (Barber and LeDrew, 1991), shoreline mapping and change detection (Tuell, 1998; Cook, 2003), and coastal water (Brzank and Heipke, 2006). In addition, both L-band and C-band InSAR imageries can help to measure relative surface water-level changes in river corridors, reservoirs, and wetlands (Alsdorf et al., 2000; Lu and Kwoun, 2008). They concluded that PALSAR can provide better results than RADARSAT-2 data with the aid of DWT. Parallel to this advancement, investigating SAR-LiDAR synergy triggered more efforts in forest monitoring (Hyde et al., 2006, 2007; Nelson et al., 2007). When a laser pulse hits the forest canopy, it can penetrate the canopy and produce more returns that may be used to identify the tree canopy structure (e.g., tree, shrub, grass, building, etc.). Feature-level fusion based on different image sources, such as SAR, LiDAR, and optical imageries, followed by a classification step, can be a good practice for target plant species identification. However, classification has to depend on the binary segmentation and supervised labeling sequentially to support the final land classification mapping. One example in the literature is the work done by Kellndorfer et al. (2010), which presented a statistical fusion of LiDAR, InSAR, and optical remote sensing data for forest stand height characterization. The recent focus in fusing SAR and LiDAR images is the retrieval of the aboveground biomass, which is a primary variable related to the carbon cycle in land ecosystems (Montesano et al., 2014). Kaasalainen et al. (2015) provided a thorough literature review of all of the state-of-the-art systems for aboveground biomass estimation. Recent major advancements taking advantage of various SAR or InSAR data fusion, feature extraction, and machine learning toward different environmental applications are summarized in Table 8.3.

8.4 SUMMARY

In this chapter, the basic concepts and principles of data fusion with different technology hubs were introduced and discussed systematically with respect to image, information, and data fusion in varying contexts. In general, image or data fusion is a synoptic term that refers to all types of possible fusion processes in remote sensing science, regardless of data types. Image fusion is one specific type of data fusion in which all input data to the fusion process are provided in the form of imageries. In remote sensing, most data fusion processes refer to image fusion, as remotely sensed data are commonly achieved in the format of digital imageries. Nevertheless, image fusion does not include all kinds of data fusion processes. New types of remote sensing systems like LiDAR provide data of the observed scene in the form of discrete data points. Hence, the fusion process of LiDAR data with other data should be referred to as data fusion instead of image fusion. As the data sources turn out to be diversified, information fusion can help and justify the complexity of the fusion process. With

TABLE 8.3

Recent Advances in Applying Microwave Data Fusion Methods for Environmental Applications

Reference	Fusion Method	Data Source	Objective	Enhanced Properties
Land Deformation, Land Cover, and Landscape Changes				
Holopainen et al. (2010)	GA and the nonparametric k-nearest neighbour (k-NN) algorithm	Airborne Laser Scanning and TerraSAR-X Radar Images	To improve the measurements of forest	Spatial Spectral
Zhang et al. (2010)	Block-regression-based fusion	Airborne X-Band SAR image and multispectral image	To improve land classification	Spatial Spectral
Kellndorfer et al. (2010)	Statistical fusion method	Airborne LVIS LIDAR, SRTM, Landsat ETM+, and ancillary data	To improve forest stand height characterization	Spatial Spectral
Lu et al. (2011)	PCA, DWT, HPF, and NMM	ALOS PALSAR and RADARSAT-2 data with Landsat	To improve the monitoring of impervious land cover in urban regions	Spatial Spectral
Reiche et al. (2013)	The MulTiFuse approach	ALOS PALSAR and Landsat data	To improve the monitoring and mapping of tropical forest degradation	
Montesano et al. (2014)	Single variable empirical models	LIDAR and SAR	To improve the measurements of forest structure gradient	Spatial Spectral
Reiche et al. (2015)	The MulTiFuse approach	Landsat NDVI and ALOS PALSAR L-band	To improve the monitoring of deforestation	Spatial Spectral Temporal
Abdikan et al. (2015)	Adjustable SAR-MS Fusion, Ehlers Fusion, High Pass Filtering, and Bayesian Data Fusion	X-band dual polarized TerraSAR-X and multispectral RapidEye data	To improve land use classification and change detection	Spatial Spectral
Shokrollahi and Ebadi (2016)	Feature and decision level fusion	PolSAR and hyperspectral images	To generate better land use and land cover map	Spatial Spectral
Xiao et al. (2016)	Hybrid feature selection method that combined the ReliefF filter approach and the genetic algorithm wrapper approach	RADARSAT-2 Fine Quad-Pol image and an HJ-1A CCD2 multispectral image	To improve land cover classification	Spatial Spectral
Water Environment				
Vespe et al. (2011)	ANN	SAR, SeaWinds on QuikSCAT, AVHRR PathFinder V5 SST product, MODIS-AQUA Chlorophyll-A	To enhance oil spill detection	Spatial Spectral

(Continued)

TABLE 8.3 (*Continued*)

Recent Advances in Applying Microwave Data Fusion Methods for Environmental Applications

Reference	Fusion Method	Data Source	Objective	Enhanced Properties
D'Addabbo et al. (2014)	Bayesian Networks for SAR/InSAR data fusion	Time series of COSMO-SkyMed stripmap SAR images	To enhance inundation detection in flooding events	Spatial Spectral Temporal
Liu et al. (2015)	PCA	Optical data by HIS transform and SAR data by wavelet transform	To enhance sea ice detection	Spatial Spectral

Note: LIDAR: Light detection and ranging; LVIS: Laser Vegetation Imaging Sensor; SRTM: Shuttle Radar Topography Mission; SAR: Synthetic aperture radar; PolSAR: Polarimetric SAR; PCA: principal component analysis; DWT: discrete wavelet transform; HPF: high-pass filter; NMM: normalized multiplication; GA: genetic algorithm; ANN: Artificial neural network; SST: sea surface temperature.

advancements in remote sensing, various new types of primary and secondary data with different specifications are available to the user community, such as hyperspectral and microwave imageries (e.g., SAR, InSAR, and PolInSAR), in addition to commonly used optical images. Therefore, developing more image/data fusion algorithms to blend different features from each individual data source into a spatially, temporally, and spectrally advantageous dataset is required in the future.

REFERENCES

Abdikan, S., Bilgin, G., Sanli, F. B., Uslu, E., and Ustuner, M., 2015. Enhancing land use classification with fusing dual-polarized TerraSAR-X and multispectral RapidEye data. *Journal of Applied Remote Sensing*, 9, 096054.

Abdikan, S., Sanli, F. B., Filiz, B. B., and Goksel, C. 2008. Fusion of SAR images (PALSAR and RADARSAT-1) with multispectral spot image: A comparative analysis of resulting images. *The International Archives of the Photogrammetry Remote Sensing and Spatial Information Sciences*, 37, 1197–1202.

Ahtonen, P., Ahton, P., Euro, M., Hallikainen, M., Solbø, S., Johansen, B., and Solheim, I. 2004. SAR and optical based algorithms for estimation of water bodies. Technical report, FloodMan Project.

Aiazzi, B., Alparone, L., Baronti, S., and Garzelli, A., 2002. Context driven fusion of high spatial and spectral resolution images based on oversampled multiresolution analysis. *IEEE Transactions on Geoscience and Remote Sensing*, 40, 2300–2312.

Alparone, L., Baronti, S., Garzelli, A., and Nencini, F., 2004. Landsat ETM+ and SAR image fusion based on generalized intensity modulation. *IEEE Transactions on Geoscience and Remote Sensing*, 42, 2832–2839.

Alsdorf, D. E., Melack, J. M., Dunne, T., Mertes, L. A. K., Hess, L. L., and Smith, L. C., 2000. Interferometric radar measurements of water-level changes on the Amazon floodplain. *Nature*, 404, 174–177.

Amarsaikhan, D. and Douglas, T., 2004. Data fusion and multisource image classification. *International Journal of Remote Sensing*, 25, 3529–3539.

Amarsaikhan, D., Blotevogel, H. H., van Genderen, J. L., Ganzorig, M., Gantuya, R., and Nergui, B., 2010. Fusing high-resolution SAR and optical imagery for improved urban land cover study and classification. *International Journal of Image and Data Fusion*, 1, 83–97.

Ardeshir Goshtasby, A. and Nikolov, S., 2007. Image fusion: Advances in the state of the art. *Information Fusion*, 8, 114–118.

Barber, D. G. and LeDrew, E. F., 1991. SAR sea ice discrimination using texture statistic: A multivariate approach. *Photogrammetric Engineering & Remote Sensing*, 57, 385–395.

Benediktsson, J. A. and Kanellopoulos, I., 1999. Classification of multisource and hyperspectral data based on decision fusion. *IEEE Transactions on Geoscience and Remote Sensing*, 37, 1367–1377.

Benediktsson, J. A., Pesaresi, M., and Arnason, K., 2003. Classification and feature extraction for remote sensing images from urban areas based on morphological transformations. *IEEE Transactions on Geoscience and Remote Sensing*, 41, 1940–1949.

Berrocal, V. J., Gelfand, A. E., and Holland, D. M., 2012. Space-time data fusion under error in computer model output: An application to modeling air quality. *Biometrics*, 68, 837–848.

Bigdeli, B., Samadzadegan, F., and Reinartz, P., 2014. A decision fusion method based on multiple support vector machine system for fusion of hyperspectral and LIDAR data. *International Journal of Image and Data Fusion*, 5, 196–209.

Bloch, I., 1996. Information combination operators for data fusion: A comparative review with classification. *IEEE Transactions on Systems, Man, and Cybernetics: Systems*, 26, 52–67.

Bourgeau-Chavez, L. L., Brunzell, S., Nolan, M., and Hyer, E. d., 2001. Analysis of SAR data for fire danger prediction in boreal Alaska. Final Report. ASF-IARC Grant NAS-98-129, 59.

Briem, G. J., Benediktsson, J. A., and Sveinsson, J. R., 2002. Multiple classifiers applied to multisource remote sensing data. *IEEE Transactions on Geoscience and Remote Sensing*, 40, 2291–2299.

Bruzzone, L., Prieto, D., and Serpico, S., 1999. A neural-statistical approach to multitemporal and multisource remote-sensing image classification. *IEEE Transactions on Geosciences and Remote Sensing*, 37, 1350–1359.

Brzank, A. and Heipke, C., 2006. Classification of LIDAR data into water and land points in coastal areas. In: *The International Archives of Photogrammetry, Remote Sensing, and Spatial Information Sciences*, Bonn, Germany, 197–202.

Bwangoy, J. R. B., Hansen, M. C., Roy, D. P., De Grandi, G., and Justice, C. O., 2010. Wetland mapping in the Congo Basin using optical and radar remotely sensed data and derived topographical indices. *Remote Sensing of Environment*, 114, 73–86.

Carrara, P., Bordogna, G., Boschetti, M., Brivio, P., Nelson, A., and Stroppiana, D., 2008. A flexible multi-source spatial-data fusion system for environmental status assessment at continental scale. *International Journal of Geographical Information Science*, 22, 781–799.

Chang, N. B., Bai, K., and Chen, C. F., 2015. Smart information reconstruction via time-space-spectrum continuum for cloud removal in satellite images. *IEEE Journal of Selected Topics in Applied Earth Observations and Remote Sensing*, 8, 1898–1912.

Chang, N.-B., Bai, K., Imen, S., Chen, C.-F., and Gao, W., 2016. Multisensor satellite image fusion and networking for all-weather environmental monitoring. *IEEE Systems Journal*, 1–17.

Chang, N.-B., Vannah, B., and Yang, Y. J., 2014a. Comparative Sensor Fusion Between Hyperspectral and Multispectral Satellite Sensors for Monitoring Microcystin Distribution in Lake Erie. *IEEE Journal of Selected Topics in Applied Earth Observations and Remote Sensing*, 7, 2426–2442.

Chang, N.-B., Vannah, B. W., Yang, Y. J., and Elovitz, M., 2014b. Integrated data fusion and mining techniques for monitoring total organic carbon concentrations in a lake. *International Journal of Remote Sensing*, 35, 1064–1093.

Chanussot, J., Mauris, G., and Lambert, P., 1999. Fuzzy fusion techniques for linear features detection in multitemporal SAR images. *IEEE Transactions on Geoscience and Remote Sensing*, 37, 2287–2297.

Chavez, P. S., Sides, S. C., and Anderson, J. A., 1991. Comparison of three different methods to merge multi resolution and multi-spectral data: Landsat TM and SPOT panchromatic. *Photogrammetric Engineering & Remote Sensing*, 57, 295–303.

Chen, B., Huang, B., and Xu, B., 2017. Multi-source remotely sensed data fusion for improving land cover classification. *ISPRS Journal of Photogrammetry and Remote Sensing*, 124, 27–39.

Chibani, Y., 2006. Additive integration of SAR features into multispectral SPOT images by means of the à trous wavelet decomposition. *ISPRS Journal of Photogrammetry and Remote Sensing*, 60, 306–314.

Choi, M., Kim, R. Y., Nam, M. R., and Kim, H., 2005. Fusion of multispectral and panchromatic satellite images using the curvelet transform. *IEEE Transactions on Geoscience and Remote Sensing*, 2, 136–140.

Coe, M., 1998. A linked global model of terrestrial hydrologic processes: simulation of the modern rivers, lakes, and wetlands. *Journal of Geophysical Research*, 103, 8885–8899.

Cook, G., 2003. Evaluating LIDAR for documenting shoreline change. In: *Proceedings of 3rd Biennial Coastal GeoTools Conference*, South Carolina, USA.

Crawford, M., 1999. Fusion of airborne polarimetric and interferometric SAR for classification of coastal environments. *IEEE Transactions on Geoscience and Remote Sensing*, 37, 1306–1313.

D'Addabbo, A., Refice, A., and Pasquariello, G., 2014. A Bayesian network approach to perform SAR/InSAR data fusion in a flood detection problem. In: *Procedings of SPIE, Vol. 9244, Image and Signal Processing for Remote Sensing*, Amsterdam, Netherlands, L. Bruzzone (Ed.), 92441A.

Dalla Mura, M., Villa, A., Benediktsson, J. A., Chanussot, J., and Bruzzone, L., 2011. Classification of hyperspectral images by using extended morphological attribute profiles and independent component analysis. *IEEE Geoscience and Remote Sensing Letters*, 8, 542–546.

Dalponte, M., Bruzzone, L., Gianelle, D., and Member, S. S., 2008. Fusion of Hyperspectral and LIDAR Remote Sensing Data for Classification of Complex Forest Areas. *IEEE Transactions on Geoscience and Remote Sensing*, 46, 1416–1427.

Dalponte, M., Bruzzone, L., and Gianelle, D., 2012. Tree species classification in the Southern Alps based on the fusion of very high geometrical resolution multispectral/hyperspectral images and LiDAR data. *Remote Sensing of Environment*, 123, 258–270.

Dey, C., Jia, X., and Fraser, D., 2008. Decision fusion for reliable flood mapping using remote sensing images. *Proceedings of Digital Image Computing: Techniques and Applications*, 5, 184–190.

Doña, C., Chang, N.-B., Caselles, V., Sánchez, J. M., Camacho, A., Delegido, J., and Vannah, B. W., 2015. Integrated satellite data fusion and mining for monitoring lake water quality status of the Albufera de Valencia in Spain. *Journal of Environmental Management*, 151, 416–426.

Du, Q., 2009. Decision fusion for classifying hyperspectral imagery with high spatial resolution. *SPIE Newsroom*, 1–3.

Du, Y., Zhang, Y., Ling, F., Wang, Q., Li, W., and Li, X., 2016. Water Bodies' Mapping from Sentinel-2 Imagery with Modified Normalized Difference Water Index at 10-m Spatial Resolution Produced by Sharpening the SWIR Band. *Remote Sensing*, 8, 354.

Duda, R. O. and Hart, P. E., 1973. *Pattern Classification and Scene Analysis*. New York: Wiley.

Durrant-Whyte, H. F., 1988. Sensor Models and Multisensor Integration. *The International Journal of Robotics Research*, 7, 97–113.

Elmenreich, W., 2002. *Sensor Fusion in Time-Triggered Systems*. Vienna University of Technology, Wein, Austria.

Fasbender, D., Brasseur, O., and Bogaert, P., 2009. Bayesian data fusion for space-time prediction of air pollutants: The case of NO_2 in belgium. *Atmospheric Environment*, 43, 4632–4645.

Fasbender, D., Radoux, J., and Bogaert, P., 2008. Bayesian data fusion for adaptable image pan sharpening. *IEEE Transactions on Geoscience and Remote Sensing*, 46, 1847–1857.

Fauvel, M., Benediktsson, J. A., Chanussot, J., and Sveinsson, J. R., 2008. Spectral and spatial classification of hyperspectral data using SVMs and morphological profiles. *IEEE Transactions on Geoscience and Remote Sensing*, 46, 3804–3814.

Fok, H. S., 2015. Data fusion of multisatellite altimetry for ocean tides modelling: A spatio-temporal approach with potential oceanographic applications. *International Journal of Image and Data Fusion*, 6, 232–248.

Fu, D., Chen, B., Wang, J., Zhu, X., and Hilker, T., 2013. An improved image fusion approach based on enhanced spatial and temporal the adaptive reflectance fusion model. *Remote Sensing*, 5, 6346–6360.

Gamba, P. and Houshmand, B., 2002. Joint analysis of SAR, LIDAR and aerial imagery for simultaneous extraction of land cover, DTM and 3D shape of buildings. *International Journal of Remote Sensing*, 23, 4439–4450.

Gao, F., Masek, J., Schwaller, M., and Hall, F., 2006. On the blending of the Landsat and MODIS surface reflectance: Predicting daily Landsat surface reflectance. *IEEE Transactions on Geoscience and Remote Sensing*, 44, 2207–2218.

Ghosh, A., Fassnacht, F. E., Joshi, P. K., and Koch, B. A., 2014. Framework for mapping tree species combining hyperspectral and LiDAR data: Role of selected classifiers and sensor across three spatial scales. *International Journal of Applied Earth Observation and Geoinformation*, 26, 49–63.

Goodenough, D. G., Chen, H., Dyk, A., and Han, T., 2005. Evaluation of multi-temporal and multi-polarization ASAR for boreal forests in Hinton. In: *Analysis of Multi-Temporal Remote Sensing Images*, Mississippi, USA.

Goodenough, D. G., Chen, H., Dyk, A., Richardson, A., and Hobart, G., 2008. Data fusion study between polarimetric SAR, hyperspectral and LiDAR data for forest information. In: *IEEE Geoscience and Remote Sensing Symposium*, 7–11 July 2008, Boston, MA, USA.

Granville, V. and Rasson, J. P., 1995. Multivariate discriminant analysis and maximum penalized likelihood density estimation. *Journal of the Royal Statistical Society*, 57(3), 501–517.

Guo, Y., Li, Y. Y., Zhu, L., Wang, Q., Lv, H., Huang, C., and Li, Y. Y., 2016. An Inversion-Based Fusion Method for Inland Water Remote Monitoring. *IEEE Journal of Selected Topics in Applied Earth Observations and Remote Sensing*, 9, 5599–5611.

Haack, B. N., Solomon, E. K., Bechdol, M. A., and Herold, N. D., 2002. Radar and optical data comparison/integration for urban delineation: A case study. *Photogrammetric Engineering and Remote Sensing*, 68, 1289–1296.

Haghighat, M. B. A., Aghagolzadeh, A., and Seyedarabi, H., 2011. Multi-focus image fusion for visual sensor networks in DCT domain. *Computers & Electrical Engineering*, 37, 789–797.

Hall, D. L. and Llinas, J., 1997. An introduction to multisensor data fusion. *Proceedings of the IEEE*, 85, 6–23.

Hardie, R. C., Eismann, M. T., and Wilson, G. L., 2004. MAP estimation for hyperspectral image resolution enhancement using an auxiliary sensor. *IEEE Transactions on Image Processing*, 13, 1174–1184.

Hazaymeh, K. and Hassan, Q. K., 2015. Fusion of MODIS and Landsat-8 surface temperature images: A new approach. *PLoS One*, 10, 1–13.

Hess, L. L., Melack, J. M., Filoso, S., and Wang, Y., 1995. Delineation of inundated area and vegetation along the Amazon floodplain with the SIR-C synthetic aperture radar. *IEEE Transactions on Geoscience and Remote Sensing*, 33, 896–904.

Hilker, T., Wulder, M. A., Coops, N. C., Linke, J., McDermid, G., Masek, J. G., Gao, F., and White, J. C., 2009. A new data fusion model for high spatial- and temporal-resolution mapping of forest disturbance based on Landsat and MODIS. *Remote Sensing of Environment*, 113, 1613–1627.

Holopainen, M., Haapanen, R., Karjalainen, M., Vastaranta, M., Hyyppä, J., Yu, X., Tuominen, S., and Hyyppä, H., 2010. Comparing accuracy of airborne laser scanning and TerraSAR-X radar images in the estimation of plot-level forest variables. *Remote Sensing*, 2, 432–445.

Hong, G., Zhang, Y., and Mercer, B., 2009. A wavelet and HIS integration method to fuse high resolution SAR and moderate resolution multispectral images. *Photogrammetric Engineering and Remote Sensing*, 75, 1–11.

Hu, J. and Li, S., 2011. Fusion of panchromatic and multispectral images using multiscale dual bilateral filter. In: *Proceedings of the 18th IEEE International Conference on Image Processing*, Brussels, Belgium, 1489–1492.

Huang, C., Chen, Y., Zhang, S., Li, L., Shi, K., and Liu, R., 2016. Surface Water Mapping from Suomi NPP-VIIRS Imagery at 30 m Resolution via Blending with Landsat Data. *Remote Sensing*, 8(8), 631, doi:10.3390/rs8080631

Huang, B. and Song, H., 2012. Spatiotemporal reflectance fusion via sparse representation. *IEEE Transactions on Geoscience and Remote Sensing*, 50, 3707–3716.

Huang, B., Wang, J., Song, H., Fu, D., and Wong, K., 2013. Generating high spatiotemporal resolution land surface temperature for urban heat island monitoring. *IEEE Geoscience and Remote Sensing Letters*, 10, 1011–1015.

Huang, X. and Zhang, L., 2010. Comparison of vector stacking, multi-SVMs fuzzy output, and multi-SVMs voting methods for multiscale VHR urban mapping. *IEEE Geoscience and Remote Sensing Letters*, 7, 261–265.

Huang, X. and Zhang, L., 2012. A multilevel decision fusion approach for urban mapping using very high-resolution multi/hyperspectral imagery. *International Journal of Remote Sensing*, 33, 3354–3372.

Huang, X., Zhang, L., and Gong, W., 2011. Information fusion of aerial images and LIDAR data in urban areas: Vector-stacking, re-classification and post-processing approaches. *International Journal of Remote Sensing*, 32, 69–84.

Huang, X., Zhang, L., and Wang, L., 2009. Evaluation of morphological texture features for mangrove forest mapping and species discrimination using multispectral IKONOS imagery. *IEEE Geoscience and Remote Sensing Letters*, 6, 393–397.

Hyde, P., Dubayah, R., Walker, W., Blair, J. B., Hofton, M., and Hunsaker, C., 2006. Mapping forest structure for wildlife habitat analysis using multi-sensor (LiDAR, SAR/InSAR, ETM+, Quickbird) synergy. *Remote Sensing of Environment*, 102, 63–73.

Hyde, P., Nelson, R., Kimes, D., and Levine, E., 2007. Exploring LiDAR–RaDAR synergy—Predicting aboveground biomass in a southwestern ponderosa pine forest using LiDAR, SAR and InSAR. *Remote Sensing of Environment*, 106, 28–38.

Imen, S., Chang, N.-B., and Yang, Y.J., 2015. Developing the remote sensing-based early warning system for monitoring TSS concentrations in Lake Mead. *Journal of Environmental Management*, 160, 73–89.

Jiang, C., Zhang, H., Shen, H., and Zhang, L., 2012. A practical compressed sensing based pan sharpening method. *IEEE Geoscience and Remote Sensing Letters*, 9, 629–633.

Jimenez, L. O., Morales-Morell, A., and Creus, A., 1999. Classification of hyperdimensional data based on feature and decision fusion approaches using projection pursuit, majority voting, and neural networks. *IEEE Transactions on Geoscience and Remote Sensing*, 37, 1360–1366.

Joshi, M. V. and Jalobeanu, A., 2010. MAP estimation for multiresolution fusion in remotely sensed images using an IGMRF prior model. *IEEE Transactions on Geoscience and Remote Sensing*, 48, 1245–1255.

Jouan, A. and Allard, Y., 2004. Land use mapping with evidential fusion of features extracted from polarimetric synthetic aperture radar and hyperspectral imagery. *Information Fusion*, 5, 251–267.

Kaasalainen, S., Holopainen, M., Karjalainen, M., Vastaranta, M., Kankare, V., Karila, K., and Osmanoglu, B., 2015. Combining LiDAR and synthetic aperture radar data to estimate forest biomass: Status and prospects. *Forests*, 6, 252–270.

Kellndorfer, J. M., Walker, W. S., LaPoint, E., Kirsch, K., Bishop, J., and Fiske, G., 2010. Statistical fusion of LiDAR, InSAR, and optical remote sensing data for forest stand height characterization: A regional-scale method based on LVIS, SRTM, Landsat ETM+, and ancillary data sets. *Journal of Geophysical Research: Biogeosciences*, 115, G00E08.

Kim, J. W., Lu, Z., Lee, H., Shum, C. K., Swarzenski, C. M., Doyle, T. W., and Baek, S. H., 2009. Integrated analysis of PALSAR/Radarsat-1 InSAR and ENVISAT altimeter for mapping of absolute water level changes in Louisiana wetland. *Remote Sensing of Environment*, 113, 2356–2365.

Leckei, D. G., 1990. Synergism of synthetic aperture radar and visible/infrared data for forest type discrimination. *Photogrammetric Engineering & Remote Sensing*, 56, 1237–1246.

Lee, D. D. and Seung, H. S., 1999. Learning the parts of objects by non-negative matrix factorization. *Nature*, 401, 788–791.

Lee, H., Shum, C. K., Yi, Y., Ibaraki, M., Kim, J.-W., Braun, A., Kuo, C.-Y., and Lu, Z., 2009. Louisiana wetland water level monitoring using retracked TOPEX/POSEIDON altimetry. *Marine Geodesy*, 32, 284–302.

Li, A., Bo, Y., Zhu, Y., Guo, P., Bi, J., and He, Y., 2013. Blending multi-resolution satellite sea surface temperature (SST) products using Bayesian maximum entropy method. *Remote Sensing of Environment*, 135, 52–63.

Li, H., Manjunath, B. S., and Mitra, S. K., 1995. Multisensor image fusion using the wavelet transform. *Graphical Models and Image Processing*, 27, 235–244.

Li, Z. and Leung, H., 2009. Fusion of multispectral and panchromatic images using a restoration based method. *IEEE Transactions on Geoscience and Remote Sensing*, 47, 1482–1491.

Liu, M., Dai, Y., Zhang, J., Zhang, X., Meng, J., and Xie, Q., 2015. PCA-based sea-ice image fusion of optical data by HIS transform and SAR data by wavelet transform. *Acta Oceanologica Sinica*, 34, 59–67.

Liu, D. and Li, J., 2016. Data field modeling and spectral-spatial feature fusion for hyperspectral data classification. *Sensor*, 16, 2146–2162.

Liu, H. and Weng, Q., 2012. Enhancing temporal resolution of satellite imagery for public health *studies*: A case study of west nile virus outbreak in los angeles in 2007. *Remote Sensing of Environment*, 117, 57–71.

Lu, Z., Dzurisin, D., Jung, H.-S., Zhang, J., and Zhang, Y., 2010. Radar image and data fusion for natural hazards characterisation. *International Journal of Image and Data Fusion*, 1, 217–242.

Lu, D., Li, G., Moran, E., Batistella, M., and Freitas, C., 2011. Mapping impervious surfaces with the integrated use of landsat thematic mapper and data: A case study in an urban–rural landscape in the brazilian amazon. *ISPRS Journal of Photogrammetry and Remote Sensing*, 66, 798–808.

Lu, Z., Kim, J. W., Lee, H., Shum, C. K., Duan, J., Ibaraki, M., Akyilmaz, O., and Read, C. H., 2009. Helmand river hydrologic studies using ALOS PALSAR InSAR and ENVISAT altimetry. *Marine Geodesy*, 32, 320–333.

Lu, Z. and Kwoun, O., 2008. RADARSAT-1 and ERS interferometric analysis over southeastern coastal Louisiana: Implication for mapping water-level changes beneath swamp forests. *IEEE Transactions on Geoscience and Remote Sensing*, 46, 2167–2184.

Lu, L., Xie, W., Zhang, J., Huang, G., Li, Q., and Zhao, Z., 2015. Woodland extraction from high-resolution CASMSAR data based on dempster-shafer evidence theory fusion. *Remote Sensing*, 7, 4068–4091.

Man, Q., Dong, P., and Guo, H., 2015. Pixel- and feature-level fusion of hyperspectral and LiDAR data for urban land-use classification. *International Journal of Remote Sensing*, 36, 1618–1644.

Mason, C. D., Speck, R., Devereux, B., Schumann, G., Neal, J., and Bates, P. D., 2010. Flood detection in urban areas using TerraSAR-X. *IEEE Transactions on Geoscience and Remote Sensing*, 48, 882–894.

Mattern, S., Raouafi, W., Bogaert, P., Fasbender, D., and Vanclooster, M., 2012. Bayesian Data Fusion (BDF) of monitoring data with a statistical groundwater contamination model to map groundwater quality at the regional scale. *Journal of Water Resource and Protection*, 4, 929–943.

Montesano, P. M., Nelson, R. F., Dubayah, R. O., Sun, G., Cook, B. D., Ranson, K. J. R., Næsset, E., and Kharuk, V., 2014. The uncertainty of biomass estimates from LiDAR and SAR across a boreal forest structure gradient. *Remote Sensing of Environment*, 154, 398–407.

Multi-Sensor Fusion. Aerosp. Def. URL http://www.hcltech.com/aerospace-and-defense/enhanced-vision-system (accessed 1.1.15).

Nelson, R. F., Hyde, P., Johnson, P., Emessiene, B., Imhoff, M. L., Campbell, R., and Edwards, W., 2007. Investigating RaDAR-LiDAR synergy in a North Carolina pine forest. *Remote Sensing of Environment*, 110, 98–108.

Niemann, K. O., Frazer, G., Loos, R., Visintini, F., and Stephen, R., 2007. Integration of first and last return LiDAR with hyperspectral data to characterize forested environments. In: *IEEE International Geoscience and Remote Sensing Symposium*, 23–28 July 2007, Barcelona, Spain.

Nunez, J., Otazu, X., Fors, O., Prades, A., Pala, V., and Arbiol, R., 1999. Multiresolution based image fusion with additive wavelet decomposition. *IEEE Transactions on Geoscience and Remote Sensing*, 37, 1204–1211.

Otazu, X., Gonzalez-Audicana, M., Fors, O., and Nunez, J., 2005. Introduction of sensor spectral response into image fusion methods. *Application to wavelet-based methods. IEEE Transactions on Geoscience and Remote Sensing*, 43, 2376–2385.

Pan, Z., Glennie, C. L., Fernandez-Diaz, J. C., Legleiter, C. J., Overstreet, B., 2016. Fusion of LiDAR orthowaveforms and hyperspectral imagery for shallow river bathymetry and turbidity sstimation. *IEEE Transactions on Geoscience and Remote Sensing*, 54, 4165–4177.

Pedergnana, M., Marpu, P. R., Dalla Mura, M., Benediktsson, J. A., and Bruzzone, L., 2012. Classification of remote sensing optical and LiDAR data using extended attribute profiles. *IEEE Journal of Selected Topics in Signal Processing*, 6, 856–865.

Pellemans, A. H. J. M., Jordans, R. W. L., and Allewijn, R., 1993. Merging multispectral and panchromatic SPOT images with respect to the radiometric properties of the sensor, *Photogrammetric Engineering & Remote Sensing*, 59, 81–87.

Pesaresi, M. and Benediktsson, J. A., 2001. A new approach for the morpho-logical segmentation of high-resolution satellite imagery. *IEEE Transactions on Geoscience and Remote Sensing*, 39, 309–320.

Petrakos, M., Benediktsson, J. A., and Kanellopoulos, I., 2001. The effect of classifier agreement on the accuracy of the combined classifier in decision level fusion. *IEEE Transactions on Geoscience and Remote Sensing*, 39, 2539–2546.

Pohl, C. and Van Genderen, J. L., 1998. Multisensor image fusion in remote sensing: Concepts, methods and applications. *International Journal of Remote Sensing*, 19, 823–854.

Pradhan, B., Jebur, M. N., Shafri, H. Z. M., and Tehrany, M. S., 2016. Data fusion technique using wavelet transform and taguchi methods for automatic landslide detection from airborne laser scanning data and QuickBird satellite imagery. *IEEE Transactions on Geoscience and Remote Sensing*, 54, 1610–1622.

Puttaswamy, S.J., Nguyen, H. M., Braverman, A., Hu, X., and Liu, Y., 2013. Statistical data fusion of multi-sensor AOD over the Continental United States. *Geocarto International*, 29, 48–64.

Ramsey, E., Rangoonwala, A., Middleton B., and Lu, Z., 2009. Satellite optical and radar data used to track wetland forest impact and short-term recovery from Hurricane Katrina. *Wetlands*, 29, 66–79.

Ranchin, T., Aiazzi, B., Alparone, L., Baronti, S., and Wald, L., 2003. Image fusion—the ARSIS concept and some successful implementation schemes. *ISPRS Journal of Photogrammetry and Remote Sensing*, 58, 4–18.

Ranchin, T. and Wald, L., 2000. Fusion of high spatial and spectral resolution images: The ARSIS concept and its implementation. *Photogrammetric Engineering & Remote Sensing*, 66, 49–61.

Reiche, J., Souza, C. M., Hoekman, D. H., Verbesselt, J., Persaud, H., and Herold, M., 2013. Feature level fusion of multi-temporal ALOS PALSAR and landsat data for mapping and monitoring of tropical deforestation and forest degradation. *IEEE Journal of Selected Topics in Applied Earth Observations and Remote Sensing*, 6, 2159–2173.

Reiche, J., Verbesselt, J., Hoekman, D., and Herold, M., 2015. Fusing Landsat and SAR time series to detect deforestation in the tropics. *Remote Sensing of Environment*, 156, 276–293.

Ross, A. A. and Govindarajan, R., 2005. Feature level fusion of hand and face biometrics. In: *Proceedings of SPIE*, the International Society for Optical Engineering, 196–204.

Saatchi, S., Halligan, K., Despain, D. G., and Crabtree, R. L., 2007. Estimation of forest fuel load from radar remote sensing. *IEEE Transactions on Geoscience and Remote Sensing*, 45, 1726–1740.

Schmidt, M., Lucas, R., Bunting, P., Verbesselt, J., and Armston, J., 2015. Multi-resolution time series imagery for forest disturbance and regrowth monitoring in Queensland, Australia. *Remote Sensing of Environment*, 158, 156–168.

Schumann, G., Bates, P. D., Horritt, M. S., Matgen, P., and Pappenberger, F., 2009. Progress in integration of remote sensing–derived flood extent and stage data and hydraulic models. *Review of Geophysics*, 47, RG4001.

Serpico, S. and Roli, F., 1995. Classification of multisensory remote-sensing images by structured neural networks. *IEEE Transactions on Geoscience, and Remote Sensing*, 33, 562–578.

Shackelford, A. K. and Davis, C. H., 2003. A hierarchical fuzzy classification approach for high-resolution multispectral data over urban areas. *IEEE Transactions on Geoscience and Remote Sensing*, 41, 1920–1932.

Shah, V. P., Younan, N. H., and King, R. L., 2008. An efficient pan-sharpening method via a combined adaptive PCA approach and contourlets. *IEEE Transactions on Geoscience and Remote Sensing*, 46, 1323–1335.

Shi, Y., Zhou, X., Yang, X., Shi, L., and Ma, S., 2015. Merging satellite ocean color data with bayesian maximum entropy method. *IEEE Journal of Selected Topics in Applied Earth Observations and Remote Sensing*, 8, 3294–3304.

Shi, W., Zhu, C., Tian, Y., and Nichol, J., 2005. Wavelet-based image fusion and assessment, *International Journal of Applied Earth Observation and Geoinformation*, 6, 241–251.

Shimoni, M., Borghys, D., Heremans, R., Perneel, C., and Acheroy, M., 2009, Fusion of PolSAR and PolInSAR data for land cover classification. *International Journal of Applied Earth Observation and Geoinformation*, 11, 169–180.

Shokrollahi, M. and Ebadi, H. J., 2016. Improving the accuracy of land cover classification using fusion of polarimetric SAR and hyperspectral images. *Journal of the Indian Society of Remote Sensing*, 44, 1017–1024.

Simard, M., Zhang, K., Rivera-Monroy, V. H., Ross, M. S., Ruiz, P. L., Castañeda-Moya, E., Twilley, R. R., and Rodriguez, E., 2006. Mapping height and biomass of mangrove forests in the everglades national park with SRTM elevation data. *Photogrammetric Engineering and Remote Sensing*, 7, 299–312.

Slatton, K., Crawford, M., and Evan, B., 2001. Fusing interferometric radar and laser altimeter data to estimate surface topography and vegetation heights. *IEEE Transactions on Geoscience and Remote Sensing*, 39, 2470–2482.

Solaiman, B., Pierce, L. E., and Ulaby, F. T., 1999. Multisensor data fusion using fuzzy concepts: Application to land-cover classification using ERS-1/JERS-1 SAR composites. *IEEE Transactions on Geoscience and Remote Sensing*, 37, 1316–1326.

Solberg, A., Brekke, C., and Husoy P. O., 2007. Oil spill detection in radarsat and envisat SAR images, *IEEE Transactions on Geoscience and Remote Sensing*, 45, 746–755.

Solberg, A., Jain, A., and Taxt, T., 1994. Multisource classification of remotely sensed data: Fusion of landsat TM and SAR images. *IEEE Transaction on Geoscience and Remote Sensing*, 32, 768–778.

Stavrakoudis, D. G., Dragozi, E., Gitas, I. Z., and Karydas, C. G., 2014. Decision fusion based on hyperspectral and multispectral satellite imagery for accurate forest species mapping. *Remote Sensing*, 6, 6897–6928.

Tewes, A., Thonfeld, F., Schmidt, M., Oomen, R. J., Zhu, X., Dubovyk, O., Menz, G., and Schellberg, J., 2015. Using rapidEye and MODIS data fusion to monitor vegetation dynamics in semi-arid rangelands in South Africa. *Remote Sensing*, 7, 6510–6534.

Tholey, N., Clandillon, S., and De Fraipont, P., 1997. The contribution of spaceborne SAR and optical data in monitoring flood events: Examples in Northern and Southern France. *Hydrological Processes*, 11, 1409–1413.

Townsend, P. A. 2002. Relationships between forest structure and the detection of flood inundation in forest wetlands using C-band SAR. *International Journal of Remote Sensing*, 23, 443–460.

Townsend, P. A. and Walsh, S. J., 1998. Modeling floodplain inundation using an integrated GIS with radar and optical remote sensing. *Geomorphology*, 21, 295–312.

Tuell, G. H., 1998. The use of high resolution synthetic aperture radar (SAR) for shoreline mapping. In: *The International Archives of Photogrammetry, Remote Sensing, and Spatial Information Sciences*, Columbus, Ohio, 592–611.

Upla, K. P., Gajjar, P. P., and Joshi, M. V., 2011. Multiresoltution fusion using contourlet transform based edge learning. In: *Procedings of IEEE International Geoscience and Remote Sensing Symptom*, Vancouver, BC, Canada, 523–526.

Van Genderen, J. L. and Pohl, C., 1994. Image fusion: Issues, techniques and applications. In: *Proceedings EARSeL Workshop of Intelligent Image Fusion*. Strasbourg, France, 18–26.

Vespe, M., Posada, M., Ferraro, G., and Greidanus, H., 2011. Data fusion of SAR derived features and ancillary information for automatic oil spill detection. *Fresenius Environmental Bulletin*, 20, 36–43.

Wald, L., 1999. Some terms of reference in data fusion. *IEEE Transactions on Geoscience and Remote Sensing*, 37, 1190–1193.

Walker, W., Kellndorfer, J., and Pierce, L., 2007. Quality assessment of SRTM C- and X-band interferometric data: Implications for the retrieval of vegetation canopy height. *Remote Sensing of Environment*, 106, 428–448.

Waltz, E., 2001. The principles and practice of image and spatial data fusion. In: Hall, D. L., Llinas, J. (Eds.), *Multisensor Data Fusion.* CRC Press, Boca Raton, FL, USA, 50–63.

Wang, H., Fan, T., Shi, A., Huang, F., and Wang, H., 2010. Fuzzy integral based information fusion for water quality monitoring using remote sensing data. *International Journal of Communications, Network and System Sciences,* 3, 737–744.

Waske, B. and Benediktsson, J. A., 2007. Fusion of support vector machines for classification of multisensor data. *IEEE Transactions on Geoscience and Remote Sensing,* 45, 3858–3866.

Wei, J., Wang, L., Liu, P., and Song, W., 2016. Spatiotemporal fusion of remote sensing images with structural sparsity and semi-coupled dictionary learning. *Remote Sensing,* 9, 21.

Weng, Q., Fu, P., and Gao, F., 2014. Generating daily land surface temperature at landsat resolution by fusing landsat and MODIS data. *Remote Sensing of Environment,* 145, 55–67.

Wu, M., Huang, W., Niu, Z., and Wang, C., 2015. Generating daily synthetic landsat imagery by combining landsat and MODIS data. *Sensors,* 15, 24002–24025.

Wu, M., Niu, Z., Wang, C., Wu, C., and Wang, L., 2012. Use of MODIS and landsat time series data to generate high-resolution temporal synthetic Landsat data using a spatial and temporal reflectance fusion model. *Journal of Applied Remote Sensing,* 6, 63507.

Xiao, Y., Jiang, Q., Wang, B., Li, Y., Liu, S., and Cui, C., 2016. Object-oriented fusion of RADARSAT-2 polarimetric synthetic aperture radar and HJ-1A multispectral data for land-cover classification. *Journal of Applied Remote Sensing,* 10, 026021.

Xie, D., Zhang, J., Zhu, X., Pan, Y., Liu, H., Yuan, Z., and Yun, Y., 2016. An Improved STARFM ith help of an unmixing-based method to generate high spatial and temporal resolution remote sensing data in complex heterogeneous regions. *Sensors,* 16, 207.

Xu, H., Guang, J., Xue, Y., de Leeuw, G., Che, Y. H., Guo, J., He, X. W., and Wang, T. K., 2015. A consistent aerosol optical depth (AOD) dataset over mainland China by integration of several AOD products. *Atmospheric Environment,* 114, 48–56.

Xu, Q., Qiu, W., Li, B., and Gao, F., 2017. Hyperspectral and panchromatic image fusion through an improved ratio enhancement. *Journal of Applied Remote Sensing,* 11, 015017.

Yang, G., Weng, Q., Pu, R., Gao, F., Sun, C., Li, H., and Zhao, C., 2016. Evaluation of ASTER-like daily land surface temperature by fusing ASTER and MODIS data during the HiWATER-MUSOEXE. *Remote Sensing,* 8, 75.

Yokoya, N., Chan, J., and Segl, K., 2016. Potential of Resolution-Enhanced Hyperspectral Data for Mineral Mapping Using Simulated EnMAP and Sentinel-2 Images. *Remote Sensing,* 8, 172.

Zhang, Y., 1999. Optimisation of building detection in satellite images by combining multispectral classification and texture filtering. *ISPRS Journal of Photogrammetry and Remote Sensing,* 54, 50–60.

Zhang, C., 2015. Applying data fusion techniques for benthic habitat mapping and monitoring in a coral reef ecosystem. *ISPRS Journal of Photogrammetry and Remote Sensing,* 104, 213–223.

Zhang, L., Huang, X., Huang, B., and Li, P.,2006. A pixel shape index coupled with spectral information for classification of high spatial resolution remotely sensed imagery. *IEEE Transactions on Geoscience and Remote Sensing,* 44, 2950–2961.

Zhang, L., Shen, H., Gong, W., and Zhang, H., 2012. Adjustable model-based fusion method for multispectral and panchromatic images. *IEEE Transactions on Systems, Man, and Cybernetics, Part B,* 42, 1693–1704.

Zhang, C. and Xie, Z., 2013. Data fusion and classifier ensemble techniques for vegetation mapping in the coastal Everglades, *Geocarto International,* 29, 228–243.

Zhang, J., Yang, J., Zhao, Z., Li, H., and Zhang, Y., 2010. Block-regression-based fusion of optical and SAR imagery for feature enhancement. *International Journal of Remote Sensing,* 31(9), 2325–2345.

Zhu, X. X. and Bamler, R., 2013. A sparse image fusion algorithm with application to pan sharpening. *IEEE Transactions on Geoscience and Remote Sensing,* 51, 2827–2836.

Zhu, X., Chen, J., Gao, F., Chen, X., and Masek, J. G., 2010. An enhanced spatial and temporal adaptive reflectance fusion model for complex heterogeneous regions. *Remote Sensing of Environment,* 114, 2610–2623.

Zitnik, M. and Zupan, B., 2015. Data Fusion by Matrix Factorization. *IEEE Transactions on Pattern Analysis and Machine Intelligence,* 37, 41–53.

Zurita-Milla, R., Clevers, J. G. P. W., and Schaepman, M. E., 2008. Unmixing-based landsat TM and MERIS FR data fusion. *IEEE Geoscience and Remote Sensing Letters,* 5, 453–457.

9 Major Techniques and Algorithms for Multisensor Data Fusion

9.1 INTRODUCTION

With the rapid advancement of space technologies and remote sensing tools/platforms such as the Earth Observing System in the National Aeronautics and Space Administration (NASA), a variety of remotely sensed data have become available in the Earth Observing System Data Information System (EOSDIS) from diversified operational satellites since the 1980s. Nevertheless, there exists a well-known trade-off between spatial, temporal, and spectral resolution of remotely sensed data in various remote sensing systems. In fact, it is always difficult to obtain one image measured by one individual instrument that has both high spatial resolution and high spectral resolution as well as high temporal resolution. For example, remotely sensed imagery from a satellite-based sensor like Landsat Thematic Mapper (TM) has a spatial resolution of 30-m with a temporal resolution of 16-days, while a Moderate Resolution Imaging Spectroradiometer (MODIS) image is 250-m (spatial resolution) on a daily (temporal resolution) basis. Thus, the remotely sensed imagery from TM has a higher spatial resolution than MODIS, but a lower temporal resolution (16-days versus 1-day). This is mainly due to competing constraints between the instantaneous field-of-view and signal-to-noise ratio limited by the detector instruments and satellite orbits.

In real-world applications, simply making use of one type of remotely sensed data may not fulfill all the requirements necessary to better monitor the changing Earth environment. For example, some satellite sensors providing multispectral images identify features spectrally rather than spatially, while other satellite sensors providing high spatial panchromatic images characterize features spatially instead of spectrally (Wang et al., 2005; Chen et al., 2008). These constraints significantly limit the broad exploration of valuable data sources in various applications, especially for those emergency response events that demand high spatial and high spectral (or high temporal) resolution simultaneously. Typically, there exist two kinds of approaches to cope with such constraints. One is to improve the capability of remote sensing instruments to make them competent for meeting the challenges of real-world applications; however, this is not only technically very difficult but also costly. The other is to blend or fuse distinct features retrieved from different data sources into one data set to enrich features of importance for complex real-world applications. Blending or fusing distinct features retrieved from different data sources turns out to be more practical for enhancing feature extraction.

Multiple observations with different characteristics from cross-mission sensors have provided unprecedented opportunities to collectively monitor the changing environment. Combining information from multiple data sources into a spatially and/or temporally regular dataset by using various modeling techniques to facilitate further data interpretation and analysis is considered a valuable tool and a common practice in remote sensing science. Regardless of whether these sensors were designed as a constellation in the beginning data fusion, or the process of blending information from multiple sources into a new data set, can be performed based on a variety of continuous data sources over the electromagnetic spectrum such as visible light, infrared light, and microwaves, as well as discrete data sources such as laser radar (e.g., LiDAR). If the input data sources for fusion are provided in terms of remotely sensed imagery, it can also be referred to as image fusion.

The basic concepts and relevant technology hubs of data fusion have been well elaborated in Chapter 8 through a thorough literature review.

Since the incipient stage of data fusion, a myriad of techniques and algorithms have been developed through various applications to advance the knowledge of data fusion. Due to different spatial, temporal, and spectral characteristics, simply using one algorithm to fuse all data sets is by no means a feasible solution. In principle, data fusion methods can be grouped into two classical categories. One is the spatiotemporal domain fusion, and the other is the transform domain fusion, each of which involves complex mathematical constructs to be used as instruments to model real-world events and to serve as an object of reasoning. The former may include Intensity-Hue-Saturation (IHS) transformation, Principal Component Analysis (PCA), and wavelet transforms; the latter can be further classified into different categories, including artificial intelligence-based fusion algorithms, probabilistic fusion algorithms, statistical fusion algorithms, and unmixing-based fusion. Some hybrid approaches exist to integrate the advantages from different data fusion methods. To improve the application potential, these recent data fusion algorithms may be grouped by function such as: (1) the enhancement of spatial resolution of multispectral and/or hyperspectral satellite imagery for terrestrial feature extraction (Chang et al., 2015), (2) the improvement of temporal resolution of satellite imagery for public health assessment (e.g., Liu and Weng, 2012), (3) the generation of advanced daily land surface temperature using extreme learning machine at Landsat resolution for climate change studies (Huang et al., 2013; Weng et al., 2014; Bai et al., 2015), (4) the optical fusion of RapidEye and MODIS imageries to explore vegetation dynamics of semi-arid rangelands (Tewes et al., 2015), (5) the fusion of multiple temporal, angular, and spectral features to improve land cover classification accuracy (e.g., Chen et al., 2017a), and (6) the exploration of the spatiotemporal distributions of environmental quality (Puttaswamy et al., 2013; Chang et al., 2014a,b; Chang et al., 2015; Xu et al., 2015a; Chang et al., 2016). Thus, the fused data or image should have enhanced spatial, temporal and/or spectral characteristics compared to each individual data set prior to the fusion process.

9.2   DATA FUSION TECHNIQUES AND ALGORITHMS

Conceptually, data fusion can be treated as a procedure that combines information from multiple data sources through a mathematical modeling framework (Figure 9.1), and the whole process can be further modeled as

$$O = F(X_1', X_2', \ldots, X_n') \tag{9.1}$$

where $(X_1', X_2', \ldots, X_n')$ denotes the transformed information from the original data files $(X_1, X_2, \ldots, X_n)$, and F represents a modeling framework to fuse the transformed information toward a final output O (Dai and Khorram, 1999). Specifically, F refers to various algorithms and techniques aiming to

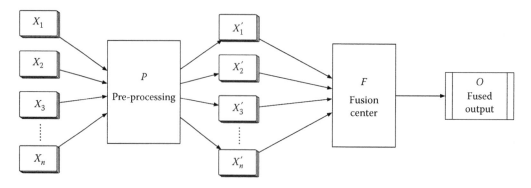

FIGURE 9.1   A general framework of data fusion.

blend information. Hence, it covers a large variety of methods. As previously discussed, image and data fusion methods can be divided into various categories, despite their fundamental principles and theories. In this section, these fundamental theories and principles lying behind each typical data fusion algorithm will be briefly introduced in addition to descriptions of their practical implementations in real-world environmental monitoring applications.

9.2.1 Pan-Sharpening

In the early stage of satellite remote sensing, in order to balance constraints between spatial, temporal, and spectral resolutions of remote sensing sensors as well as to keep satellites' weight, cost, bandwidth, and complexity down, a few sensors with limited capabilities were deployed onboard satellites. To enhance the Earth observing capabilities, satellite sensors providing high resolution panchromatic (PAN) images are commonly deployed in addition to sensors offering common medium/low resolution multispectral (MS) images. Configuring satellite sensors in such a way is a relatively common practice. For example, the well-known Landsat 7 incorporates sensors providing six 30-m resolution multispectral bands and a 60-m thermal infrared band together with a 15-m resolution panchromatic band. Such band combinations are commonly bundled in satellite remote sensing, and there exist a variety of similar satellite sensors onboard Earth observation platforms such as SPOT, QuickBird, OrbView, IKONOS, WorldView, and GeoEye, which commonly include both coarse spatial resolution MS bands and a single fine spatial resolution PAN band.

Pan-sharpening, shorthand for panchromatic sharpening, is a pixel-level fusion technique aimed at fusing a coarser spatial resolution MS image with a finer spatial resolution PAN image to create a new dataset that has the spectral properties of MS with the spatial resolution of PAN images. It aims to increase the spatial resolution of the MS to match that of the PAN images, while preserving the initial spectral information simultaneously (Amro et al., 2011; Vivone et al., 2015). In other words, the PAN image is used to sharpen the MS image; "to sharpen" means "to improve" the spatial resolution of the MS image. A generic pan-sharpening process typically encompasses the following procedures, including (1) up-sampling and interpolation of MS images, (2) alignment and co-registration between MS and PAN images, and (3) the core fusion steps (Figure 9.2).

With the increasing availability of high-resolution remotely sensed images, the demand for pan-sharpened data is continuously growing. As an important method to enhance images by improving the spatial resolution of the MS images, pan-sharpening enables us to perform a large

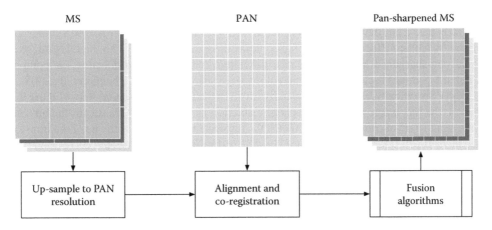

FIGURE 9.2 A schematic illustration of a generic pan-sharpening procedure. MS and PAN denote multispectral and panchromatic images, respectively.

variety of remote sensing tasks that require images with both high spatial and spectral resolutions, such as simple visual image interpretation (Laporterie-Déjean et al., 2005), change detection (Souza et al., 2003), object recognition and classification (Mohammadzadeh and Zoej, 2006; Aguilar et al., 2013), and higher-level product generation and advanced manipulation (Günlü et al., 2014).

A myriad of image fusion algorithms has been developed for meeting different pan-sharpening objectives over the past decades. In the public domain, there exists an enormous amount of pan-sharpening techniques and methods, which even date back to the 1980s. For example, the Intensity Modulation (IM) integration method was proposed for the integration of Landsat Multispectral Scanner (MSS) and Seasat Synthetic Aperture Radar (SAR) images by modulating the intensity of each pixel of the MSS channels with the value of the corresponding pixel of the SAR image (Wong and Orth, 1980). The actual pan-sharpening emerged after the launch of the French SPOT satellite system in February of 1986 when PAN images became available. Meanwhile, the paradigm term for image fusion shifted from the original "integration" to "merge" by taking advantage of IHS transformation (Welch and Ehlers, 1987; Tu et al., 2004). Since then, a variety of methods have been proposed for sharpening MS images with PAN data, such as High Pass Filtering (HPF) (Chavez and Bowell, 1988), Brovey Transform (BT) (Chavez Jr. et al., 1991; Bovolo et al., 2010), PCA (González-Audícana et al., 2004; Yang et al., 2007, 2012b) and Gram-Schmidt (GS) method (Laben and Brower, 2000; Aiazzi et al., 2006; Dalla Mura et al., 2015). In addition to these classical methods, pan-sharpening can also be performed by making use of methods like wavelet transform (Zhou et al., 1998), contourlet transform (Amro and Mateos, 2010), and generalized Laplacian pyramid (Aiazzi et al., 2002), as well as statistical methods such as spatially adaptive fusion methods (Aiazzi et al., 2002), Bayesian-based approaches (Fasbender et al., 2008), and many others (Wang et al., 2014).

Previous studies regarding various pan-sharpening methods attempted to classify them into distinct categories based on principles used in the fusion process. For example, Ranchin and Wald (2000) classified them into three groups including projection method, substitution method, and relative spectral contribution method, and Schowengerdt (2006) further classified these methods briefly into spatial domain and spectral domain. In general, it is hard to find a universal classification of methods in the literature (Aiazzi et al., 2009; Amro et al., 2011). According to Amro et al. (2011), the pan-sharpening methods can be grouped into the following typical categories.

9.2.1.1   Component Substitution

The component substitution (CS) methods are classical approaches for pan-sharpening which work through projecting the MS image into another space, assuming that this CS output isolates the spatial structure from the spectral information. Subsequently, the transformed MS image can be enhanced by replacing the component that contains the spatial structure with the PAN image. Then the final pan-sharpened MS image can be generated by reversely transforming the component-substituted MS image back to the original color space (Shettigara, 1992; Thomas et al., 2008; Amro et al., 2011). Specifically, a CS method usually involves the following essential steps (Figure 9.3): (1) up-sample the MS image to the same spatial resolution as the PAN, (2) transform the up-sampled MS image into a set of components, (3) match the histogram of the PAN with one of the transformed components of MS (i.e., aiming to reduce possible distortion as the greater the correlation between the PAN image and the replaced component, the lower the distortion introduced by the data fusion process), (4) replace that component with the histogram matched PAN image, and (5) reversely transform all components (with the substituted) back to the original color space to obtain a pan-sharpened MS image. Methods that work in such a way to sharpen MS images (Figure 9.4a) include the commonly used IHS (Carper et al., 1990), PCA (Chavez and Kwarteng, 1989) (Figure 9.4b), GS (Farebrother, 1974; Laben and Brower, 2000) (Figure 9.4c), and BT (Chavez Jr. et al., 1991; Bovolo et al., 2010) (Figure 9.4d), because all these methods involve performing a different transformation

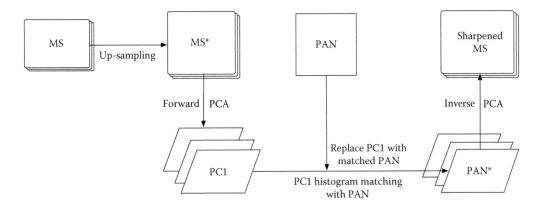

FIGURE 9.3 Pan-sharpening with PCA transform.

on the MS image. Among these methods, the GS algorithm was found to excel over other techniques due to its better performance in spectral preservation and spatial injection (Aiazzi et al., 2009).

Compared to other methods, CS techniques are global, meaning they operate in the same way on the whole image. Hence, techniques belonging to this category enable us to render the spatial details into the final product (Aiazzi et al., 2007). In general, CS techniques are fast and easy

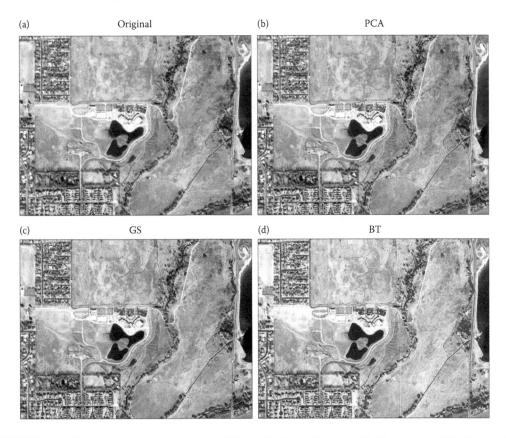

FIGURE 9.4 Pan-sharpened images using different fusion techniques. (a) One scene of Multispectral Imaging (MSI) image over Boulder, Colorado, USA, (b) pan-sharpened with PCA method, (c) pan-sharpened with GS method, and (d) pan-sharpened with BT method. Both the MSI (with a spatial resolution of 2.8 m) and PAN (with a spatial resolution of 0.7 m) images were acquired by instruments onboard the QuickBird satellite.

to implement and provide sharpened images with good visual/geometrical quality in most cases (Aiazzi et al., 2007). On the other hand, the global operating capability makes CS techniques incapable of accounting for local dissimilarities between the PAN and MS images originating from the spectral mismatch between the PAN and MS channels of the instruments, which may produce significant spectral distortions (Thomas et al., 2008). In order to find one component that is most suitable for substitution, statistical tests or weighted measures are incorporated to adaptively select an optimal component for substitution and transformation, and such methods are known as adaptive component substitution (Shah et al., 2008; Rahmani et al., 2010). A more detailed description of the main CS methods can be found in Vivone et al. (2015).

9.2.1.2 Relative Spectral Contribution

The relative spectral contribution (RSC) family can be considered a variant of the CS family, as a linear combination of the spectral bands is used instead of substitution. The process involves (1) up-sampling the MS image to the same spatial resolution as the PAN image, (2) matching the histogram of the PAN with each up-sampled MS band, and (3) pan-sharpening each up-sampled MS band by a linear combination of the spectral information at each pixel (i, j) through the following mathematical computations:

$$HRMS_b(i, j) = \frac{MS_b^h(i, j) \times PAN(i, j)}{\sum_b MS_b^h(i, j)} \tag{9.2}$$

where $HRMS_b$ denotes the pan-sharpened MS image and MS_b^h represents the high resolution (up-sampled) MS image at band b ($b = 1, 2, \ldots, B$).

Methods falling in the RSC family include BT and IM. BT is a simple method of merging data from different sensors based on the chromaticity transform, aiming at increasing the visual contrast in the high and low ends of the data's histogram, with an assumption that the spectral range spanned by the PAN image is the same as that being covered by the multispectral channels (Gillespie et al., 1987). Because only three bands are involved, a pan-sharpened image with BT has great spectral distortions and hence is not suitable for pixel-based classification as the pixel values are changed drastically (Vijayaraj et al., 2004) (Figure 9.4d). The IM method can be performed by replacing the summation of all MS bands in Equation 9.2 with the intensity component of the IHS transformation (Vijayaraj et al., 2004). As reported in the literature, the IM method may result in color distortions if the spectral range of the intensity replacement (or modulation) image is different from the spectral range covered by the three bands used in the color composition (Alparone et al., 2004). Comparisons between CS and RSC methods have been discussed in detail in Tu et al. (2001) and Wang et al. (2005).

9.2.1.3 High Frequency Injection

The concept of high frequency injection (HFJ) is to transfer the high-frequency details extracted from the PAN image into the up-sampled MS images by applying spatial filtering techniques (Tsai, 2003). An early idea for such an approach was initialized by Chavez (1986), who aimed to extract the spectral information from the Landsat TM and combine it with the spatial information from a data set with a much higher spatial resolution. Relevant methods include HPF (Chavez and Bowell, 1988) and high pass modulation (HPM) (Schowengerdt, 2006), which work through the following steps: (1) un-sample the MS image according to the resolution of the PAN image, (2) perform low pass filtering on the PAN image, (3) derive a high frequency image by subtracting the filtered image from the original PAN, and (4) obtain a pan-sharpened image by adding the high frequency image to each band of the MS. Further investigations indicate that the HFJ methods have superior performance compared with many other pan-sharpening methods such as those in the CS family due to the HFJ methods' lower spectral distortion and edge noises (Wald et al., 1997).

9.2.1.4 Multi-Resolution Transformation

Spatial transformation methods like convolution and Fourier transform have long been considered a valuable tool for extracting the spatial information of remote sensing images. In order to access spatial information over a wide range of scales or directions from local to global in addition to time and frequency, methods such as Laplacian pyramids (Burt and Adelson, 1983; Aiazzi et al., 2002), wavelet transform (Zhou et al., 1998), contourlet transform (Amro and Mateos, 2010), and curvelets transform (Dong et al., 2015a) can also be applied as a data representation beside these two extreme transformations. The basic assumption of such methods is that the activity level of source images can be measured by the decomposed coefficients in a selected transform domain (Liu et al., 2017a). Among these methods, wavelet transform has been widely used in pan-sharpening fusion schemes, which provides a framework to decompose images into a hierarchy with a decreasing degree of resolution, separating detailed spatial information between successive levels (Burt and Adelson, 1983). The essence of such a method is to decompose MS and PAN images into different components with variation scales so that spatial details can be enhanced in the sharpened MS images (Zhang, 2010).

Similar to CS and HFJ methods, multi-resolution transformation (MRT)-based data fusion can also be performed at the pixel level, but MRT methods enable the spectral adjustment of PAN images to be compatible with each MS band prior to the data fusion process while maintaining the original colors and statistical parameters (Zhou et al., 1998; Chang et al., 2016). Despite such advancements, MRT approaches may still exhibit certain spatial distortions (e.g., ringing artifacts) which decrease the visual quality of the fused data (Masi et al., 2016). The pan-sharpening process using wavelet and/or contourlet transform can be summarized as follows (Amro et al., 2011): (1) forward transform PAN and MS images using a sub-band and directional decomposition such as the subsampled or non-subsampled wavelet or contourlet transform, (2) apply a fusion rule onto the transform coefficients, and (3) obtain the pan-sharpened image by performing the inverse transform.

9.2.1.5 Statistical and Probabilistic Methods

In addition to the aforementioned methods, pan-sharpening can also be realized by making use of a set of methods that exploits the statistical characteristics of the MS and PAN images. By taking advantage of the substantial redundancy existing in the PAN data and the local correlation between PAN and MS images, Price (1987, 1999) successfully combined PAN and MS imageries from dual-resolution satellite instruments; the method was further improved by Park and Kang (2004) and named after the spatially adaptive algorithm. Compared to the original method, the spatially adaptive algorithm features adaptive insertion of information according to the local correlation between the two images, which sharpens the MS images and prevents spectral distortion as much as possible. By using a nonlocal parameter optimization scheme, MS bands are also sharpened by evaluating band-dependent generalized intensities (Garzelli et al., 2008; Garzelli, 2015). Additionally, Bayesian framework has also been broadly used for pan-sharpening by considering the available prior knowledge about the expected characteristics of the pan-sharpened image, which renders the problem of pan-sharpening into a probabilistic framework (Fasbender et al., 2008). These statistical methods have also been extensively used in fusing other types of remotely sensed imagery in addition to pan-sharpening; more details related to such data fusion methods will be introduced in the following subsections.

In recent years, several new kinds of pan-sharpening methods have been developed, emerging as a new branch in this field (e.g., Toet, 1989; Zhang and Guo, 2009; Yang and Li, 2010; Liang et al., 2012; Liu et al., 2017b). To date, there exists a large variety of methods and algorithms available for pan-sharpening, and comparisons of these methods have been well discussed in the literature (Chavez et al., 1991; Wang et al., 2005; Ehlers et al., 2010; Amro et al., 2011; Vivone et al., 2015). In most cases, pan-sharpening methods were applied to fuse MS and PAN images for the enhancement of the spatial details of MS while preserving the spectral characteristics. Nevertheless, with the availability of hyperspectral images, pan-sharpening has been also exploited for data fusion of PAN and hyperspectral data (Licciardi et al., 2012; Loncan et al., 2015). It is clear that conventional methods are not suitable for this task due to possible issues such as nonsimultaneous acquisition, co-registration of the data, and different spatial coverages and resolutions (Vivone et al., 2015).

9.2.2 STATISTICAL FUSION METHODS

As previously discussed, the core process of data fusion can be treated as one statistical modeling procedure, aiming to compute a new data value for each pixel by performing certain mathematical operations on the given input. In such a context, the data fusion based on CS methods can be considered one special case, as substitutions are performed rather than complex mathematical computations. On the other hand, although the missing spatial details of the low-resolution MS image can be enhanced with the aid of high-resolution PAN images simply by making use of methods such as CS and MRT, possible spectral distortions are introduced into the fused image if the PAN image is not exactly equivalent to the structural component of the low-resolution MS image (Zhong et al., 2016). In many cases, the data fusion can be also performed based on two MS images and/or hyperspectral images rather than simply based on MS and PAN images. More advanced methods should be used to model the relevant relations between information from each single sensor to account for data heterogeneities and spatial dependencies, in particular the feature extraction for multitemporal change detection over different terrestrial and aquatic environments (Chang et al., 2014b; Doña et al., 2015; Imen et al., 2015).

Generally, data fusion methods taking advantage of any statistical framework to characterize the inherent relationships between different features of the input images can all be referred to as statistical data fusion. Methods falling in this category always involve a statistical modeling process that plays an essential and critical role in addressing spatial, temporal, and spectral heterogeneities arising from instrumental, observational, and algorithmic differences between distinct inputs. To date, a large number of statistical modeling algorithms have been developed and used to fuse various remote sensing data/images in real-world applications. In general, these approaches can be broadly grouped into the following categories.

9.2.2.1 Regression-Based Techniques

Regression approaches have been used to fuse multiple datasets synergistically by constructing a regressed model between the target output and various data inputs, especially for cases with diversified data sources. For instance, by constructing a set of multiple linear regression models, the spatiotemporal distribution of mud content was clearly mapped to monitor the sediment grain-size of intertidal flats in the Westerschelde of the Netherlands through the data fusion of information from both space-borne microwave (SAR) and optical/shortwave infrared remote sensing data (van der Wal and Herman, 2007). Similarly, Srivastava et al. (2013) applied a multiple linear regression approach to fuse soil moisture data from the satellite soil moisture dedicated mission (SMOS) and WRF-NOAH Land Surface Model (WRF-NOAH LSM) to improve the accuracy of soil moisture deficit estimation. Based on a regression tree analysis, Kellndorfer et al. (2010) succeeded in fusing various remote sensing data (from LiDAR, InSAR, and Landsat) for forest stand height characterization by modeling the relationship between LiDAR measured canopy height and a suite of potential predictor variables derived from InSAR and optical datasets. Data fusion based on such methods, to some extent, can be considered a convolution of multiple data sources toward a synergistic output with improved accuracy or enhanced features.

9.2.2.2 Geostatistical Approaches

As the remotely sensed data are often provided in terms of 2-dimensional imagery with detailed spatial information associated, spatial relationships between various images can be modeled and incorporated to aid in data fusion. Differing from other types of data fusion methods, geostatistical solutions provide another family of statistical fusion approaches by explicitly incorporating the concept of spatial analysis (e.g., spatial correlation). Kriging, one classic geostatistical approach that is commonly used in spatial interpolation applications, has been also explored as a potential tool for image fusion objectives (Pardo-Igúzquiza et al., 2006; Atkinson et al., 2008; Meng et al., 2010; Sales et al., 2013). The theoretical basis of Kriging was proposed by Georges Matheron in 1960, and the basic idea is to predict the value of a function at a given location by computing a weighted average of the known values of the function near that point. To a larger extent, Kriging can be also considered a kind of regression analysis based on a single random field rather than multiple observations. To date, a variety of Kriging estimators has been developed in addition to the simple one, including Ordinary Kriging, Universal Kriging, Indicator Kriging, Disjunctive Kriging, Co-Kriging, and Kriging with External Drift. Descriptions of such estimators have been detailed in Mitchell (2007).

Due to the significant advantage of preserving spectral properties of the observed coarse resolution images, Kriging-based methods have been widely used in image fusion applications for downscaling purposes (Nishii et al., 1996; Pardo-Igúzquiza et al., 2006; Chatterjee et al., 2010; Meng et al., 2010; Nguyen et al., 2012; Puttaswamy et al., 2013; Sales et al., 2013). For instance, with the aid of a traditional universal Kriging approach and a spatial statistical data fusion scheme, Puttaswamy et al. (2013) successfully merged two satellite-based daily Aerosol Optical Depth (AOD) measurements with ground-based observations over the continental United States to improve spatial coverage and overall accuracy. As summarized in Sales et al. (2013), the primary advantages of Kriging over other techniques are mainly due to its capacity in accounting for: (1) pixel size differences between two inputs with different spatial resolutions by considering the sensor's point spread function, (2) the spatial correlation structure of the attribute values at image pixels, (3) the preservation of the spectral characteristics of the imagery, and (4) the utilization of all spectral information in the data fusion procedure.

In recent years, a set of more advanced Kriging methods was successfully developed and applied for various image downscaling and/or pan-sharpening purposes. For instance, Meng et al. (2010) developed an image fusion scheme based on Regression Kriging (RK) to fuse multitemporal high-resolution satellite images by taking consideration of the correlation between response variables (i.e., the image to be fused) and predictor variables (i.e., the image with finer spatial resolution), spatial autocorrelation among pixels in the predictor images, and the unbiased estimation with minimized variance. In 2013, Sales et al. (2013) proposed a method of Kriging with External Drift (KED) to downscale five 500-m MODIS pixel bands to match two 250-m pixel bands. Later in 2015, a new method termed Area-To-Point Regression Kriging (ATPRK) was proposed, also aiming to downscale 500-m MODIS bands 3–7 to a resolution of 250-m (Wang et al., 2015a). The ATPRK method involves two essential steps: (1) regression modeling aiming to incorporate fine spatial resolution ancillary data, and (2) Area-To-Point Kriging (ATPK)-based residual downscaling (i.e., downscaling coarse residuals to the desired fine spatial resolution) (Figure 9.5). ATPRK enables the preservation of the spectral properties of the original coarse data and the extension of the spectral properties to other supplementary data (Wang et al., 2015a, 2017a). Based on ATPRK, two more extended methods, termed adaptive ATPRK (Wang et al., 2016) and spectral–spatial adaptive ATPRK (Zhang et al., 2017), were proposed for pan-sharpening and MODIS image downscaling purposes, respectively, with improvements focusing mainly on enhancing spectral fidelity and spatial details in the data fusion process.

9.2.2.3 Spatiotemporal Modeling Algorithms

In most cases, the input data for data fusion are always provided in terms of remotely sensed images at different spatial and temporal scales, for instance, fine temporal resolution but coarse spatial resolution MODIS images and coarse temporal resolution but fine spatial resolution Landsat data. In order to generate data with both fine spatial and temporal resolutions, the relevant spatial and temporal differences must be well modeled and accounted for in the data fusion process. Hence,

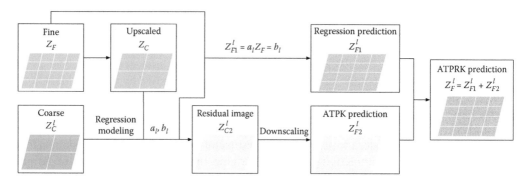

FIGURE 9.5 A schematic illustration of the use of the ATPRK algorithm to fuse coarse and fine images. (Adapted from Wang, Q., Shi, W., and Atkinson, P. M., 2016. *ISPRS Journal of Photogrammetry and Remote Sensing*, 114, 151–165.)

techniques and algorithms aimed at modeling spatial and temporal relationships between the input data sources for meeting fusion objectives are often categorized as Spatio-Temporal Data Fusion (STDF) methods. In recent years, a large number of STDF methods have been developed under different assumptions and application contexts. Among them, the Spatial and Temporal Adaptive Reflectance Fusion Model (STARFM) is the most well-known (Gao et al., 2006). STARFM is a typical image-pair based approach that was originally developed with the intention of fusing daily MODIS land surface reflectance (1,000-m) with 16-day Landsat TM reflectance (30-m) to generate synthetic Landsat-like imagery with a spatial resolution of 30-m on a daily basis (Figure 9.6).

Typically, STARFM involves the following four essential steps: (1) coarse MODIS images are first reprojected and resampled to the fine Landsat imagery in order to have the identical spatial resolution and georeference frame, (2) a moving window with fixed size is used to slide over the Landsat imagery to identify spectrally similar neighboring pixels, (3) an optimal weight is calculated for each neighbor pixel relying on spectral, temporal, and spatial differences, and (4) the surface reflectance of the central pixel of the moving window is computed in a modeling framework. The key is to model spatiotemporal variations between the up-sampled coarse image and the observed fine image to account for the relevant spectral and temporal differences. More specifically, STARFM works by relying on at least one coarse-fine image pair on temporally close days as well as one

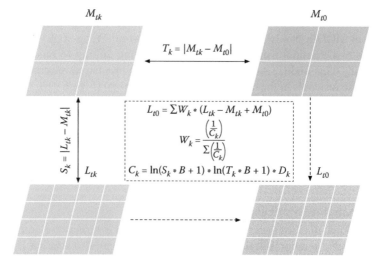

FIGURE 9.6 A schematic illustration of the STARFM algorithm based on one coarse-coarse and one coarse-fine image pair for data fusion.

coarse image to obtain a fused image with fine spatial resolution on the prediction day (e.g., Figure 9.6). Such a method assumes that pixels of the same class or adjacent pixels should have similar pixel values, and hence the temporal variations of the high spatial resolution image in one moment can be predicted by matching the neighboring similar pixels at different temporal phases.

Mathematically, the modeling framework can be considered a linearly weighted combination of the coarse temporal variations added to the available fine spatial resolution image, as the weights are determined by spectral difference, temporal difference, and location distance (Wang et al., 2017b). Regarding Figure 9.6, the prediction process can be modeled as (Gao et al., 2006):

$$L(x_{w/2}, y_{w/2}, t0) = \sum_{i=1}^{w}\sum_{j=1}^{w} W_{ij}(L(x_i, y_j, tk) - M(x_i, y_j, tk) + M(x_i, y_j, t0)) \tag{9.3}$$

where $L(x_{w/2}, y_{w/2}, t0)$ is a predicted pixel value in Landsat-like imagery for the time $t0$. w denotes the size of the moving window (e.g., a default of 1,500 m in STARFM) and hence $(x_{w/2}, y_{w/2})$ denotes the central pixel in this moving window. $L(x_i, y_j, tk)$ and $M(x_i, y_j, tk)$ are the observed Landsat and up-sampled MODIS reflectance values at time tk, which are used as base images to characterize spectral differences between two distinct sensors. $M(x_i, y_j, t0)$ is the up-sampled MODIS reflectance value at the window location (x_i, y_j) observed at $t0$. As shown in Equation 9.3, the core of the modeling is to determine weight W_{ij} for each neighboring pixel (x_i, y_j), which is calculated as a normalized reverse distance of combined weighting function C_{ij}.

$$W_{ij} = (1/C_{ij}) / \sum_{i=1}^{w}\sum_{j=1}^{w} (1/C_{ij}) \tag{9.4}$$

where C_{ij} denotes the combined weighting calculated based on spectral (S_{ij}) and temporal (T_{ij}) differences as well as spatial distance (D_{ij}),

$$C_{ij} = S_{ij} * T_{ij} * D_{ij} \tag{9.5}$$

$$S_{ij} = |L(x_i, y_j, tk) - M(x_i, y_j, tk)| \tag{9.6}$$

$$T_{ij} = |M(x_i, y_j, tk) - M(x_i, y_j, t0)| \tag{9.7}$$

$$D_{ij} = 1.0 + \frac{\sqrt{(x_{w/2} - x_i)^2 + (y_{w/2} - y_j)^2}}{A} \tag{9.8}$$

where A is a constant that defines the relative importance of spatial distance to the spectral and temporal distance; hence, a smaller value of A yields a larger dynamic range of D_{ij} (Gao et al., 2006). It is clear that STARFM works by fully relying on modeling statistical relations between image pairs at the pixel level, without performing complex transformation or substitution.

Due to its adaptive nature, STARFM has been extensively used in practice to aid in environmental surveillance by providing fused images with enhanced spatial, temporal, and spectral characteristics, such as land cover change detection (Hilker et al., 2009b; Walker et al., 2012), evapotranspiration mapping (Cammalleri et al., 2014), flood mapping (Zhang et al., 2014), water quality mapping (Chang et al., 2014b; Doña et al., 2015; Imen et al., 2015), and public health studies (Liu and Weng, 2012). Despite the salient capacity in generating fused data with both high spatial resolution and frequency coverage, it should be noted that there exist several limitations associated with STARFM, as reported in the literature (Hilker et al., 2009a; Zhu et al., 2010). First, STARFM is inadequate for predicting abrupt land cover changes that are not recorded in any baseline fine resolution image. This is mainly due to the fact that the predicted Landsat-like image is a linear weighted combination of the baseline fine resolution image and the modeled temporal changes. Second, STARFM fails to handle directional dependence of reflectance as the method simply treats the input data as a purely statistical number. Last, STARFM does not perform well on heterogeneous landscapes because temporal changes are derived from pure, homogeneous patches of land cover at the MODIS pixel scale.

In order to solve these limitations, a set of STARFM-like or STARFM-based methods was proposed in recent years. In 2009, Hilker et al. (2009a) developed a Spatial and Temporal Adaptive Algorithm for Mapping Reflectance CHange (STAARCH) to fuse Landsat and MODIS reflectance data for forest disturbance mapping purposes. By taking advantage of the tasseled cap transform of two or more image pairs, STAARCH enables the detection of spatial and temporal changes with a high level of detail and thus improves the final accuracy of the synthetic images. In 2010, an Enhanced STARFM (ESTARFM) was developed by making use of the observed reflectance trend between two points in time and spectral unmixing theory, aiming to improve the final prediction accuracy in heterogeneous landscapes (Zhu et al., 2010). A simple performance comparison of these three methods can be found in Gao et al. (2015). To address the spatial autocorrelation issue embedded in ESTARFM, Fu et al. (2013) modified a similar pixel selection scheme and thus derived a new method termed ESTARFM.

Aside from the methods discussed above, many other available techniques may also fall into this category due to their statistical nature. Random weighting, for instance, was used to fuse multi-sensor observations by adopting the weights of each individual sensor to obtain a final optimal weight distribution (Gao et al., 2011). Similarly, Bisquert et al. (2015) proposed a simple and fast data fusion scheme based on a weighted average of two input images (i.e., one coarse-fine image pair), considering their temporal validity to the image to be fused. Recently, dictionary-pair learning was also applied to STDF (e.g., Huang and Song, 2012; Song and Huang, 2013; Wei et al., 2016). Huang and Song (2012) developed a Sparse-Representation-based Spatiotemporal Reflectance Fusion Model (SPSTFM) to model temporal changes between coarse-fine image pairs, with the trained dictionary used to predict the unknown fine spatial resolution reflectance based on the modeled temporal changes among coarse images. Such a method was further extended to perform image fusion simply based on one image pair (Song and Huang, 2013; Chen et al., 2017b). In addition to remote sensing applications, statistical fusion methods are also a valuable tool in the biomedical domain. For example, by using a convex optimization method known as semidefinite programming (Vandenberghe and Boyd, 1996), a set of kernel-based heterogeneous descriptions of the same set of genes was fused together to provide better visualization and understanding of the gene structure (Lanckriet et al., 2004).

9.2.3 UNMIXING-BASED FUSION METHODS

The unmixing approach commonly uses a linear mixing model to analyze mixed pixels, which assumes that the spectrum of a mixed pixel is a linear combination of the pure spectra of the components present in that pixel weighted by their fractional coverage (Settle and Drake, 1993; Zurita-Milla et al., 2009). Mathematically, the spectral property $C(t,b)$ of a pixel with coarse resolution (known as a mixed pixel) at spectral band b at date t can be modeled as a weighted summation of the mean spectral properties of different endmembers (with a total number of n) derived from a fine resolution image $\overline{F(c,t,b)}$ and their abundances $A(c,t)$, regardless of errors arising from other external impacts such as atmospheric effects.

$$C(t,b) = \sum_{c=1}^{n} \overline{F(c,t,b)}\, A(c,t) + \varepsilon \qquad (9.9)$$

If there exists *a priori* knowledge about both the components that might be present in a given scene and their pure spectra, their sub-pixel proportions can then be retrieved through a linear mixing model in a process known as spectral unmixing (Adams et al., 1995). Due to its sound physical basis and effectiveness in analyzing mixed pixels, spectral unmixing has been widely used in the remote sensing community for various applications, such as the quantification of land cover change at sub-pixel scales (Kresslerl and Steinnocher, 1999; Haertel et al., 2004). It is clear that the performance of spectral unmixing relies largely on the quality of the *a priori* knowledge with respect to the scene composition and its pure spectral quality. Hence, spectral unmixing is preferable for hyperspectral images, whereas it cannot be performed on panchromatic images (Zurita-Milla et al., 2009).

One of the most important tasks of image fusion is to improve the spatial resolution of coarse resolution images for spatial detail enhancements; this process is often referred to as downscaling. The linear mixing model can also be used to downscale the spectral information of the coarse image to the spatial resolution identical to the fine image, and this application is commonly known as spatial unmixing (Zhukov et al., 1999; Zurita-Milla et al., 2008, 2011). Spatial unmixing is different from spectral unmixing, however. The former aims to estimate the class endmembers within each coarse pixel based on the known class proportions that are calculated from the fine spatial resolution image, and the latter is designed to estimate the class proportions within the coarse pixel based on the class endmembers that are predetermined by either endmember extraction or reference to supervised information (Wang et al., 2017b). To be more specific, spatial unmixing does not require *a priori* knowledge of the main components present in the low spatial resolution image, as the input (pure signal) and spectral unmixing is in fact the output of spatial unmixing (Zurita-Milla et al., 2009). Therefore, spatial unmixing can even be applied with mixed pixels or a small number of spectral bands in coarse resolution images.

Figure 9.7 depicts a theoretical workflow of the unmixing-based data fusion algorithm, which primarily consists of the following four steps (Zhukov et al., 1999; Zurita-Milla et al., 2008, 2011; Gevaert and García-Haro, 2015; Zhu et al., 2016): (1) define endmembers at coarse resolution by clustering the input fine resolution data, (2) calculate endmember proportions of each coarse pixel, (3) unmix the coarse pixels within a given moving window based on the computed endmember fractions in the previous step, and (4) assign the unmixed spectral value to fine pixels. There exists

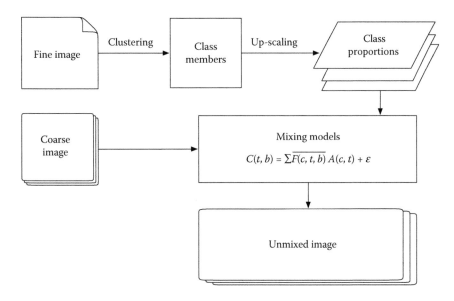

FIGURE 9.7   A schematic illustration of spatial unmixing-based fusion algorithm.

a big assumption associated with spatial unmixing methods, that is, no land-cover/land-use changes occur during the period as the endmember fractions remain constant for each coarse image (Wang et al., 2017b).

An early framework of spatial unmixing was initialized by Zhukov et al. (1999), by developing a multisensor, multi-resolution technique to fuse low- and high-resolution images acquired at different times for a synergetic interpretation. This method was further enhanced by performing various optimizations to the processes prior to unmixing in order to gain possible accuracy improvements via: (1) accounting for both spatial distance and spectral dissimilarity between pixels to tune weights assigned to the linear mixing model (Busetto et al., 2008), (2) optimizing the number of endmembers and the size of the moving window used to solve the unmixing equations (Zurita-Milla et al., 2008), and (3) making use of a high spatial resolution land use database (Zurita-Milla et al., 2009).

Recently, Amorós-López et al. (2011) developed a regularized spatial unmixing method to downscale MEdium Resolution Imaging Spectrometer (MERIS) full resolution images by making use of Landsat/TM images; they added a regularization term to the cost function of the spatial unmixing to limit large deviations of the unmixed pixels for each class. This method was later used for crop monitoring by generating a series of Landsat-like MERIS images (Amorós-López et al., 2013), and was further improved to downscale MODIS data by incorporating prior class spectra (Xu et al., 2015b). In 2012, Wu et al. (2012) proposed a Spatial and Temporal Data Fusion Model (STDFM) to generate Landsat-like surface reflectance from MODIS by considering both the spatial variation and the nonlinear temporal change information. In STDFM, the prediction was estimated through unmixing endmember reflectance at both input and prediction date and then adding the estimated change back to the base fine-resolution image under the assumption that the temporal variation properties of each land-cover class are constant (Wu et al., 2012). That is,

$$F(t0,b) = F(tk,b) + \Delta U(\Delta tk,b) \tag{9.10}$$

where

$$\Delta U(\Delta tk,b) = U(t0,b) - U(tk,b) \tag{9.11}$$

Here, $F(t0, b)$ denotes the predicted fine spatial resolution image for band b at time $t0$ based on one base image at time tk. U is the unmixed image and ΔU represents the variation between two unmixed images.

To fuse coarse resolution hyperspectral and fine resolution multispectral data, Yokoya et al. (2012) developed a method termed coupled nonnegative matrix factorization unmixing, with the main principle of combining abundance matrices from both hyperspectral and multispectral data after alternately applying unmixing to them. Later in 2013, an enhanced version of STDFM (ESTDFM) was proposed that incorporated a patch-based ISODATA classification method as well as making use of a sliding window to better determine the endmember in order to better fuse Landsat and MODIS surface reflectance (Zhang et al., 2013). In 2015, Gevaert and García-Haro (2015) proposed a method called the Spatial and Temporal Reflectance Unmixing Model (STRUM) which incorporated the main features of STARFM in order to obtain temporally stable synthetic imagery at Landsat spatial resolution. Similarly, by integrating spectral unmixing analysis with a thin plate spline interpolator, Zhu et al. (2016) developed a method termed Flexible Spatiotemporal DAta Fusion (FSDAF) for fusing satellite images with different resolutions. The FSDAF method features minimum input data requirements, heterogeneous landscapes accommodations, and the capability to predict both gradual change and land cover type change (Zhu et al., 2016).

9.2.4 Probabilistic Fusion Methods

As claimed by Zhang and Huang (2015), all the CS and MRT methods can be derived from a Bayesian fusion framework by adjusting a weight parameter to balance contributions from the spatial injection and spectral preservation models. This claim strongly indicates the strength and superiority of the Bayesian approach in advancing the data fusion modeling process. Here, data fusion methods taking advantage of probabilistic theories like Bayesian are all referred to as probabilistic fusion methods. In addition to the traditional Bayesian method (Solberg et al., 1994; Mascarenhas et al., 1996; Fasbender et al., 2008, 2009), some other well-known methods such as maximum likelihood (Xu et al., 2015a), Markova random field (Nishii, 2003), random forest (Liu et al., 2014), and Dempster-Shafer evidence theory (Bendjebbour et al., 2001; Lu et al., 2015) that work in a probabilistic framework have been widely used in data fusion processes. A great advantage of performing data fusion using probability theory is its ability to model the problem in a probabilistic framework without restricting modeling hypotheses (Fasbender et al., 2008).

In data fusion frameworks, the Bayesian theory enables us to take into account the uncertainties in the input multi-scale data. The straightforward formulation of such a data fusion method is often fast and easy to implement. In a Bayesian fusion paradigm, Bayesian Maximum Entropy (BME), one of the nonlinear spatiotemporal geostatistical methods that can theoretically blend multi-scale data from different sources with different accuracies (Li et al., 2013b), has been widely used to fuse related data from multiple sources in a probabilistic manner with a non-Gaussian distribution hypothesis (Christakos and Li, 1998). A basic workflow to fuse multiple data making use of BME can be found in Li et al. (2013a,b) and Tang et al. (2016). Theoretically, in BME fusion framework, data with uncertainties (e.g., images at coarser resolution) are routinely considered soft (or fuzzy) data that can be expressed by probability distribution, whereas the accurate data (e.g., images at finer resolution) are treated as hard data with a probability of 1 (Li et al., 2013b). Hence, both types of data can be effectively integrated based on a Bayesian paradigm.

Mathematically, the BME process to blend multiple observations for a new estimation $\widehat{x_k}$ at the location (x, y) at time t can be expressed as

$$\widehat{x_k} = \int x_k f^*(x_k | x_{soft}, x_{hard}) d_{x_k} \tag{9.12}$$

where $f^*(x_k | x_{soft}, x_{hard})$ denotes a posterior Probability Density Function (PDF) over the spatiotemporal adjacent pixel observations (Li et al., 2013b). x_{soft} and x_{hard} are probabilistic soft data and hard data, respectively. Specifically, the hard data refer to those data obtained with high accuracy; for instance, ground-based instruments measured temperature at sparse monitoring stations. In contrast, the soft data refer to data sets with uncertainties, such as satellite retrievals, model simulated results, assimilated data sets, and so on. Because the posterior PDF at the estimation point is derived from the prior PDF in the Bayesian rule when soft data are involved, the posterior PDF can be written as

$$f^*(x_k | x_{soft}, x_{hard}) = \frac{f(x_{soft}, x_{hard}, x_k)}{f(x_{soft}, x_{hard})} = \frac{f_G(x_{map})}{f(x_{soft}, x_{hard})} \tag{9.13}$$

where $f(x_{soft}, x_{hard})$ is *a priori* PDF at the spatiotemporal adjacent pixels and f denotes a detailed distribution function, for example, Gaussian distribution. $f_G(x_{map})$ is the joint PDF that can be attained by maximizing the entropy under the given constraint of the general knowledge G, which may consist of physical laws, scientific theories, logical principles, and summary statistics (Li et al., 2013b).

As elucidated in Li et al. (2013a) and Tang et al. (2016), the entropy H in BME framework can be defined as

$$H = -\int f_G(x_{map})\log f_G(x_{map})d_{x_{map}} \tag{9.14}$$

Introducing the Lagrange multipliers λ_α maximizes the entropy H equivalent to maximize the following relation:

$$L[f_G(x_{map})] = -\int f_G(x_{map})\log f_G(x_{map})d_{x_{map}}$$
$$-\sum_\alpha^N \lambda_\alpha \left[\varphi_\alpha(x_{map})f_G(x_{map})d_{x_{map}} - \overline{\varphi_\alpha(x_{map})}\right] \tag{9.15}$$

where $L[f_G(x_{map})]$ is the objective function for maximizing entropy, $\varphi_\alpha(x_{map})$ is a set of functions x_{map} such as the mean and covariance moments, and $\overline{\varphi_\alpha(x_{map})}$ is the expected value of $\varphi_\alpha(x_{map})$. Based on Equation 9.15, the maximum entropy solution for $f_G(x_{map})$ can be attained by setting the partial derivatives to zero and solving the system of equations with respect to the λ_α

$$f_G(x_{map}) = \frac{\exp\left[\sum_\alpha^N \lambda_\alpha \varphi_\alpha(x_{map})\right]}{\int \exp\left[\sum_\alpha^N \lambda_\alpha \varphi_\alpha(x_{map})\right]d_{x_{map}}} \tag{9.16}$$

The final estimation of $\widehat{x_k}$ can be finally attained by replacing $f_G(x_{map})$ in Equations 9.12 and 9.13 (Tang et al., 2016).

Because of its effectiveness in accounting for differences between multiple data sources, BME has been widely used to integrate *a priori* knowledge and uncertain data to enrich the subjective information and then consider their uncertainties in deriving much more objective results. For example, Li et al. (2013b) succeeded in fusing thermal infrared and microwave sea surface temperature (SST) products to improve the accuracy, spatial resolution, and completeness of satellite SST products. A similar framework was also used to fuse leaf area index (Li et al., 2013a), chlorophyll-a concentration (Shi et al., 2015), and Aerosol Optical Depth (AOD) (Christakos and Li, 1998). Similarly, Fasbender et al. (2009) developed a space–time prediction scheme to account for secondary information sources contributing to the variation of nitrogen dioxide concentration based on a Bayesian data fusion framework. Such a fusion framework can also be applied for pan-sharpening (Fasbender et al., 2008). Likewise, by making use of maximum likelihood, Xu et al. (2015a) proposed a data fusion framework to fuse multiple AOD products aiming to create a consistent AOD dataset with improved spatial coverage.

Markov Random Field (MRF) is a set of random variables with a Markov property described by an undirected graph which has a representation of dependencies similar to Bayesian networks but with an undirected nature. Thus, to some extent, an MRF can represent certain dependencies that a Bayesian network cannot. Because of this capability, MRF has been widely used to model various low- to mid-level tasks in image processing and computer vision (Li, 2009), for example, data fusion. For instance, Xu et al. (2011) developed an MRF-based data fusion framework to fuse multispectral images by incorporating the contextual constraints via MRF models into the fusion model. Similarly, Sun et al. (2013) developed an MRF-based gradient domain image fusion framework, in which the salient structures of input images were fused in the gradient domain and then the final fused image was reconstructed by solving a Poisson equation which forces the gradients of the fused image to be close to the fused gradients. In the framework, an MRF model was applied to accurately estimate region-based fusion weights for the salient objects or structures. Similarly, in order to fuse SAR images and optical imagery to estimate the nutrient fertility (total inorganic nitrogen) in coastal waters, Liu et al. (2014) developed a data fusion framework by making use of the random forest method. In the fusion framework, the random forest algorithm was used to merge four different input variables from the SAR and optical imagery to generate a new dataset (i.e., total inorganic nitrogen).

With the development of probabilistic theory, more advanced methods such as evidential belief reasoning can also be applied in probabilistic fusion framework. Dempster-Shafer (D-S) evidence theory, initiated in Dempster's work (Dempster, 1968) and mathematically formalized by Shafer (Shafer, 1976), has long been considered a generalization to the Bayesian theory and a popular tool to deal with uncertainty and imprecision (Khaleghi et al., 2013). Different from traditional Bayesian methods, D-S theory introduces the notion of assigning belief and plausibility to possible measurement hypotheses along with the required combination rule in order to fuse them (Khaleghi et al., 2013).

Mathematically, if we consider X to be all possible states of a system and 2^X to be all possible subsets of X, D-S theory will assign a belief mass m rather than a probability mass to each element E of 2^X to represent possible propositions with respect to the system state X. Here, the belief mass function m constrains the following two properties (Khaleghi et al., 2013):

$$m(\varnothing) = 0 \tag{9.17}$$

$$\sum_{E \in 2^X} m(E) = 1 \tag{9.18}$$

where $m(E)$ represents the level of evidence or confidence in E. If $m(E) > 0$, then the subset E of 2^X is referred to as a focal element (Zhou et al., 2013). Based on the mass function m, a probability range can be obtained for a given E

$$bel(E) \leq P(E) \leq pl(E) \tag{9.19}$$

Here, $bel(E)$ and $pl(E)$ refer to the belief of E and the plausibility of E, respectively, which can be calculated as

$$bel(E) = \sum_{B \in E} m(B) \tag{9.20}$$

$$pl(E) = \sum_{B \cap E \neq \varnothing} m(B) \tag{9.21}$$

If there exist two information sources with a belief mass function of m_1 and m_2, this information can then be fused based on Dempster's rule of combination (Khaleghi et al., 2013)

$$m_{1,2}(E) = (m_1 \oplus m_2)(E) = \frac{\sum_{B \cap C = E \neq \varnothing} m_1(B) m_2(C)}{1 - K} \tag{9.22}$$

where $m_{1,2}$ is the joint belief mass function, which should follow $m_{1,2}(\varnothing) = 0$. K represents the amount of conflict between the sources, which can be computed as

$$K = \sum_{B \cap C = \varnothing} m_1(B) m_2(C) \tag{9.23}$$

It is clear that D-S theory differs from traditional Bayesian inference by allowing each source to contribute information with different levels of detail. In other words, D-S theory is more flexible (Khaleghi et al., 2013). Due to this advantage, D-S theory has become a promising approach to fuse multisensor observations in recent years. For example, in order to better map and monitor woodland resources, Lu et al. (2015) proposed a D-S evidence theory-based fusion framework, aiming to fuse classification results from polarimetric synthetic aperture radar (PolSAR) (classified in a supervised manner) and interferometric SAR (InSAR) (classified in an unsupervised manner) data. The results indicated that the fused products had higher accuracy (95%) than that of each individual data set before fusion (88% and 90%). In general, probabilistic fusion methods could yield better performance compared to other simple statistical fusion approaches due to their abilities to reduce uncertainty in the fusion modeling process.

9.2.5 Neural Network-Based Fusion Methods

The critical step in data fusion process is modeling the relevant relations between various data inputs at different scales. Linear models have been widely used to establish relations between different data inputs because of their simplicities and effectiveness. Despite such advantages, it is noticeable there exist deficiencies related to the fact that many real-world systems are not consistent with such a linear assumption nonlinear. It is always difficult to obtain an accurate result by simply making use of linear models. For example, linear mixing models are often used to unmix coarse resolution images by referring to calculated endmember fractions from the fine resolution images, under an assumption that the spectral property of a coarse pixel is a linear weighted summation of various endmembers within this pixel. In other words, the relationship between them has already been prescribed to such a linear combination, which may in turn limit a full exploration toward higher prediction accuracy.

To cope with these deficiencies, nonlinear methods, such as artificial intelligence-based approaches, can be of great help. Among them, the most well-known approach is the Artificial Neural Network (ANN). ANNs represent a set of decision processing models inspired by the a biological neural network that is a series of interconnected neurons whose activation defines a recognizable linear pathway acting as decision aggregates to global decisions (Figure 9.8).

$$net_i = \sum_j \omega_{i,j} x_j + b_i \tag{9.24}$$

$$O_i = f(net_i) \tag{9.25}$$

where ω denotes the weights between two connected neurons that can be determined by various learning methods. x represents various inputs and b is the bias added to the weighted inputs. By applying an activation function f (e.g., typically a sigmoid function of $(1 + \exp(-k))^{-1}$) to the net input, the output O can be obtained. Because of its nature, ANN enables the capture and representation of complex input/output relationships through a learning process without making assumptions regarding data distribution or the nature of the relation between inputs and outputs. Additionally, the learned empirical knowledge can be saved and further used for future data

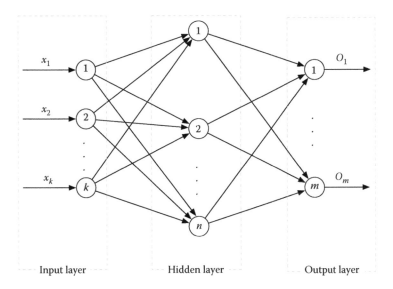

FIGURE 9.8 An example of a typical neural network architecture with one single hidden layer.

prediction purposes. As the learning process can be solved in a global optimization manner, the neural network approach is error tolerant and relatively insensitive to noise. All these features make the neural network approach a powerful tool with sound modeling capabilities, which is capable of generating solutions to complex problems beyond formal descriptions in spite of the drawback that neural networks are often considered black box models (Del Carmen Valdes and Inamura, 2001).

Because of the inherent distribution-free advantage, the neural network approach has long been considered a valuable tool for the advancement of data fusion techniques. Specifically, the neural network approach has the inherent ability to integrate information from multisource data and to identify complex relationships between mixed pixels (Dai and Khorram, 1999). This unique distribution-free feature makes the neural network approach excel among statistical approaches because it avoids exploring a predetermined relationship posed on the fusion of multisource data and thus is adaptable to numerous data inputs. At present, a large variety of neural-scale-based fusion approaches are available for data fusion.

By using a Back-Propagation Neural-network (BPN) and a Polynomial Neural Network (PNN), distinct features from infrared (IR) multispectral surveillance data were fused to provide excellent discrimination applicability between heavy and light space objects (Haberstroh and Kadar, 1993). The basic principle is to learn patterns embedded in each input, then data fusion is often performed on the feature- and/or decision-level. Similar data fusion frameworks taking advantage of BPN were also used to analyze multitemporal change of land cover characterization (Dai and Khorram, 1999), spatial resolution enhancement (Valdés Hernández and Inamura, 2000), and pan-sharpening (Del Carmen Valdes and Inamura, 2001). Apart from the conventional BPN, a set of more complex neural networks, such as Radial Basis Function (RBF)-based neural network (Chen and Li, 2006), pulse-coupled neural network (Li et al., 2006), Genetic Algorithm and Self-Organizing Feature Map (GA-SOFM) integrated artificial neural network (Yang et al., 2010), and MultiLayer Perceptron (MLP) neural network (Del Frate et al., 2014), has been applied to fuse multisensor, multifrequency (e.g., optical and microwave) remote sensing images in various practices.

With the recent development of theory in neural networks, several more advanced neural networks have been proposed and applied to advance data fusion practices. For instance, Huang et al. (2015) developed a Deep Neural Network (DNN)-based image fusion framework for pan-sharpening purposes by fully taking advantage of the feature learning ability of DNN. The proposed DNN-based data fusion framework mainly consists of three procedures, including (1) patch extraction to generate the training set, (2) DNN training, and (3) final high-resolution image reconstruction. According to Huang et al. (2015), once a training set $\{x_p^i, y_p^i\}_{i=1}^N$ (N is the total number of training image patches) is extracted from the high-resolution PAN image ($\{x_p^i\}_{i=1}^N$) and the synthetic PAN image of the up-sampled low-resolution MS image ($\{y_p^i\}_{i=1}^N$), a DNN model can be trained by using the Modified Sparse Denoising Autoencoder (MSDA) algorithm to learn the relations between coarse and fine images. Mathematically, the feedforward functions of MSDA can be defined as:

$$h(y_p^i) = s(Wy_p^i + b) \tag{9.26}$$

$$\hat{x}(y_p^i) = s(W'h(y_p^i) + b') \tag{9.27}$$

where s is the sigmoid activation function, and W (or W') and b (or b') are the encoding (or decoding) weights and biases, respectively. $h(y_p^i)$ represents the hidden layer's activation, while $\hat{x}(y_p^i)$ denotes the reconstructed input that is an approximation of x_p^i. Parameters $\Theta = \{W, W', b, b'\}$ can then be trained by minimizing the following cost function with a Lagrangian approach:

$$L\left(\left\{x_p^i, y_p^i\right\}_{i=1}^N; \Theta\right) = \frac{1}{N} \sum_{i=1}^N \left\|x_p^i - \hat{x}\left(y_p^i\right)\right\|_2^2 + \frac{\lambda}{2}\left(\|W\|_F^2 + \|W'\|_F^2\right) + \beta KL(\hat{\rho} \| \rho) \tag{9.28}$$

where λ and β are two balancing parameters determined by cross validation. As indicated, the second term in the right is a weight decay term, and the last term $KL(\hat{\rho} \| \rho)$ is the Kullback–Leibler divergence, which is a sparsity term forcing $\hat{\rho}$ (the average activation function of the hidden layer) to approximate ρ:

$$KL(\hat{\rho} \| \rho) = \rho \log \frac{\rho}{\hat{\rho}} + (1 - \rho) \log \frac{1 - \rho}{1 - \hat{\rho}} \tag{9.29}$$

$$\hat{\rho} = \frac{1}{N} \sum_{i=1}^N h\left(y_p^i\right) \tag{9.30}$$

After training an MSDA, the successive layer is trained in turn by taking the hidden activation values $h(x_p^i)$ and $h(y_p^i)$ of the previous layer as the input in order to construct a deep learning network architecture; such a scheme is thus termed stacked MSDA (S-MSDA) (Huang et al., 2015). The experimental results show that such a data fusion scheme outperforms other pan-sharpening methods in terms of both visual perception and numerical measures. Similarly, Dong et al. (2015b) proposed a further deep learning method for single image super-resolution through an end-to-end mapping between the coarse- and fine-resolution images by making use of a deep Convolutional Neural Network (CNN). Compared to standard neural networks, CNN can yield better results with less training time because of its unique structures (Jing et al., 2017). Likewise, CNN can also be used for pan-sharpening (Masi et al., 2016; Zhong et al., 2016) as well as fusion of multispectral and hyperspectral images (Palsson et al., 2017).

Compared to other fusion methods, a neural-scale-based fusion scheme could yield higher accuracy for the fusion output, especially in dealing with complex data sources as no distribution priori are guaranteed. The reason can be largely attributed to the nonlinear, nonparametric, and distribution-free characteristics associated with neural scale methods. Despite the effectiveness of such methods, a possible disadvantage that should be noted is that the complex modeling process may require even more time than simple approaches such as linear regression or weighted summation.

9.2.6 Fuzzy Set Theory-Based Fusion Methods

Fuzzy set theory, initialized by Zadeh (1965) with the concept of partial membership to a set, has been found to provide a simple, yet powerful mathematic tool to model the vagueness and ambiguity in complex systems by imitating the human reasoning ability to reach conclusions with imprecise, incomplete, and not even totally reliable information (Solaiman et al., 1999; Melgani, 2004). Because of this distinct feature, fuzzy set theory has been used in a wide range of domains to process unprecise or uncertain data. In recent years, fuzzy-based data fusion methods were also developed and applied for various decision fusion schemes. For instance, Tupin et al. (1999) developed a fuzzy fusion approach to combine several structure detectors to facilitate SAR image classification. Likewise, a fuzzy-based multisensor data fusion classifier was developed for land cover classification by integrating multisensor and contextual information in a single and homogeneous framework (Solaiman et al., 1999). The integration was realized by iteratively updating the fuzzy membership maps corresponding to different thematic classes and by using spatial contextual membership information based on different fuzzy contextual classification rules. Similar works can also be found in the literature, for example, a fuzzy fusion of spectral, spatial, and temporal contextual information approaches for the classification of multisensor remote sensing images (Melgani, 2004), a fuzzy decision rule-based fusion for the classification of urban remote sensing images (Fauvel et al., 2006), a Choquet fuzzy integral-based fusion scheme to estimate the chlorophyll-a concentration in a water body (Wang et al., 2010), and a fuzzy K-nearest neighbor classification system to fuse hyperspectral and LiDAR data (Bigdeli et al., 2014a). Results from these studies indicate that the incorporation of fuzzy set theory in fusion framework could yield higher classification accuracy.

9.2.7 Support Vector Machine-Based Fusion Methods

In addition to the aforementioned approaches, accounting for uncertainties other high-level nonlinear, nonparametric modeling techniques can also be applied for data fusion purposes, such as SVM. SVM, well-known in the fields of machine learning and pattern recognition, works as a nonparametric classifier approach which aims to discriminate inputs into two separate classes by fitting an optimal linear separating hyperplane to the training samples in a multidimensional feature space (Vapnick, 1998). A detailed description of the general concept of SVM and a brief summary of SVM theory can be found in Burges (1998) and Waske and Benediktsson (2007), respectively. Waske and Benediktsson (2007) presented a decision fusion scheme to combine multisensor data with the aid of two SVMs, with one used for classification and the other for decision fusion. Different from previous decision fusion frameworks which simply fused the final classification outputs, the original outputs of each SVM discriminant function were fused instead. With a similar framework, hyperspectral and LiDAR data were successfully fused in the decision level by making use of a SVM-based classifier fusion system (Bigdeli et al., 2014b).

9.2.8 Evolutionary Algorithms

Essentially, the modeling framework of image/data fusion can be largely considered as an optimization problem, as the core is to find an optimal solution to the data fusion models so as to fuse multiple data sources seamlessly. Therefore, evolutionary algorithms can be used as promising approaches in helping solve complex data fusion problems. In the evolutionary computing domain, there exists a variety of evolutionary algorithms with different optimization strategies, such as Genetic Algorithm (GA) (Fraser, 1962), evolutionary programming (Fogel and Fogel, 1996), Particle Swarm Optimization (PSO) (Kennedy and Eberhart, 1995), and so forth. Among these algorithms, GA and PSO are two popular methods commonly used in the remote sensing domain to deal with various complicated optimization and search problems, thus both of them will be introduced in the following sections to delineate how they help in the image/data fusion process.

GA, one of the most well-known evolutionary algorithms, works based on the principle inspired by the mechanism of natural selection-more specifically, the biological genetics and evolution in Darwin's theory. GA has been demonstrated to possess a sound ability to solve many optimization and search problems due to its intelligent and nonlinear nature (Fraser, 1962). Generally, GA works in a manner of simulating "survival of the fittest" among individuals by evolving generation-by-generation to search for a better solution to an optimization problem (Melanie, 1998). In GA, the evolution starts from a population of randomly generated individuals through an iterative process, while each individual represents a possible solution to the problem that can be mutated and altered based on certain rules (e.g., selection, crossover, and mutation). In each iteration, the population of all individuals refers to a generation that is arranged in a structure analogous to chromosomes. To make the optimization process evolve at each step, GA will randomly select a set of individuals from the current population and use them as parents to produce children for the next generation. Although randomly selected, GA does not work in a purely random manner in practice. Instead, the randomness is constrained with the former generation information in order to direct the optimization into the search space with better performance. After evolving through a suite of generations, the population will finally evolve toward an optimal solution. More details with respect to the foundations of GA can be found in Rawlins (1991).

In general, GA works on a principle analogous to gene structure which behaves like chromosomes to evolve toward an optimal solution. Essentially, it is associated with the following foundations: (1) individuals within a population will compete with each other, (2) more offspring will be generated from those individuals winning the competition, (3) "good" genes (individuals) will propagate throughout the population to produce even better offspring (i.e., "survival of the fittest"), and (4) each successive generation will be more adaptable to the environment than the previous one. These unique properties fully resemble the process to search an optimal solution for a given optimization problem within a search space, as solutions with poor performance will be discarded in a stepwise manner. As opposed to the classical derivative-based optimization algorithm, GA has two distinct advantages: (1) GA generates a population of points rather than a single point at each iteration, and thus it is more efficient, and (2) GA uses a random generator to select the next population rather than a deterministic computation, which in turn renders robustness of the results.

Because of its capability and effectiveness in quickly and efficiently reaching a reasonable solution to a complex problem by searching through a large and complex search space, in image/ data fusion domain, GA has been widely used in various fusion schemes to optimize either the fusion inputs or fused outputs toward the improvement of the final accuracy (Mumtaz et al., 2008; Lacewell et al., 2010). For instance, in order to maintain the detail and edge information from multisensor images, Li and Peng (2015) proposed a GA-aided cellular neural network based image fusion scheme, in which the GA algorithm was specifically applied for parameter optimization purposes. Analogously, a real-valued GA model was used to provide an optimum injection of PAN information into the MS images to improve the pan-sharpening performance (Garzelli and Nencini, 2006). Similar integrations can be also found in the literature, for example, GA and ANN (Peng and Dang, 2010), GA and morphological filters (Addesso et al., 2012), and so on.

The PSO algorithm was originally proposed in 1995 and was inspired by the flocking and schooling patterns of birds and fish (Kennedy and Eberhart, 1995). Analogous to GA, PSO is also a population-based optimization method that is commonly used to identify the optimal solution for evolutionary pathway problems. However, the PSO uses a different strategy as opposed to other algorithms to search for a similar optimal solution. In PSO, the optimization is performed based on a swarm of particles, and each particle represents a possible solution to the optimization problem in the search space. For each particle, there are two critical aspects associated with it, one is position and the other is velocity. Generally, the position of a particle is mainly influenced by the personal best position of this particle and the global best position (best position among the whole particles) in the swarm (Chen and Leou, 2012).

It is clear that each particle can evolve by learning from the previous experiences of all particles in the swarm, even the particle itself. During the evolving process, the position and velocity of each particle are updated iteratively unless the optimum solution is attained while the potential solutions in each iteration will be updated as well until reaching a maximum number of iterations or a global optimum result (Kusetogullari et al., 2015). As claimed in Chen and Leou (2012), PSO is mainly featured by the following advantages as compared with other evolutionary algorithms: (1) Each particle in the swarm represents a potential solution that is "flying" within the hyperspace with a velocity. (2) Each particle can learn from the previous experiences of all particles as the swarm and particles have their own memories. This is the unique feature of PSO which does not exist in any other evolutionary algorithm. (3) Only primitive mathematical operations are required by PSO as the following states can be updated automatically.

As an effective off-line optimization method, PSO has been widely used in image/data fusion applications. For instance, Wang et al. (2015b) adopted a PSO algorithm to obtain the optimal weights that can be used for better pan-sharpening MS images by the PAN image, as the optimization was equivalent to the maximization of the radiometric similarities between the MS and the up-scaled PAN images. The results indicated that the PSO-aided fusion scheme showed superior performance to popular component substitution-based fusion methods (Wang et al., 2015b). Saeedi and Faez (2011) performed a similar work by integrating a multi-objective PSO algorithm and the shiftable contourlet transform for image pan-sharpening. Likewise, Raghavendra et al. (2011) successfully developed two distinct image fusion schemes by making use of PSO to find an optimal strategy to fuse sub-band coefficients of visible and near infrared images so as to improve the face verification performance. Many other applications can be also found in the literature, such as Siddiqui et al. (2011); Ahmed et al. (2013); Gharbia et al. (2016), and so on.

In addition to image/data fusion, PSO has been also used in remote sensing to deal with other optimization problems, including image classification (Ding and Chen, 2012; Soliman et al., 2012), feature extraction (dimensionality reduction) (Chang et al., 2014a; Yang et al., 2012a), image enhancement (super-resolution) (Erturk et al., 2014), change detection (Kusetogullari et al., 2015), and so on. In recent years, a set of extended PSO algorithms have been developed to advance the optimization process, such as adaptive PSO (Zhan et al., 2009), orthogonal PSO (Zhan et al., 2011), accelerated PSO (Yang et al., 2011), and so on.

9.2.9 Hybrid Methods

There is no doubt that each data fusion method has its own strengths and weaknesses. There is no single data fusion approach sufficient to fulfill all requirements, however; in other words, each method is more or less limited by some constraints (Chang et al., 2016). Hence, it is always advisable to integrate several data fusion techniques with different capabilities to cope with the relevant drawbacks from each individual method toward a more accurate or better result. Such an integrated data fusion framework is commonly referred to as hybrid data fusion. Because of the complementary strengths from various techniques, hybrid data fusion methods always yield better results than a single approach alone. Therefore, hybrid data fusion methods have been widely used in practice by remote sensing domains to solve various complex problems.

In order to enhance the classification accuracy of hyperspectral data, an integrated feature and decision level fusion framework was developed based on a set of algorithms, including projection pursuit, majority voting, max rule, min rule, and average rule (Jimenez et al., 1999). The experimental practices demonstrate that projection pursuit-based feature fusion and majority voting-based data fusion algorithms enable the extraction of more information from the training samples while avoiding the problem of overfitting and the Hughes phenomenon, and in turn the final classification accuracy is improved (Jimenez et al., 1999). Based on the concept of high pass filtering, González-Audícana et al. (2004) developed a new fusion scheme to fuse MS and PAN images based on the multi-resolution wavelet decomposition. In this fusion scheme, the improved IHS and PCA methods were both used to inject the spatial details of the PAN image into the MS one. Further comparisons indicate that the proposed hybrid fusion scheme yielded merged images with improved quality with respect to those obtained by standard IHS, PCA, and standard wavelet-based fusion methods. Similarly, a hybrid wavelet-artificial intelligence fusion approach was developed to downscale MODIS land surface temporal data based on Landsat data in which the wavelets were used mainly to capture the properties of the signals in different directions (Moosavi et al., 2015).

Due to the powerful optimization capacity, GA has been widely used in data fusion schemes in conjunction with other fusion methods to improve the accuracy of fused outputs. In order to better retain the detail and edge information of the original images, a novel multisource image fusion scheme was developed based on cellular neural networks and GA, and the GA method was used to determine optimal parameters for the neural network fusion framework (Li and Peng, 2015). In addition to these complex methods, a hybrid fusion framework can also be attained by integrating several statistical approaches. For example, Wang et al. (2017b) developed an enhanced spatiotemporal data fusion framework aiming to increase prediction accuracy in fusing MODIS and Landsat images. To achieve this goal, two separate data fusion methods, ATPRK and STARFM, were used successively to fuse 250-m bands with 500-m bands of MODIS to produce the interim 250-m MODIS data while STARFM was used to blend the 250-m fused MODIS with 30-m Landsat images. Such a data fusion scheme enabled the capture of more information for abrupt changes and heterogeneous landscapes than simply using the original 500-m MODIS data, thus in turn increasing the accuracy of spatiotemporal fusion predictions (Wang et al., 2017b).

Although hybrid approaches could yield better performance in image fusion practices compared to using a single approach alone, comprehensive knowledge and understanding of each method are always required to guarantee accurate modeling and integration of these separate methods in the data fusion framework; otherwise, an improper integration could result in even more erroneous results. Generally, the degree of success for data fusion is often case-dependent. Nevertheless, there exist several niches in performing data fusion in order to better serve the users' needs (Pohl and Van Genderen, 1998). In an attempt to confirm such niches, there are a few issues that deserve our attention.

First, users should clearly know the primary objective of data fusion. More specifically, they should know the features that are to be enhanced. For instance, in meteorological applications, users may require data having frequent and repetitive coverage with relatively low spatial resolution; in contrast, users aiming to perform land cover change detection or urban mapping may demand an image with a higher spatial resolution, and military surveillance users may need to have images with both high spatial and temporal resolutions.

The second issue with which users should be familiar is related to the selection of the data for data fusion. At present, an enormous amount of satellite-based data is available, but with different characteristics which depend largely on the sensor capability, orbit of the platform, imaging geometry, ultimate resolution, and even bias levels and atmospheric constraints such as cloud cover and sun angle (Ardeshir Goshtasby and Nikolov, 2007; Nguyen et al., 2009; Castanedo, 2013). Thus, these factors need to be carefully considered in selecting proper data selection for the data fusion process, because a good data set can make the data fusion more successful. Third, the users also have to take into account the need for data pre-processing like up-sampling and registration, depending on which processing strategy one needs to follow.

Furthermore, the users should determine which technique to use for the problems at hand to explore the full advantage of the data available as different methods may yield distinct outputs. Last but not least, proper quality evaluation measures or criteria are necessary to assess the accuracy of the data fusion process. After data fusion, the accuracy of the fused outputs should be well assessed to guarantee a quality assured result. A set of statistical measures that can be used to assess the performance of data fusion results will be introduced in the next chapter.

9.3 SUMMARY

In this chapter, a set of common image and data fusion algorithms was introduced by providing details related to the underlying principles and theories of each method. STARFM, an advanced multisensor data fusion algorithm originally proposed to create Landsat-like MODIS imagery by enhancing the spatial resolution of MODIS data, was elaborated here. In addition to STARFM, data fusion techniques making use of pan-sharpening, neural-network-based theory, probability theory, evolutionary algorithms, statistical approach, image unmixing, and many others were described as well. In general, data fusion approaches can be applied not only to enhance spatial, temporal, and/or spectral resolution of the input remotely sensed imagery, but also to improve prediction accuracy and to optimize data quality and spatial coverage simultaneously.

REFERENCES

Adams, J. B., Sabol, D. E., Kapos, V., Filho, R. A., Roberts, D., Smith, M. O., and Gillespie, A. R., 1995. Classification of multispectral images based on fractions and endmembers: Applications to land-cover change in the Brazilian Amazon. *Remote Sensing of Environment*, 52, 137–152.

Addesso, P., Conte, R., Longo, M., Restaino, R., and Vivone, G., 2012. A pansharpening algorithm based on genetic optimization of Morphological Filters. In: *2012 IEEE International Geoscience and Remote Sensing Symposium*, 5438–5441.

Aguilar, M. A., Saldaña, M. M., and Aguilar, F. J., 2013. GeoEye-1 and WorldView-2 pan-sharpened imagery for object-based classification in urban environments. *International Journal of Remote Sensing*, 34, 2583–2606.

Ahmed, T., Singh, D., Gupta, S., and Raman, B., 2013. Partice swarm optimization based fusion of MODIS and PALSAR images for hotspot detection. In: *2013 IEEE International Conference on Microwave and Photonics (ICMAP)*, 1–6.

Aiazzi, B., Baronti, S., Alparone, L., and Garzelli, A., 2002. Context-driven fusion of high spatial and spectral resolution images based on oversampled multiresolution analysis. *IEEE Transactions on Geoscience and Remote Sensing*, 40, 2300–2312.

Aiazzi, B., Baronti, S., Lotti, F., and Selva, M., 2009. A comparison between global and context-adaptive pansharpening of multispectral images. *IEEE Geoscience and Remote Sensing Letters*, 6, 302–306.

Aiazzi, B., Baronti, S., and Selva, M., 2007. Improving component substitution pansharpening through multivariate regression of MS+Pan data. *IEEE Transactions on Geoscience and Remote Sensing*, 45, 3230–3239.

Aiazzi, B., Baronti, S., Selva, M., and Alparone, L., 2006. Enhanced Gram-Schmidt spectral sharpening based on multivariate regression of MS and pan data. In: *2006 IEEE International Symposium on Geoscience and Remote Sensing*, 3806–3809.

Alparone, L., Facheris, L., Marta, S., Baronti, S., Cnr, I., Panciatichi, V., Garzelli, A., and Nencini, F., 2004. Fusion of multispectral and SAR images by intensity modulation. In: *Procceedings 7th International Conference on Information Fusion*, 637–643.

Amorós-López, J., Gómez-Chova, L., Alonso, L., Guanter, L., Moreno, J., and Camps-Valls, G., 2011. Regularized multiresolution spatial unmixing for ENVISAT/MERIS and landsat/TM image fusion. *IEEE Geoscience and Remote Sensing Letters*, 8, 844–848.

Amorós-López, J., Gómez-Chova, L., Alonso, L., Guanter, L., Zurita-Milla, R., Moreno, J., and Camps-Valls, G., 2013. Multitemporal fusion of Landsat/TM and ENVISAT/MERIS for crop monitoring. *International Journal of Applied Earth Observation and Geoinformation*, 23, 132–141.

Amro, I. and Mateos, J., 2010. Multispectral image pansharpening based on the contourlet transform. In: *Information Optics and Photonics: Algorithms, Systems, and Applications*, 247–261.

Amro, I., Mateos, J., Vega, M., Molina, R., and Katsaggelos, A. K., 2011. A survey of classical methods and new trends in pansharpening of multispectral images. *EURASIP Journal on Advances in Signal Processing*, 2011, 79.

Ardeshir Goshtasby, A., and Nikolov, S., 2007. Image fusion: Advances in the state of the art. *Information Fusion*, 8, 114–118.

Atkinson, P. M., Pardo-Igúzquiza, E., and Chica-Olmo, M., 2008. Downscaling cokriging for super-resolution mapping of continua in remotely sensed images. *IEEE Transactions on Geoscience and Remote Sensing*, 46, 573–580.

Bai, Y., Wong, M., Shi, W.-Z., Wu, L.-X., and Qin, K., 2015. Advancing of land surface temperature retrieval using extreme learning machine and spatio-temporal adaptive data fusion algorithm. *Remote Sensing*, 7, 4424–4441.

Bendjebbour, A., Delignon, Y., Fouque, L., Samson, V., and Pieczynski, W., 2001. Multisensor image segmentation using Dempster-Shafer fusion in Markov fields context. *IEEE Transactions on Geoscience and Remote Sensing*, 39, 1789–1798.

Bigdeli, B., Samadzadegan, F., and Reinartz, P., 2014a. Feature grouping-based multiple fuzzy classifier system for fusion of hyperspectral and LIDAR data. *Journal of Applied Remote Sensing*, 8, 83509.

Bigdeli, B., Samadzadegan, F., and Reinartz, P., 2014b. A decision fusion method based on multiple support vector machine system for fusion of hyperspectral and LIDAR data. *International Journal of Image and Data Fusion*, 5, 196–209.

Bisquert, M., Bordogna, G., Bégué, A., Candiani, G., Teisseire, M., and Poncelet, P., 2015. A Simple fusion method for image time series based on the estimation of image temporal validity. *Remote Sensing*, 7, 704–724.

Bovolo, F., Bruzzone, L., Capobianco, L., Garzelli, A., Marchesi, S., and Nencini, F., 2010. Analysis of the effects of pansharpening in change detection on VHR images. *IEEE Geoscience and Remote Sensing Letters*, 7, 53–57.

Burges, J. C., 1998. A tutorial on support vector machines for pattern recognition. *Data Mining and Knowledge Discovery*, 2, 121–167.

Burt, P. J. and Adelson, E. H., 1983. The laplacian pyramid as a compact image code. *IEEE Transactions on Communications*, 31, 532–540.

Busetto, L., Meroni, M., and Colombo, R., 2008. Combining medium and coarse spatial resolution satellite data to improve the estimation of sub-pixel NDVI time series. *Remote Sensing of Environment*, 112, 118–131.

Cammalleri, C., Anderson, M. C., Gao, F., Hain, C. R., and Kustas, W. P., 2014. Mapping daily evapotranspiration at field scales over rainfed and irrigated agricultural areas using remote sensing data fusion. *Agricultural & Forest Meteorology*, 186, 1–11.

Carper, W. J., Lillesand, T. M., and Kiefer, R. W., 1990. The use of intensity-hue-saturation transformations for merging SPOT panchromatic and multispectral image data. *Photogrammetric Engineering and Remote Sensing*, 56, 459–467.

Castanedo, F., 2013. A review of data fusion techniques. *Science World Journal*, 2013, 1–19.

Chang, N.-B., Bai, K., Imen, S., Chen, C.-F., and Gao, W., 2016. Multisensor satellite image fusion and networking for all-weather environmental monitoring. *IEEE Systems Journal*, 1–17.

Chang, Y.-L., Liu, J.-N., Chen, Y.-L., Chang, W.-Y., Hsieh, T.-J., and Huang, B., 2014a. Hyperspectral band selection based on parallel particle swarm optimization and impurity function band prioritization schemes. *Journal of Applied Remote Sensing*, 8, 84798.

Chang, N.-B., Vannah, B. W., Yang, Y. J., and Elovitz, M., 2014b. Integrated data fusion and mining techniques for monitoring total organic carbon concentrations in a lake. *International Journal of Remote Sensing*, 35, 1064–1093.

Chang, Y.-L., Wang, Y. C., Fu, Y.-S., Han, C.-C., Chanussot, J., and Huang, B., 2015. Multisource Data Fusion and Fisher Criterion-Based Nearest Feature Space Approach to Landslide Classification. *IEEE Journal of Selected Topics in Applied Earth Observations and Remote Sensing*, 8, 576–588.

Chatterjee, A., Michalak, A. M., Kahn, R. a., Paradise, S. R., Braverman, A. J., and Miller, C. E., 2010. A geostatistical data fusion technique for merging remote sensing and ground-based observations of aerosol optical thickness. *Journal of Geophysical Research*, 115, D20207.

Chavez, P., 1986. Digital merging of Landsat TM and digitized NHAP data for 1: 24,000- scale image mapping. *Photogrammetric Engineering and Remote Sensing*, 52–10, 1637–1646.

Chavez, P. S. and Bowell, J. A., 1988. Comparison of the spectral information content of Landsat Thematic Mapper and SPOT for three different sites in the Phoenix, Arizona region. *Photogrammetric Engineering and Remote Sensing*, 54, 1699–1708.

Chavez, P. S. and Kwarteng, A. Y., 1989. Extracting spectral contrast in Landsat thematic mapper image data using selective principal component analysis. *Photogrammetric Engineering and Remote Sensing*, 55, 339–348.

Chavez Jr., P. S., Sides, S. C., and Anderson, J. A., 1991. Comparison of three different methods to merge multiresolution and multispectral data: Landsat TM and SPOT panchromatic. *Photogrammetric Engineering and Remote Sensing*, 57, 295–303.

Chen, B., Huang, B., and Xu, B., 2017a. Multi-source remotely sensed data fusion for improving land cover classification. *ISPRS Journal of Photogrammetry and Remote Sensing*, 124, 27–39.

Chen, B., Huang, B., and Xu, B., 2017b. A hierarchical spatiotemporal adaptive fusion model using one image pair. *International Journal of Digital Earth*, 10, 639–655.

Chen, H.-Y. and Leou, J.-J., 2012. Multispectral and multiresolution image fusion using particle swarm optimization. *Multimedia Tools and Applications*, 60, 495–518.

Chen, Y. W. and Li, B. Y., 2006. Remote sensing image fusion based on adaptive RBF neural network. In: *International Conference on Neural Information Processing*, 314–323.

Chen, S., Su, H., Zhang, R., Tian, J., and Yang, L., 2008. The tradeoff analysis for remote sensing image fusion using expanded spectral angle mapper. *Sensors*, 8, 520–528.

Christakos, G. and Li, X., 1998. Bayesian maximum entropy analysis and mapping: A farewell to kriging estimators? *Mathematical Geology*, 30, 435–462.

Dai, X. and Khorram, S., 1999. Data fusion using artificial neural networks: A case study on multitemporal change analysis. *Computers, Environment and Urban Systems*, 23, 19–31.

Dalla Mura, M., Vivone, G., Restaino, R., Addesso, P., and Chanussot, J., 2015. Global and local gram-schmidt methods for hyperspectral pansharpening. In: *2015 IEEE International Geoscience and Remote Sensing Symposium (IGARSS)*, 37–40.

Del Carmen Valdes, M., and Inamura, M., 2001. Improvement of remotely sensed low spatial resolution images by back-propagated neural networks using data fusion techniques. *International Journal of Remote Sensing*, 22, 629–642.

Del Frate, F., Latini, D., Picchiani, M., Schiavon, G., and Vittucci, C., 2014. A neural network architecture combining VHR SAR and multispectral data for precision farming in viticulture. In: *2014 IEEE Geoscience and Remote Sensing Symposium*, 1508–1511.

Dempster, A. P., 1968. A generalization of bayesian inference. *Journal of the Royal Statistical Society Series B*, 30, 205–247.

Ding, S. and Chen, L., 2012. Spectral and spatial feature classification of hyperspectral images based on particle swarm optimisation. *International Journal of Innovative Computing and Applications*, 4, 233–242.

Doña, C., Chang, N.-B., Caselles, V., Sánchez, J. M., Camacho, A., Delegido, J., and Vannah, B. W., 2015. Integrated satellite data fusion and mining for monitoring lake water quality status of the Albufera de Valencia in Spain. *The Journal of Environmental Management*, 151, 416–426.

Dong, C., Loy, C. C., He, K., and Tang, X., 2015b. Image super-resolution using deep convolutional networks. *IEEE Transactions on Pattern Analysis and Machine Intelligence*, 38, 295–307.

Dong, L., Yang, Q., Wu, H., Xiao, H., and Xu, M., 2015a. High quality multi-spectral and panchromatic image fusion technologies based on curvelet transform. *Neurocomputing* 159, 268–274.

Ehlers, M., Klonus, S., Johan Åstrand, P., and Rosso, P., 2010. Multi-sensor image fusion for pansharpening in remote sensing. *International Journal of Image and Data Fusion*, 1, 25–45.

Erturk, A., Gullu, M. K., Cesmeci, D., Gercek, D., and Erturk, S., 2014. Spatial resolution enhancement of hyperspectral images using unmixing and binary particle swarm optimization. *IEEE Geoscience and Remote Sensing Letters*, 11, 2100–2104.

Farebrother, R. W., 1974. Algorithm AS 79: Gram-schmidt regression. *Applied Statistics*, 23, 470.

Fasbender, D., Brasseur, O., and Bogaert, P., 2009. Bayesian data fusion for space-time prediction of air pollutants: The case of NO_2 in Belgium. *Atmospheric Environment*, 43, 4632–4645.

Fasbender, D., Radoux, J., and Bogaert, P., 2008. Bayesian data fusion for adaptable image pansharpening. *IEEE Transactions on Geoscience and Remote Sensing*, 46, 1847–1857.

Fauvel, M., Chanussot, J., and Benediktsson, J. A., 2006. Decision fusion for the classification of urban remote sensing images. *IEEE Transactions on Geoscience and Remote Sensing*, 44, 2828–2838.

Fogel, D. B. and Fogel, L. J., 1996. An Introduction to Evolutionary Programming, In: *Artificial Evolution. AE 1995. Lecture Notes in Computer Science*, 21–33.

Fraser, A. S., 1962. Simulation of genetic systems. *Journal of Theoretical Biology*, 2, 329–346.

Fu, D., Chen, B., Wang, J., Zhu, X., and Hilker, T., 2013. An improved image fusion approach based on enhanced spatial and temporal the adaptive reflectance fusion model. *Remote Sensing*, 5, 6346–6360.

Gao, F., Hilker, T., Zhu, X., Anderson, M., Masek, J., Wang, P., and Yang, Y., 2015. Fusing landsat and MODIS data for vegetation monitoring. *IEEE Geoscience and Remote Sensing Magazine*, 3, 47–60.

Gao, F., Masek, J., Schwaller, M., and Hall, F., 2006. On the blending of the Landsat and MODIS surface reflectance: predicting daily Landsat surface reflectance. *IEEE Transactions on Geoscience and Remote Sensing*, 44, 2207–2218.

Gao, S., Zhong, Y., and Li, W., 2011. Random weighting method for multisensor data fusion. *IEEE Sensors Journal*, 11, 1955–1961.

Garzelli, A., 2015. Pansharpening of multispectral images based on nonlocal parameter optimization. *IEEE Transactions on Geoscience and Remote Sensing*, 53, 2096–2107.

Garzelli, A. and Nencini, F., 2006. PAN-sharpening of very high resolution multispectral images using genetic algorithms. *International Journal of Remote Sensing*, 27, 3273–3292.

Garzelli, A., Nencini, F., and Capobianco, L., 2008. Optimal MMSE pan sharpening of very high resolution multispectral images. *IEEE Transactions on Geoscience and Remote Sensing*, 46, 228–236.

Gevaert, C. M. and García-Haro, F. J., 2015. A comparison of STARFM and an unmixing-based algorithm for Landsat and MODIS data fusion. *Remote Sensing of Environment*, 156, 34–44.

Gharbia, R., Baz, A. H. El, and Hassanien, A. E., 2016. An adaptive image fusion rule for remote sensing images based on the particle swarm optimization. In: *2016 IEEE International Conference on Computing, Communication and Automation (ICCCA)*, 1080–1085.

Gillespie, A. R., Kahle, A. B., and Walker, R. E., 1987. Color enhancement of highly correlated images. II. Channel ratio and "chromaticity" transformation techniques. *Remote Sensing of Environment*, 22, 343–365.

González-Audícana, M., Saleta, J. L. J. L., Catalán, R. G., García, R., Gonzalez-Audicana, M., Saleta, J. L. J. L., Catalan, R. G., and Garcia, R., 2004. Fusion of multispectral and panchromatic images using improved IHS and PCA mergers based on wavelet decomposition. *IEEE Transactions on Geoscience and Remote Sensing*, 42, 1291–1299.

Günlü, A., Ercanlı, İ., Sönmez, T., and Başkent, E. Z., 2014. Prediction of some stand parameters using pan-sharpened IKONOS satellite image. *European Journal of Remote Sensing*, 47, 329–342.

Haberstroh, R. and Kadar, I., 1993. Multispectral data fusion using neural networks. In: *1955, Signal Processing, Sensor Fusion, and Target Recognition II*, 65–75.

Haertel, V., Shimabukuro, Y. E., and Almeida, R., 2004. Fraction images in multitemporal change detection. *International Journal of Remote Sensing*, 25, 5473–5489.

Hilker, T., Wulder, M. A., Coops, N. C., Linke, J., McDermid, G., Masek, J. G., Gao, F., and White, J. C., 2009a. A new data fusion model for high spatial- and temporal-resolution mapping of forest disturbance based on Landsat and MODIS. *Remote Sensing of Environment*, 113, 1613–1627.

Hilker, T., Wulder, M. A., Coops, N. C., Seitz, N., White, J. C., Gao, F., Masek, J. G., and Stenhouse, G., 2009b. Generation of dense time series synthetic Landsat data through data blending with MODIS using a spatial and temporal adaptive reflectance fusion model. *Remote Sensing of Environment*, 113, 1988–1999.

Huang, B. and Song, H., 2012. Spatiotemporal reflectance fusion via sparse representation. *IEEE Transactions on Geoscience and Remote Sensing*, 50, 3707–3716.

Huang, B., Wang, J., Song, H., Fu, D., and Wong, K., 2013. Generating high spatiotemporal resolution land surface temperature for urban heat island monitoring. *IEEE Geoscience and Remote Sensing Letters*, 10, 1011–1015.

Huang, W., Xiao, L., Wei, Z., Liu, H., and Tang, S., 2015. A new pan-sharpening method with deep neural networks. *IEEE Geoscience and Remote Sensing Letters*, 12, 1037–1041.

Imen, S., Chang, N.-B., and Yang, Y. J., 2015. Developing the remote sensing-based early warning system for monitoring TSS concentrations in Lake Mead. *Journal of Environmental Management*, 160, 73–89.

Jimenez, L. O., Morales-Morell, A., and Creus, A., 1999. Classification of hyperdimensional data based on feature and decision fusion approaches using projection pursuit, majority voting, and neural networks. *IEEE Transactions on Geoscience and Remote Sensing*, 37, 1360–1366.

Jing, L., Wang, T., Zhao, M., and Wang, P., 2017. An adaptive multi-sensor data fusion method based on deep convolutional neural networks for fault diagnosis of planetary gearbox. *Sensors*, 17, 414.

Kellndorfer, J. M., Walker, W. S., Lapoint, E., Kirsch, K., Bishop, J., and Fiske, G., 2010. Statistical fusion of lidar, InSAR, and optical remote sensing data for forest stand height characterization: A regional-scale method based on LVIS, SRTM, Landsat ETM+, and ancillary data sets. *Journal of Geophysical Research: Biogeosciences* 115, 1–10.

Kennedy, J. and Eberhart, R., 1995. Particle swarm optimization. In: *Proceedings of ICNN'95-International Conference on Neural Networks*. 1942–1948.

Khaleghi, B., Khamis, A., Karray, F. O., and Razavi, S. N., 2013. Multisensor data fusion: A review of the state-of-the-art. *Information Fusion*, 14, 28–44.

Kresslerl, F. P. and Steinnocher, K. T., 1999. Detecting land cover changes from NOAA-AVHRR data by using spectral mixture analysis. *International Journal of Applied Earth Observation and Geoinformation*, 1, 21–26.

Kusetogullari, H., Yavariabdi, A., and Celik, T., 2015. Unsupervised change detection in multitemporal multispectral satellite images using parallel particle swarm optimization. *IEEE Journal of Selected Topics in Applied Earth Observations and Remote Sensing*, 8, 2151–2164.

Laben, C. and Brower, B., 2000. Process for enhancing the spatial resolution of multispectral imagery using pan-sharpening. United States Patent, 6.

Lacewell, C. W., Gebril, M., Buaba, R., and Homaifar, A., 2010. Optimization of image fusion using genetic algorithms and discrete wavelet transform. In: *Proceedings of the IEEE 2010 National Aerospace & Electronics Conference*, 116–121.

Lanckriet, G. R. G., De Bie, T., Cristianini, N., Jordan, M. I., and Noble, W. S., 2004. A statistical framework for genomic data fusion. *Bioinformatics*, 20, 2626–2635.

Laporterie-Déjean, F., de Boissezon, H., Flouzat, G., and Lefèvre-Fonollosa, M.-J., 2005. Thematic and statistical evaluations of five panchromatic/multispectral fusion methods on simulated PLEIADES-HR images. *Information Fusion*, 6, 193–212.

Li, S. Z., 2009. *Markov Random Field Modeling in Image Analysis*. Computer Science Workbench. Springer-Verlag, London, UK.

Li, A., Bo, Y., and Chen, L., 2013a. Bayesian maximum entropy data fusion of field-observed leaf area index (LAI) and Landsat Enhanced Thematic Mapper Plus-derived LAI. *International Journal of Remote Sensing*, 34, 227–246.

Li, A., Bo, Y., Zhu, Y., Guo, P., Bi, J., and He, Y., 2013b. Blending multi-resolution satellite sea surface temperature (SST) products using Bayesian maximum entropy method. *Remote Sensing of Environment*, 135, 52–63.

Li, M., Cai, W., and Tan, Z., 2006. A region-based multi-sensor image fusion scheme using pulse-coupled neural network. *Pattern Recognition Letters*, 27, 1948–1956.

Li, J. and Peng, Z., 2015. Multi-source image fusion algorithm based on cellular neural networks with genetic algorithm. *Optik-International Journal for Light and Electron Optics*, 126, 5230–5236.

Liang, J., He, Y., Liu, D., and Zeng, X., 2012. Image fusion using higher order singular value decomposition. *IEEE Trans. Image processing*, 21, 2898–2909.

Licciardi, G. A., Khan, M. M., Chanussot, J., Montanvert, A., Condat, L., Jutten, C., 2012. Fusion of hyperspectral and panchromatic images using multiresolution analysis and nonlinear PCA band reduction. *EURASIP Journal on Advances in Signal Processing*, 2012, 207.

Liu, Z., Chai, Y., Yin, H., Zhou, J., and Zhu, Z., 2017a. A novel multi-focus image fusion approach based on image decomposition. *Information Fusion*, 35, 102–116.

Liu, Y., Chen, X., Peng, H., and Wang, Z., 2017b. Multi-focus image fusion with a deep convolutional neural network. *Information Fusion*, 36, 191–207.

Liu, M., Liu, X., Li, J., Ding, C., and Jiang, J., 2014. Evaluating total inorganic nitrogen in coastal waters through fusion of multi-temporal RADARSAT-2 and optical imagery using random forest algorithm. *International Journal of Applied Earth Observation and Geoinformation*, 33, 192–202.

Liu, H., Weng, Q., 2012. Enhancing temporal resolution of satellite imagery for public health studies: A case study of west nile virus outbreak in Los Angeles in 2007. *Remote Sensing of Environment*, 117, 57–71.

Loncan, L., De Almeida, L. B., Bioucas-Dias, J. M., Briottet, X., Chanussot, J., Dobigeon, N., Fabre, S. et al., 2015. Hyperspectral pansharpening: A review. *IEEE Geoscience and Remote Sensing Magazine*, 3, 27–46.

Lu, L., Xie, W., Zhang, J., Huang, G., Li, Q., and Zhao, Z., 2015. Woodland extraction from high-resolution CASMSAR data based on Dempster-Shafer evidence theory fusion. *Remote Sensing*, 7, 4068–4091.

Mascarenhas, N. D. A., Banon, G. J. F., and Candeias, A. L. B., 1996. Multispectral image data fusion under a Bayesian approach. *International Journal of Remote Sensing*, 17, 1457–1471.

Masi, G., Cozzolino, D., Verdoliva, L., and Scarpa, G., 2016. Pansharpening by convolutional neural networks. *Remote Sensing*, 8, 594.

Melanie, M., 1998. *An Introduction to Genetic Algorithms*. The MIT Press, Cambridge, MA, USA.

Melgani, F., 2004. Classification of multitemporal remote-sensing images by a fuzzy fusion of spectral and spatio-temporal contextual information. *International Journal of Pattern Recognition and Artificial Intelligence*, 18, 143–156.

Meng, Q., Borders, B., and Madden, M., 2010. High-resolution satellite image fusion using regression kriging. *International Journal of Remote Sensing*, 31, 1857–1876.

Mitchell, H. B., 2007. *Multi-Sensor Data Fusion: An Introduction*. Springer, Berlin, Heidelberg, Germany.

Mohammadzadeh, A. and Zoej, J. V., 2006. Road extraction based on fuzzy logic and mathematical morphology from pan-sharpened ikonos images. *Photogrammetric Record*, 21, 44–60.

Moosavi, V., Talebi, A., Mokhtari, M. H., Shamsi, S. R. F., and Niazi, Y., 2015. A wavelet-artificial intelligence fusion approach (WAIFA) for blending Landsat and MODIS surface temperature. *Remote Sensing of Environment*, 169, 243–254.

Mumtaz, A., Majid, A., and Mumtaz, A., 2008. Genetic algorithms and its application to image fusion. In: *2008 IEEE the 4th International Conference on Emerging Technologies*, 6–10.

Nguyen, H., Cressie, N., and Braverman, A., 2012. Spatial statistical data fusion for remote sensing applications. *Journal of the American Statistical Association*, 107, 1004–1018.

Nguyen, H., Franke, K., and Petrovic, S., 2009. Optimizing a class of feature selection measures. In: *NIPS 2009 Workshop on Discrete Optimization in Machine Learning: Submodularity, Sparsity & Polyhedra (DISCML)*, Vancouver, Canada.

Nishii, R., 2003. A markov random field-based approach to decision-level fusion for remote sensing image classification. *IEEE Transactions on Geoscience and Remote Sensing*, 41, 2316–2319.

Nishii, R., Kusanobu, S., and Tanaka, S., 1996. Enhancement of low spatial resolution image based on high resolution bands. *IEEE Transactions on Geoscience and Remote Sensing*, 34, 1151–1158.

Palsson, F., Sveinsson, J. R., Ulfarsson, M. O., 2017. Multispectral and hyperspectral image fusion using a 3-D-convolutional neural network. *IEEE Geoscience and Remote Sensing Letters*, 14, 639–643.

Pardo-Igúzquiza, E., Chica-Olmo, M., and Atkinson, P. M., 2006. Downscaling cokriging for image sharpening. *Remote Sensing of Environment*, 102, 86–98.

Park, J. H. and Kang, M. G., 2004. Spatially adaptive multi-resolution multispectral image fusion. *International Journal of Remote Sensing*, 25, 5491–5508.

Peng, X. and Dang, A., 2010. Hybrid genetic algorithm (GA)-based neural network for multispectral image fusion. In: *2010 IEEE International Geoscience and Remote Sensing Symposium*, 496–498.

Pohl, C. and Van Genderen, J. L., 1998. Multisensor image fusion in remote sensing: Concepts, methods and applications. *International Journal of Remote Sensing*, 19, 823–854.

Price, J. C., 1987. Combining panchromatic and multispectral imagery from dual resolution satellite instruments. *Remote Sensing of Environment*, 21, 119–128.

Price, J. C., 1999. Combining multispectral data of differing spatial resolution. *IEEE Transactions on Geoscience and Remote Sensing*, 37, 1199–1203.

Puttaswamy, S. J., Nguyen, H. M., Braverman, A., Hu, X., and Liu, Y., 2013. Statistical data fusion of multi-sensor AOD over the continental United States. *Geocarto International*, 29, 48–64.

Raghavendra, R., Dorizzi, B., Rao, A., and Hemantha Kumar, G., 2011. Particle swarm optimization based fusion of near infrared and visible images for improved face verification. *Pattern Recognition*, 44, 401–411.

Rahmani, S., Strait, M., Merkurjev, D., Moeller, M., and Wittman, T., 2010. An adaptive IHS pan-sharpening method. *IEEE Geoscience and Remote Sensing Letters*, 7, 746–750.

Ranchin, T. and Wald, L., 2000. Fusion of high spatial and spectral resolution images: The ARSIS concept and its implementation. *Photogrammetric Engineering and Remote Sensing*, 66, 49–61.

Rawlins, G. J. E. *(editor)*, 1991. *Foundations of Genetic Algorithms*. Morgan Kaufmann, San Mateo, CA.

Saeedi, J. and Faez, K., 2011. A new pan-sharpening method using multiobjective particle swarm optimization and the shiftable contourlet transform. *ISPRS Journal of Photogrammetry and Remote Sensing*, 66, 365–381.

Sales, M. H. R., Souza, C. M., and Kyriakidis, P. C., 2013. Fusion of MODIS images using kriging with external drift. *IEEE Transactions on Geoscience and Remote Sensing*, 51, 2250–2259.

Schowengerdt, R. A., 2006. *Remote Sensing: Models and Methods for Image Processing*. Academic Press, Cambridge, Massachusetts, USA; 3 edition (September 11, 2006)

Siddiqui, A. B., Arfan Jaffar, M., Hussain, A., and Mirza, A. M., 2011. Block-based pixel level multi-focus image fusion using particle swarm optimization. *International Journal of Innovative Computing, Information and Control*, 7, 3583–3596.

Settle, J. J. and Drake, N. A., 1993. Linear mixing and the estimation of ground cover proportions. *International Journal of Remote Sensing*, 14, 1159–1177.

Shafer, G., 1976. *A Mathematical Theory of Evidence*. Princeton University Press, Victoria.

Shah, V. P., Younan, N. H., and King, R. L., 2008. An efficient pan-sharpening method via a combined adaptive PCA approach and contourlets. *IEEE Transactions on Geoscience and Remote Sensing*, 46, 1323–1335.

Shettigara, V. K., 1992. A generalized component substitution technique for spatial enhancement of multispectral images using a higher resolution data set. *Photogrammetric Engineering and Remote Sensing*, 58, 561–567.

Shi, Y., Zhou, X., Yang, X., Shi, L., and Ma, S., 2015. Merging satellite ocean color data with Bayesian maximum entropy method. *IEEE Journal of Selected Topics in Applied Earth Observations and Remote Sensing*, 8, 3294–3304.

Solaiman, B., Pierce, L. E., and Ulaby, F. T., 1999. Multisensor data fusion using fuzzy concepts: Application to land-cover classification using ERS-1/JERS-1 SAR composites. *IEEE Transactions on Geoscience and Remote Sensing*, 37, 1316–1326.

Solberg, A. H. S., Jain, A. K., and Taxt, T., 1994. Multisource classification of remotely sensed data: fusion of Landsat TM and SAR images. *IEEE Transactions on Geoscience and Remote Sensing*, 32, 768–778.

Soliman, O. S., Mahmoud, A. S., and Hassan, S. M., 2012. Remote sensing satellite images classification using support vector machine and particle swarm optimization. In: *2012 IEEE Third International Conference on Innovations in Bio-Inspired Computing and Applications*, 280–285.

Song, H. and Huang, B., 2013. Spatiotemporal satellite image fusion through one-pair image learning. *IEEE Transactions on Geoscience and Remote Sensing*, 51, 1883–1896.

Souza, C. M., Firestone, L., Silva, L. M., and Roberts, D., 2003. Mapping forest degradation in the Eastern Amazon from SPOT 4 through spectral mixture models. *Remote Sensing of Environment*, 87, 494–506.

Srivastava, P. K., Han, D., Rico-Ramirez, M. A., Al-Shrafany, D., and Islam, T., 2013. Data fusion techniques for improving soil moisture deficit using SMOS satellite and WRF-NOAH land surface model. *Water Resources Management*, 27, 5069–5087.

Sun, J., Zhu, H., Xu, Z., and Han, C., 2013. Poisson image fusion based on Markov random field fusion model. *Information Fusion*, 14, 241–254.

Tang, Q., Bo, Y., and Zhu, Y., 2016. Spatiotemporal fusion of multiple-satellite aerosol optical depth (AOD) products using Bayesian maximum entropy method. *Journal of Geophysical Research: Atmospheres*, 121, 4034–4048.

Tewes, A., Thonfeld, F., Schmidt, M., Oomen, R. J., Zhu, X., Dubovyk, O., Menz, G., and Schellberg, J., 2015. Using RapidEye and MODIS data fusion to monitor vegetation dynamics in semi-arid rangelands in South Africa. *Remote Sensing*, 7, 6510–6534.

Thomas, C., Ranchin, T., Wald, L., and Chanussot, J., 2008. Synthesis of multispectral images to high spatial resolution: A critical review of fusion methods based on remote sensing physics. *IEEE Transactions on Geoscience and Remote Sensing*, 46, 1301–1312.

Toet, A., 1989. A morphological pyramidal image decomposition. *Pattern Recognition, Pattern Recognition Letters*, 9, 255–261.

Tsai, V. J. D., 2003. Frequency-based fusion of multiresolution images. In: *2003 IEEE International Geoscience and Remote Sensing Symposium*, 3665–3667.

Tu, T.-M., Huang, P. S., Hung, C.-L., and Chang, C.-P., 2004. A fast intensity–hue–saturation fusion technique with spectral adjustment for IKONOS imagery. *IEEE Geoscience and Remote Sensing Letters*, 1, 309–312.

Tu, T.-M., Su, S.-C., Shyu, H.-C., and Huang, P. S., 2001. A new look at IHS-like image fusion methods. *Information Fusion*, 2, 177–186.

Tupin, F., Bloch, I., and Maitre, H., 1999. A first step toward automatic interpretation of SAR images using evidential fusion of several structure detectors. *IEEE Transactions on Geoscience and Remote Sensing*, 37, 1327–1343.

Valdés Hernández, M. del C., and Inamura, M., 2000. Spatial resolution improvement of remotely sensed images by a fully interconnected neural network approach. *IEEE Transactions on Geoscience and Remote Sensing*, 38, 2426–2430.

Vandenberghe, L. and Boyd, S., 1996. Semidefinite Programming. *Society for Industrial and Applied Mathematics*, 38, 49–95.

van der Wal, D. and Herman, P. M. J., 2007. Regression-based synergy of optical, shortwave infrared and microwave remote sensing for monitoring the grain-size of intertidal sediments. *Remote Sensing of Environment*, 111, 89–106.

Vapnick, V. N., 1998. *Statistical Learning Theory*. Wiley, New York.

Vijayaraj, V., O'Hara, C. G., and Younan, N. H., 2004. Quality analysis of pansharpened images. In: *2004 IEEE International Geoscience and Remote Sensing Symposium*, 1–4.

Vivone, G., Alparone, L., Chanussot, J., Dalla Mura, M., Garzelli, A., Licciardi, G. A., Restaino, R., and Wald, L., 2015. A critical comparison among pansharpening algorithms. *IEEE Transactions on Geoscience and Remote Sensing*, 53, 2565–2586.

Wald, L., Ranchin, T., and Mangolini, M., 1997. Fusion of satellite images of different spatial resolutions: Assessing the quality of resulting images. *Photogrammetric Engineering and Remote Sensing*, 63, 691–699.

Walker, J. J., de Beurs, K. M., Wynne, R. H., and Gao, F., 2012. Evaluation of Landsat and MODIS data fusion products for analysis of dryland forest phenology. *Remote Sensing of Environment*, 117, 381–393.

Wang, H., Fan, T., Shi, A., Huang, F., and Wang, H., 2010. Fuzzy integral based information fusion for water quality monitoring using remote sensing data. *International Journal of Communications, Network and System Sciences*, 3, 737–744.

Wang, Q., Shi, W., and Atkinson, P. M., 2016. Area-to-point regression kriging for pan-sharpening. *ISPRS Journal of Photogrammetry and Remote Sensing*, 114, 151–165.

Wang, Q., Shi, W., Atkinson, P. M., and Wei, Q., 2017a. Approximate area-to-point regression kriging for fast hyperspectral image sharpening. *IEEE Journal of Selected Topics in Applied Earth Observations and Remote Sensing*, 10, 286–295.

Wang, Q., Shi, W., Atkinson, P. M., and Zhao, Y., 2015a. Downscaling MODIS images with area-to-point regression kriging. *Remote Sensing of Environment*, 166, 191–204.

Wang, W., Jiao, L., and Yang, S., 2014. Fusion of multispectral and panchromatic images via sparse representation and local autoregressive model. *Information Fusion*, 20, 73–87.

Wang, W., Jiao, L., and Yang, S., 2015b. Novel adaptive component-substitution-based pan-sharpening using particle swarm optimization. *IEEE Geoscience and Remote Sensing Letters*, 12, 781–785.

Wang, Q., Zhang, Y., Onojeghuo, A. O., Zhu, X., and Atkinson, P. M., 2017b. Enhancing spatio-temporal fusion of MODIS and Landsat data by incorporating 250 m MODIS data. *IEEE Journal of Selected Topics in Applied Earth Observations and Remote Sensing*, 10, 4116–4123.

Wang, Z., Ziou, D., Armenakis, C., Li, D., and Li, Q., 2005. A comparative analysis of image fusion methods. *IEEE Transactions on Geoscience and Remote Sensing*, 43, 1391–1402.

Waske, B. and Benediktsson, J. A., 2007. Fusion of support vector machines for classification of multisensor data. *IEEE Transactions on Geoscience and Remote Sensing*, 45, 3858–3866.

Wei, J., Wang, L., Liu, P., and Song, W., 2016. Spatiotemporal fusion of remote sensing images with structural sparsity and semi-coupled dictionary learning. *Remote Sensing*, 9, 21.

Welch, R. and Ehlers, M., 1987. Merging multiresolution SPOT HRV and Landsat TM data. *Photogrammetric Engineering and Remote Sensing*, 53, 301–303.

Weng, Q., Fu, P., and Gao, F., 2014. Generating daily land surface temperature at Landsat resolution by fusing Landsat and MODIS data. *Remote Sensing of Environment*, 145, 55–67.

Wong, F. H. and Orth, R., 1980. Registration of SEASAT/LANDSAT composite images to UTM coordinates, In: Proceedings of the Sixth Canadian Syinposium on Remote Sensing, 161–164.

Wu, M., Niu, Z., Wang, C., Wu, C., and Wang, L., 2012. Use of MODIS and Landsat time series data to generate high-resolution temporal synthetic Landsat data using a spatial and temporal reflectance fusion model. *Journal of Applied Remote Sensing*, 6, 63507.

Xu, M., Chen, H., and Varshney, P. K., 2011. An image fusion approach based on Markov random fields. *IEEE Transactions on Geoscience and Remote Sensing*, 49, 5116–5127.

Xu, H., Guang, J., Xue, Y., de Leeuw, G., Che, Y. H., Guo, J., He, X. W., and Wang, T. K., 2015a. A consistent aerosol optical depth (AOD) dataset over mainland China by integration of several AOD products. *Atmospheric Environment*, 114, 48–56.

Xu, Y., Huang, B., Xu, Y., Cao, K., Guo, C., and Meng, D., 2015b. Spatial and temporal image fusion via regularized spatial unmixing. *IEEE Geoscience and Remote Sensing Letters*, 12, 1362–1366.

Yang, X.-S., Deb, S., and Fong, S., 2011. Accelerated particle swarm optimization and support vector machine for business optimization and applications. In: *Networked Digital Technologies: Third International Conference, NDT 2011*, 53–66. Macau, China, July 11–13, 2011.

Yang, H., Du, Q., and Chen, G., 2012a. Particle swarm optimization-based hyperspectral dimensionality reduction for urban land cover classification. *IEEE Journal of Selected Topics in Applied Earth Observations and Remote Sensing*, 5, 544–554.

Yang, B. and Li, S., 2010. Multifocus image fusion and restoration with sparse representation. *IEEE Transactions on Instrumentation and Measurement*, 59, 884–892.

Yang, G., Pu, R., Huang, W., Wang, J., and Zhao, C., 2010. A novel method to estimate subpixel temperature by fusing solar-reflective and thermal-infrared remote-sensing data with an artificial neural network. *IEEE Transactions on Geoscience and Remote Sensing*, 48, 2170–2178.

Yang, S., Wang, M., and Jiao, L., 2012b. Fusion of multispectral and panchromatic images based on support value transform and adaptive principal component analysis. *Information Fusion*, 13, 177–184.

Yang, S., Zeng, L., Jiao, L., and Xiao, J., 2007. Fusion of multispectral and panchromatic images using improved GIHS and PCA mergers based on contourlet. In: *Proceedings of the SPIE*, Volume 6787, Multispectral Image Processing.

Yokoya, N., Yairi, T., and Iwasaki, A., 2012. Coupled nonnegative matrix factorization unmixing for hyperspectral and multispectral data fusion. *IEEE Transactions on Geoscience and Remote Sensing*, 50, 528–537.

Zadeh, L. A., 1965. Fuzzy sets. *Information and Control*, 8, 338–353.

Zhan, Z. H., Zhang, J., Li, Y., and Chung, H. S.-H., 2009. Adaptive particle swarm optimization. *IEEE Transactions on Systems, Man, and Cybernetics, Part B (Cybernetics)*, 39, 1362–1381.

Zhan, Z.-H., Zhang, J., Li, Y., and Shi, Y.-H., 2011. Orthogonal learning particle swarm optimization. *IEEE Transactions on Evolutionary Computation*, 15, 832–847.

Zhang, J., 2010. Multi-source remote sensing data fusion: Status and trends. *International Journal of Image and Data Fusion*, 1, 5–24.

Zhang, Y., Atkinson, P. M., Ling, F., Wang, Q., Li, X., Shi, L., and Du, Y., 2017. Spectral–spatial adaptive area-to-point regression Kriging for MODIS image downscaling. *IEEE Journal of Selected Topics in Applied Earth Observations and Remote Sensing*, 10, 1883–1896.

Zhang, Q. and Guo, B. L., 2009. Multifocus image fusion using the nonsubsampled contourlet transform. *Signal Processing*, 89, 1334–1346.

Zhang, H. and Huang, B., 2015. A new look at image fusion methods from a Bayesian perspective. *Remote Sensing*, 7, 6828–6861.

Zhang, W., Li, A., Jin, H., Bian, J., Zhang, Z., Lei, G., Qin, Z., and Huang, C., 2013. An enhanced spatial and temporal data fusion model for fusing Landsat and MODIS surface reflectance to generate high temporal Landsat-like data. *Remote Sensing*, 5, 5346–5368.

Zhang, F., Zhu, X., and Liu, D., 2014. Blending MODIS and Landsat images for urban flood mapping. *International Journal of Remote Sensing*, 35, 3237–3253.

Zhong, J., Yang, B., Huang, G., Zhong, F., and Chen, Z., 2016. Remote sensing image fusion with convolutional neural network. *Sensing and Imaging*, 17, 10.

Zhou, J., Civco, D. L., and Silander, J. A., 1998. A wavelet transform method to merge Landsat TM and SPOT panchromatic data. *International Journal of Remote Sensing*, 19, 743–757.

Zhou, J., Liu, L., Guo, J., and Sun, L., 2013. Multisensor data fusion for water quality evaluation using Dempster-Shafer evidence theory. *International Journal Of Distributed Sensor Networks*, 2013, 1–6.

Zhu, X., Chen, J., Gao, F., Chen, X., and Masek, J. G., 2010. An enhanced spatial and temporal adaptive reflectance fusion model for complex heterogeneous regions. *Remote Sensing of Environment*, 114, 2610–2623.

Zhu, X., Helmer, E. H., Gao, F., Liu, D., Chen, J., and Lefsky, M. A., 2016. A flexible spatiotemporal method for fusing satellite images with different resolutions. *Remote Sensing of Environment*, 172, 165–177.

Zhukov, B., Oertel, D., Lanzl, F., and Reinhäckel, G., 1999. Unmixing-based multisensor multiresolution image fusion. *IEEE Transactions on Geoscience and Remote Sensing*, 37, 1212–1226.

Zurita-Milla, R., Clevers, J. G. P. W., and Schaepman, M. E., 2008. Unmixing-based landsat TM and MERIS FR data fusion. *IEEE Geoscience and Remote Sensing Letters*, 5, 453–457.

Zurita-Milla, R., Clevers, J. G. P. W., Van Gijsel, J. A. E., and Schaepman, M. E., 2011. Using MERIS fused images for land-cover mapping and vegetation status assessment in heterogeneous landscapes. *International Journal of Remote Sensing*, 32, 973–991.

Zurita-Milla, R., Kaiser, G., Clevers, J. G. P. W., Schneider, W., and Schaepman, M. E., 2009. Downscaling time series of MERIS full resolution data to monitor vegetation seasonal dynamics. *Remote Sensing of Environment*, 113, 1874–1885.

10 System Design of Data Fusion and the Relevant Performance Evaluation Metrics

10.1 INTRODUCTION

Since the 1990s, numerous image and data fusion algorithms have been developed to deal with different practical issues based on synergistic remote sensing data with improved spatial, temporal, and spectral resolution. As more remote sensing data become available, there often exist multiple solutions through different data fusion alternatives in dealing with a unique problem. With the recent advancements in remote sensing technologies and data fusion theories, decisions in regard to which solution is most suitable for the problem at hand is a critical challenge. The two intertwined questions that need to be addressed are: (1) what remote sensing data should be used to fit in the prescribed data fusion framework, and (2) which data fusion method should be applied to deal with the given problem based on the available data sources. Two real-world cases are illustrated in the next section to provide a holistic understanding of this intertwined argument as well as bring self-awareness to the system design of plausible data fusion frameworks.

Following the boom in development of various data fusion algorithms, as described in previous chapter, performance evaluation of data fusion techniques in a quantitative manner has become essential to the proper assessment of the data fusion knowledge base. Since the initialization of data fusion techniques in the mid-1990s, quality assurance approaches have been developed and applied to evaluate the performance of the employed data fusion techniques. Performance evaluation metrics thus play a critical role in determining the effectiveness of the data fusion outputs. Broadly, quality assurance approaches used in the data fusion domain can be classified into two categories, either subjective or objective. When taking image fusion into account, the most direct way to assess the quality of image fusion is to perform a simple visual analysis on the fused image in which a visual plausible judgement always indicates good quality of the fused output and vice versa. However, quality assurance through such a method is subjective, as the final determined quality depends largely on the visual perception of the observers as well as the viewing conditions. Different judgements could be made under distinct viewing conditions by the same observer, and different observers may give conflicting conclusions for the same target image even under identical viewing conditions due to the different visual perception (or sensitivity) of each individual observer.

At an early stage of data fusion, many attempts were directed to derive certain quality assurance measures based on a thorough understanding of the human visual system. Despite the past endeavor, however, none of these measures have been attained since the visual system has not been well understood (Wang et al., 2008). In addition, in most cases, the fused images exhibit a certain level of information loss or distortions that are hard to detect through visual perception alone, in particular for hyperspectral images in which the information is abundant. Moreover, it is also time-consuming and labor-intensive to check a huge number of fused images visually. Therefore, a general, intuitive, and objective performance measure is essential for providing fair assessments of the fused outputs in order to compare different data fusion techniques with a similar purpose.

To date, a suite of statistical (or mathematical) indicators and metrics has been developed or adapted to evaluate the quality of data fusion outputs in a quantitative manner. Compared to the traditional visual analysis and other qualitative approaches, statistical measures are more attractive. First of all, statistical measures are able to provide a more general and objective assessment of the fused outputs by calculating a single statistic value based on a fixed equation to indicate the quality of the target image. Meanwhile, statistical measures enable us to perform a comparison of the proposed data fusion techniques directly and intuitively toward a robust final conclusion. Last but not least, statistical measures are independent of observers and viewing conditions. In other words, such measures can be used in a more general and universal manner on various targets without taking consideration of external factors.

This chapter is organized to demonstrate two system design exercises for practical data fusion applications to provide a holistic comprehension of the two intertwined questions mentioned above. Following that logic flow, a suite of performance evaluation metrics is introduced mathematically before the final summary of this chapter.

10.2 SYSTEM DESIGN OF SUITABLE DATA FUSION FRAMEWORKS

In order to provide holistic comprehension of how an appropriate data fusion framework may be designed to be most suitable for solving the unique problem at hand, this section demonstrates two cases of system design for practical data fusion applications. The logic flows and basic principles lying behind the scene are detailed in each demonstrated case, guiding the manner of selecting data sources and fusion methods and organizing a proper data fusion framework toward possible implementations. For comparison, one case considers forest species discrimination and distribution mapping (pixel-level fusion) and the other case deals with landslide detection (decision-level fusion).

10.2.1 System Design for Data Fusion—Case 1

Case 1 is designed to conduct multispectral and hyperspectral data fusion to aid in forest species discrimination and distribution mapping. First, an understanding of forest dynamics in a particular area is of critical importance in local ecosystem conservation. Mongolia has few forest resources in contrast to other countries, hence a detailed investigation of the species composition and distribution in Mongolia is essential for creating efficient forest preservation and management policies. Due to the large size of this country, ground-level field surveys for the assessment of tree species distribution is by no means an easy task. Such work is often laborious, expensive, and time-consuming. There is no doubt that remote sensing techniques emerge as promising tools for dealing with such challenging tasks. Toward such a goal, multisource data fusion-aided feature extraction should be applied, with data fusion aiming to combine the comparative advantages of each data source to gain better spatial, temporal, and spectral resolution for accurate species discrimination.

In order to construct an appropriate data fusion framework for such a task, the first step is to determine what data sources should be used. With the fast advancement of remote sensor technologies, a suite of Very High Resolution (VHR) optical sensors (like SPOT, QuickBird, and IKONOS), hyperspectral sensors, SAR, and even LiDAR mounted on either air-borne or space-borne platforms, have become available for meeting various research challenges. Regarding forest mapping, VHR remotely sensed imageries, such as QuickBird, IKONOS, SPOT, and Landsat, have proven more valuable than older generation sensors in discriminating various forest species (Wang et al., 2004b). In addition, hyperspectral imageries taken from satellite/air-borne sensors (like AVIRIS and Hyperion) have been recognized as particularly useful data for discriminating different species of the same genus due to their great potential for characterizing forest species distribution, leaf area, and stand biomass with increased spectral information (Martin et al., 1998; Clark et al., 2005; Dalponte et al., 2008; Stavrakoudis et al., 2012). However, it is not viable when using the limited spectral information provided by VHR multispectral images. Therefore, it is valuable to fuse VHR multispectral images with hyperspectral imageries to explore the spectral differences of tree species for discrimination and mapping purposes. For instance, a fine spatial resolution Landsat data set (30-m, 7 channels in

0.45−2.35 μm) can be fused with hyperspectral MERIS (300-m, 15 channels in 412.5−900 nm) data to gain better spatial and spectral resolution for species discrimination.

Once the data sources for data fusion are determined, the next critical step is to construct the data fusion framework by selecting an optimal data fusion method. As described in Chapter 9, numerous data fusion approaches are available at present. A common form of data fusion based on multispectral and hyperspectral imageries is similar to pan-sharpening performed on multispectral imagery that incorporates co-registration and resampling at the pixel level; there are three approaches to be considered in this context, including pixel-level fusion such as spectral unmixing, feature-level fusion such as image stacking, and decision-level fusion.

In order to determine which data fusion approach should be used, we need to first have a clear idea of the advantages of each individual algorithm. The spectral unmixing-based data fusion approach works mainly by relying on linear mixing model to combine two images acquired at different spatial resolutions, aiming to downscale a low spatial resolution image co-registered with a high spatial resolution dataset. Such a linear mixing model assumes that the spectrum of a mixed pixel is a linear combination of the pure spectra of the components present in that pixel weighted by their fractional coverage (Settle and Drake, 1993). Thus, it is possible to use this linear mixing model to retrieve their sub-pixel proportions based on the given spectra information of tree species (i.e., pure spectra). Despite the sound physical basis of the linear mixing model in analyzing the mixed pixel, resampling hyperspectral imagery with hundreds of bands into the resolution of a VHR multispectral imagery is often challenging.

The second method of data fusion is the so-called vector or image stacking approach, which is performed at the feature level by aggregating input vectors from multiple sources to produce a new composite feature space for subsequent classification purposes. Such applications may include, but are not limited to, the combination of multispectral or hyperspectral data with SAR (Hong et al., 2009; Byun et al., 2013) or with LiDAR information (Dalponte et al., 2008; García et al., 2011; Man et al., 2015) as well as fusion with new textural features derived from the initial image (Fauvel et al., 2008) or features obtained through object-based image analysis (Huang and Zhang, 2010, 2012) based on a single source of primary information.

The last approach is decision-level fusion, which often employs an independent classifier for each data source and then combines the outputs from all the classifiers using a hard or soft fusion operator for final mapping. For this reason, such a method is also commonly referred to as classifier fusion (Fauvel et al., 2006; Bigdeli et al., 2014a,b). With respect to hyperspectral data, decision-level fusion can be used to deal with the high dimensionality of hyperspectral data by performing independent classification in their native resolution for each band data source with multiple trained classifiers on different feature subsets (Ceamanos et al., 2010; Dalponte et al., 2012; Bigdeli et al., 2014a). A variety of classifiers, such as Bayesian reasoning, Dempster-Shafer (D-S) evidence reasoning, voting system, cluster analysis, fuzzy set theory, artificial neural networks, and the entropy method, can be used for decision-level fusion. Although the computational burden and data volume produced will be significantly reduced, such a data fusion framework is always complex to some extent.

According to the simple comparison of the three kinds of data fusion approaches detailed above, it is noticeable that spatial unmixing is more appropriate for fusing Landsat and MERIS data sets for the following two reasons. First, spatial unmixing has a good physical basis when using linear mixing model for dealing with mixed pixels than any other method. Second, the pixel-level fusion framework is easy to construct and implement due to its simple and straightforward modeling process. A spatial unmixing-based data fusion framework can then be constructed to generate Landsat-like MERIS images for forest species discrimination and distribution mapping. An illustrative data fusion framework is presented in Figure 10.1 and can be briefly summarized as follows. First, the fine spatial resolution Landsat image is segmented into K classes by making use of an unsupervised clustering algorithm (i.e., data mining), and the membership of each Landsat pixel to the clusters is also calculated. Next, these posterior probabilities at Landsat resolution are used to calculate the abundance or proportion for each MERIS pixel. Then, each MERIS pixel is

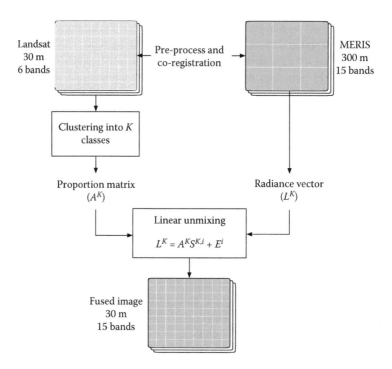

FIGURE 10.1 An illustrative sketch of using spatial unmixing-based fusion framework to fuse Landsat and MERIS imageries.

unmixed in a window through linear mixture equations. Finally, the fused image is obtained by assigning the estimated spectra to the corresponding Landsat pixels, which are estimated from the MERIS endmembers of each land cover weighted by the Landsat posterior class probabilities. The use of these posteriors from the Landsat image guarantees the spectral variability of the land cover classes within each MERIS pixel footprint in the fused image. Similar schemes can be found in the literature (Zurita-Milla et al., 2008, 2011).

It is clear that the spatial unmixing can be easily attained by solving the linear equation shown in Figure 10.1, where the radiance vector L^K of the mixed pixel is contributed by a set of sub-pixels that have an identical spatial distribution (or proportion, denoted as A^K therein) to that of the fine resolution image collectively. Based on this assumption, the radiance vector of each sub-pixel (denoted as $S^{K,i}$) can be obtained by solving the linear regression equation in a least square manner.

10.2.2 SYSTEM DESIGN FOR DATA FUSION—CASE 2

Case 2 is designed to conduct decision-level fusion between SAR and optical remote sensing imageries to aid in landslide detection. As we know, Japan is a country significantly prone to landslide after storms and flooding. Following the philosophy in Case 1, there are two critical factors associated with the data fusion framework for monitoring the vagaries of the environmental scenes in the storm events regarding landslides; that is, we must consider what data sources and data fusion algorithms should be used to solve the problem at hand. For landslide detection, the typical method is mainly based on change detection, which can be performed by using multitemporal land cover indicators such as Normalized Difference Vegetation Index (NDVI) of the region of interest after geometric correction. An object-based method can also be applied for change detection of land cover for landslide detection in which multi-scale image segmentation is commonly applied to group pixels with similar spectral or textural characteristics. With such efforts, landslides can be identified according to some geomorphic and geographical features.

Nevertheless, numerous constraints hinder landslide detection methods in practical applications, including the data sources, pre-processing requirements, size of the study region, the types of landslides, semi-empirical parameter values, and landscape complexity. Among these factors, one critical limitation is the availability and quality of data sources. For instance, optical remote sensing data are not functional for some atmospheric/environmental conditions, because optical remote sensors cannot penetrate through clouds and haze. By contrast, microwave data can be used as an alternative since microwave sensors enable us to acquire data in all weather conditions. However, microwave data are often provided in a lower temporal resolution, which makes it difficult to use for a quick disaster response. In this context, fusing microwave data (e.g., Synthetic Aperture Radar [SAR]) with optical satellite imageries would certainly enhance the overall efficacy and robustness of landslide detection and prediction.

Despite various data fusion schemes found in the literature for the possible fusion of optical and microwave data such as intensity modulation (Alparone et al., 2004b), band ratio (Zhang et al., 2010), wavelet transform (Hong et al., 2009), object-based fusion (Ban et al., 2010; Ban and Jacob, 2013), and classifier-based fusion (Waske and Benediktsson, 2007; Bao et al., 2012), constructing an appropriate fusion framework to fuse SAR and optical imagery for a practical condition is still a challenging task. This is due to the fact that SAR and multispectral data are different in essence due to different spectral, structure, and electromagnetic ranges. Regarding landslide detection and prediction in a highly heterogeneous region in Japan, a decision-level data fusion framework would be more appropriate for the improvement of land cover classification accuracy, as it benefits from both spectral and textural features of multispectral and structural features of SAR data through a dimensionality reduction and Support Vector Machine (SVM) classification system. Figure 10.2 demonstrates an illustrative sketch of such a decision-level data fusion framework.

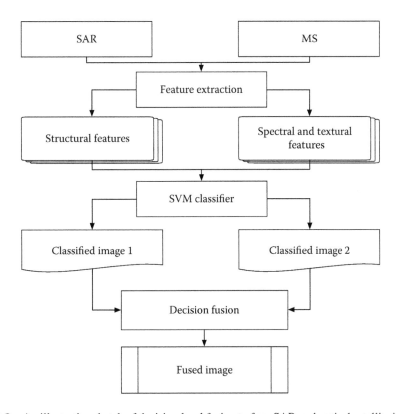

FIGURE 10.2   An illustrative sketch of decision-level fusion to fuse SAR and optical satellite imageries.

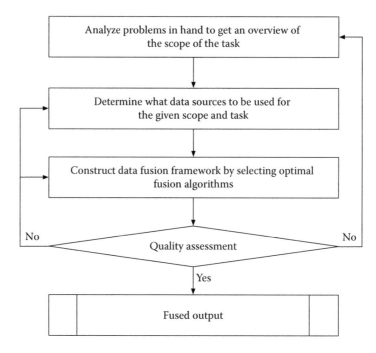

FIGURE 10.3   The system design of a data fusion framework.

10.2.3   The Philosophy for System Design of Data Fusion

Creating an appropriate data fusion framework in a practical way always requires prior knowledge in domain science in order to bridge the gaps between data sources and data fusion algorithms to be applied and analyze the problem at hand through a holistic approach. The logic flow toward the construction of a data fusion framework can be generally summarized in a system of systems engineering manner, as illustrated in Figure 10.3. First, we must analyze problems at hand to determine the scope of data fusion for a specific case, that is, what features do we need and at what scale? The second step is to determine what data sources should be applied in order to best serve our needs. At present, there exist various remote sensing data; hence choosing a good data source is necessary to facilitate effective and efficient processing and modeling. Next, we must construct a data fusion framework by selecting an appropriate fusion algorithm suitable for the given data sources and tasks. Finally, the performance of the fusion scheme should be evaluated by applying some performance assessment metrics to confirm the effectiveness and robustness of the proposed data fusion framework based on different given data sources if more than one needs to be considered.

10.3   PERFORMANCE EVALUATION METRICS FOR DATA FUSION

As described in the introduction section, the performance of data fusion techniques and/or the quality of the fused outputs can be assessed either in a qualitative or quantitative manner. Qualitative evaluation mainly refers to assessment of the performance manually through visual analysis, and hence such a method is subjective. In contrast, quantitative assessments are often performed based on a set of statistical measures that indicate the relative quality of the fused outputs, which makes the performance evaluation more objective than its counterpart. In the following subsections, both qualitative and quantitative evaluation methods are delineated in association with proposed statistical metrics that can be used for performance evaluation in the data fusion domain.

TABLE 10.1

Qualitative Measures for Image Quality Assessment

Grade	Absolute Measure	Relative Measure
1	Excellent	Performs the best overall
2	Good	Better than the average level
3	Fair	Average level
4	Poor	Lower than the average level
5	Very poor	Performs the least overall

10.3.1 QUALITATIVE ANALYSIS

As a basic performance evaluation method, visual analysis is the simplest but most direct qualitative approach for assessing the quality of fused images. In general, several absolute and relative measures are available to the observers for reference, which they can use to render their final decision regarding the image quality (Table 10.1) (Shi et al., 2005). Such an evaluation can be performed on one image as a whole, although it can be also conducted on one specific feature (e.g., colors or shapes) in the fused image. With respect to the latter, a simple weighted mean can be further applied to each feature set to reach a final overall quality judgment. Such a method is also referred to as the mean opinion score method, as an overall score may be given to the target image (Carnec et al., 2003). Although visual analysis enables us to provide an intuitive method for image quality assessment, such a method is fully subjective and depends highly on the experience of observers. In other words, different assessment results could be given by different observers, even with the same target image. In addition, the viewing conditions may impact the observers' judgement (Wald et al., 1997).

10.3.2 QUANTITATIVE ANALYSIS

In the past decades, many objective analysis methods have been proposed to evaluate the quality of fused outputs, mainly through mathematical modeling approaches. Such methods are also referred to as quantitative analysis, as opposed to qualitative analysis. Generally, performance evaluation via quantitative analysis is used to assess the spectral and/or spatial similarities between the fused image and the reference images (if applicable) based on several predefined statistical indicators. These indicators are in turn referred to as performance evaluation metrics (Jagalingam and Hegde, 2015). In the following subsections, several widely-used performance evaluation metrics are discussed mathematically. In general, these metrics and indicators can be divided into two distinct categories: without reference image and with reference image, by considering whether or not a reference image is required in calculating these metrics.

10.3.2.1 Without Reference Image

In most cases, finding a reference image at a specific time with spatial, temporal, and spectral characteristics similar to the fused image is difficult, which is the reason why we need to perform data fusion to create such a fused image. In this context, a quality measure simply based on the fused image is highly anticipated for quality assurance of the fused image. In other words, none of the reference images are required. These indicators can be defined as below.

1. *Standard Deviation*: Standard Deviation (*SD*) was first proposed at the end of the 19th century (Pearson, 1894), and was used to replace an earlier alternative for the same idea, like mean error. In general, *SD* is a measure used to quantify the amount of variations or dispersions in a set of data values (Bland and Altman, 1996). Therefore, in the data fusion domain, *SD* can be applied to the fused image to estimate the degree of deviation of the

grey levels of each pixel to the grey level average of the whole image to represent the contrast among pixels in the image (Shi et al., 2005). Mathematically, for the fused image X with a size of $N \times M$, the SD value can be calculated as

$$SD = \sqrt{\frac{1}{N \times M} \sum_{i=1}^{N} \sum_{j=1}^{M} (x_{i,j} - \bar{x})^2} \tag{10.1}$$

$$\bar{x} = \frac{1}{N \times M} \sum_{i=1}^{N} \sum_{j=1}^{M} x_{i,j} \tag{10.2}$$

where $\bar{x}$ denotes the arithmetic mean of pixel values of the image. The SD value is nonnegative, and a value of zero implies that all pixels in the image have an identical value; otherwise, a positive SD will be obtained. With respect to a fused image, a large SD often indicates a heterogeneous image (i.e., pixel values with large variations) and vice versa.

2. *Information Entropy*: Information entropy, more specifically the Shannon entropy, was proposed by Claude E. Shannon in 1948 to provide an absolute limit on the best possible average length of lossless encoding or compression of an information source, which can be treated as the expected value (mean) of the information contained in each message (Shannon, 1948). In image processing, information entropy is commonly used as a statistical measure of information randomness (or richness) to characterize the texture of the target image; this process is also commonly referred to as image entropy (Shi et al., 2005; Tsai et al., 2008). Mathematically, information entropy, often denoted as H, can be calculated as

$$H = -\sum P \log_2 P \tag{10.3}$$

where P is the probability mass function which contains the histogram counts of pixel values. Essentially, image entropy uses Shannon's formula to calculate the cumulative probability or frequency of individual gray levels and outputs, a value that quantifies the amount of information per pixel. This value is given as the number of bits per pixel and is limited by the image quantization. For instance, an 8-bit image has a maximum information content per pixel of 8-bits (Roberts et al., 2008). Theoretically, an image that is perfectly flat (e.g., has the same data values) will have an entropy of zero, whereas images with more dynamic values should have large entropy values. Consequently, information entropy can be used as a good indicator for assessing the quality of the fused image by evaluating the increase of information content (a high entropy value implies the fused image has rich information content).

3. *Erreur Relative Globale Adimensionnelle de Synthèse*: In French, *Erreur relative globale adimensionnelle de synthèse* (ERGAS), also known as the relative dimensionless global error in synthesis, is considered a good indicator for depicting the level of spectral distortions in the fused image through comparison to the original image before fusion. Toward this goal, this index needs to account for scale issues (i.e., change of resolution) between the fused images and the original image before fusion. To address this issue, ERGAS incorporates a scale adjustment factor, which is defined as follows (Wald, 2000):

$$ERGAS = 100 \frac{h}{l} \sqrt{\frac{1}{N} \sum_{k=1}^{N} \left(\frac{RMSE(k)}{\mu(k)} \right)^2} \tag{10.4}$$

where h and l are the spatial resolutions of the fine and coarse images, respectively. N is the total number of bands in the coarse image, $\mu(k)$ is the spectral mean of the kth original band of the coarse image, and $RMSE(k)$ is the root-mean-square error between the kth band of the original coarse image and the fused image.

Theoretically, *ERGAS* never reaches a zero value, because the bands that were used for the calculation of this index have slightly different characteristics (band centers and bandwidths). Despite this, *ERGAS* can be used as a good indicator to assess the quality of the fused images by computing the quality of the fused image in terms of the normalized average error of each band of the processed image (Zurita-Milla et al., 2008). The lower the value of *ERGAS* (closer to zero), the fewer spectral distortions are introduced into the fused image compared to the original image. Due to the advantage of no reference image, *ERGAS* has been widely used in the literature to assess the quality of fused images, especially for pan-sharpened images (e.g., Sheng et al., 2008; Zurita-Milla et al., 2008; Chen and Leou, 2012; Xu et al., 2015).

4. *Mutual Information*: Mutual information (*MI*) is a basic concept of information theory that measures the statistical dependence between two random variables and the amount of information that one variable contains about the other. *MI* was first applied by Qu et al. (2002) as an information measure for evaluating image fusion performance, based on the theory that *MI* describes the similarity of the image intensity distributions of the corresponding image pair (Qu et al., 2002). As a similarity measure, *MI* enables us to represent how much information is transformed into the fused images from both input images. Since no assumption is made regarding the nature of the relation between the image intensities in the input images, *MI* is thus very general and can be applied automatically without prior processing.

If A and B represent two source images for fusion while F is the fused image, the amount of information transformed from A and B to F can be modeled as

$$MI = I_{FA}(f,a) + I_{FB}(f,b) \tag{10.5}$$

where

$$I_{FA}(f,a) = \sum_{f,a} P_{FA}(f,a) \log_2 \frac{P_{FA}(f,a)}{P_F(f)P_A(a)} \tag{10.6}$$

$$I_{FB}(f,b) = \sum_{f,b} P_{FB}(f,b) \log_2 \frac{P_{FB}(f,b)}{P_F(f)P_B(b)} \tag{10.7}$$

In the above equations, $P_A(a)$, $P_B(b)$, and $P_F(f)$ are histograms of images A, B, and F. $P_{FA}(f,a)$ and $P_{FB}(f,b)$ are the joint histograms of F and A and F and B, respectively (Qu et al., 2002). MI works under the assumption that the fused image should contain the important information from two source images, and a higher MI value thus implies better performance of the fusion process (Mumtaz et al., 2008).

10.3.2.2 With Reference Image

The best way to assess the quality or accuracy of one fused output is to directly compare the output with the "true" data, and such "true" data are often called reference data. However, as stated previously, it is always difficult to find reference data for direct comparisons. In practical conditions, for example in image fusion, we may use an image obtained in other nearby temporal phases as a representative. By taking advantage of reference images, the performance of fusion schemes can be evaluated by calculating statistical metrics to indicate spectral or spatial similarities between the fused output and the reference image. In contrast to those metrics without reference images, statistical measures based on reference images have been widely used in the literature due to their efficacy and higher accuracy in demonstrating the quality of the fused output.

1. *Mean Bias*: In image processing, mean bias can be defined as the difference between the expected values of two images. In image fusion, mean bias can be used to estimate the spectral distortions between the reference and fused images (González-Audícana et al., 2004). In order to cancel out the spectral differences due to temporal intervals, mean bias is often calculated in the form of relative difference as

$$Bias = \frac{\overline{X'} - \overline{X}}{\overline{X}} = \frac{1}{N \times M} \sum_{j=1}^{N} \sum_{i=1}^{M} \frac{x'_{i,j} - x_{i,j}}{x_{i,j}} \qquad (10.8)$$

where $x_{i,j}$ and $x'_{i,j}$ denote the spectral values (or intensity) of the reference image and the fused image at the pixel (i, j). $\overline{X}$ and $\overline{X'}$ are the mean spectral values of the reference image and the fused image, respectively. Obviously, a perfect value of bias is zero. The larger the bias value, the more severe the spectral distortions in the fused image. Bias estimated in the current form mainly emphasizes the overall deviations but ignores the details because positive and negative differences between pixels can cancel each other out and result in a low bias value (even zero) despite significant spectral distortions. In this context, an absolute form can be calculated to avoid such a problem (Shi et al., 2005), for instance

$$Bias = \frac{1}{N \times M} \sum_{j=1}^{N} \sum_{i=1}^{M} \frac{|x'_{i,j} - x_{i,j}|}{x_{i,j}} \qquad (10.9)$$

2. *Correlation Coefficient*: Correlation coefficient is calculated to quantify a type of correlation and dependence between two or more values. In most cases, correlation coefficient refers to the Pearson product-moment correlation coefficient. Also known as r, R, or Pearson's r; it is commonly used as a measure of the strength and direction of the linear relationship between two variables, which is defined as the covariance of the variables divided by the product of their standard deviations, given by

$$r = \frac{\sum_{j=1}^{n} \sum_{i=1}^{m} (x_{i,j} - \overline{X})(x'_{i,j} - \overline{X'})}{\sqrt{\sum_{j=1}^{n} \sum_{i=1}^{m} (x_{i,j} - \overline{X})^2 (x'_{i,j} - \overline{X'})^2}} \qquad (10.10)$$

Clearly, the r value varies from -1 to 1; the closer to 1 (or -1), the higher the correlation between two variables. A positive R means two variables co-vary in phase (both increase or decrease) whereas a negative implies the opposite. An r value of zero means these two variables are almost independent from each other. In image fusion, R is commonly calculated to indicate the spectral similarity between the fused and the reference images. The larger the r, the fewer spectral distortions in the fused image.

3. *Root-Mean-Squared Error*: Root-Mean-Squared Error (RMSE or rmse), also known as root-mean-squared deviation, is a scale-dependent measure commonly used to estimate the difference between values predicted by a model and the values actually observed (Hyndman and Koehler, 2006). In addition, RMSE has been widely used to assess the performance of image fusing techniques by measuring the deviation of the pixel values of the fused image in contrast to the reference image (Ranchin and Wald, 2000). Given the

reference image X and the fused image X' with a size of $N \times M$, the RMSE between them can be calculated as

$$rmse = \sqrt{\frac{\sum_{i=1}^{N}\sum_{j=1}^{M}(x_{i,j} - x'_{i,j})^2}{N \times M}} \tag{10.11}$$

Due to the mean-square nature, the value of RMSE is nonnegative, with a perfect value of zero. The larger the RMSE value, the more severe the spectral distortions in the fused image relative to the reference one. In most cases, RMSE is used in conjunction with r to better depict the similarities between the reference and the fused images.

4. *Relative Average Spectral Error*: In image processing, the Relative Average Spectral Error (RASE) is an analog of RMSE, which is commonly expressed in percentage to characterize the average performance of a method in the considered spectral bands (Ranchin and Wald, 2000). Mathematically, the RASE is defined as

$$RASE = \frac{100}{M}\sqrt{\frac{1}{N}\sum_{i=1}^{N}RMSE^2(X_i)} \tag{10.12}$$

where M represents the mean radiance of N original spectral images X. RASE is considered to be independent of the spectral bands (i.e., wavelength). For instance, the RASE calculated based on SPOT-XS band 2 and band 3 yield similar quality (Wald, 2000). In other words, this indicator seems to be more universal in practical conditions. Hence, it has been frequently used to evaluate the quality of pan-sharpened images.

5. *Structural Similarity Index*: The Structural SIMilarity (SSIM) index was developed by Wang et al. (2004a) for assessing image similarity. As a perceptual criterion for assessing image similarity, SSIM adequately reflects the distortion perceived by a human observer (Amorós-López et al., 2010). When it is applied to evaluate the fused image, it enables the comparison of the local patterns of pixel intensities between the reference and fused images (Jagalingam and Hegde, 2015). Mathematically, SSIM is defined as

$$SSIM(x,y) = \frac{(2\mu_x\mu_y + C_1)(2\sigma_{xy} + C_2)}{(\mu_x^2 + \mu_y^2 + C_1)(\sigma_x^2 + \sigma_y^2 + C_2)} \tag{10.13}$$

where, C_1 is a constant that is included to avoid instability when $\mu_x^2 + \mu_y^2$ is close to zero and C_2 is a constant that is included to avoid instability when $\sigma_x^2 + \sigma_y^2$ is close to zero. μ_x and μ_y denote mean intensities of the reference and the fused images while σ_x and σ_y denote standard deviations. For the SSIM index, the values closer to 1 indicate good preservation of spatial details with a score of 1 when both images are identical (Wang et al., 2004a).

6. *Spectral Angle Mapper*: The Spectral Angle Mapper (SAM) is motivated by the observation that changes in illumination conditions due to factors such as shadows, slope variation, sun position, and light cloud, result in changes in only the magnitude of a pixel's vector, rather than the direction (Harvey et al., 2002). Hence, SAM has been widely used to evaluate the spectral fidelity of fused images by calculating the spectral angles between the reference image and the fused image to measure spectral distortions (Kruse et al., 1993). Mathematically, SAM can be expressed as

$$SAM \triangleq \arccos\left(\frac{\langle x_{i,j} \cdot x'_{i,j}\rangle}{\|x_{i,j}\|\|x'_{i,j}\|}\right) \tag{10.14}$$

where $x_{i,j}$ and $x'_{i,j}$ denote the spectral values at the pixel (i, j) in the reference and the fused images. Commonly, SAM can be measured in either degrees or radians and is usually averaged over the whole image to yield a global measurement of spectral distortion (Nencini et al., 2007). The smaller the value of SAM, the less spectral distortion. A SAM value of 0 indicates the absence of spectral distortion, but possible radiometric distortion (Zhang et al., 2009).

7. *Universal Image Quality Index*: To replace traditional error summation methods, Wang and Bovik (2002) proposed a universal index to evaluate image similarities, which was also applied to validate image fusion techniques. This universal image quality index is often abbreviated as Universal Image Quality Index (UIQI), and is designed to model any image distortion as a combination of three factors: loss of correlation coefficient, radiometric distortion, and contrast distortion (Wang and Bovik, 2002). Therefore, this index does not treat image degradation as errors but rather as structural distortion, so it evaluates the spectral distortion and structural distortion simultaneously (Roberts et al., 2008). Let X and X' denote the reference and the fused images, respectively; the UIQI defined by Wang and Bovik (2002) can be calculated as

$$UIQI = \frac{4\sigma_{XX'}\overline{X}\cdot\overline{X'}}{(\sigma_X^2 + \sigma_{X'}^2)[(\overline{X})^2 + (\overline{X'})^2]} \tag{10.15}$$

where $\sigma_{XX'}$ is the covariance between the reference image X and the fused image X'. $\overline{X}$ and $\overline{X'}$ are the mean spectral values of the reference image and the fused image while σ_X^2 and $\sigma_{X'}^2$ are the variances of the reference and the fused image, respectively.

More specifically, the above equation can be further rewritten as the following, with three distinct factors incorporated:

$$UIQI = \frac{\sigma_{XX'}}{\sigma_X\sigma_{X'}} \cdot \frac{2\overline{X}\cdot\overline{X'}}{(\overline{X})^2 + (\overline{X'})^2} \cdot \frac{\sigma_X\sigma_{X'}}{(\sigma_X^2 + \sigma_{X'}^2)} \tag{10.16}$$

The first right component is the correlation coefficient between two images, the second term measures the mean radiometric distortion, and the third depicts the contrast distortion (Wang and Bovik, 2002). The value of UIQI varies between −1 and 1. The closer the UIQI value is to 1, the less distortion is observed in the fused images, and a perfect value of 1 is obtained if two images are identical. Experimental results have shown that the UIQI performs significantly better than the commonly used mean-squared error when facing various image distortions (Yang et al., 2007; Zhang et al., 2009; Wei et al., 2015).

8. Q_4 *Index*: The Q_4 index is an extension of UIQI, with improvements in using quaternions or hypercomplex numbers to account for both radiometric and spectral measurements (Amorós-López et al., 2013). Hence it is also used as a similarity measure to depict distortions between a distorted image and a reference image. Similar to UIQI, the Q_4 index can be mathematically defined as

$$Q_4 = \frac{\sigma_{z1z2}}{\sigma_{z1}\sigma_{z2}} \cdot \frac{2|\overline{z_1}|\cdot|\overline{z_2}|}{|\overline{z_1}|^2 + |\overline{z_2}|^2} \cdot \frac{2\sigma_{z1}\sigma_{z2}}{\sigma_{z1}^2 + \sigma_{z2}^2} \tag{10.17}$$

where $z_1 = a_1 + ib_1 + jc_1 + kd_1$ denotes the four-band reference image and z_2 is analogous to z_1 but for the fused image. a, b, c, and d denote the radiance values of a given image pixel in the four bands, typically acquired in the B, G, R, and NIR wavelengths, while i, j, and k are imaginary units. σ_{z1} and σ_{z2} are the square roots of the variances of z_1 and z_2.

Q_4 enables the characterization of spectral distortions and inaccuracies in spatial enhancement. Unfortunately, Q_4 cannot be utilized to evaluate images with a number of bands greater than four, as the hypercomplex numbers with more than four components do not exhibit the same correlation properties as quaternions (Alparone et al., 2004a). Q_4 varies in the range of [0, 1], with an ideal value of 1. Higher Q_4 scores correspond to better spectral information preservation since the Q_4 index measures the similarity between two images (Huang et al., 2014). The closer the value of the Q_4 index is to 1, the higher the spectral quality of the fusion product (Shi et al., 2011). Commonly, Q_4 were calculated as averages on $N \times N$ blocks, which obtained the final result by averaging all the values over the whole image.

9. *Peak Signal-to-Noise Ratio*: Peak Signal-to-Noise Ratio (PSNR) is an engineering term regarding the ratio between the maximum possible power of a signal and the power of corrupting noise that affects the fidelity of its representation. In image fusion, PSNR is commonly applied to assess the radiometric distortion of the fused image compared to the reference one, assuming that distortion is only caused by additive signal-independent noise (Damera-Venkata et al., 2000; Zhang et al., 2009; Yuhendra et al., 2012). For the fused image X' and the reference image X of size $N \times M$, PSNR can be calculated as (Damera-Venkata et al., 2000)

$$PSNR = 10 \log_{10} \left(\frac{(\max(X'))^2}{\sum_{i=1}^{M} \sum_{j=1}^{N} (x(i,j) - x'(i,j))^2} \right) \tag{10.18}$$

where $x(i,j)$ and $x'(i,j)$ are pixel values of the reference image X and the fused image X', respectively. $\max(X')$ denotes the maximum pixel value of the fused image (e.g., 255 for 8-bit images). In general, a larger PSNR value indicates better performance of the fusion scheme. However, this is not true for all cases, since it only works when comparing results from the same codec (or codec type) and same content (Huynh-Thu and Ghanbari, 2008).

10.4   SUMMARY

In this chapter, the systematic work flow of constructing an appropriate data fusion framework to deal with real-world issues is demonstrated through the illustration of two practical system planning applications. With this in mind, creating an optimal data fusion framework in a practical way always requires domain knowledge to analyze the problem at hand in order to determine what data sources and what data fusion algorithms should be woven together for applications. Once a data fusion framework is created for the given problem, a performance evaluation should also be employed to assess the quality of the fused outputs. The quality of the fused outputs can be evaluated in either a qualitative or quantitative manner. Compared to the qualitative analysis, quantitative analysis often yields better accuracy by making use of various statistical metrics. Based on whether or not the reference image is required, these quality evaluation metrics can be broadly divided into two different categories as discussed in this chapter. Overall, the quality of the fused outputs should be carefully evaluated based on various quality assessment metrics to ensure the quality of the fused images.

REFERENCES

Alparone, L., Baronti, S., Garzelli, A., and Nencini, F., 2004a. A global quality measurement of pan-sharpened multispectral imagery. *IEEE Geoscience and Remote Sensing Letters*, 1, 313–317.

Alparone, L., Facheris, L., Marta, S., Baronti, S., Cnr, I., Panciatichi, V., Garzelli, A., and Nencini, F., 2004b. Fusion of multispectral and SAR images by intensity modulation. In: *Procceedings of the 7th International Conference on Information Fusion*, Philadelphia, Pennsylvania, USA, 637–643.

Amorós-López, J., Gómez-Chova, L., Alonso, L., Guanter, L., Zurita-Milla, R., Moreno, J., and Camps-Valls, G., 2013. Multitemporal fusion of landsat/TM and ENVISAT/MERIS for crop monitoring. *International Journal of Applied Earth Observation and Geoinformation*, 23, 132–141.

Amorós-López, J., Gómez-Chova, L., Guanter, L., Alonso, L., Moreno, J., and Camps-Vails, G., 2010. Multi-resolution spatial unmixing for MERIS and landsat image fusion. In: 2010 IEEE *International Geoscience and Remote Sensing Symposium (IGARSS)*, Honolulu, Hawaii, USA, 3672–3675.

Ban, Y., Hu, H., and Rangel, I. M., 2010. Fusion of quickbird MS and RADARSAT SAR data for urban land-cover mapping: object-based and knowledge-based approach. *International Journal of Remote Sensing*, 31, 1391–1410.

Ban, Y. and Jacob, A., 2013. Object-based fusion of multitemporal multiangle ENVISAT ASAR and HJ-1B multispectral data for urban land-cover mapping. *IEEE Transactions on Geoscience and Remote Sensing*, 51, 1998–2006.

Bao, C., Huang, G., and Yanga, S., 2012. Application of fusion with SAR and optical images in land use classification based on SVM. In: *ISPRS-International Archives of the Photogrammetry, Remote Sensing and Spatial Information Sciences*, Melborune, Australia, 11–14.

Bigdeli, B., Samadzadegan, F., and Reinartz, P., 2014a. Feature grouping-based multiple fuzzy classifier system for fusion of hyperspectral and LIDAR data. *Journal of Applied Remote Sensing*, 8, 83509.

Bigdeli, B., Samadzadegan, F., and Reinartz, P., 2014b. A decision fusion method based on multiple support vector machine system for fusion of hyperspectral and LIDAR data. *International Journal of Image and Data Fusion*, 5, 196–209.

Bland, J. M. and Altman, D. G., 1996. Measurement error. *British Medical Journal*, 313, 744.

Byun, Y., Choi, J., and Han, Y., 2013. An area-based image fusion scheme for the integration of SAR and optical satellite imagery. *IEEE Journal of Selected Topics in Applied Earth Observations and Remote Sensing*, 6, 2212–2220.

Carnec, M., Le Callet, P., and Barba, D., 2003. An image quality assessment method based on perception of structural information. In: *2003 IEEE International Conference on Image Processing (Cat. No.03CH37429), Barcelona, Spain, III-185-8*.

Ceamanos, X., Waske, B., Benediktsson, J. A., Chanussot, J., Fauvel, M., and Sveinsson, J. R., 2010. A classifier ensemble based on fusion of support vector machines for classifying hyperspectral data. *International Journal of Image and Data Fusion*, 1, 293–307.

Chen, H.-Y. and Leou, J.-J., 2012. Multispectral and multiresolution image fusion using particle swarm optimization. *Multimedia Tools and Applications*, 60, 495–518.

Clark, M. L., Roberts, D. A., and Clark, D. B., 2005. Hyperspectral discrimination of tropical rain forest tree species at leaf to crown scales. *Remote Sensing of Environment*, 96, 375–398.

Dalponte, M., Bruzzone, L., and Gianelle, D., 2008. Fusion of hyperspectral and LIDAR remote sensing data for classification of complex forest areas. *IEEE Transactions on Geoscience and Remote Sensing*, 46, 1416–1427.

Dalponte, M., Bruzzone, L., and Gianelle, D., 2012. Tree species classification in the Southern Alps based on the fusion of very high geometrical resolution multispectral/hyperspectral images and LiDAR data. *Remote Sensing of Environment*, 123, 258–270.

Damera-Venkata, N., Kite, T. D., Geisler, W. S., Evans, B. L., and Bovik, A. C., 2000. Image quality assessment based on a degradation model. *IEEE Transactions on Image Processing*, 9, 636–650.

Fauvel, M., Benediktsson, J. A., Chanussot, J., and Sveinsson, J. R., 2008. Spectral and spatial classification of hyperspectral data using SVMs and morphological profiles. *IEEE Transactions on Geoscience and Remote Sensing*, 46, 3804–3814.

Fauvel, M., Chanussot, J., and Benediktsson, J. A., 2006. Decision fusion for the classification of urban remote sensing images. *IEEE Transactions on Geoscience and Remote Sensing*, 44, 2828–2838.

García, M., Riaño, D., Chuvieco, E., Salas, J., and Danson, F. M., 2011. Multispectral and LiDAR data fusion for fuel type mapping using support vector machine and decision rules. *Remote Sensing of Environment*, 115, 1369–1379.

González-Audícana, M., Saleta, J. L., Catalán, R. G., and García, R., 2004. Fusion of multispectral and panchromatic images using improved IHS and PCA mergers based on wavelet decomposition. *IEEE Transactions on Geoscience and Remote Sensing*, 42, 1291–1299.

Harvey, N. R., Theiler, J., Brumby, S. P., Perkins, S., Szymanski, J. J., Bloch, J. J., Porter, R. B., Galassi, M., and Young, A. C., 2002. Comparison of GENIE and conventional supervised classifiers for multispectral image feature extraction. *IEEE Transactions on Geoscience and Remote Sensing*, 40, 393–404.

Hong, G., Zhang, Y., and Mercer, B., 2009. A Wavelet and IHS integration method to fuse high resolution SAR with moderate resolution multispectral images. *Photogrammetric Engineering & Remote Sensing*, 75, 1213–1223.

Huang, X. and Zhang, L., 2010. Comparison of vector stacking, multi-SVMs fuzzy output, and multi-SVMs voting methods for multiscale VHR urban mapping. *IEEE Geoscience and Remote Sensing Letters*, 7, 261–265.

Huang, X. and Zhang, L., 2012. A multilevel decision fusion approach for urban mapping using very high-resolution multi/hyperspectral imagery. *International Journal of Remote Sensing*, 33, 3354–3372.

Huang, X., Wen, D., Xie, J., and Zhang, L., 2014. Quality assessment of panchromatic and multispectral image fusion for the ZY-3 satellite: From an information extraction perspective. *IEEE Geoscience and Remote Sensing Letters*, 11, 753–757.

Huynh-Thu, Q. and Ghanbari, M., 2008. Scope of validity of PSNR in image/video quality assessment. *Electronics Letters*, 44, 800.

Hyndman, R. J. and Koehler, A. B., 2006. Another look at measures of forecast accuracy. *International Journal of Forecasting*, 22, 679–688.

Jagalingam, P. and Hegde, A. V., 2015. A review of quality metrics for fused image. *Aquatic Procedia*, 4, 133–142.

Kruse, F., Lefkoff, A., Boardman, J., Heidebrecht, K., Shapiro, A., Barloon, P., and Goetz, A., 1993. The spectral image processing system (SIPS)—interactive visualization and analysis of imaging spectrometer data. *Remote Sensing of Environment*, 44, 145–163.

Man, Q., Dong, P., and Guo, H., 2015. Pixel- and feature-level fusion of hyperspectral and lidar data for urban land-use classification. *International Journal of Remote Sensing*, 36, 1618–1644.

Martin, M., Newman, S., Aber, J., and Congalton, R. G., 1998. Determining forest species composition using high spectral resolution remote sensing data. *Remote Sensing of Environment*, 65, 249–254.

Mumtaz, A., Majid, A., and Mumtaz, A., 2008. Genetic algorithms and its application to image fusion. In: *2008 4th IEEE International Conference on Emerging Technologies*, Rawalpindi, Pakistan, 6–10.

Nencini, F., Garzelli, A., Baronti, S., and Alparone, L., 2007. Remote sensing image fusion using the curvelet transform. *Information Fusion*, 8, 143–156.

Pearson, K., 1894. On the dissection of asymmetrical frequency curves. *Philosophical Transactions of the Royal Society A*, 185, 71–110.

Qu, G., Zhang, D., and Yan, P., 2002. Information measure for performance of image fusion. *Electronics Letters*, 38, 313.

Ranchin, T. and Wald, L., 2000. Fusion of high spatial and spectral resolution images: The ARSIS concept and its implementation. *Photogrammetric Engineering & Remote Sensing*, 66, 49–61.

Roberts, J. W., Van Aardt, J., and Ahmed, F., 2008. Assessment of image fusion procedures using entropy, image quality, and multispectral classification. *Journal of Applied Remote Sensing*, 2, 1–28.

Settle, J. J. and Drake, N. A., 1993. Linear mixing and the estimation of ground cover proportions. *International Journal of Remote Sensing*, 14, 1159–1177.

Shannon, C. E., 1948. A mathematical theory of communication. *Illinois Research*, 27, 379–423.

Sheng, Z., Shi, W. Z., Liu, J., and Tian, J., 2008. Remote sensing image fusion using multiscale mapped LS-SVM. *IEEE Transactions on Geoscience and Remote Sensing*, 46, 1313–1322.

Shi, A., Xu, L., Xu, F., and Huang, C., 2011. Multispectral and panchromatic image fusion based on improved bilateral filter. *Journal of Applied Remote Sensing*, 5, 53542.

Shi, W., Zhu, C., Tian, Y., and Nichol, J., 2005. Wavelet-based image fusion and quality assessment. *International Journal of Applied Earth Observation and Geoinformation*, 6, 241–251.

Stavrakoudis, D. G., Galidaki, G. N., Gitas, I. Z., and Theocharis, J. B., 2012. A genetic fuzzy-rule-based classifier for land cover classification from hyperspectral imagery. *IEEE Transactions on Geoscience and Remote Sensing*, 50, 130–148.

Tsai, D.-Y., Lee, Y., and Matsuyama, E., 2008. Information entropy measure for evaluation of image quality. *Journal of Digital Imaging*, 21, 338–347.

Wald, L., 2000. Quality of high resolution synthesised images: is there a simple criterion? In: Third conference on *Fusion of Earth Data: Merging Point Measurements, Raster Maps and Remotely Sensed Images*, Sophia Antipolis, France, 99–103.

Wald, L., Ranchin, T., and Mangolini, M., 1997. Fusion of satellite images of different spatial resolutions: Assessing the quality of resulting images. *Photogrammetric Engineering & Remote Sensing*, 63, 691–699.

Wang, Z. and Bovik, A. C., 2002. A universal image quality index. *IEEE Signal Processing Letters*, 9, 81–84.

Wang, Z., Bovik, A. C., Sheikh, H. R., and Simoncelli, E. P., 2004a. Image quality assessment: From error visibility to structural similarity. *IEEE Transactions on Image Processing*, 13, 600–612.

Wang, Q., Shen, Y., and Jin, J., 2008. Performance evaluation of image fusion techniques. In: *Image Fusion: Algorithms and Applications*, Stathaki, T. (Ed.). Academic Press, Cambridge, Massachusetts, USA, 469–492.

Wang, L., Sousa, W. P., Gong, P., and Biging, G. S., 2004b. Comparison of IKONOS and QuickBird images for mapping mangrove species on the Caribbean coast of Panama. *Remote Sensing of Environment*, 91, 432–440.

Waske, B. and Benediktsson, J. A., 2007. Fusion of support vector machines for classification of multisensor data. *IEEE Transactions on Geoscience and Remote Sensing*, 45, 3858–3866.

Wei, Q., Bioucas-Dias, J., Dobigeon, N., and Tourneret, J.-Y., 2015. Hyperspectral and multispectral image fusion based on a sparse representation. *IEEE Transactions on Geoscience and Remote Sensing*, 53, 3658–3668.

Xu, Y., Huang, B., Xu, Y., Cao, K., Guo, C., and Meng, D., 2015. Spatial and temporal image fusion via regularized spatial unmixing. *IEEE Geoscience and Remote Sensing Letters*, 12, 1362–1366.

Yang, S., Zeng, L., and Jiao, L., 2007. Fusion of multispectral and panchromatic images using improved GIHS and PCA mergers based on contourlet. In: Proc. of SPIE 6787, *Multispectral Image Processing,* Wuhan, China, 67871J–67871J–7.

Yuhendra, Alimuddin I., Sumantyo J. T. S., and Kuze H., 2012. Assessment of pan-sharpening methods applied to image fusion of remotely sensed multi-band data. *International Journal of Applied Earth Observation and Geoinformation*, 18, 165–175.

Zhang, J., Yang, J., Zhao, Z., Li, H., and Zhang, Y., 2010. Block-regression based fusion of optical and SAR imagery for feature enhancement. *International Journal of Remote Sensing*, 31, 2325–2345.

Zhang, Y., De Backer, S., and Scheunders, P., 2009. Noise-resistant wavelet-based Bayesian fusion of multispectral and hyperspectral images. *IEEE Transactions on Geoscience and Remote Sensing*, 47, 3834–3843.

Zurita-Milla, R., Clevers, J. G. P. W., and Schaepman, M. E., 2008. Unmixing-based landsat TM and MERIS FR data fusion. *IEEE Geoscience and Remote Sensing Letters*, 5, 453–457.

Zurita-Milla, R., Clevers, J. G. P. W., Van Gijsel, J. A. E., and Schaepman, M. E., 2011. Using MERIS fused images for land-cover mapping and vegetation status assessment in heterogeneous landscapes. *International Journal of Remote Sensing*, 32, 973–991.

Part IV

Integrated Data Merging, Data Reconstruction, Data Fusion, and Machine Learning

11 Cross-Mission Data Merging Methods and Algorithms

11.1 INTRODUCTION

Remote sensing for environmental monitoring is challenging due to many dynamic environmental factors such as aerosols, sun glint, clouds, and others during image collection (Gordon and Wang, 1994; Moore et al., 2000; Wang and Bailey, 2001). It is thus apparent that obtaining full, clear images over the study area with a single sensor observation is not an easy job. It is always difficult to collect comprehensive information of a target at the ground level based on a single instrument's observations, and this is especially true in the tropical and subtropical regions where heavy cloud cover is not unusual in all seasons. Since the late 1980s, more digital image data have become available from diversified operational Earth Observing System (EOS) satellites. Since the 1990s, combining information from different satellite-based image data with multispatial, multitemporal, and multispectral resolutions into a single dataset by using various modeling techniques has been considered a valuable tool for better earth observation. Thus, the concept of image fusion was initialized in the early 1990s (Shen, 1990; Vornberger and Blindschadler, 1992; Pohl and van Genderen, 1995; Gross and Schott, 1998). In the early 2000s, Luo et al. (2002) formalized the principles and practices of image fusion and data fusion to improve spatial resolution.

Cross-mission satellite sensors with similar characteristics provide a synergistic opportunity to improve spatial and temporal coverage of a study region by constellating a suite of satellites conceptually into a single team that may enable us to flexibly produce a quality unified product with no or little data gap, namely data merging (Gregg, 1998; Gregg and Woodward, 1998). Starting with the data merger activities undertaken by the National Aeronautics and Space Administration's (NASA) Sensor Intercomparison and Merger for Biological and Interdisciplinary Studies (SIMBIOS) project in 2002, the remote sensing community has been experiencing a regime shift to create scientific quality remote sensing data encompassing measurements from multiple satellite missions (Fargion and McClain, 2002). Such a regime shift from "traditional image processing" to "data/ image fusion" and to "data merging" has resulted in significant advancements in numerous signal/ image processing applications.

Data merging has the potential to eliminate incompatibility resulting from sensors, such as scan angle dependencies, as well as seasonal and spatial trends in data. Data merging can produce merged products that have better spatial and temporal coverage than either individual mission does. In most cases, the selection of the satellite sensors for data merging depends on the characteristics of orbit, platform, imaging geometry, and ultimate resolution for applications over the spatial, temporal, and spectral domains (Ardeshir Goshtasby and Nikolov, 2007; Zhang, 2015). Due to different design principles for platforms, instruments, and retrieval algorithms, the data consistency among cross-mission sensors needs to be verified after bias removal and modification for band shifting in most cases leading to produce a cohesive data record. In addition to improving spatial coverage, data merging can also be applied to enhance temporal coverage given different types of data gaps in differing temporal scales associated with different individual mission over specified time period. This is exactly the case when generating a long-term coherent data record at the decadal or multi-decadal scale for a dedicated assessment to improve cross-sensor continuity of satellite-derived products with the aid of some unique temporal aliasing algorithms (Hu and Le, 2014; Barnes and Hu, 2015).

In response to the need for real world applications, the most applicable data merging scheme is to consider the primary radiometric products, such as the normalized water-leaving radiance or remote sensing reflectance. These primary radiometric products provide a data science foundation for deriving the rest of the higher-level data products to support a variety of environmental monitoring (D'Alimonte et al., 2008). These temporal aliasing algorithms often do not exist and thus require being tailored case by case depending on which and how many satellites are going to coalesce. In any circumstance, effective data merging based on a set of primary radiometric data products must minimize or eliminate discrepancies resulting from collective differences associated with instrumental (e.g., different sensor design, center wavelengths, and band widths), algorithmic (e.g., calibration, atmospheric correction, aerosol models), and temporal (e.g., local overpassing time) aspects embedded in these cross-mission products. For example, an optically-based technique was applied to compute radiometric quantities at the desired center wavelengths by solving an inverse radiative transfer problem to adjust spectral distortion due to mismatched center wavelengths over different products (Mélin and Zibordi, 2007).

To deal with the issues above, advancements with the aid of data merging in the last two decades can be classified into the four categories below.

11.1.1 Data Merging with Bio-Optical or Geophysical Models

In addition to the analytical merging models mentioned above, bio-optical or geophysical methods were introduced to merge multiple ocean color sensor observations in the early 2000s. For example, a semi-analytical bio-optical ocean color data merging model was developed to generate global retrievals of three bio-geochemically relevant variables based on the normalized water-leaving radiance from Sea-Viewing Wide Field-of-View Sensor (SeaWiFS) and Moderate Resolution Imaging Spectroradiometer (MODIS) (Maritorena and Siegel, 2005). These water quality constituents include the chlorophyll concentration, the combined dissolved and detrital absorption coefficient, and the particulate backscattering coefficient related to inorganic suspended solids in a water body. By applying the same bio-optical merging model, observations from SeaWiFS, MODIS, and Medium Resolution Imaging Spectrometer (MERIS) were further merged to create a coherent long-term time series of ocean color products under the NASA Ocean Color MEaSUREs and the ESA GlobColour projects (Maritorena et al., 2010). Improvement in global coverage in merged data compared to individual missions (2002–2009) was used as a performance index. Average and standard deviation of coverage for daily and multiple-day composite imagery for SeaWiFS, Aqua, and MERIS may be compared with each other when the three sensors are merged for the 2002–2009 period. Further, a complete error budget for the final merged data products was developed which took the sources of uncertainty associated with input water-leaving radiances into account and provided different uncertainty levels for the merged output products.

11.1.2 Data Merging with Machine Learning Techniques

In the early 2000s, a machine learning technique was used to reduce discrepancies among chlorophyll-a (chl-a hereafter) concentrations derived from MODIS onboard the Terra satellite and SeaWiFS to produce consistent global ocean color coverage on a daily basis (Kwiatkowska and Fargion, 2003). This is because data from these two sensors were revealed to show apparent discrepancies caused by differences in sensor design, calibration, processing algorithms, and from the rate of change in the atmosphere and ocean within a specified time difference between sensor imaging of the same regions on the ground (Kwiatkowska and Fargion, 2003). In this case, Support Vector Machines (SVMs) were used to bring MODIS data to the SeaWiFS representation after resolving convoluted data relationships between the two sensors. Merged SeaWiFS data were considered to exemplify a consistent ocean color baseline with the aid of SVMs to assimilate a variety of bio-optical, atmospheric, viewing geometry, and ancillary information.

11.1.3 DATA MERGING WITH STATISTICAL TECHNIQUES

The noisy and highly dynamic nature of differences between cross-mission sensors often makes empirical methods difficult to apply. To obtain the observations of low chl-*a* concentrations, for example, statistical methods such as multivariate linear regression were also utilized to remove systematic biases among sensors by projecting the MODIS and SeaWiFS observations onto the *in situ* measurements. This correction scheme improved the consistency between cross-mission sensors with the salient effects at blue bands (i.e., 412 and 443 nm). Similar approaches were also used to create scientifically reliable ocean color climate data records (Gregg and Casey, 2010) as well as post-launch radiometric recalibration (Gregg et al., 2009).

11.1.4 DATA MERGING WITH INTEGRATED STATISTICAL AND MACHINE LEARNING TECHNIQUES

Although data fusion can synergistically advance the spatial and temporal resolution of image data, possible data gaps may still exist due to either internal (e.g., instrument mechanical failure) or external (e.g., cloud contamination) factors hindering the remote sensing for Earth's surface observation. To cope with these gaps, methods such as the Spectral Information Adaptation and Synthesis Scheme (SIASS) based on the Quantile-Quantile (Q-Q) adjustment and extreme learning machine (ELM) were developed for improving spatiotemporal coverage through data merging toward reconstruction via the time–space–spectrum continuum in satellite images (Bai et al., 2015; Chang et al., 2015). In SIASS, three cross-mission sensors, including MODIS-Aqua, MODIS-Terra, and Visible Infrared Imaging Radiometer Suite (VIIRS) onboard Suomi-NPP, provide the synergistic opportunity to improve spatial and temporal coverage by merging their observations with the aid of signal processing techniques and machine learning tools (i.e., ELM). Shi et al. (2015) proposed to use the Bayesian Maximum Entropy (BME) method for merging chlorophyll-*a* (chl-*a*) concentration data obtained by the SeaWiFS onboard Orbview-2, the MERIS onboard ENVISAT, and the MODIS onboard Aqua. In their study, MODIS chl-*a* concentration on the current day was considered as the baseline with which the BME model was used to link baseline data with other chl-*a* concentrations collected on previous days by MERIS and SeaWiFS. Root-mean-squared error and correlation between *in situ* chl-*a* measurements and the BME-merged chl-*a* from 1-day data and 3-day data were generated and compared, respectively, for performance evaluation. Besides, common observations between successive generations of sensors during the overlapped time period provide a few synergistic opportunities for sensor inter-comparisons and data merging. Similarly, Bai et al. (2016) developed a modified Q-Q adjustment for data merging of two different total column ozone (TCO) time series datasets by removing apparent cross-mission TCO biases between Ozone Monitoring Instrument (OMI) and Ozone Mapping and Profiler Suite (OMPS). Common observations during the overlapped time period of 2012–2015 were employed to characterize and quantify the systematic bias between OMI and OMPS through the modified bias correction method (Bai et al., 2016). This led to the creation of a single long-term coherent TCO record that can be applied for the assessment of ozone layer variability in the past decades. For selected appropriate ocean color products for data merging, the current and scheduled ocean color sensors are summarized in Tables 11.1 and 11.2.

The above mentioned approaches, whether analytical, statistical, or empirical, are all capable of reducing discrepancies among cross-mission sensors for consistency improvement to some extent. Recent progress in the artificial intelligence-based algorithms have motivated advanced feature extraction with higher accuracy. Nevertheless, these artificial intelligence-based methods still have some pros and cons. On one hand, the artificial intelligence-based algorithms need abundant labeled samples for learning, classification, and pattern recognition in addition to the radiometric products, such as satellite solar zenith angle, viewing geometry, and other specific parameters such as aerosol optical depth, ozone amount, and water vapor, which are not always available (Bai et al., 2015). On the other hand, those artificial intelligence-based algorithms are highly dependent on the *in situ* measurements for cross calibration, which limits the broader real world applications of these artificial intelligence-based algorithms on a large scale (Bai et al., 2015).

These *in situ* measurement-based cross calibration schemes can only reduce the systematic bias resulting from the algorithmic and instrumental differences among satellite sensors. In other words, they cannot remove the location-dependent biases that are intimately related to local factors such as hydrodynamic or meteorological factors. For example, the bias between two cross-mission sensors would not be the same at two different locations in one lake if one sensor is located in the outflow with strong advection of water mass while the other is located in the center of the lake with small hydrodynamic disturbances. This bias is also hard to quantify owing to the stochastic nature embedded in the temporal differences among sensors. In addition, previous approaches only considered merging observations at the common bands, while giving up some valuable spectral information across some neighboring bands during the data merging process. For example, no observation is recorded at 531 nm wavelength by VIIRS while observations are provided in the MODIS product associated with this band. Thus, spectral information at 531 nm should be reconstructed from available neighboring bands in VIIRS data to preserve the spectral characteristics at this wavelength while merging MODIS with VIIRS (Lee et al., 2014). Although previous band-shifting methods can address such an issue, it still requires using the complex bio-optical model on some occasions (Mélin and Sclep, 2015). For the purpose of demonstration, the following sections illustrate the SIASS algorithm to show how data merging may be carried out to improve spatiotemporal coverage of remote sensing images after removing both the systematic and the location-dependent biases in a cloudy aquatic environment.

11.2 THE SIASS ALGORITHM

SIASS, which is an integrated statistical and machine learning method for data merging, took advantage of a few synergistic opportunities to show how the barriers described above at the end of Section 11.1 can be overcome holistically. The primary strength of SIASS rests upon the generalized scheme developed for bias correction between any pairs of cross-mission sensors with similar design characteristics (Bai et al., 2015). The secondary strength is that with the aid of SIASS, cross-mission discrepancies can be removed without using any *in situ* measurements as long as sensors' common observations can be available. Compared to other data merging methods, SIASS is able to not only eliminate incompatibilities between the common bands, but also reconstruct spectral information for those mismatched bands (i.e., bands at different wavelengths) among sensors.

Merging cross-mission ocean color reflectance products derived from MODIS-Aqua/Terra and VIIRS-NPP to generate consistent ocean color reflectance products for further feature extraction is the major objective in SIASS (Table 11.3). The essence is to calibrate all observed reflectance data collected from different sensors to the same level and then merge them together to improve spatial and/or temporal coverage. The hypothesis under SIASS is that the observed reflectance differences among cross-mission sensors are mainly the result of the systematic bias and the location dependent bias. This implies the SIASS algorithm needs to remove systematic bias existing between cross-mission products caused by: (1) instrumental difference, (2) wavelength and band width, and (3) algorithmic and calibration bias, as well as location dependent bias caused by: (1) different sensor overpassing times, (2) biological processes, and (3) external forcing such as wind and water mass advection. Practical implementation of the SIASS method was assessed by merging cross-mission ocean color reflectance observations collected from MODIS and VIIRS over Lake Nicaragua during the time period of 2012–2015 and demonstrated in the following subsections.

11.2.1 Data Merging via SIASS

SIASS, a novel bias correction scheme developed by Bai et al. (2015), was designed to improve the spatial coverage of ocean color reflectance for better monitoring and understanding of aquatic

TABLE 11.1

Current Ocean Color Sensors

Sensor/Data Link	Agency	Satellite	Launch Date	Swath (km)	Spatial Resolution (m)	Bands	Spectral Coverage (nm)	Orbit
COCTS CZI	CNSA (China)	HY-1B (China)	April 11, 2007	2400	1100	10	402–12,500	Polar
				500	250	4	433–695	
GOCI	KARI/KIOST (South Korea)	COMS	June 26, 2010	2500	500	8	400–865	Geostationary
HICO	ONR, DOD, and NASA	JEM-EF Int. Space Stn.	September 18, 2009	50 km Selected Coastal Scenes	100	124	380–1000	51.6° 15.8 orbits p/d
MERSI	CNSA (China)	FY-3A (China)	May 27, 2008	2400	250/1000	20	402–2155	Polar
MERSI	CNSA (China)	FY-3B (China)	November 5, 2010	2400	250/1000	20	402–2155	Polar
MERSI	CNSA (China)	FY-3C (China)	September 23, 2013	2400	250/1000	20	402–2155	Polar
MODIS-Aqua	NASA (USA)	Aqua (EOS-PM1)	May 4, 2002	2330	250/500/1000	36	405–14,385	Polar
MODIS-Terra	NASA (USA)	Terra (EOS-AM1)	December 18, 1999	2330	250/500/1000	36	405–14,385	Polar
OCM-2	ISRO (India)	Oceansat-2 (India)	September 23, 2009	1420	360/4000	8	400–900	Polar
VIIRS	NOAA (USA)	Soumi NPP	October 28, 2011	3000	375/750	22	402–11,800	Polar

TABLE 11.2

Scheduled Ocean Color Sensors

Sensor/Data Link	Agency	Satellite	Launch Date	Swath (km)	Spatial Resolution (m)	Bands	Spectral Coverage (nm)	Orbit
OLCI	ESA/EUMETSAT	Sentinel 3A	June 2015	1270	300/1200	21	400–1020	Polar
COCTS CZI	CNSA (China)	HY-1C/D (China)	2015	2900 1000	1100 250	10 10	402–12,500 433–885	Polar
SGLI	JAXA (Japan)	GCOM-C	2016	1150–1400	250/1000	19	375–12,500	Polar
COCTS CZI	CNSA (China)	HY-1E/F	2017	2900 1000	1100 250	10 4	402–12,500 433–885	Polar
HIS	DLR (Germany)	EnMAP	2017	30	30	242	420–2450	Polar
OCM-3	ISRO (India)	Oceansat–3	2017	1400	360/1	13	400–1010	Polar
OLCI	ESA/EUMETSAT	Sentinel—3B	2017	1265	260	21	390–1040	Polar
VIIRS	NOAA/NASA (USA)	JPSS—1	2017	3000	370/740	22	402–11,800	Polar
Multispectral Optical Camera	INPE/CONAE	SABIA-MAR	2018	200/2200	200/1100	16	380–11,800	Polar
GOCI-II	KARI/KIOST (South Korea)	GeoKompsta 2B	2018	1200×1500 TBD	250/1000	13	412–1240 TBD	Geostationary
OCI	NASA	PACE	2018	*	*	*	*	Polar
HYSI—VNIR	ISRO (India)	GISAT—1	*(planned)	250	320	60	400–870	Geostationary (35.786 km) at 93.5∞E
OES	NASA	ACE	>2020	TBD	1000	26	350–2135	Polar
Coastal Ocean Color Imaging Spec (Name TBD)	NASA	GEO—CAPE	>2022	TBD	250–375	155 TBD	340–2160	Geostationary
VSWIR Instrument	NASA	HyspIRI	>2022	145	60	10 nm contiguous bands	380–2500	LEO, Sun Sync

TABLE 11.3

Characteristics of MODIS and VIIRS Ocean Color Remote Sensing Reflectance Products

	MODIS	VIIRS
Wavelengths (nm)	412: 405–420	410: 402–422
	443: 438–448	443: 436–454
	469: 459–479	486: 478–488
	488: 483–493	551: 545–565
	531: 526–536	671: 662–682
	547: 546–556	
	555: 545–565	
	645: 620–670	
	667: 662–672	
	678: 673–683	
Temporal resolution	Daily	Daily
Spatial resolution (m)	1,000	750
Local crossing time	MODIS-Terra: 10:30 AM	1:30 PM
	MODIS-Aqua: 1:30 PM	
Data format	HDF	NetCDF
Data access	NASA OBPG	NASA OBPG

Note: The Level-2 ocean color reflectance data of MODIS-Terra, MODIS-Aqua, and VIIRS used in this study were acquired from the NASA Ocean Biology Processing Group (OBPG).

dynamics, based on a set of cross-mission ocean color reflectance products collected in an overlapped time period from three satellite sensors, namely MODIS-Aqua, MODIS, Terra, and VIIRS-NPP. SIASS can be applied for the case with more than three satellite sensors, however. With this SIASS algorithm, biases across missions can be quantified and then removed to improve consistency among cross-mission sensor observations.

Figure 11.1 (Bai et al., 2015) shows such an adaptive bias correction scheme to remove the systematic bias and the location-dependent bias among sensors sequentially via the proper integration of: (1) singular value decomposition (SVD) (Businger and Golub, 1969; Nash, 1990), (2) original Q-Q adjustment (Wilk and Gnanadesikan, 1968), (3) empirical mode decomposition (EMD) (Drazin, 1992), and (4) ELM (Huang et al., 2006). There is no doubt that in the beginning of the SIASS algorithm, one of these satellite sensors must be chosen as the baseline among MODIS-Aqua/Terra and VIIRS-NPP, while the rest are complementary, in order to merge multiple satellite data sets. This means observations collected by the rest of the satellite sensors must be projected onto the corresponding observations at the baseline sensor level so as to be consistent with the baseline observations toward data merging. To ease the discussion in the selection of the baseline sensor in the following subsections, paired comparisons of MODIS-Aqua and MODIS-Terra, are abbreviated as AT; MODIS-Aqua and VIIRS-NPP are abbreviated as AV; and MODIS-Terra and VIIRS-NPP are abbreviated as TV. The selection of the baseline sensor can be made possible by using three predetermined criteria. These three criteria for the selection of the baseline sensor across such three pairwise settings (i.e., AT, AV, and TV) include: (1) any single sensor which can show the highest number of clear pixels should have a comparative advantage, (2) any paired sensors that may contribute to more clear pixels with higher coverage ratio (to be formally defined in Section 11.2.1.5) may gain more comparative advantage, and (3) the smaller the difference between ocean color reflectance observations at the common band among sensors, the lower relative biases

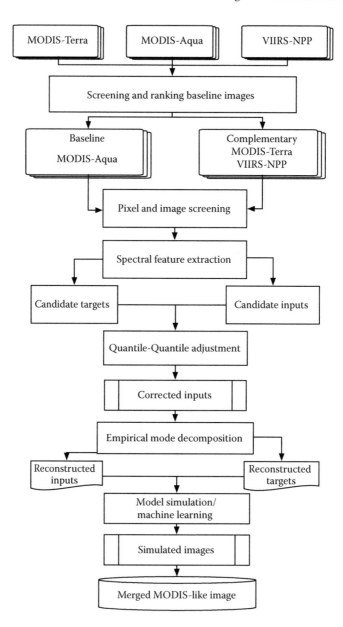

FIGURE 11.1 Schematic flowchart of the cross-mission ocean color reflectance merging scheme with SIASS.

between sensors. The criteria for screening and selection of the baseline sensor will be explained in greater detail in Section 11.3.

Reflectance values (abbreviated as Rrs herein) of VIIRS-NPP and MODIS-Terra have to be mapped onto the level of MODIS-Aqua in case MODIS-Aqua can be selected as the baseline in this demonstration (Figure 11.1). Following this basic procedure (Figure 11.1), SIASS was proposed to eliminate biases between cross-mission sensor observations for wavelengths in the synchronized bands while reconstructing spectral information for wavelengths of mismatched bands. Whereas the systematic bias can be processed via Q-Q adjustment, the location-dependent bias can be removed via EMD. ELM is finally employed to perform the reconstruction of satellite images via nonlinear prediction and deal with band shifting issue if any. Figure 11.2 shows the schematic of SIASS when VIIRS images are merged with Aqua images. They are described in detail below.

11.2.1.1 SVD

In remote sensing, a pixel is usually defined as a squared area to pinpoint any location on Earth with unique geographic information such as latitude and longitude. Clear pixels associated with both baseline and complementary satellites are subject to SVD for noise removal at first before the Q-Q adjustment can be applied. The SVD of a matrix is usually referred to as the final and best factorization of a matrix:

$$A = WDP^{T} \qquad (11.1)$$

where W and P is an orthogonal matrix and D is a diagonal matrix. In the decomposition, A can be any matrix. Note that if A is a symmetric positive definite, its eigenvectors are orthogonal and can be presented as $A = QDQ^{T}$, a special case of SVD in which $P = W = Q$. For a more general A, the SVD requires two different matrices, W and P. We also know that A might not be the same as WDP^{T} because of noise embedded in the signals, which provides the opportunity to help filter out the random bias in remote sensing images. The reference time series that is generated for the subsequent Q-Q adjustment is the first principal component of all matchups in the overlapped study time period and regions (see the middle of Figure 11.2).

11.2.1.2 Q-Q adjustment

During the procedure for systematic bias correction, each clear pixel within the overlapped time period that has been denoised may be screened out for the construction of a reference time series database for both the baseline and complementary satellite sensors. Assume that the reference time series database for each pixel may have a good memory effect on the hidden patterns of systematic bias. Based on such a reference time series database, the cumulative distribution function (CDF) can be generated for statistical consideration of systematic bias correction. Given the observations collected from the selected baseline and complementary satellites, pixels associated with complementary satellites have to be corrected up to the corresponding levels of the baseline satellite. By picking up the candidate pixels collected by the complementary satellites for adjustment one by one over the entire study region, the new individual CDF comprised of all candidate pixel values over the study period can also be generated. When addressing each pixel on a rolling basis across the two CDFs based on the following equations, the bias correction for the entire study region may be achieved stepwise by using the Q-Q adjustment thereby eliminating systematic bias. These two CDFs are exactly the blue CDF (candidate) and the red CDF (baseline) in the middle of Figure 11.2.

Mathematically, for each pixel over a geographical grid point common to cross-mission sensors, the cross-mission ocean color reflectance bias at wavelength λ (denoted as $\Delta Rrs(\lambda)$ hereafter) is assumed to mainly consist of two portions: the sensor-dependent systematic bias ($\Delta Rrs^{SAT}(\lambda)$) and the location-dependent bias ($\Delta Rrs^{LCT}(\lambda)$).

Therefore, the two fundamental equations below lay down the foundation of SIASS:

$$< Rrs(\lambda) > = Rrs(\lambda) + < \Delta Rrs(\lambda) > \qquad (11.2)$$

$$< \Delta Rrs(\lambda) > = < \Delta Rrs^{SAT}(\lambda) > + < \Delta Rrs^{LCT}(\lambda) > \qquad (11.3)$$

where $< >$ denotes the ensemble mean.

Generally, $\Delta Rrs^{SAT}(\lambda)$ is mainly related to the instrumental and algorithmic differences among sensors. Ideally, this bias is nearly consistent and thus can be quantified through the sensor

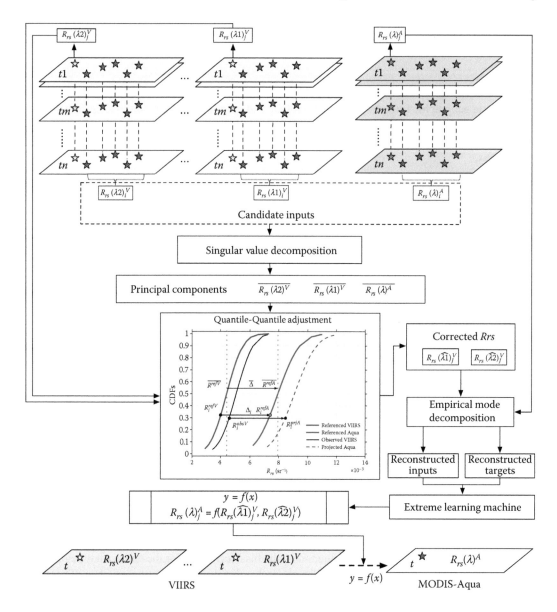

FIGURE 11.2 A schematic illustration of the SIASS method when VIIRS images are merged with Aqua images.

inter-comparisons. The simplest way to characterize this bias is to calculate the average differences through matchup analysis, which can be calculated as

$$\Delta Rrs^{SAT}(\lambda) = \frac{1}{N}\sum_{i=1}^{N}\left(Rrs_i^{SAT1}(\lambda) - Rrs_i^{SAT2}(\lambda)\right) \tag{11.4}$$

$$\left\langle \Delta Rrs^{SAT}(\lambda)\right\rangle = \frac{1}{T}\sum_{t=1}^{T}\left(Rrs_t^{SAT}(\lambda)\right) \tag{11.5}$$

where i denotes the number of matchups among sensors on the same date and t is the time. Following Equations 11.6 through 11.14, each pixel time series can be fully corrected to eliminate the systematic bias through the Q-Q adjustment (middle of Figure 11.2 or Figure 11.3).

Based on the Q-Q adjustment method (Figure 11.3), one observation from VIIRS (i.e., Rrs_i^{obsV}) can be calibrated to the MODIS-Aqua level by adding the associated cross-mission bias (ΔRrs, formulated in (11.2)) between these two sets of sensor observations pixel by pixel, which can be modeled as

$$Rrs_i^{prjA} = Rrs_i^{obsV} + \Delta Rrs_i \tag{11.6}$$

where Rrs_i^{prjA} is the projected observation after bias correction, and i is ith percentiles in CDFs of Rrs^{obsV}. In Figure 11.3, referenced VIIRS time series data stand for common observations of VIIRS after the treatment of spectral feature extraction during the overlapped time period with Aqua. Similarly, referenced Aqua time series stand for common observations of MODIS-Aqua after the treatment of spectral feature extraction during the overlapped time period with VIIRS. Observed VIIRS time series data stand for the original VIIRS observations with no further treatment. Projected Aqua time series data stand for projected data at the Aqua level from the VIIRS observations. In ΔRrs_i, the associated systematic bias at the ith percentiles (ΔRrs_i^{SAT}) can be expressed as the sum of the mean radiometric shift ($\overline{\Delta}$) plus the corresponding deviation Δ_i'

$$\Delta Rrs_i^{SAT} = g\overline{\Delta} + f\Delta_i', \tag{11.7}$$

where

$$\Delta_i = Rrs_i^{refA} - Rrs_i^{refV} \tag{11.8}$$

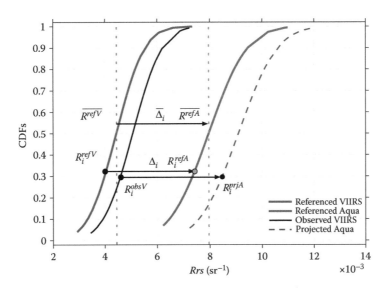

FIGURE 11.3 An illustrative example of the Q-Q adjustment method for calibrating cross-mission sensors observations for each relevant band. Observations from one sensor (solid black contour) can be calibrated to the other sensor level (dashed black contour) by removing associated biases, which can be characterized from common differences between cross-mission sensors observations during the overlapped time period (blue and red contour). (Bai, K., Chang, N. B., and Chen, C. F., 2015. *IEEE Transactions on Geoscience and Remote Sensing*, 54, 311–329.)

$$\bar{\Delta} = \frac{1}{N} \sum_{i=1}^{N} \Delta_i = \overline{Rrs^{refA}} - \overline{Rrs^{refV}} \tag{11.9}$$

$$\Delta_i' = \Delta_i - \bar{\Delta} \tag{11.10}$$

$$g = \frac{\left(\sum\limits_{i=1}^{N} Rrs_i^{obsV} \right) / N}{\left(\sum\limits_{i=1}^{N} Rrs_i^{refV} \right) / N} = \frac{\overline{Rrs^{obsV}}}{\overline{Rrs^{refV}}} \tag{11.11}$$

$$f = \frac{IQR_{Rrs^{obsV}}}{IQR_{Rrs^{refV}}} \tag{11.12}$$

$$IQR_{Rrs^{obsV}} = Rrs^{obsV} \big|_{p=75\%} - Rrs^{obsV} \big|_{p=25\%} \tag{11.13}$$

$$IQR_{Rrs^{refV}} = Rrs^{refV} \big|_{p=75\%} - Rrs^{refV} \big|_{p=25\%} \tag{11.14}$$

Terms $IQR_{Rrs^{obsV}}$ and $IQR_{Rrs^{refV}}$ in Equations 11.12 through 11.14 are the interquartile ranges of the observed and referenced observations, respectively, calculated as the differences between the 75th ($p = 75\%$) and 25th ($p = 25\%$) percentiles.

In this study, Rrs^{refA} and Rrs^{refV} were derived from the matchups of common observations of both sensors' observations. The basic underlying theory is that induced discrepancies due to instrumental and algorithmic differences may be memorized along with the time, which can be characterized from the long-term historical observations (Figure 11.2, bottom panel). From Equations 11.6 through 11.14, it is observed that ΔRrs_i^{SAT} is mainly modulated by g and f; if $g = f = 1$, ΔRrs_i^{SAT} would be a special case in which only Δ_i is added to Rrs_i^{obsV} without any further adjustment. In practice, because of different instrumental response functions and the degradation of sensors over time, this bias is not always consistent. Therefore, the offset of the average differences cannot eliminate discrepancies among cross-mission sensors, and could even result in unintended new biases. To avoid uncertainty that might result from outliers in observations, the median values of Rrs^{refA} and Rrs^{refV} instead of the ensemble mean were used to calculate the average radiometric shift $\bar{\Delta}$ in Equation 11.9. Similar values were applied in Equation 11.11 to calculate the factor g as well.

It is anticipated that systematic bias ΔRrs^{SAT} at all common bands among sensors can be greatly mitigated by adopting the Q-Q adjustment method. Given that this Q-Q adjustment method is based on observations, such a method can be applicable for synthesizing and generating spectral information for mismatched bands, such as the band at 531 nm wavelengths between MODIS and VIIRS. For instance, observations from two neighboring bands of VIIRS (i.e., 486 nm and 551 nm) should be used as the baseline information for possible image reconstruction to achieve this goal.

11.2.1.3 EMD

According to Figure 11.2, spectral feature extraction via EMD is designed to retrieve the major spectral information polishing referenced datasets for both baseline and complementary satellite sensors, respectively. Given any observed dataset, the systematic bias can be eliminated by the Q-Q adjustment globally based on two relevant referenced datasets. Even with the Q-Q adjustment, time series of ocean color reflectance from different sensors at one particular pixel on the same date might not correspond in the adjusted dataset due to the highly dynamic nature of aquatic or atmospheric environments. Locational bias may be corrected further to deal with the unexpected fluctuations.

To some extent, these high frequency fluctuations might be referred to as white noise and could be eliminated to better characterize long-term relationships. For this purpose, EMD was chosen to reconstruct the projected time series in this study. Unlike the well-known Fourier transform, which requires linear and stationary data to avoid energy spreading in the energy-frequency domain, EMD can decompose any nonlinear and nonstationary time series into a finite and often small number of intrinsic mode functions (IMFs) that allow well-behaved Hilbert transforms (Huang et al., 1998). EMD can thus be applied to analyze some nonlinear and nonstationary processes in dynamic environments because of the decomposition capability of IMFs based on the local characteristic time scale of the data that may be deemed as a combination of fast and slow oscillations. The location-dependent bias can then be removed from the measurements in the aquatic or atmospheric environment to address those synoptic biases.

Such a decomposition scheme is designed based on the direct extraction of the energy associated with various intrinsic time scales, and the IMFs are extracted level by level from the highest frequency to the lowest frequency in which local oscillations represent the highest frequency riding on the corresponding lower frequency part of the data. This extraction procedure will continue until no complete oscillation can be extracted in the residual (Huang et al., 1998). In other words, the first IMF is the time series with the highest frequency oscillations extracted and thus can usually be regarded as the white noise to be removed. After the removal of white noise, a new time series data set can be reconstructed by adding all remnant IMFs together other than the first one. The reconstructed time series instead of the projected time series data set is thus applied to characterize the location-dependent relationships among sensors. The mathematical algorithm can be shown below.

If $X(t)$ represents the time series after the Q-Q adjustment, the procedure of extracting an IMF, called sifting, can be described as follows to: (1) identify all the local extrema in the test data, (2) connect all the local maxima by a cubic spline line as the upper envelope, and (3) repeat the procedure for the local minima to produce the lower envelope. The upper and lower envelopes should cover all the data between them, and the average of these two envelopes is denoted as m_1; subtracting the m_1 from the original time series $X(t)$ we can obtain the first component h_1:

$$X(t) - m_1 = h_1 \tag{11.15}$$

Component h_1 in Equation 11.5 should satisfy the definition of an IMF. After the first round of sifting, a crest may become a local maximum and a valley a local minimum. New extrema generated in this way reveal the proper modes lost in the initial sifting. In the next step, h_1 is treated as new input data:

$$h_1 - m_{11} = h_{11} \tag{11.16}$$

After repeated sifting up to k times, h_1 becomes an IMF, that is

$$h_{1(k-1)} - m_{1k} = h_{1k} \tag{11.17}$$

Then, h_{1k} is designated as the first IMF component of the data:

$$c_1 = h_{1k} \tag{11.18}$$

Here, a normalized precipitation time series is used as an example (Figure 11.4) to illustrate the decomposition process from the highest frequency to the residual.

11.2.1.4 ELM

After removing the systematic bias, location-dependent bias $\Delta Rrs^{LCT}(\lambda)$ in Equation 11.3 must also be addressed. As previously described, $\Delta Rrs^{LCT}(\lambda)$ at one geographic grid could be different from

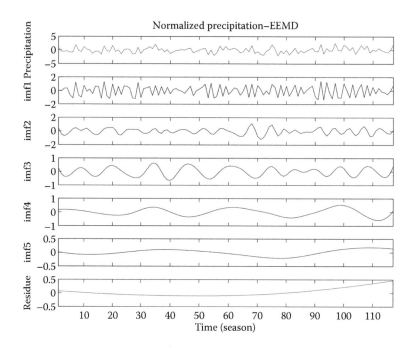

FIGURE 11.4 IMFs of a normalized precipitation time series produced by EMD.

the others; therefore, relationships within such pairwise settings (i.e., AV and AT) in this context should be examined individually for each particular pixel to complete the location-dependent bias correction. In the SIASS algorithm, a fast machine learning tool, ELM, was utilized to establish complex relationships among sensors that can be used to estimate the location-dependent bias at each geographical grid or pixel after the white noise can be removal by EMD.

To know more about ELM, we may start with the general mathematical architecture of a feedforward neural network model (Figure 11.5). In this formulation, we assume a training set $\{(x_i, y_i) \mid x_i \in R^n, y_i \in R^m, i = 1, 2, 3, \ldots, N\}$ with N training samples will be used to establish n-input and m-output for function generalization. Assuming L number of hidden nodes and the Single Layer

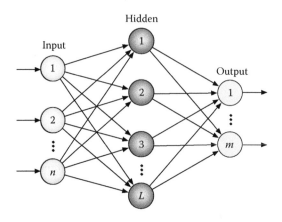

FIGURE 11.5 The architecture of a feedforward neural network with n input units, L hidden units, and m output units.

Forward Network (SLFN), the simplest form of a standard artificial neural network model, then the activation function $g(x)$ can be modeled as

$$\sum_{i=1}^{L} v_i g_i(x_j) = \sum_{i=1}^{L} v_i g(a_i \cdot x_j + b_i) \quad (j=1, 2, 3, \ldots, N) \tag{11.19}$$

where $\mathbf{a_i} = [a_{i1}, a_{i2}, \ldots, a_{in}]^T$ $(i = 1,2,3, \ldots, L)$ is the weight vector connecting the ith hidden node with the input nodes; and $\mathbf{v_i} = [v_{i1}, v_{i2}, \ldots, v_{in}]^T$ $(i = 1,2,3, \ldots, L)$ is the weight vector connecting the ith hidden node with the output nodes. Here, b_i is the bias of the ith hidden node and $g(x)$ is the activation function, including sine and cosine radial basis, and other non-regular functions as well as sigmoidal function.

These equations show the forward calculation process of SLFN to generate the predictions Y for N number of training inputs, the general expression of which is (Figure 11.5):

$$HV = Y \tag{11.20}$$

Here, H is the hidden layer output matrix of the neural network and can be calculated as:

$$H = \begin{bmatrix} g(a1 \cdot x1 + b1) & \cdots & g(aL \cdot x1 + bL) \\ \vdots & \ddots & \vdots \\ g(a1 \cdot xN + b1) & \cdots & g(aL \cdot xN + bL) \end{bmatrix}_{N \times L} \tag{11.21}$$

$$V = \begin{bmatrix} v_1^T \\ \vdots \\ v_L^T \end{bmatrix}_{L \times m} \quad Y = \begin{bmatrix} y_1^T \\ \vdots \\ y_N^T \end{bmatrix}_{L \times m} \tag{11.22}$$

After this process is complete, then the backpropagation (BP) algorithm is run to train the SLFN, which then tries to find the specific sets of $(a_i, b_i, i = 1, 2, 3, \ldots, L)$ to minimize the difference between approximations and targets (Huang et al., 2006):

$$\| H(a1, \ldots, aL, b1, \ldots, bL)\hat{}V - Y \| = \min\left(\| H(a1, \ldots, aL, b1, \ldots, bL)V - Y \|\right)|_{a_i, b_i, V} \tag{11.23}$$

The process of minimizations seeks the gradient descent-based algorithms that adjust the weights and biases through iteration of backward propagation, expressed as

$$W_k = W_{k-1} - \eta \frac{dE(W)}{dW} \tag{11.24}$$

where η is the learning rate and E is the error left in each predictive iteration. These gradient descent-based algorithms are useful for a vast number of problems and applications, but they have iterative learning steps and so are slower than is required. They also have problems associated with over-fitting and local minima instead of global minima.

ELM is an effective feed forward neural network that tends toward a global optimum (Huang et al., 2006). Empirical results reveal that ELM and its variants are more effective and accurate than other state-of-the-art learning algorithms for enhancing the learning pace and accuracy for SLFN. Unlike the customary learning algorithms that require altering and tuning a considerable number of parameters in SLFN, the input weights and hidden layer biases in ELM can be randomly allocated if the initiation functions are vastly differentiable with the hidden output layer remaining unaltered (Huang et al., 2006). Subsequently, the preparation is comparable to finding a base standard arrangement for the straight framework $HV = Y$ because the yield weights, V, are the main variable of SLFN to be tuned. So the ELM Equation 11.23 now becomes,

$$|| H(a1,...,aL,b1,...,bL)\hat{}V - Y || = \min(|| H(a1,...,aL,b1,...,bL)V - Y ||)|_V \qquad (11.25)$$

The linear framework can be attempted with the guidance of the least square method (LSM), and the output weights are analytically decided by

$$V = H^{\dagger}Y \qquad (11.26)$$

where $H^{\dagger}$ is the Moore-Penrose generalized inverse of H (Huang et al., 2006).

11.2.1.5 Performance Evaluation

A few indexes must be selected to evaluate the performance of the data merging scheme. To meet this goal, the average monthly data coverage ratio (i.e., $\overline{CR}$) can be calculated as the ensemble mean of daily clear pixels coverage ratio (CR) over the lake each month

$$CR = 100 \times \frac{N_{clear}}{N_{total}} \qquad (11.27)$$

$$\overline{CR} = \frac{1}{T} \sum_{t=1}^{T} CR_t \qquad (11.28)$$

where N_{clear} is the number of clear water pixels (i.e., having data value) and N_{total} is the total number of water pixels over the study area. Similarly, the average monthly mean absolute bias ($\overline{MAB}$) can be calculated as the average of the mean absolute bias (MAB) between clear pixel matchups among sensors:

$$MAB = 100 * \frac{1}{N} \sum_{i=1}^{N} | Rrs_i^{SAT1} - Rrs_i^{SAT2} | \qquad (11.29)$$

$$\overline{MAB} = \frac{1}{T} \sum_{t=1}^{T} MAB_t \qquad (11.30)$$

11.3 ILLUSTRATIVE EXAMPLE FOR DEMONSTRATION

11.3.1 STUDY AREA

Lake Nicaragua was chosen as the study area for demonstration due to its unique location in a tropical region where heavy clouds persist during wet seasons (Figure 11.6). This lake, which is the largest freshwater lake in Central America, has an area of 8,264 km². The two general seasons are the dry season from November to April and the wet season from May to October. Heavy cloud cover is phenomenal during the wet season, with an annual average cloud cover of nearly 70% (Chang et al., 2015). As the lake has been considered as a future drinking water source by the Nicaragua government and several other Central American countries, water quality monitoring in terms of some biophysical parameters to characterize infrastructure conditions and pollution impacts in this lake on a near real-time basis is essential.

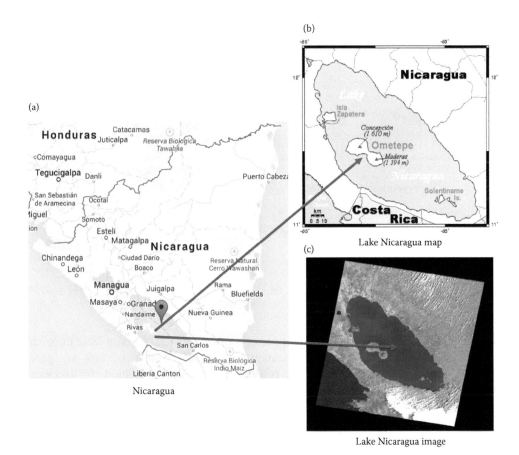

FIGURE 11.6 The study region—Lake Nicaragua map (a) with a blowout detail (b) and a satellite image (c). (National Automatous University of Nicaragua) (© Google, map data © 2016 Google).

The statistical summary of the number of clear sky days associated with MODIS-Terra and MODIS-Aqua in 2013 for each pixel in Lake Nicaragua is shown in Figure 11.7. The central part of the lake is mostly clear in an observed year. The time series of percentage of cloud (i.e., cloudy pixels/total pixels, POC) in daily MODIS-Terra and MODIS-Aqua images over Lake Nicaragua can be expressed by Figure 11.8. Summer time (wet season) is mostly cloudy. In 2013, it is indicative that MODIS-Terra collected better images with much fewer cloudy pixels in the central part of the lake. Cloudy pixels can be recovered by using SIASS to support better environmental monitoring at the ground level.

11.3.2 BASELINE SENSOR SELECTION

Traditional manual sampling campaigns of water quality monitoring are laborious, time consuming, and costly and these campaigns are incapable of capturing vast spatial variability holistically. To provide holistic insight for water quality management with a large spatial coverage, daily satellite ocean color products should be considered to provide near real-time water quality monitoring. At present, MODIS aboard the Terra (1999-present, MODIS-Terra hereafter) and the Aqua (2002-present, MODIS-Aqua hereafter) as well as VIIRS aboard the S-NPP (2011-present) are orbiting the Earth with a capability of providing the ocean color products and monitoring the inland waters as well. The features of these three sensors have been presented in Table 11.1. All three sensors can map the Earth on a daily basis at a moderate spatial resolution. The general

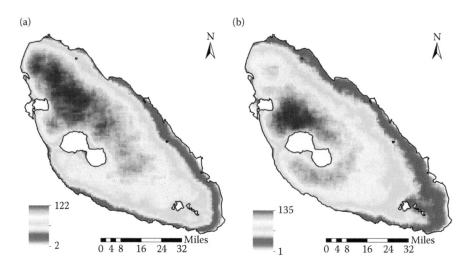

FIGURE 11.7 Number of clear-sky days in 2013 for each pixel in daily (a) Terra images and (b) Aqua images over Lake Nicaragua.

performance and prediction accuracy of ocean color products derived based on these three sensors via pre-processing, enhancement, transformation, and classification algorithms are generally comparable with each other. This niche triggers a unique synergistic opportunity for data merging. In the following part of this section, the SIASS algorithm will be applied to process remotely sensed images of MODIS-Terra, MODIS-Aqua, and VIIRS-NPP for data merging.

The potential benefits from merging cross-mission sensor observations can be evaluated in terms of *CR* and *MAB* (Figures 11.9 and 11.10) leading to screening and selecting the baseline sensor. In this data merging scheme, observations from MODIS-Aqua were reserved without any correction or computation. VIIRS observations were first merged with those of MODIS-Aqua, and the merged products were further merged with observations of MODIS-Terra (i.e., denoted as ATV in Figures 11.9 and 11.10). However, data merging may take place the other way around for those grids having no information (i.e., no data value due to cloud impact) in MODIS-Aqua when there are grid values in VIIRS-NPP and MODIS-Terra observations. In this case, those value-missing MODIS-Aqua grid values were projected from the other two sensors' observations and then merged to fill in the gap.

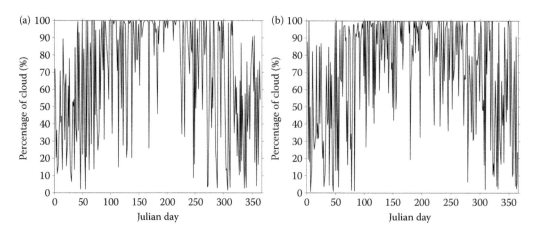

FIGURE 11.8 Time series of POC derived from daily (a) Terra and (b) Aqua images over Lake Nicaragua.

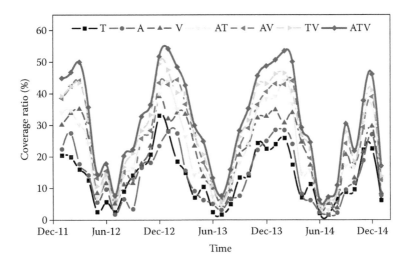

FIGURE 11.9 Comparison of the average monthly data coverage ratio over Lake Nicaragua during 2012–2015 at different fusion scheme. T: Terra; A: Aqua; V: VIIRS; AT: Aqua+Terra; AV: Aqua +VIIRS; TV: Terra+VIIRS; ATV: Aqua+Terra+VIIRS. (Bai, K., Chang, N. B., and Chen, C. F., 2015. *IEEE Transactions on Geoscience and Remote Sensing*, 54, 311–329.)

In Figure 11.9, the $\overline{CR}$ of merged products over Lake Nicaragua is nearly twice as much as that of any single sensor observation. Regarding the $\overline{CR}$ of a single sensor mission, VIIRS-NPP has a slightly larger $\overline{CR}$ than MODIS, especially during dry seasons (November–April). Meanwhile, indicative of a larger CR, the $\overline{CR}$ of merged products between any two sensors such as VIIRS and MODIS would be more competitive. This can be inferred from the larger $\overline{CR}$ of AV and TV compared to that of AT in Figure 11.9. Overall, it seems likely that VIIRS has priority over MODIS to be chosen as the baseline satellite sensor. Nevertheless, as shown in Figure 11.10, comparisons of $\overline{MAB}$ between any two sensors indicate that better agreement can be confirmed by selecting MODIS-Aqua as the baseline satellite sensor. This choice can be deduced from the smaller $\overline{MAB}$ of AV and AT when comparing with that of TV. When MODIS-Aqua can be regarded as the baseline

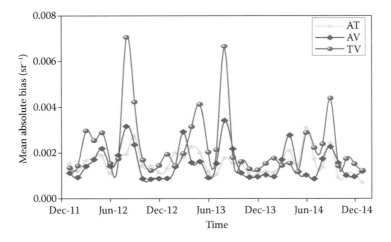

FIGURE 11.10 Monthly mean absolute bias (*MAB*) of observed ocean color reflectance at 443 nm among sensors. Combinations of sensors are identified by the letter associated with each individual sensor: Terra (T), Aqua (A), VIIRS (V).

sensor, the other two sensors' observations are required to be projected onto the MODIS-Aqua level. This setting may avoid showing extreme biases between MODIS-Terra and VIIRS-NPP (e.g., large biases of TV; see Figure 11.10). Therefore, observations derived from MODIS-Aqua were selected as the baseline satellite sensor for merging observations between VIIRS-NPP and MODIS-Terra successively over Lake Nicaragua based on: (1) spectral (i.e., number of available bands) difference, (2) overpassing time differences (Table 11.1), (3) the statistics of daily clear pixels coverage ratio (i.e., CR), and (4) relative biases (i.e., MAB) between sensors collectively.

11.3.3 Systematic Bias Correction

In this section, the Q-Q adjustment method was applied to remove systematic bias among three sensors. As described in Section 11.2, historical time series of ocean color reflectance Rrs^{SAT} were defined as references for systematic bias ΔRrs_i^{SAT} characterization. During the 2012–2015 overlapped time period, all available observations of each sensor over Lake Nicaragua were applied to derive the relevant reference time series from which the CDFs for the Q-Q adjustment can be computed for each relevant band. Matchups between MODIS-Aqua and VIIRS bands were first extracted from both historical time series, respectively, denoted as pixels in Figure 11.11. For this purpose, all historical images were ranked from high to low based on the number of matchups between both sensor observations for each date. Yet the number of matchups decreased over time as more observations recorded on different days were included. Note that sufficient matchups should be sampled to formalize the CDFs to better characterize the systematic bias among sensors.

However, when there is band shift, the SIASS algorithm should be advanced to merge observations collected from the neighboring bands. For instance, when dealing with the band at MODIS 531 nm wavelengths, observations collected from two VIIRS bands at 486 nm and 551 nm wavelengths pre-processed by using the SVD were used simultaneously to generate the projected spectral information of 531 nm at the MODIS-Aqua level jointly.

As mentioned above, principal components (PCs) of these observations associated with each band were computed via SVD to reduce the uncertainties embedded in the chaotic raw time series data. Statistics indicate that the first PC (PC1) of these observations can explain almost 95% of the total variance. Thus, PC1 (i.e., the first principal component) of each sensor's observations should be applied as the historical reference in the Q-Q adjustment for systematic bias characterization. Because the bias correction scheme is entirely data dependent, large outliers in the reference time series would thus introduce higher uncertainties and end up with new biases during the Q-Q adjustment. Therefore, a quality control approach became necessary for building a quality-assured correction scheme.

Let us choose the case that requires band shifting as an example. After calculating the PC1 of each set of observations (e.g., V1-PC, V2-PC, A-PC in Figure 11.11), the absolute bias between the V-PC and A-PC was computed accordingly (Figure 11.12):

$$Bias_i = 100 \times \frac{|V\text{-}PC_i - A\text{-}PC_i|}{A\text{-}PC_i} \tag{11.31}$$

where i is the number of days in the extracted discrete time series.

For quality control purposes, data points with values larger than the total average plus one standard deviation (i.e., >95% in CDFs of the PC bias) of the whole bias time series may be screened out. Then the associated data points in the PC time series of those corresponding days with large bias may be discarded. Based on the comparisons of the derived PCs of each sensor in Figure 11.12, it is noticeable that the long-term trends are still similar to each other although the local fluctuations of each sensor's observations are different. These final reference time series data are applicable

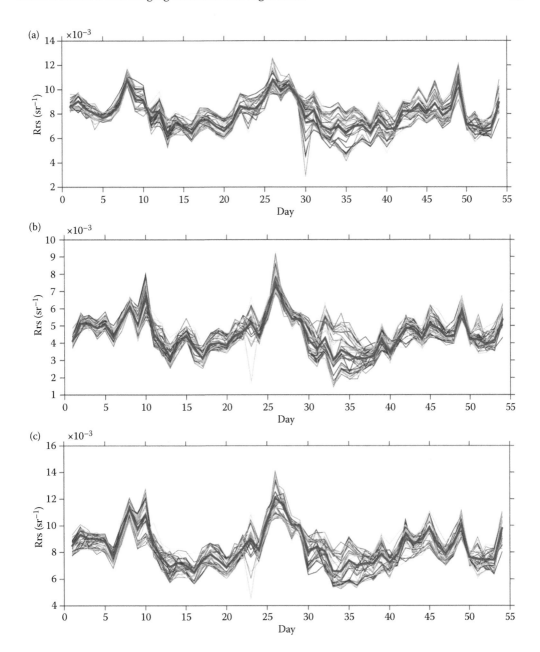

FIGURE 11.11 The first principal component (PC1, thick blue line) and associated ocean color reflectance time series (thin lines) of MODIS-Aqua (531 nm (a)) and VIIRS (488 nm (b) and 551 nm (c)).

to the band shifting after the Q-Q adjustment. It is worthwhile to mention that following such screening steps and quality control measures (Figures 11.11 and 11.12), only 54 days of observations were reserved for further systematic bias characterization. It is thus indicative that there is an urgent need for data merging over this region to improve the relevant spatial and temporal coverage of observations.

Once the referenced time series (i.e., PC time series) data are created, the Q-Q adjustment method may take place for systematic bias correction among sensors. Following Equations 11.6 through 11.14, observations collected by one sensor can be calibrated to the other based on the

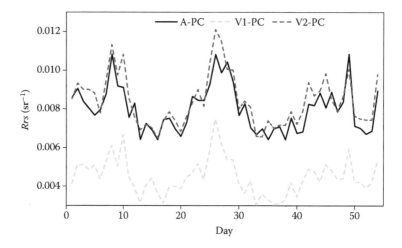

FIGURE 11.12 Comparison of PC1 (A-PC = PC1 of MODIS-Aqua; V1-PC = PC1 of VIIRS-NPP at 488 nm; V2-PC = PC1 of VIIRS-NPP at 551 nm).

referenced time series data. Taking the band shifting case as an example into account continuously, a comparison of $Rrs^{VIIRS}(488)$ and $Rrs^{VIIRS}(551)$ before and after being projected to $Rrs^{AQUA}(531)$ indicates that the systematic bias among sensors can be largely removed by adopting the Q-Q adjustment method (Figure 11.13a). This can be evidenced by the fact that the CDF of projected time series (PRJ_A1) is almost overlapped with that of $Rrs^{AQUA}(531)$ after bias correction

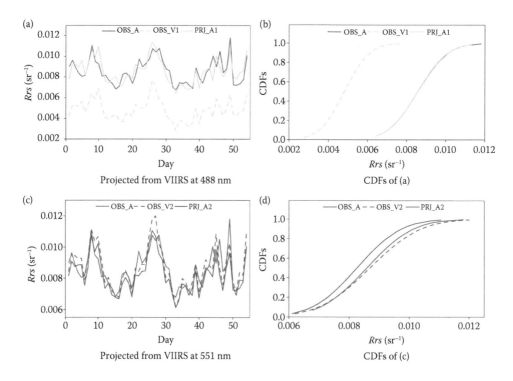

FIGURE 11.13 Comparisons of ocean color reflectance before and after systematic bias correction. (a) Time series of observed $Rrs^{VIIRS}(488)$ (OBS-V1), $Rrs^{AQUA}(531)$ (OBS-A), and projected time series (PRJ_A1) from OBS-V1; (b) CDFs of each time series in (a); (c) same as in (a) but from $Rrs^{VIIRS}(551)$ (OBS-V2); (d) CDFs of each time series in (c).

(Figure 11.13b) (i.e., the red and solid green curves). In fact, the projected time series data are not generated by a simple transition from the original raw data through adding the average differences between two referenced time series data. This can be evidenced by the fact that the differences are not consistent between time series before and after systematic bias correction as shown in Figure 11.13c. During the Q-Q adjustment, minor corrections were made for observations with small values while major corrections were made for large values at band 551 nm of VIIRS, reflecting an insightful correction logic (Bai et al., 2015). For instance, larger biases might occur at those peak values between VIIRS-NPP and MODIS-Aqua observations (Figure 11.13c). This is driven by the different instrumental responses or algorithmic differences in dealing with extreme values associated with different sensors (Bai et al., 2015). Therefore, it is advantageous to use such an adaptive bias correction scheme (i.e., Q-Q adjustment) for removing systematic biases among cross-mission sensor observations.

11.3.4 LOCATION-DEPENDENT BIAS CORRECTION

Once the systematic bias can be eliminated for each pixel value, it is the right time for possible removal of the inherent location-dependent bias at each grid. Due to the highly dynamic nature of aquatic environments in this illustrative example, the original reflectance time series with fluctuations randomly distributed are always nonlinear and nonstationary and such fluctuations are similar to white noise. According to Section 11.2, building a reliable model for the derivation of relationships between the projected time series after removing the systematic bias and the genuine time series with the removal of location-dependent time series is difficult, regardless of which method is chosen. In view of such complexity, EMD, which is a modern signal processing tool, was employed first to remove fluctuations screened out by the first IMF. It is known that the first IMF has the highest frequency in the sense that removing these high frequency signals may not affect the long-term variability of the original signals. Rather, removing these high frequency signals would improve the stability and efficiency in model generalization when conducting the location-dependent bias. Comparisons of the time series before and after reconstruction via EMD clearly show that the reconstructed time series data are much more representative (i.e., smoother) after removing those high frequency fluctuations than the original projected time series. It is noticeable that the reconstructed time series still maintain their original long-term variability (Figure 11.14) (Bai et al., 2015).

ELM is then chosen to establish relationships for removing the possible location-dependent bias (i.e., ΔRrs^{LCT}) from the reconstructed time series. In this context, reconstructed MODIS-Aqua time series data were defined as targets while reconstructed time series of the projected observations (from VIIRS or MODIS-Terra) time series data were used as inputs for ELM (Bai et al., 2015). We randomly screened 70% of these reconstructed time series data for ELM training, while the rest of data were used for testing purposes. The stopping criterion for the training process was created based on the coefficient of determination (i.e., R^2) between the predicted values and the associated target values during the training stage. The training process continues until the R^2 value ends up at 0.8. Once a robust ELM model is generated, it may then be used for prediction based on the given new observations for the remaining 30% of data. To avoid random simulation and improve robustness, multiple models were simulated via 30 trials for each particular pixel.

11.3.5 SPECTRAL INFORMATION SYNTHESIS

To verify the efficacy of the ELM model in SIASS, we continue our effort to deal with the mismatched bands via experimental analyses conducted for spectral information synthesis (e.g., VIIRS-NPP 488 nm and 551 nm) whereas spectral adjustment at the common bands would be

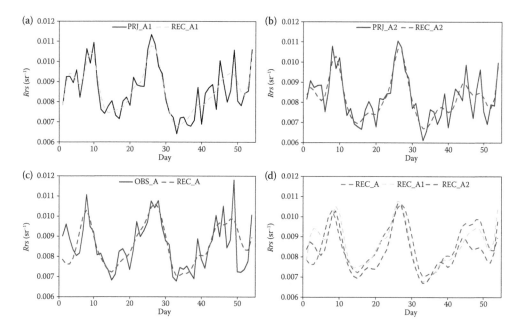

FIGURE 11.14 Comparisons of the time series before and after reconstruction. Here, PRJ denotes projected time series in Figure 11.13 (A1 is associated with VIIRS 488 nm and A2 is associated with VIIRS 551 nm) while REC denotes reconstructed time series. (a) Reconstructed from the projected time series I (PRJ_A1), (b) reconstructed from the projected time series II (PRJ_A2), (c) reconstructed from Aqua 531 nm (OBS_A), (d) reconstructed time series. (Bai, K., Chang, N. B., and Chen, C. F., 2015. *IEEE Transactions on Geoscience and Remote Sensing*, 54, 311–329.)

deemed relatively straightforward. To achieve this goal, the validation scheme was performed by applying SIASS to reconstruct 400 observed MODIS-Aqua ocean color reflectance grids in one image based on associated VIIRS-NPP and MODIS-Terra observations collected on January 10, 2014. These 400 data points (i.e., pixel values) were extracted from a 20 × 20 clear scene in each sensor observation of MODIS-Terra. Spectral information of two adjacent projected VIIRS-NPP bands with wavelengths of 488 nm and 551 nm were synthesized to generate observations at MOIDS-Aqua 531 nm wavelengths level (Figure 11.15). Before applying SIASS, salient biases were identified between VIIRS-NPP and MODIS-Aqua observations. While the case of VIIRS-NPP at 551 nm slightly overestimated the reflectance value of MODIS-Aqua at 531 nm, VIIRS-NPP at 488 nm largely underestimated the reflectance value of MODIS-Aqua at 531 nm. Inconsistent biases between observations associated with MODIS-Aqua 531 nm and VIIRS-NPP 551 nm also reveal the complexity of bias correction among sensors, which are driven by the nonlinear and nonstationary nature of biases among sensors (Bai et al., 2015). These inconsistencies might be due to instrumental responses, algorithmic differences, calibration uncertainties, local overpassing time differences of satellite sensors, and aquatic dynamics (Bai et al., 2015).

However, the robust ELM model can work well and the synthesized time series data based on VIIRS-NPP 488 nm and 551 nm largely avoids relevant issues as mentioned above. The synthesized spectral information from neighboring bands with the aid of the ELM model preserves the unique spectral characteristics of these mismatched bands. Fair agreement can be observed between sensor measurements and reconstructed spectral information at MODIS-Aqua 531 nm level (Figure 11.15c), leading to confirmation of the efficacy of the SIASS method. By applying SIASS, merged ocean color reflectance products at these three distinctive wavelengths were generated (Figure 11.16). Prior to the development of SIASS, this weakness could largely be overcome by using complex

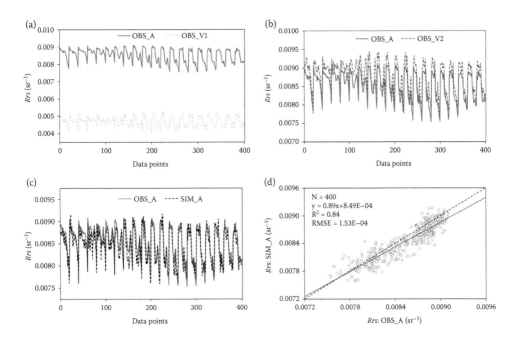

FIGURE 11.15 Comparisons between the observed and simulated data for 400 pixels at 531 nm on January 10, 2014. OBS_V1 denotes the observations from VIIRS at 488 nm and OBS_V2 for 551 nm, OBS_A are the observations from MODIS-Aqua at 531 nm. (a) Aqua 531 nm vs. VIIRS 488 nm, (b) Aqua 531 nm vs. VIIRS 551 nm, (c) Aqua 531 nm vs. synthesized, (d) scatter plot of (c).

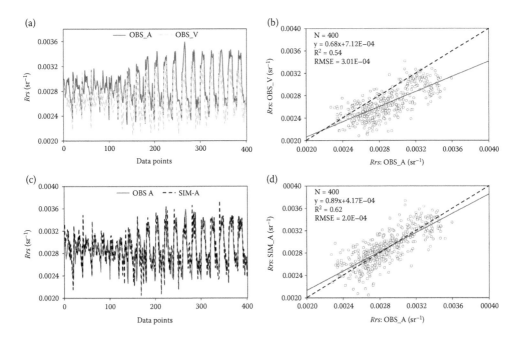

FIGURE 11.16 Comparisons between the observed and simulated data for 400 pixels at 443 nm on January 10, 2014. Observations from VIIRS at 443 nm were used to simulate the associated spectral information at MODIS-Aqua 443 nm. (a) Aqua 443 nm vs. VIIRS 443 nm, (b) scatter plot of (a), (c) Aqua 443 nm vs. simulated, (d) scatter plot of (c).

bio-optical (i.e., geophysical model) models or band-shifting processes (Maritorena and Siegel, 2005; Maritorena et al., 2010; Lee et al., 2014; Mélin and Sclep, 2015). This advantage makes SIASS stand out among other similar bias correction schemes given that SIASS may preserve some valuable spectral information while working for the mismatched bands among sensors (Bai et al., 2015).

To verify the anticipated role of ELM in this SIASS algorithm, the Mean Relative Difference (*MRD*) was calculated pairwise for ocean color reflectance observations between the observed data collected from MODIS-Aqua or either of the two other complementary satellites as well as the simulated data.

$$MRD = 100 \times \frac{1}{N} \sum_{i=1}^{N} \frac{Rrs_i^{SAT1} - Rrs_i^{SAT2}}{Rrs_i^{SAT2}} \qquad (11.32)$$

where Rrs_i^{SAT2} is the ocean color reflectance observations from MODIS-Aqua; Rrs_i^{SAT1} denotes observations from MODIS-Terra or VIIRS-NPP or simulated ocean color reflectance; and N is the number of samples in each satellite observation. The associated *MRD* values derived from the results (Figures 11.15 through 11.18) were comparatively summarized (Table 11.4) (Bai et al., 2015).

Before applying SIASS, apparent inconsistences were observed between MODIS-Aqua and the other two complementary satellites, especially at 443 nm (Table 11.4). Whereas VIIRS-NPP significantly underestimated ocean color reflectance observations, MODIS-Terra significantly overestimated these values relative to MODIS-Aqua. The largest correction effect was confirmed at 443 nm, with an MRD <1% after applying SIASS. Finally, these maps in Figure 11.19 show that the spatial coverage ratio of MODIS-Aqua observations at the 443 nm wavelength over Lake Nicaragua was significantly improved by merging with VIIRS-NPP and MODIS-Terra successively.

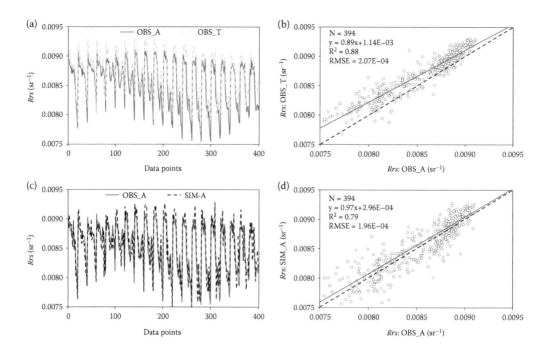

FIGURE 11.17 Comparisons between the observed and simulated data for 400 pixels at 531 nm on January 10, 2014. Observations from VIIRS at 443 nm were used to simulate the associated spectral information at MODIS-Aqua 531 nm. (a) Aqua 531 nm vs. Terra 531 nm, (b) Scatter plot of (a), (c) 531 nm vs. simulated, (d) scatter plot of (c).

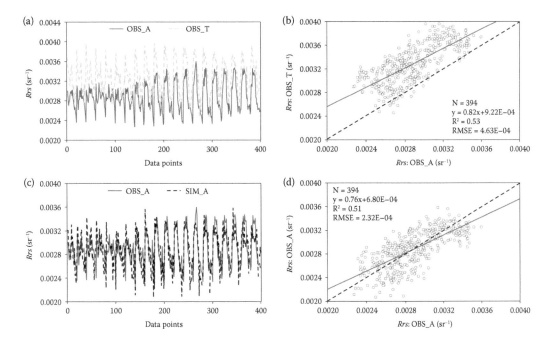

FIGURE 11.18 Comparisons between the observed and simulated data for 400 pixels at 443 nm on January 10, 2014. Observations from MODIS-Terra at 443 nm were used to simulate the associated spectral information at MODIS-Aqua 443 nm. (a) Aqua 443 nm vs. Terra 443 nm, (b) scatter plot of (a), (c) Aqua 443 nm vs. simulated, (d) scatter plot of (c).

TABLE 11.4

Comparisons of *MRD* between Pairwise Ocean Color Reflectance From the Cross-Mission Observations Before and After Spectral Adaptation with SIASS

Sensor Pair	AV		AT	
	Before (%)	After (%)	Before (%)	After (%)
531 nm	44.96/2.7[a]	0.98	1.88	0.23
443 nm	−7.31	0.53	14.21	0.02

Source: Bai, K., Chang, N. B., and Chen, C. F., 2015. *IEEE Transactions on Geoscience and Remote Sensing*, 54, 311–329.

[a] Two bands of VIIRS (488 nm/551 nm) observations (Figure 11.15).

11.4 SUMMARY

In this chapter, by taking advantage of the common time period of observations temporally between successive generations of satellite sensors, an adaptive method with an integrated statistics and machine learning technique, SIASS, was proposed to remove the systematic and location-dependent biases between cross-mission ocean color sensors for observation merging purposes. With the aid of the Q-Q adjustment method and a machine-learning–based correction scheme, SIASS was able to remove the instrumental- and algorithmic-related systematic biases as well as location-dependent bias simultaneously. Compared to previous methods using complex bio-optical models and scarce

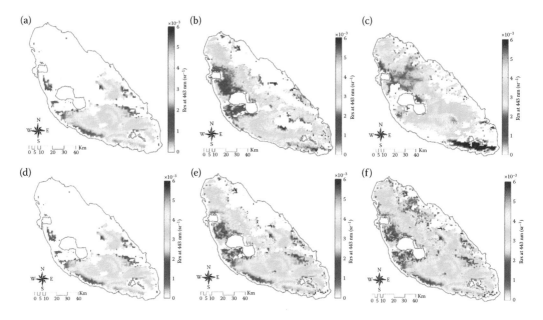

FIGURE 11.19 Comparisons of MODIS-Aqua ocean color reflectance before and after merging with VIIRS and MODIS-Terra at 443 nm on December 5, 2013. The original observations of each sensor are shown in the upper panel while the merged are shown in the bottom panel. (a) MODIS-Aqua, (b) VIIRS-NPP, (c) MODIS-Terra, (d) MODIS-Terra, (e) MODIS-Aqua fused with VIIRS-NPP, (f) MODIS-Aqua fused with VIIRS-NPP and MODIS-Terra. (Bai, K., Chang, N. B., and Chen, C. F., 2015. *IEEE Transactions on Geoscience and Remote Sensing*, 54, 311–329.)

in situ measurements for possible bias correction, SIASS is more adaptive as it solely relies on common observations from cross-mission sensors.

REFERENCES

Ardeshir Goshtasby, A. and Nikolov, S., 2007. Image fusion: Advances in the state of the art. *Information Fusion*, 8, 114–118.

Bai, K., Chang, N. B., and Chen, C. F., 2015. Spectral information adaptation and synthesis scheme for merging cross-mission consistent ocean color reflectance observations from MODIS and VIIRS. *IEEE Transactions on Geoscience and Remote Sensing*, 54, 311–329.

Bai, K., Chang, N. B., Yu, H., and Gao, W., 2016. Statistical bias correction for creating coherent total ozone record from OMI and OMPS observations. *Remote Sensing of Environment*, 182, 150–168.

Barnes, B. B. and Hu, C., 2015. Cross-sensor continuity of satellite-derived water clarity in the Gulf of Mexico: Insights into temporal aliasing and implications for long-term water clarity assessment. *IEEE Transactions on Geoscience and Remote Sensing*, 53, 1761–1772.

Businger, P. A. and Golub, G. H., 1969. Algorithm 358: Singular value decomposition of a complex matrix. *Communications of the ACM*, 12, 564–565.

Chang, N. B., Bai, K., and Chen, C. F., 2015. Smart information reconstruction via time-space-spectrum continuum for cloud removal in satellite images. *IEEE Journal of Selected Topics in Applied Earth Observations*, 99, 1–19.

D'Alimonte, D., Zibordi, G., and Mélin, F., 2008. A statistical method for generating cross-mission consistent normalized water-leaving radiances. *IEEE Transactions on Geoscience and Remote Sensing*, 46, 4075–4093.

Drazin, P. G., 1992. *Nonlinear Systems*. Cambridge University Press, Cambridge, USA.

Fargion, G. and McClain, C. R., 2002. *SIMBIOS Project 2001 Annual Report, 2002. NASA Tech Memo. 2002–210005*. NASA Goddard Space Flight Center, Greenbelt, MD.

Gordon, H. R. and Wang, M., 1994. Retrieval of water-leaving radiance and aerosol optical thickness over the oceans with SeaWiFS: A preliminary algorithm. *Applied Optics*, 33, 443–52.

Gregg, W. W., 1998. Coverage opportunities for global ocean color in a multimission era. *IEEE Transactions on Geoscience and Remote Sensing*, 36, 1620–1627.

Gregg, W. W. and Casey, N. W., 2010. Improving the consistency of ocean color data: A step toward climate data records. *Geophysical Research Letters*, 37, L04605.

Gregg, W. W., Casey, N. W., O'Reilly, J. E., and Esaias, W. E., 2009. An empirical approach to ocean color data: Reducing bias and the need for post-launch radiometric re-calibration. *Remote Sensing of Environment*, 113, 1598–1612.

Gregg, W. W. and Woodward, R. H., 1998. Improvements in coverage frequency of ocean color: Combining data from SeaWiFS and MODIS. *IEEE Transactions on Geoscience and Remote Sensing*, 36, 1350–1353.

Gross, H. N. and Schott, J. R., 1998. Application of spectral mixture analysis and image fusion techniques for image sharpening. *Remote Sensing of Environment*, 63, 85–94.

Hu, C. and Le, C., 2014. Ocean color continuity from VIIRS measurements over Tampa Bay. *IEEE Geoscience and Remote Sensing Letters*, 11, 945–949.

Huang, N. E., Shen, Z., Long, S. R., Wu, M. C., Shih, H. H., Zheng, Q., Yen, N.-C., Tung, C. C., and Liu, H. H., 1998. The empirical mode decomposition and the Hilbert spectrum for nonlinear and non-stationary time series analysis. *Proceedings of the Royal Society of London. Series A, Mathematical and Physical Sciences*, 454, 903–995.

Huang, G.-B., Zhu, Q.-Y., and Siew, C.-K., 2006. Extreme learning machine: Theory and applications. *Neurocomputing*, 70, 489–501.

Kwiatkowska, E. J. and Fargion, G. S., 2003. Application of machine-learning techniques toward the creation of a consistent and calibrated global chlorophyll concentration baseline dataset using remotely sensed ocean color data. *IEEE Transactions on Geoscience and Remote Sensing*, 41, 2844–2860.

Lee, Z., Shang, S., Hu, C., and Zibordi, G., 2014. Spectral interdependence of remote-sensing reflectance and its implications on the design of ocean color satellite sensors. *Applied Optics*, 53, 3301–3310.

Luo, R. C., Yih, C. C., and Su, K. L., 2002. Multisensor fusion and integration: approaches, applications, and future research directions. *IEEE Sensors Journal*, 2, 107–119.

Maritorena, S., D'Andon, O. H. F., Mangin, A., and Siegel, D. A., 2010. Merged satellite ocean color data products using a bio-optical model: Characteristics, benefits and issues. *Remote Sensing of Environment*, 114, 1791–1804.

Maritorena, S. and Siegel, D. A., 2005. Consistent merging of satellite ocean color data sets using a bio-optical model. *Remote Sensing of Environment*, 94, 429–440.

Mélin, F. and Sclep, G., 2015. Band shifting for ocean color multi-spectral reflectance data. *Optics Express*, 23, 2262.

Mélin, F. and Zibordi, G., 2007. Optically based technique for producing merged spectra of water-leaving radiances from ocean color remote sensing. *Applied Optics*, 46, 3856–3869.

Moore, K. D., Voss, K. J., and Gordon, H. R., 2000. Spectral reflectance of whitecaps: Their contribution to water-leaving radiance. *Journal of Geophysical Research: Oceans*, 105, 6493–6499.

Nash, J. C., 1990. The singular-value decomposition and its use to solve least-squares problems. In: *Compact Numerical Methods for Computers: Linear Algebra and Function Minimisation*, Adam Hilger (Ed.), Bristol, England, CRC Press, Boca Raton, Florida, USA, 30–48.

Pohl, C. and van Genderen, J. L., 1995, Image fusion of microwave and optical remote sensing data for map updating in the Tropics. In: *Proceedings of SPIE 2579, Image and Signal Processing for Remote Sensing II*, Paris, France, 2–10.

Shen, S. S., 1990. Summary of types of data fusion methods utilized in workshop papers. In: *Multisource Data Integration in Remote Sensing, Proceedings of Workshop*, Maryland, USA, *NASA Conference Publication 3099*, 145–149.

Shi, Y., Zhou, X., Yang, X., Shi, L., and Ma, S., 2015. Merging satellite ocean color data with Bayesian maximum entropy method. *IEEE Journal of Selected Topics in Applied Earth Observations and Remote Sensing*, 8, 3294–3304.

Vornberger, P. L. and Blindschadler, R. A., 1992. Multi-spectral analysis of ice sheets using co-registered SAR and TM imagery. *International Journal of Remote Sensing*, 13, 637–645.

Wang, M. and Bailey, S. W., 2001. Correction of sun glint contamination on the SeaWiFS ocean and atmosphere products. *Applied Optics*, 40, 4790–4798.

Wilk, M. B. and Gnanadesikan, R., 1968. Probability plotting methods for the analysis for the analysis of data. *Biometrika*, 55, 1–17.

Zhang, C., 2015. Applying data fusion techniques for benthic habitat mapping and monitoring in a coral reef ecosystem. *ISPRS Journal of Photogrammetry and Remote Sensing*, 104, 213–223.

12 Cloudy Pixel Removal and Image Reconstruction

12.1 INTRODUCTION

It is known that approximately 50% to 75% of the world is cloud covered (NASA, 2016). Over the past 20 years, abundant space- and air-borne remote sensors on board platforms have provided a variety of remotely sensed imageries that greatly help in understanding the Earth's environment. Since most of the remotely sensed imageries are retrieved from the visible and infrared spectral ranges, these collected imageries are highly vulnerable to the presence of clouds and cloud shadows when trying to observe the Earth's surface. This is especially true during wet seasons or across middle and low latitude regions where evaporation may cause heavy cloud cover most of the time (Asner, 2001). Whereas clouds provide critical information about the water cycle related to the meteorological and hydrological studies and lead to better understanding of weather phenomena and water movement, clouds and cloud shadows are regarded as contamination which cause major issues in many environmental monitoring and earth observation programs if they are not solely deigned to explore the clouds themselves in meteorology (Melgani, 2006; Eckardt et al., 2013).

Clouds and cloud shadows corrupt satellite images such as Landsat and Moderate Resolution Imaging Spectroradiometer (MODIS) images when viewing the ground level of the Earth. Some special features such as high brightness, low contrast, and low frequency in thin clouds have deeply influenced the quality of the visible bands when observing the ground or sea level. For example, cloud coverage at Lake Nicaragua in Central America became an issue when the water quality monitoring had to be carried out on a long-term basis (Figure 12.1). To recover or reconstruct these cloudy pixels, there is a need to detect the exact cloudy area in the images.

The threshold method and the supervised classification method were developed for cloud mask in recent decades. In the threshold method, the characteristic graph is built using bands sensitive to clouds. A straight forward example consists of a target image and a reference image, both of which require image registration and digital number normalization. Note that in order to get good detection results, the reference and the target images are required to have similar spatial resolution. The coarse cloud mask may be extracted through the threshold method based on the blue band information and further refined using the cloud transition edge index along the edge of the coarse cloud mask. The threshold method is more popular because of its simplicity and effectiveness. The threshold method was applied for the Advanced Very High Resolution Radiometer (AVHRR) data (Saunders, 1986; Di Vittorio and Emery, 2002), MODIS (Platnick et al., 2003), and Landsat Top of Atmosphere (TOA) reflectance and Brightness Temperature (BT) followed by using Potential Cloud Pixels (PCPs) (Zhu and Woodcock, 2012). The threshold method has some limitations, however. First, the threshold method might fail in snow and ice regions because snow and ice have reflection characteristics similar to cloud in visible bands. Second, the threshold method might fail to detect translucent thin clouds. The supervised classification method may overcome these limitations by using some machine learning tools for differentiation between cloudy and non cloudy pixels. For example, the Artificial Neural Networks (ANN) may help with cloud detection on the SPOT images (Jang, et al., 2006). It is also possible to use the visible bands of the whole sky images as inputs to train the MultiLayer Perceptron (MLP) neural networks and Support Vector Machine (SVM) for automatic cloud detection (Taravat et al., 2015). Furthermore, the visual saliency features may be used to automatically generate some basic saliency maps of clouds followed by employing a random forest classifier trained to distinguish clouds (Hu et al., 2015).

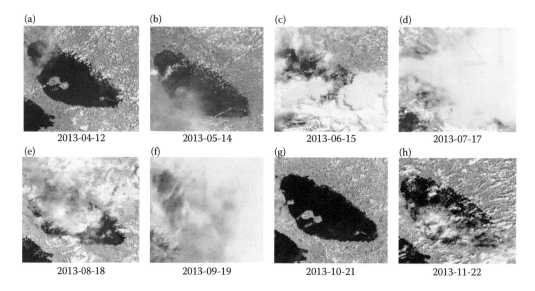

FIGURE 12.1   Landsat 8 Operational Land Imager (OLI) true color images over Lake Nicaragua in 2013.

Besides automatic detection of cloud and shadow, cloud removal and pixel reconstruction research is in an acute need for satellite imageries in order to reduce cloud interference in feature extraction and content-based mapping when processing original, merged or fused images. In Chapter 4, we discussed gap filling or cloudy pixel reconstruction methods based on spatial information, temporal information, and both. For cloud removal, the approaches developed in the past few decades can be classified into five categories (Lin et al., 2013, 2014): inpainting-based, multispectral information-based, multitemporal information-based, auxiliary sensor, and hybrid methods. Their definitions and comparisons are summarized below as well as in Table 12.1. Pros and cons of methods for cloudy pixel recovery are listed in Table 12.2.

- *Inpainting-based method*: This method uses the cloudy image itself by exploiting the remaining clear parts inside a cloud-covered region to interpolate the cloudy regions.
- *Multispectral information-based method*: This method builds relationships between cloudy bands and cloud-free auxiliary bands to reconstruct missing pixel values in which histogram matching, least square fitting, and wavelet reconstruction were applied.
- *Multitemporal information-based method*: This method exploits a cloud-free auxiliary image from different dates to estimate cloudy pixels for the same area.
- *Auxiliary sensor-based*: This method uses the principle of combining information collected by different satellite sensors to reconstruct cloudy pixels.
- *Hybrid method*: The hybrid method may combine any two methods or even all three methods to exhibit higher application potential in dealing with a dynamic environment with heavy cloud.

However, there is no single method that may complete cloud removal and missing pixel reconstruction. The hybrid algorithms take advantage of synergistic capacity from different methods to fill in gaps caused by clouds and cloud shadows, as discussed by the following three categories below.

1. Cloudy pixel reconstruction method based on spatial and/or multispectral information: This method uses cloud-free pixels from the image itself to predict the values for contaminated areas. Approaches can vary and include using focal mean (Long et al., 1999), geostatistics

TABLE 12.1

Comparisons of Methods for Cloudy Pixel Recovery

Method	Algorithm	Niche	Reference
Inpainting-based method	The bandelet transform and the multiscale geometrical grouping	To incorporate geometrical flow to restore the cloudy parts	Maalouf et al. (2009)
	The propagation of the spectro–geometrical information	To incorporate both geometrical flow and spectral information from the clear parts of the image to restore the cloudy parts	Lorenzi et al. (2011)
	The co-kriging method	To combine both temporal and spatial information for the best estimates	Addink (1999)
Multispectral-based method	The SAR images taken at the same day	To have prior information about contaminated zone structure	Zhu et al. (2016)
	The haze optimized transformation for visible bands	To radiometrically adjust their cloud contamination	Zhang et al. (2002)
	The haze optimized transformation for NIR band over the shallow water environment	To predict the spatial distribution of haze intensity in each visible band by establishing a linear regression model for aquatic zones	Ji (2008)
	The restoration of the cloud contaminated area of aqua MODIS band 6	To incorporate histogram matching with local least squares fitting between bands 6 and 7	Rakwatin et al. (2009)
	The empirical and radiative transfer model	To recover thin clouds in visible bands based on optical information from visible bands and NIR band	Lv et al. (2016)
Multitemporal information-based method	The regression trees to estimate pixel values contaminated by clouds and cloud shadows	To create a reference cloud-free scene from different date	Helmer and Ruefenacht (2005)
	The contextual prediction method	To reconstruct cloudy pixels by establishing the relationship between cloudy image and reference images	Melgani (2006)
	The post-reconstruction method as an improvement of the contextual reconstruction approach	To reconstruct cloudy pixels by using spectral and spatial correlations when identifying the relevant images	Benabdelkader and Melgani (2008)
	The three-step method by (1) image enhancement, (2) base map selection with the least cloudy pixels, (3) multiscale wavelet-based fusion method to deal with the tradition zone	To generate cloud-free satellite images by combining multitemporal images of the cloud-free areas	Tseng et al. (2008)
	The closest spectral fit method	To reconstruct cloud and cloud-shadow contaminated pixels based on their similar free cloud pixel values	Meng et al. (2009)

(Continued)

TABLE 12.1 (*Continued*)

Comparisons of Methods for Cloudy Pixel Recovery

Method	Algorithm	Niche	Reference
	The NSPI/GNSPI method	To remove strips produced by the failure of scan-line corrector of the Landsat 7 Enhanced Thermal Mapper Plus (ETM+) sensor	Chen et al. (2011), Zhu et al. (2012)
	The information cloning method	To clone information from clear patches to the corresponding cloudy patches based on temporal correlation of mutitemporal images	Lin et al. (2012, 2014)
	The SVR with new combinations of new kernel functions	To perform a compression in the number of values needed for a good reconstruction	Lorenzi et al. (2013)
	The MRF global function method	To build the pixel-offset based spatiotemporal MRF global function to find the most suitable similar pixel to replace the cloudy area and guided spatially by a reference image	Cheng et al. (2014)
	The random measurement cloud removal algorithm	To synthetize the cloud contaminated zones via time series through sparse reconstruction	Cerra et al. (2016)
	The sparse representation via multitemporal dictionary learning	To reconstruct cloudy images with the multitemporal auxiliary image with dictionary learning	Xu et al. (2016)
Auxiliary sensor-based method	The semi-physical fusion method	To estimate Landsat reflectance for gap filling based on MODIS BRDF and Albedo product as well as Landsat available data	Roy et al. (2008)
	The fusion of a HRI using two types of auxiliary images, i.e., LRI and a SAR image	To incorporate two dictionary pairs are trained jointly: One pair is generated from the HRI and LRI gradient image patches, and the other is generated from the HRI and SAR gradient image patches	Huang et al. (2015)

DN: digital number; SAR: Synthetic Aperture Radar; HRI: high-resolution image; LRI: low-resolution image; BRDF: Bidirectional Reflectance Distribution Function; NIR: near infrared; SVR: Support Vector Regression; GNSPI: Geostatistical Neighborhood Similar Pixel Interpolator; MRF: Markov random fields.

TABLE 12.2

Pros and Cons of Methods for Cloudy Pixel Recovery

	Pros	Cons
Inpainting-based method	• It can give a visually presumable result with computational efficiency • The algorithm is simple	• It is only applicable for small gaps • It is limited in reconstructing cloudy pixels lying in the edge between two land cover types • It has issues of low accuracy and lack of restoring information • This limitation appears oftentimes in heterogeneous landscapes
Multispectral information-based method	• This method has high accuracy for thin clouds	• Multispectral images are needed in gap filling • It cannot work well with heavy clouds
Multitemporal information-based method	• This method makes use of the multitemporal acquisitions of remote sensing images collected by the same sensor or similar sensors	• It has low accuracy in a highly dynamic environment
Auxiliary sensor-based method	• This method may incorporate multiple inputs from different satellite sensors	• This method is limited by the spatial resolution difference, temporal coherence, and spectral comparability between sensors

methods (Zhang et al., 2007) or nearest neighbor (Chen et al., 2011) to perform gap filling. There are two ways to potentially improve the effectiveness of cloud removal. One is to include the multichannel or multiband contextual information, and the other is to utilize the texture information retrieved from microwave images. Both expanded the inpainting-based method into the use of multi-spectral information. For instance, Benabdelkader and Melgani (2008) utilized the contextual spatio-spectral post reconstruction of cloud contaminated images with the aid of multichannel information. Microwave images are not affected by cloud, while cloud is always a problem for optical remote sensing data. Therefore, there is a synergistic opportunity to make use of microwave images to reconstruct optical remote sensing data. The ALOS satellite has both microwave and optical sensors with similar resolution, providing such a synergistic opportunity. With this understanding, Hoan and Tateishi (2009) developed a method for removing cloud in ALOS-AVNIR2 optical remote sensing images based on ALOS-PALSAR data. In their study, cloud detection was conducted based on a combination of Total Reflectance Radiance Index (TRRI) and Cloud-Soil Index (CSI). Eckardt et al. (2013) further presented a new, synergistic approach using optical and synesthetic aperture radar (SAR) data to amend the missing pixels via a Closest Feature Vector (CFV). The methodology was applied to mono-temporal, multifrequency SAR data from TerraSAR-X (X-Band), ERS (C-Band), and ALOS Palsar (L-Band) in concert with multispectral remote sensing Landsat and SPOT 4 images for the assessment of the image restoration performance (Figure 12.2).

2. Cloudy pixel reconstruction method based on multitemporal and spatial information: This method is designed to replace contaminated pixels with cloud-free pixels from a reference image collected at a different time. Wang et al. (1999) adopted wavelet transformation to incorporate complementary information into the composite from multitemporal images via data fusion. In addition, the regression relationship and/or histogram matching between the

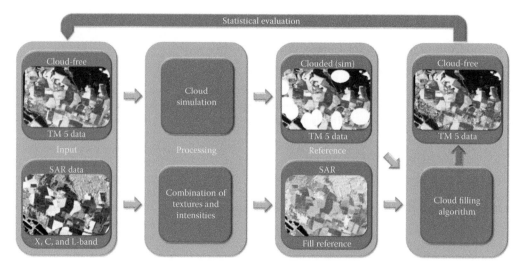

FIGURE 12.2 Workflow of cloud removal using optical and SAR data. (Eckardt, R. et al., 2013. *Remote Sensing*, 5, 2973–3006.)

target image and a reference image may be retrieved to generate a cloud-free image (Helmer and Ruefenacht, 2005). Martinuzzi et al. (2007) developed a semi-automated method to mask clouds and shadows in Landsat ETM+ imagery for a variety of landscapes. Roy et al. (2008) developed a semi-physical data fusion approach that used the MODIS Bidirectional Reflectance Distribution Function (BRDF)/Albedo land surface characterization product and Landsat ETM+ data to predict ETM+ reflectance on the same date. Meng et al. (2009) developed a Closest Spectral Fit (CSF) method to remove clouds and cloud shadows from remotely sensed optical images by taking advantage of spectral similar pixels detected from multitemporal images. An analog method is the Neighborhood Similar Pixel Interpolator (NSPI) proposed by Chen et al. (2011) that was used to fill gaps in Landsat ETM+ SLC-off images. By taking advantage of a similar principle, Jin et al. (2013) developed an automated cloud and shadow detection and filling algorithm by using two-date Landsat imagery. Because this method depends solely on the reference image, the quality of the filled value could be heavily affected by any difference between the reference and target images, which can be caused by phenology, pre-processing techniques, spatial resolution, atmospheric conditions, and disturbances.

3. Cloudy pixel reconstruction method based on inpainting, multitemporal, and/or multispectral information: Melgani (2006) developed the Expectation–Maximization (EM) algorithm for cloud removal based on multitemporal multispectral images with respect to spatiotemporal context via SVMs. With the availability of a reference image, Gu et al. (2011) first used spatial, temporal, and spectral information in MODIS imageries simultaneously for pixel reconstruction of thin and heavy clouds. The key success is tied to analyzing the cloud spectral characters derived from the thirty-six bands of MODIS data, indicating that the spectral reflections of ground and cloud are different in various MODIS bands. The newly developed image reconstruction algorithms, such as SMart Information Reconstruction (SMIR) (Chang et al., 2015), Self-Organizing Map (SOM), and information cloning are relatively complicated. Whereas the SOM requires making use of complementary remote sensing observations (Jouini et al., 2013), information cloning requires using the global optimization scheme (Lin et al., 2013). Lin et al. (2013) proposed the information cloning algorithm, assuming that land covers change insignificantly over a short period of time. The patch-based information reconstruction is mathematically formulated as a Poisson

equation and solved using a global optimization process linking cloud-free patches to their corresponding cloud contaminated patches for image reconstruction. Jouini et al. (2013) presented a method able to fill large data gaps in satellite chlorophyll (CHL) images, based on the principle of SOM classification methods. In SOM, MODIS ocean color band enables us to measure the distribution of chlorophyll (the main pigment of phytoplankton) at the surface of the ocean on a daily basis with a spatial resolution of 1 km at nadir. However, the effective data retrieval represents only a minor fraction of the total observed pixels, because observations are often heavily disturbed by clouds, thick aerosol layers or sun glint. This limitation strongly reduces the potential applications of ocean color observations. The neural classification method makes use of complementary oceanic remote sensing observations, including sea surface temperature (SST) and sea surface height (SSH) to supplement the information. SOM relies on two assumptions: (1) a state of the ocean can be locally defined by its values of SST, SSH, and CHL, and (2) a codebook of possible (SST, SSH, CHL) situations, if large enough. Therefore, each situation of SOM is characterized by the values of the three variables CHL, SST, and SSH over a 3×3 spatial window surrounding the pixel at three successive times over specified the spatiotemporal domain to account for the spatiotemporal context. It leads to the reconstruction of incomplete situations of CHL even in the condition of heavy clouds. SMIR will be introduced in the end of this chapter in greater detail by employing integrated pattern recognition, big data analytics, and machine learning technique (Chang et al., 2015).

12.2 BASICS OF CLOUD REMOVAL FOR OPTICAL REMOTE SENSING IMAGES

12.2.1 SUBSTITUTION APPROACHES

Mitchell et al. (1977) started developing a distortion model to deal with cloud impact and proposed a filter to reduce the impact of cloudy pixels on Landsat satellite imageries. Recovery of true ground reflectance from the recorded images is not possible when the cloud cover is too thick to allow ground reflectance information to reach the sensor (Chanda and Majumder, 1991). For this reason, the algorithms proposed early in the 1980s and 1990s are applicable only when obscuration is driven by atmospheric haze and thin cloud. Liu and Hunt (1984) proposed a revised distortion model to remove thin clouds and Chanda and Majumder (1991) further generated an iterative scheme for the removal of thin cloud from Landsat imageries. Following the same trends, Cihlar and Howarth (1994) and Simpson and Stitt (1998) achieved thin cloud removal for AVHRR imageries. These preliminary approaches were designed mainly based on simple substitution and interpolation to support the cloud removal. Substitution approaches are used to fill in the missing values of cloudy pixels with the cloud-free pixels in the neighborhood collected either in the same time period or from previous time periods with spatial relationships, and are mainly applied to homogeneous landscapes of a less dynamic nature (Long et al., 1999; Sprott, 2004; Tseng et al., 2008; Lin et al., 2014). Among these substitution methods, one representative method is the CSF method proposed by Meng et al. (2009), which works by taking advantage of spectral similarity to reconstruct cloud contaminated pixels by filling in those pixels with data values of spectrally similar pixels. A schematic illustration of the CSF method is shown in Figure 12.3. Generally, two main conceptions in CSF may be highlighted. First, the location-based one-to-one correspondence and spectral-based closest fit are defined and the location-based one-to-one correspondence was applied to identify pixels with the same locations in both base image and reference images. Then the spectral-based closest fit was identified and applied to determine the most similar pixels in an image (Meng et al., 2009). Analogous methods can be also found in the literature, linking cloud-free pixels with cloudy pixels based on two optical satellite images or even optical and microwave images, that is, SAR data (Hoan and Tateishi, 2009; Eckardt et al., 2013; Mill et al., 2014).

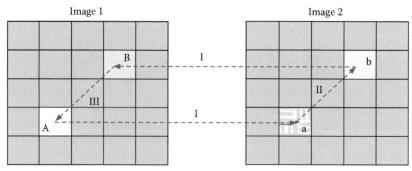

I: Location-based one-to-one correspondence   II: Spectral-based closet fit

III: Data value replacement

FIGURE 12.3 A schematic illustration of the closest spectral fit (CSF) method. A: cloudy pixel to be reconstructed; a: non-cloudy pixel at the same location of A; b: spectral similar pixel to a; B: pixel at the same location of b.

12.2.2 Interpolation Approaches

Although substitution techniques enable us to reconstruct the information of cloudy or value-missing pixels efficiently, there is a big invariant or homogeneous assumption associated with such methods. Specifically, cloudy pixels should be spectrally most similar to their neighborhoods (homogeneous nature) or have similar spatial pattern with the reference image (e.g., CSF). Such an assumption may not always hold because of inherent time- and space-dependent differences. To address this issue, interpolation approaches establishing the relationships between cloudy and cloud-free pixels through either linear or nonlinear models can be applied to simulate and predict the values for cloudy pixels (Warner et al., 2001; Moody et al., 2005; Wang et al., 2005; Benabdelkader and Melgani, 2008). For a single image, spatial interpolation is utilized to remove clouds by selecting neighboring pixels as candidate inputs for interpolation or prediction, because the reflectance value of one pixel is usually dependent on the reflectance values of associated pixels in the neighborhood. In addition, temporal interpolation provides a possible way to fill in the missing pixel values due to cloud cover through historical time series data. Nevertheless, hybrid approaches also show some integrative ability to mitigate and eliminate cloud contamination through different combinations over spectral, space, and time domain (Gafurov and Bárdossy, 2009; Paudel and Andersen, 2011; Xia et al., 2012; Lin et al., 2013). For instance, Wang et al. (2005) developed an algorithm with three steps. They first de-noise the two images with the method of Wiener filter to wipe off the primary noise. Then they adopt the method of Poisson Matting to segment the edge of the area covered by cloud and use wavelet analysis to restore the area originally occupied by cloud; this is followed by the use of a B-spline based model to repair the residual missing pixels.

In practice, substitution and simple interpolation methods may fail to reconstruct cloudy pixels at some extreme environments, in particular with high dynamic natures, for example, water bodies. Figure 12.4 shows two remote sensing reflectance images collected by MODIS on Terra and Aqua over Lake Nicaragua on January 19, 2012. While the cloud impact on the Aqua image is mainly salient on the top (north part) of the lake, the cloud impact on the Terra image is mainly present at the right (east side) of the lake. Ideally, such an image pair should be perfect data sources for cloud removal to implement substitution. Nonetheless, spectral differences between them are salient. As shown in Figure 12.5a, remote sensing reflectance values between these two images are poorly related to each other, with a correlation coefficient value of 0.32. Additionally, the comparison via relative difference reveals significant inconsistencies between them (Figure 12.5b). Further comparisons between Terra-based reflectance values collected at January 19 and 21, 2012, also reveal such an

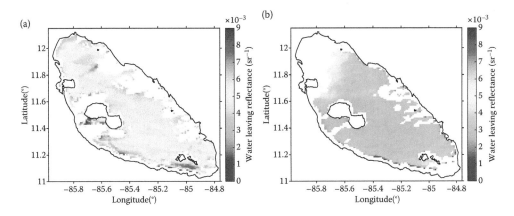

FIGURE 12.4 Water-leaving reflectance of MODIS blue band on January 19, 2012. (a) Terra image and (b) Aqua image.

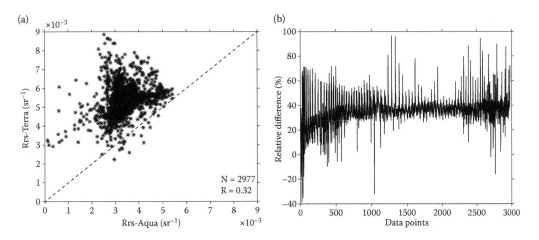

FIGURE 12.5 Comparison between Terra and Aqua reflectance values of blue band (443 nm) on January 19, 2012. Note that the spectral difference calculated between the pixel values (i.e., reflectance value) of a likely pair of satellites, such as Terra and Aqua, is associated with a given spectral condition (or simulated pixel values between Terra and Aqua reflectance values associated with a given spectral condition), and the equation is $100 \times$ (Terra pixel value- Aqua pixel value)/Terra pixel value. (a) comparison of reflectance values and (b) comparison via relative difference.

effect (Figure 12.6). Therefore, substitution and linear interpolation based on multitemporal images would fail in such a highly dynamic water environment.

12.3 CLOUD REMOVAL WITH MACHINE LEARNING TECHNIQUES

The finding in the last section encourages the inclusion of spatial, spectral, and temporal information simultaneously in order to take advantage of all merits associated with these three dimensions. In addition, more advanced learning algorithms should be used to deal with complex relations between different data sources, especially for those with highly nonlinear and nonstationary natures. As a branch of machine learning models, ANN-based models provide a suite of powerful tools for data classification, feature extraction, and pattern recognition. A popular form of ANN model is the well-known Backward Propagation Neural-network (BPN). Among multilayered neural networks, BPN is a supervised learning method to solve non-linear problems where outputs come close to target

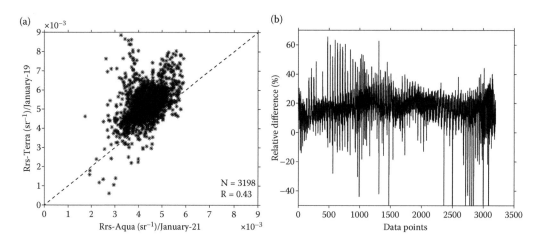

FIGURE 12.6 Comparison of Terra reflectance values of blue band (443 nm) collected on January 19 and 21, 2012. (a) comparison of reflectance values and (b) comparison via relative difference.

outputs for a training set containing a pattern (Rumelhart et al., 1986; Werbos, 1990). These ANN-based models have a common architecture of interconnected hidden layers in the middle to connect the single input and output layers, providing learning capacity which might have long computational time for learning and over-fitting issues (Rumelhart et al., 1986; Werbos, 1990).

In practice, most cloud removal or image reconstruction methods work relying on complementary images to detect spectral similar pixels for possible substitution or interpolation. However, complementary images may also be unavailable in some extreme situations, for example, over tropical regions where cloud cover is much higher. In such context, more advanced methods simply based on one kind of image will be advisable to reconstruct the missing information. Here, an advanced cloud removal method, termed as SMIR, designed based on the strength of ELM with fast computing capability (Chang et al., 2015), will be introduced. The essence of SMIR is to restore the information of value missing pixels based on the current available information in each image. The schematic flowchart of the SMIR is presented in Figure 12.7 which emphasizes the memory effect of the long-term data base the helps the cloudy pixel reconstruction through a time-space-spectrum continuum. The reason to use ELM is mainly due to its fast learning speed against other machine learning tools, such as BPNs, to quickly scan over the whole memory and extract relevant information spontaneously for image reconstruction. It is similar to a memory retrieval in a biological system. The usage of ELM enables us to deal with big data to better depict the inherent relations hidden behind numerous images, given that the computational time required by ELM is much smaller than that of ANN. In this way, the memory effect embedded in SMIR can be largely enhanced with the aid of a big database that may record all relevant pixel values in the past few decades. In general, the SMIR mainly includes four crucial steps: (1) image acquisition and pre-processing, (2) cloud detection, (3) input selection (i.e., selecting candidate images and candidate pixels), and (4) prediction (Chang et al., 2015).

In this study, daily MODIS-Terra water-leaving surface reflectance data are utilized to reduce the cloud impact. This product, measured at 10 different narrow bands with 1 km spatial resolution, is provided by NASA's ocean color science team. These data can be downloaded from their archives (http://oceancolor.gsfc.nasa.gov). Statistics, including the number of valid images in each year over Lake Nicaragua through 2003–2013 and the cloud cover information based on the Percentage of Cloud (POC) as defined in Chapter 11, are listed in Table 12.3, which shows severe cloud cover of Lake Nicaragua with an averaged POC of more than 70% throughout the year. For certain days, the lake is fully covered by clouds within 5 years of daily composition (i.e., the same date of the year), meaning that POC is 100%. With an 11-year time series of MODIS-Terra ocean color images, the

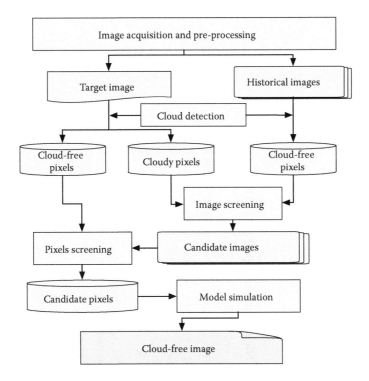

FIGURE 12.7   Schematic flowchart of the SMIR method.

averaged POC can be reduced from 38% based on cloud statistics of 5-year to 18% based on cloud statistics of 11-year. The time series plots of the POC with the 5- and the 11-year cloud statistics database of Lake Nicaragua are shown in Figure 12.8. Severe cloud cover mainly takes place from May to October during the wet season whereas less cloud cover is found from November to April during the dry season in Nicaragua.

Since the core of SMIR is to generalize a set of prediction models for each target pixel for data value simulation, the selection of training inputs is thus of critical importance. First, the candidate images at target pixels are screened out as input for the training process. By taking one cloudy pixel in the MODIS-Terra ocean color image collected on January 5, 2013, as an example (Figure 12.9), the pre-candidate images are screened out from the historical time series by selecting images whose pixels are cloud-free at the same location (Chang et al., 2015). Then, the overlap ratio between the target cloudy image (i.e., image to be recovered) and each pre-candidate image is calculated. The candidate pixels are obtained by choosing collaborating cloud-free pixels in those pre-candidate images based on one arbitrary overlapping ratio defined by the users. (Figure 12.9).

For each cloudy pixel, different training inputs and corresponding targets may be selected, leading to the establishment of different prediction models through the machine learning tools. In this study, the learning performance of ELMs, described in Chapter 10, is compared with the counterpart of the conventional BPN algorithms. The source code of basic ELM algorithm, in MATLAB® version, was adopted from http://www.ntu.edu.sg/home/egbhuang/. The ANN tools used in this study are available from the Neural Network ToolBox™ provided by the MathWorks in MATLAB environment. Unlike ELMs, which are independent of applications, ANNs can be fitted to the target problem and suitable ANNs should be selected beforehand. In our case (i.e., information recovery), the main problem falls into the function approximation category and thus several proper BPN training algorithms are needed for comparison with ELM. These BPN training algorithms include but are not limited to the Levenberg-Marquardt algorithm (trainlm) (Hagan and

TABLE 12.3

Statistics of MODIS-Terra Ocean Color Images with Full Coverage Over Lake Nicaragua through 2003–2013

Year		2003	2004	2005	2006	2007	2008	2009	2010	2011	2012	2013	C5 2009–2013	C11 2003–2013
								Annual						
No. of images		293	291	296	303	291	303	300	292	290	290	283	366	366
POC (%)	Min.	0.08	0.04	0.35	0.73	0.18	0.15	0.33	0.04	0.04	0.26	1.5	0.01	0
	Max.	100	100	100	100	100	100	100	100	100	100	100	100	97.44
	Avg.	71.87	71.23	75.5	75.55	75.31	73.51	72.53	74.36	72.65	70.3	70.26	37.59	17.53

Source: Chang, N. B., Bai, K., and Chen, C. F., 2015. *IEEE Journal of Selected Topics in Earth Observations and Remote Sensing*, 99, 1–19.
C5: Composited climatologic images with the latest 5 years (2009–2013) images, C11: same as C5 but with the past 11 years (2003–2013) images.

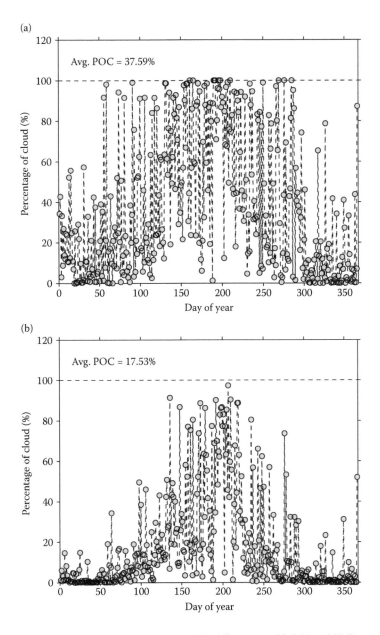

FIGURE 12.8 Time series of the cloud coverage over Lake Nicaragua with 5 (a) and 11 (b) years composition. (Chang, N. B., Bai, K., and Chen, C. F., 2015. *IEEE Journal of Selected Topics in Earth Observations and Remote Sensing*, 99, 1–19.)

Menhaj, 1994), the scaled conjugate gradient algorithm (trainscg) (Gordon and Wang, 1994), and the Bayesian regulation algorithm (trainbr) (Møller, 1993).

In this study, the performance of all the training algorithms applied for nonlinear model simulation was evaluated based on a common case of Lake Nicaragua which is a cloudy lake (Chang et al., 2015). Data values of 50 cloudy pixels in the MODIS-Terra ocean color image collected on January 5, 2013, were first recorded and then masked as cloudy pixels (i.e., no data value given to those pixels). The proposed SMIR method with different training algorithms was implemented to reconstruct the missing pixel values, respectively. In order to have a fair comparison, the same inputs and targets in association with the same number of hidden neurons

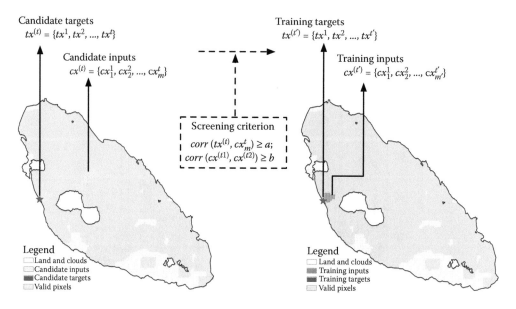

FIGURE 12.9 Schematic view of the selection of candidate pixels in SMIR algorithm. (Chang, N. B., Bai, K., and Chen, C. F., 2015. *IEEE Journal of Selected Topics in Earth Observations and Remote Sensing*, 99, 1–19.)

(i.e., 10) were utilized for each learning algorithm. Meanwhile, early stopping criterion was defined. Considering an ANN model with only one hidden layer in the ANN-based cloud removal algorithm, different results from different trials for band 443 nm may be a good example to showcase the learning capacity in terms of different training inputs (Figure 12.10). In this figure, it is obvious that different results were obtained with different generalized models generated from different training inputs. In other words, 70% randomly selected data were used to generalize one model to predict the left figure, and then another model was simulated based on newly selected inputs to predict the right figure. It is noted that the results derived based on two different training inputs differ from each other, in turn suggesting the need to use an ensemble mean rather than one individual result.

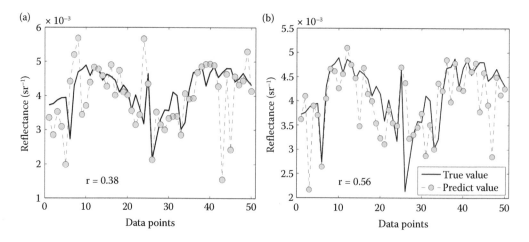

FIGURE 12.10 Results for blue band (443 nm) with ANN-based cloud removal algorithm based on different training inputs. (a) results with input I and (b) results with input II.

For all applications except training with trainbr (due to large computational time), 10 trials of simulation were performed for each pixel. The final results were obtained by averaging all these 10 trials. In other words, the ensemble mean was used rather than each individual simulated result. This is mainly due to the fact that the prediction accuracy of the neural network-based algorithm depends partially on the structure of the neural network model. The performance of the proposed algorithm is characterized by the training time (T_cpu), the correlation coefficient (r), and the Root Mean Squared Error ($rmse$) between the true data value and predicted value (Chang et al., 2015), which is calculated by:

$$rmse = \sqrt{\frac{\sum_{i=1}^{N}(\hat{y}_i - y_i)^2}{N}} \tag{12.1}$$

where $\hat{y}_i$ and y_i are the simulated and true data values, respectively. Statistical results are compared in Table 12.4.

Results reveal that the ELM outperforms the other three traditional BPN algorithms, especially in terms of training time. This shows that the ELM can be generalized and implemented with much faster training speed than other training algorithms (i.e., at least 50 times faster). With respect to the training accuracy, the ELM performs slightly better than the other three algorithms. The differences of $rmse$ and r seem anti-salient due to the limited pixels chosen for an experimental test. However, there is no guarantee that these differences should be much bigger in real world applications with larger datasets since ELM might not be able to handle a large number of outliers and success in the training stage does not imply the same success in the testing and validation stages. These comparisons in a small scale confirm that ELM is more competitive in terms of both training speed and prediction accuracy. Furthermore, the scaled conjugate gradient algorithm (trainscg) outperforms the other two BPN algorithms, especially in terms of the training speed. Considering this, the trainscg may be employed as the control and compared against the performance of ELM in the operational assessment of SMIR.

TABLE 12.4

Comparisons between Different Training Algorithms in Conjunction with Single Hidden Layer Feedforward Networks

	T_cpu (s)	$rmse$ (sr^{-1})	r (Pearson)
ELM	**10.8**	**7.2E-04**	**0.72**
trainscg	508.4	7.6E-04	0.71
trainlm	4324.2	7.7E-04	0.69
trainbr[a]	16863.5	9.4E-04	0.54

Source: Chang, N. B., Bai, K., and Chen, C. F., 2015. *IEEE Journal of Selected Topics in Earth Observations and Remote Sensing*, 99, 1–19.

Note: The correlation coefficient (r), root mean squared error (rmse) and training time (T_cpu: second) were employed to evaluate the simulation performance. Same inputs and targets in association with same number of hidden neurons (10) were utilized for each learning algorithm.

Each indicator with best performance is highlighted in bold.

[a] Due to large computational time, only 1 trial simulation was predicted with the trainbr.

In the ANN models, the number of hidden layers and neurons (nodes) are of critical importance in defining network architecture. It has been proven that feedforward networks with one hidden layer and enough neurons can solve any finite input-output mapping problem. Since ELMs belong to the category of SLFNs, the number of hidden neurons is the only parameter that decides the architecture of the neural network. According to Huang and Babri (1998),

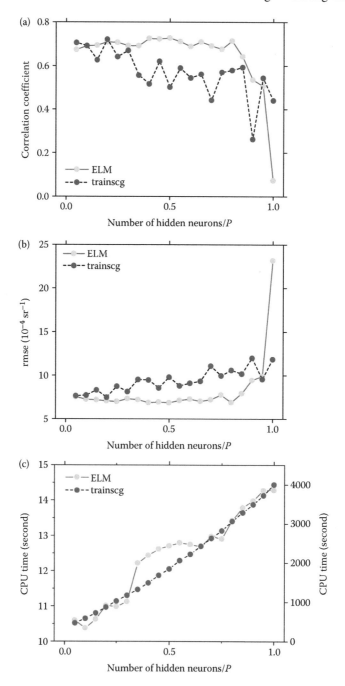

FIGURE 12.11 Performance of ELM and trainscg varying with the number of hidden neurons. P is the number of candidate images in this study. (a) Correlation coefficient, (b) root mean square error, and (c) CPU time (left axis for ELM and right axis for trainscg).

Narasimha et al. (2008), Sheela and Deepa (2013), and Wang et al. (2014), the number of hidden neurons given in the network is highly related to the learning capacity of the neural network. Varying criteria have been proposed to define the upper and lower bounds of the number of hidden neurons. In this study, the criterion proposed by Sheela and Deepa (2013) and Wang et al. (2014) was applied to determine the suitable bounds of the number of hidden neurons used in the feedforward networks:

$$L \leq P \leq \frac{L(N+M)}{M} \tag{12.2}$$

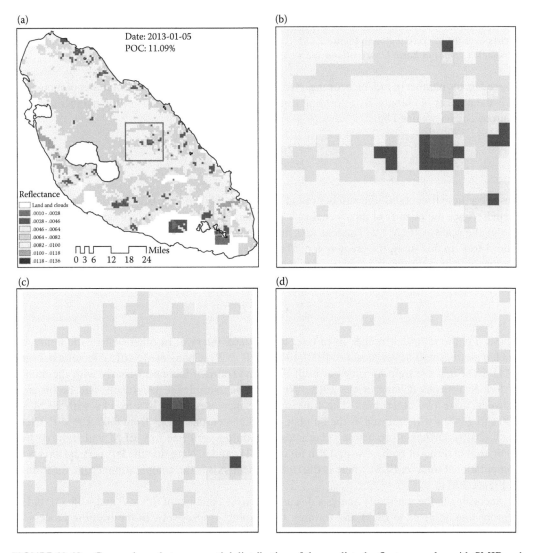

FIGURE 12.12 Comparisons between spatial distribution of the predicted reflectance value with SMIR and true data value. (a) Location of the selected region for reconstruction (outlined with red rectangle, 20 × 20 pixels). (b) Spatial distribution of the observed MODIS-Terra reflectance data at 531 nm on January 5, 2013. (c) Simulated value with ELM-based SMIR. (d) Simulated value with trainscg-based SMIR. (Chang, N. B., Bai, K., and Chen, C. F., 2015. *IEEE Journal of Selected Topics in Earth Observations and Remote Sensing*, 99, 1–19.)

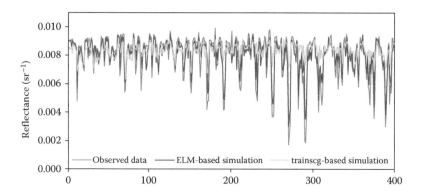

FIGURE 12.13   Pixel-wise comparison of remote sensing reflectance values between observed and simulated scenes over the selected region in Figure 12.12. (Chang, N. B., Bai, K., and Chen, C. F., 2015. *IEEE Journal of Selected Topics in Earth Observations and Remote Sensing*, 99, 1–19.)

where L is the number of hidden neurons, P is the number of samples, N is the number of inputs, and M is the number of outputs. By reforming Equation 12.2, bounding limits of the number of hidden neurons can be obtained by:

$$\frac{P \times M}{N + M} \leq L \leq P \qquad (12.3)$$

Based on these limitations, the number of hidden neurons should vary between the lower and upper bounds. Since the number of inputs is always much larger than the number of outputs (i.e., 1 in this study), the value of $M/(N + M)$ is close to 0, meaning that the number of hidden neurons can change from 1 to P. In this study, the prediction performance of the ELM with a varying number of hidden neurons was investigated and compared to the trainscg as is shown in Figure 12.11. The number of hidden neurons is defined as

$$L = \eta \times P \qquad (12.4)$$

where η is the ratio that ranges from 0.05 to 1 at 0.05 interval. In other words, the number of hidden neurons will be given as $L = [0.05 \times P, 0.1 \times P, \ldots, P]$ (Chang et al., 2015). Results in Figure 12.11 show that the training times of both ELM and trainscg increase linearly with the number of hidden neurons (Chang et al., 2015). Nevertheless, significant differences can be observed between these two algorithms. The training time increases from 10 s to 14 s with ELM and 500 s to 4,000 s with trainscg as the number of hidden neurons varies from $0.05 \times P$ to P (Chang et al., 2015). Different prediction accuracies in training performed by ELM and trainscg were compared in terms of both *rmse* and *r* in Figure 12.11a and b. Note that the left vertical axis is the computational time for ELM whereas the right vertical axis is the computational time for trainscg in Figure 12.11c.

According to the sensitivity analysis discussed above, a number of $0.5 \times P$ and $0.05 \times P$ hidden neurons can be given to ELM and trainscg for training, respectively. In the training process, an r-squared (R^2) value of 0.9 was given as the stopping criterion. In other words, only the trained model that predicts the unseen validating data with an r^2 value reaching 0.9 or even higher will be considered; otherwise, this model will not be used. To assess the general performance of SMIR in reconstructing the value missing pixels, a sampled region with 20×20 cloud-free pixels was first masked and then the SMIR method was used to simulate these masked values for demonstration (Figure 12.12). Specifically, data values in the selected region were masked as cloudy pixels after reserving their original values. Next, the SMIR method was applied to predict those values.

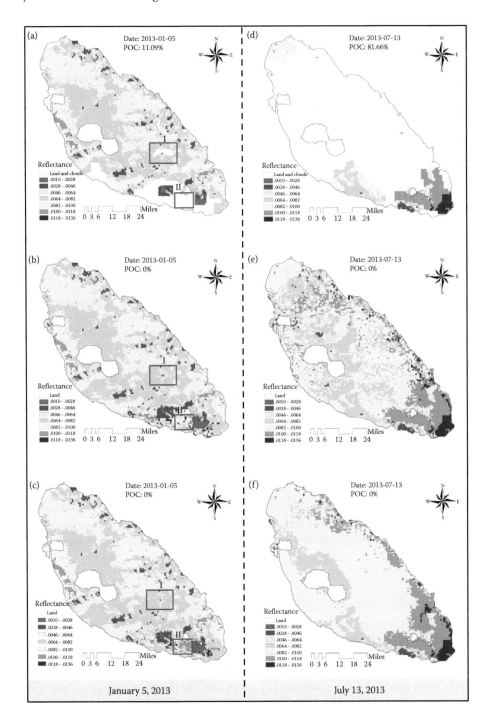

FIGURE 12.14 Comparisons of MODIS-Terra reflectance images at 531 nm with small (January 5, 2013) and severe (July 13, 2013) cloud cover before and after cloud removal with different training algorithms utilized by SMIR. POC denotes percentage of clouds. Note the spectral differences shown in the outlined regions (red rectangles) in the reconstructed images. (a) Cloudy image with slight cloud cover; (b) reconstructed image with ELM-based SMIR; (c) reconstructed image with trainscg-based SMIR; (d) cloudy image with severe cloud cover; (e) reconstructed image with ELM-based SMIR; and (f) reconstructed image with trainscg-based SMIR. (Chang, N. B., Bai, K., and Chen, C. F., 2015. *IEEE Journal of Selected Topics in Earth Observations and Remote Sensing*, 99, 1–19.)

For comparison purpose, the learning algorithm of trainscg was also used. The final predicted results of both learning algorithms are presented in Figure 12.12c and d, respectively. It is clear that the ELM-based method outperforms the trainscg-based counterpart, which is strongly evidenced by the spatial patterns of these two simulated scenes. Pixel-wise comparison of data values are shown in Figure 12.13. In general, both methods can predict the local variability of data values among these pixels. Nonetheless, the trainscg-based method failed to predict the local minimum whereas the ELM-based method holistically performed well, in particular in predicting the local minimums.

In Figure 12.14, under different cloud cover conditions, both ELM-based and trainscg-based SMIR methods are performed to reconstruct the cloudy pixel value of MODIS-Terra water-leaving reflectance images. Only the observed data were utilized as inputs in the modeling practices regardless of how many simulation runs were carried out to deal with the issue of error correction. The first scenario analysis with a light cloud impact was performed for the images collected on January 5, 2013, with cloud cover of 11.09% over Lake Nicaragua (see the left panel of Figure 12.14). It shows that the SMIR method is capable of reconstructing all missing values caused by cloud impact with reasonable results. In contrast to the trainscg, ELM works much faster for a total of 866 data values being reconstructed. While the ELM-based SMIR method spent only 0.54 hour, trainscg-based SMIR method took 4.6 hours for the completion of the same work. Overall, ELM shows much better prediction accuracy than trainscg in this real world application. ELM-based SMIR may successfully produce a continuous patch as opposed to the scattered patch generated by the trainscg-based SMIR method in region I (see the left panel in Figure 12.14). Furthermore, significant differences can be observed between patches in region II, where larger reflectance values are predicted by trainscg-based than ELM-based SMIR method (see the left panel in Figure 12.14).

In this study, a severe scenario was also considered for the images collected on July 13, 2013, with cloud cover of 81.66% over Lake Nicaragua (see the right panel in Figure 12.14). Clear pixel values are distributed in the northern and southern parts of the lake forming three different patches. In this case, the ELM-based method reconstructed all 6,376 data points at a speed of 3.15 hours while the trainscg-based method spent more than 60 hours to finish the same task. However, both ELM and trainscg had poor prediction accuracy due to insufficient training data and weak relationships between far reaching pixels. Within these 6,376 points, only half are really reconstructed through the machine learning method, and the rest of the pixels are either inferred from the adjacent pixels in the same images or drawn from images collected a certain time period ago. To overcome these drawbacks, more historical time series data points collected from other data sources, such as MODIS-Aqua, can be added for possible enhancement of the memory effect for the better learning capacity.

12.4   SUMMARY

In this chapter, some traditional and advanced cloud removal algorithms in terms of inpainting-based, multitemporal-based, multispectral-based or hybrid methods were discussed sequentially. With the use of machine learning tools, such as ANN and ELM, the most recent algorithm, namely SMIR, was highlighted. This novel method takes advantage of the learning power embedded in ELM and BPN algorithms to characterize the complicated relationships between the cloud-free pixels and cloudy pixels with the aid of memory effect from the historical time series over spatial and spectrum domains. Thus, it is feasible to perform well even in highly dynamic and heterogeneous environments where other approaches might fail.

REFERENCES

Addink, E. A., 1999. A comparison of conventional and geostatistical methods to replace clouded pixels in NOAA–AVHRR images. *International Journal of Remote Sensing*, 20, 961–977.
Asner, G. P., 2001. Cloud cover in Landsat observations of the Brazilian Amazon. *International Journal of Remote Sensing*, 22, 3855–3862.

Benabdelkader, S. and Melgani, F., 2008. Contextual spatio-spectral post-reconstruction of cloud-contaminated images. *IEEE Geoscience and Remote Sensing Letters*, 5, 204–208.

Cerra, D., Bieniarz, J., and Beyer, F., 2016. Cloud removal in image time series through sparse reconstruction from random measurements. *IEEE Journal of Selected Topics in Earth Observations and Remote Sensing*, 9, 1–14.

Chanda, B. and Majumder, D., 1991. An iterative algorithm for removing the effects of thin cloud cover from Landsat imagery. *Mathematical Geology*, 23, 853–860.

Chang, N. B., Bai, K., and Chen, C. F., 2015. Smart information reconstruction via time-space-spectrum continuum for cloud removal in satellite images. *IEEE Journal of Selected Topics in Earth Observations and Remote Sensing*, 99, 1–19.

Chen, J., Zhu, X., Vogelmann, J. E., Gao, F., and Jin, S., 2011. A simple and effective method for filling gaps in Landsat ETM+SLC-Off images. *Remote Sensing of Environment*, 115, 1053–1064.

Cheng, Q., Shen, H., Zhang, L., Yuan, Q., and Zeng, C., 2014. Cloud removal for remotely sensed images by similar pixel replacement guided with a spatio-temporal MRF model. *ISPRS Journal of Photogrammetry and Remote Sensing*, 92, 54–68.

Cihlar, J. and Howarth, J., 1994. Detection and removal of cloud contamination from AVHRR images. *IEEE Transactions on Geoscience and Remote Sensing*, 32, 583–589.

Di Vittorio, A. and Emery, W., 2002. An automated, dynamic threshold cloud-masking algorithm for daytime AVHRR images over land. *IEEE Transactions on Geoscience and Remote Sensing*, 40, 1682–1694.

Eckardt, R., Berger, C., Thiel, C., and Schmullius, C., 2013. Removal of optically thick clouds from multi-spectral satellite images using multi-frequency SAR data. *Remote Sensing*, 5, 2973–3006.

Gafurov, A. and Bárdossy, A., 2009. Cloud removal methodology from MODIS snow cover product. *Hydrology and Earth System Sciences*, 13, 1361–1373.

Gordon, H. and Wang, M., 1994. Retrieval of water-leaving radiance and aerosol optical thickness over the oceans with SeaWiFS: A preliminary algorithm. *Applied Optics*, 33, 443–452.

Gu, L., Ren, R., and Zhang, S., 2011. Automatic cloud detection and removal algorithm for MODIS remote sensing imagery. *Journal of Software*, 6, 1289–1296.

Hagan, M. T. and Menhaj, M. B., 1994. Training feedforward networks with the Marquardt algorithm. *IEEE Transactions on Neural Networks*, 5, 989–993.

Helmer, E. and Ruefenacht, B., 2005. Cloud-free satellite image mosaics with regression trees and histogram matching. *Photogrammetric Engineering and Remote Sensing*, 71, 1079–1089.

Hoan, N. T. and Tateishi, R., 2009. Cloud removal of optical image using SAR data for ALOS applications. *Journal of The Remote Sensing Society of Japan*, 29, 410–417.

Hu, X., Wang, Y. and Shan, J. 2015. Automatic recognition of cloud images by using visual saliency features. *IEEE Geoscience and Remote Sensing Letters*, 12, 1760–1764.

Huang, G. and Babri, H., 1998. Upper bounds on the number of hidden neurons in feedforward networks with arbitrary bounded nonlinear activation functions. *IEEE Transactions on Neural Networks*, 9, 224–229.

Huang, B., Li, Y., Han, X., Cui, Y., Li W., and Li, R., 2015. Cloud removal from optical satellite imagery with SAR imagery using sparse representation. *IEEE Transactions on Geoscience and Remote Sensing*, 12, 1046–1050.

Jang, J. D., Viau, A., Anctil, F. and Bartholomé, E. 2006. Neural network application for cloud detection in SPOT VEGETATION images. *International Journal of Remote Sensing*, 27, 719–736.

Ji, C. Y., 2008. Haze reduction from the visible bands of LANDSAT TM and ETM + images over a shallow water reef environment. *Remote Sensing of Environment*, 112, 1773–1783.

Jin, S., Homer, C., Yang, L., Xian, G., Fry, J., Danielson, P., and Townsend, P. A., 2013. Automated cloud and shadow detection and filling using two-date Landsat imagery in the USA. *International Journal of Remote Sensing*, 34, 1540–1560.

Jouini, M., Lévy, M., Crépon, M., and Thiria, S., 2013. Reconstruction of satellite chlorophyll images under heavy cloud coverage using a neural classification method. *Remote Sensing of Environment*, 131, 232–246.

Lin, C. H., Lai, K. H., Chen Z. B., and Chen, J. Y., 2014. Patch-based information reconstruction of cloud-contaminated multi-temporal images. *IEEE Transactions on Geoscience and Remote Sensing*, 52, 163–174.

Lin, C. H., Tsai, P. H., Lai, K. H., and Chen, J. Y., 2013. Cloud removal from multitemporal satellite images using information cloning. *IEEE Transactions on Geoscience and Remote Sensing*, 51, 232–241.

Liu, Z. K. and Hunt, B. R., 1984. A new approach to removing cloud cover from satellite imagery. *Computer Vision, Graphics and Image Processing*, 25, 252–256.

Long, D. G., Remund, Q. P., and Daum, D. L., 1999. A cloud-removal algorithm for SSM/I data. *IEEE Transactions on Geoscience and Remote Sensing*, 37, 54–62.

Lorenzi, L., Melgani, F., and Mercier, G., 2011. Multiresolution inpainting for reconstruction of missing data in VHR images. *International Geoscience and Remote Sensing Symposium (IGARSS)*, 8, 531–534.

Lorenzi, L., Mercier, G., and Melgani, F., 2013. Support vector regression with kernel combination for missing data reconstruction. *IEEE Geoscience and Remote Sensing Letters*, 10, 367–371.

Lv, H., Wang, Y., and Shen, Y., 2016. An empirical and radiative transfer model based Algorithm to remove thin clouds in visible bands. *Remote Sensing of Environment*, 179, 183–195.

Maalouf, A., Carre, P., and Augereau, B., 2009. A bandelet-based inpainting technique for clouds removal from remotely. *IEEE Transactions on Geoscience and Remote Sensing*, 47, 2363–2371.

Martinuzzi, S., Gould, W. A., and Ramos González, O.M., 2007. Creating cloud-free landsat ETM+ data sets in tropical landscapes: Cloud and cloud-shadow removal. *General Technical Report IITF-GTR-32*. International Institute of Tropical Forestry (IITF), U.S. Department of Agriculture (USDA), Forest Service. Puerto Rico: IITF.

Melgani, F., 2006. Contextual reconstruction of cloud-contaminated multitemporal multispectral images. *IEEE Transactions on Geoscience and Remote Sensing*, 44, 442–455.

Meng, Q., Borders, B., Cleszewski, C., and Madden, M., 2009. Closest spectral fit for removing clouds and cloud shadows. *Photogrammetric Engineering & Remote Sensing*, 75, 569–576.

Mill, S., Ukaivbe, D., and Zhu, W., 2014. Clouds and cloud shadows removal from infrared satellite images in remote sensing system. In: *ASEE 2014 Zone 1 Conference*, April 3–5, 2014, University of Bridgeport, Bridgeport, CT, USA.

Mitchell, O., Delp, E. III, and Chen, P., 1977. Filtering to remove cloud cover in satellite imagery. *IEEE Transactions on Geoscience Electronics*, 15, 137–141.

Møller, M., 1993. A scaled conjugate gradient algorithm for fast supervised learning. *Neural Networks*, 6, 525–533.

Moody, E. G., King, M. D., Platnick, S., Schaaf, C. B., and Gao, F., 2005. Spatially complete global spectral surface albedos: Value-added datasets derived from terra MODIS land products. *IEEE Transactions on Geoscience and Remote Sensing*, 43, 144–157.

Narasimha, P., Manry, M., and Maldonado, F., 2008. Upper bound on pattern storage in feedforward networks. *Neurocomputing*, 71, 3612–3616.

NASA, 2016. www.nasa.gov/vision/earth/lookingatearth/icesat_light.html, accessed by June 2016.

Paudel, K. P. and Andersen, P., 2011. Monitoring snow cover variability in an agropastoral area in the Trans Himalayan region of Nepal using MODIS data with improved cloud removal methodology. *Remote Sensing of Environment*, 115, 1234–1246.

Platnick, S., Platnick, S., King, M. D., Ackerman, S., Menzel, P., Baum, B. A., Riedi, J. C., and Frey, R. A. 2003. The MODIS cloud products: Algorithms and examples from Terra. *IEEE Transactions on Geoscience and Remote Sensing*, 41, 459–473.

Rakwatin, P., Takeuchi, W., and Yasuoka, Y., 2009. Restoration of Aqua MODIS Band 6 using histogram matching and local least squares fitting. *IEEE Transactions on Geoscience and Remote Sensing*, 47, 613–627.

Roy, D. P., Ju, J., Lewis, P., Schaaf, C., Gao, F., Hansen, M., and Lindquist, E., 2008. Multi-Temporal MODIS–Landsat data fusion for relative radiometric normalization, gap filling, and prediction of Landsat data. *Remote Sensing of Environment*, 112, 3112–3130.

Rumelhart, D., Hinton, G., and Williams, R., 1986. Learning internal representations by error propagation. In: Rumelhart, D. E. and McClelland, J. L. (Eds.), *Parallel Distributed Processing* (Vol. 1). MIT Press, Cambridge, MA, USA.

Saunders, R. W., 1986. An automated scheme for the removal of cloud contamination from AVHRR radiances over western Europe. *International Journal of Remote Sensing*, 7, 867–886.

Sheela, K. and Deepa, S., 2013. Review on methods to fix number of hidden neurons in neural networks. *Mathematical Problems in Engineering*, 2013, 1–11.

Simpson, J. and Stitt, J., 1998. A procedure for the detection and removal of cloud shadow from AVHRR data over land. *IEEE Transactions on Geoscience and Remote Sensing*, 36, 880–897.

Sprott, J. C., 2004. A method for approximating missing data in spatial patterns. *Computers & Graphics*, 28, 113–117.

Taravat, A., Del Frate, F., Cornaro, C., and Vergari, S., 2015. Neural networks and support vector machine algorithms for automatic cloud classification of whole-sky ground-based images. *IEEE Geoscience and Remote Sensing Letters*, 12, 666–670.

Tseng, D. C., Tseng, H. T., and Chien, C. L., 2008. Automatic cloud removal from multi-temporal SPOT images. *Applied Mathematics and Computation*, 205, 584–600.

Wang, B., Ono, A., Muramatsu, K., and Fujiwara, N., 1999. Automated detection and removal of clouds and their shadows from Landsat TM images. *IEICE Transactions on Information and Systems*, 82, 453–460.

Wang, Z., Jin, J., Liang, J., Yan, K., and Peng, Q., 2005. A new cloud removal algorithm for multi-spectral images. In: *Proceedings of SPIE*, Vol. 6043, MIPPR 2005: SAR and Multispectral Image Processing, Wuhan, China.

Wang, J., Wang, C., and Chen, C. L. P., 2014. The bounded capacity of fuzzy neural networks (FNNs) via a new fully connected neural fuzzy inference system (F-CONFIS) with its applications. *IEEE Transactions on Fuzzy Systems*, 22, 1373–1386.

Warner, J. X., Gille, J. C., Edwards, D. P., Ziskin, D. C., Smith, M. W., Bailey, P. L., and Rokke, L., 2001. Cloud detection and clearing for the earth observing system Terra satellite Measurements of Pollution in the Troposphere (MOPITT) Experiment. *Applied Optics*, 40, 1269–1284.

Werbos, P. J., 1990. Back propagation through time: What it does and how to do it. *Proceedings of the IEEE*, 78, 1550–1560.

Xia, Q., Gao, X., Chu, W., and Sorooshian, S., 2012. Estimation of daily cloud-free, snow-covered areas from MODIS based on variational interpolation. *Water Resources Research*, 48, W09523.

Xu, M., Jia, X., and Pickering, M., 2016. Cloud removal based on sparse representation via multitemporal dictionary learning. *IEEE Transactions on Geoscience and Remote Sensing*, 54, 2998–3006.

Zhang, Y., Guindon, B., and Cihlar, J., 2002. An image transform to characterize and compensate for spatial variations in thin cloud contamination of Landsat images. *Remote Sensing of Environment*, 82, 173–187.

Zhang, C., Li, W., and Travis, D., 2007. Gaps-fill of SLC-off Landsat ETM plus satellite image using a geostatistical approach. *International Journal of Remote Sensing*, 28, 5103–5122.

Zhu, X., Liu, D., and Chen, J., 2012. A new geostatistical approach for filling gaps in Landsat ETM+ SLC-off images. *Remote Sensing of Environment*, 124, 49–60.

Zhu, Z. and Woodcock, C. E., 2012. Object-based cloud and cloud shadow detection in Landsat imagery. *Remote Sensing of Environment*, 118, 83–94.

Zhu, C., Zhao, Z., and Zhu, X., 2016. Cloud removal for optical images using SAR structure data. In: *Proceedings of International Conference on Signal Processing (ICSP)*, November 6–10, 2016, Chengdu, China, 1872–1875.

13 Integrated Data Fusion and Machine Learning for Intelligent Feature Extraction

13.1 INTRODUCTION

13.1.1 BACKGROUND

Earth observation and environmental monitoring taken from one individual satellite sensor commonly have restricted spatial, temporal or spectral coverage and resolution due to the inherent limitations of signal-to-noise ratio, view angle, sensor characteristics, platform and payload design, and orbital features in remote sensing. Since the late 1980s, many more digital image data have become available from diversified operational Earth Observing System (EOS) satellites in the United States and others around the globe. Combining information from different satellite-based images with different spatial, temporal, and spectral resolutions into a single dataset by using various image processing, information retrieval, and data science techniques has been considered a valuable tool in remote sensing science. The concept of image fusion was initialized in the 1990s (Shen, 1990; Pohl and van Genderen, 1998). The process of blending information from multiple sources into a new dataset is commonly referred to as image or data fusion, which aims to produce new data products with higher spatial or temporal resolution or both while preserving the spectral properties embedded in the individual source image as much as possible. In the early 2000s, Waltz (2001) and Luo et al. (2002) formalized the principles and practices of image and spatial data fusion. These efforts are prone to be in concert with traditional sensor fusion and information fusion techniques depending on the synergistic opportunities in various applications. Image and data fusion are thus considered a common practice of blending information from multiple image or data sources into a spatially, temporally, and spectrally advantageous dataset to facilitate advanced interpretation and decision making (Fok, 2015). Because of this advancement, a variety of image and data fusion technologies have been extensively applied in order to deepen the insight of environmental monitoring and management.

From Chapter 11, it is known that merging remote-sensing images for environmental monitoring is based on images collected from different satellites with similar temporal or spatial coverage rather than their temporal or spatial resolution. The essence of multisensor data fusion that is mainly aimed to improve temporal or spatial resolution or both is related to extracting valuable features embedded in different images/data and retrieving expanded information with the aid of data science techniques. There are over 40 data science techniques that may be generally grouped into either parametric or non-parametric methods at present in which machine learning and data mining techniques are part of them. The availability of satellite images with varying spectral, spatial, and temporal resolution and the advancements of data science techniques, such as statistical inference, pattern recognition, and machine learning or data mining techniques, provide us with a wealth of research niches for complex image or data fusion toward performing intelligent feature extraction and generating sound content-based mapping. In most cases, the key to initial success is tied to the selection of the satellite sensors for data fusion, which depends on the characteristics of orbit, platform, and imaging geometry, as well as the capability of feature extraction and classifier selection to meet the goals of required coverage, resolution, and information retrieval simultaneously (Ardeshir and Nikolov, 2007; Nguyen, 2009; Castanedo, 2013).

13.1.2 THE PATHWAY OF DATA FUSION

Numerous data fusion approaches were proposed and employed for different applications from pixel to feature to decision level from the early 2000s onwards (Wang et al., 2005; Nikolakopoulos, 2008; Abdikan et al., 2014). At the pixel level, for instance, Landsat satellite imageries have high spatial resolution, but such an application suffers from a long overpass interval of 16 days. Free coarse resolution sensors with a daily repeat cycle, such as Moderate Resolution Imaging Spectroradiometer (MODIS), are incapable of providing detailed environmental information because of low spatial resolution. This discrepancy can be harmonized by using the data, image, and/or sensor fusion opportunities in which the high spatial resolution Landsat and the high temporal resolution MODIS imageries may be fused to optimally produce better images with both high spatial and temporal resolutions in support of better feature extraction (Chang et al., 2014a,b). In most cases, data fusion methods are applied to minimize the gaps between spatial and temporal scale and aid in subsequent feature extraction pixel by pixel based on multispectral or hyperspectral images. This method is commonly referred to as the spatial-temporal data fusion approach at the pixel level. Along this line, image and data fusion technologies can be further expanded at feature or decision level and advanced by involving microwave remote sensing technologies, providing additional monitoring capacity in texture and topographical dimensions in almost all weather conditions (Reiche et al., 2013; Abdikan et al., 2014). Whether the success in various applications can be warranted depends on the subsequent feature extraction and classifier selection, however, with different learning capacities.

In principle, supervised learning provides better classification than its unsupervised alternative (Du, 2009). Whereas an unsupervised classifier can generally recognize spectrally homogeneous areas well, a supervised classifier may result in more accurate classification for some heterogeneous areas (Du, 2009). As the accuracy of individual classifiers is limited, fusion of classifiers for better classification of multisensor data offered the opportunity of a new endeavor for data fusion at the feature or decision level (Petrakos et al., 2001; Waske and Benediktsson, 2007). Thus, fusing supervised and unsupervised approaches may yield complementary effects, leading to better classification performance at different levels. With this combination, the impact from trivial spectral variations may be alleviated during the unsupervised classification and subtle differences between spectrally similar pixels may not be overly exaggerated during the supervised classification (Du, 2009). Notwithstanding, data fusion may be performed either by multiple classifiers at the feature level or classifier fusion at the decision level based on a fusion operator to synthesize the results acquired from a few single-source classifiers. In the literature, these fusion operators were designed based on the classifiers' soft outputs (either probabilistic or fuzzy). They can be conducted through the use of complex fusion operators with Genetic Algorithm (GA) self-organizing neuro-fuzzy multilayered classifiers (Mitrakis et al., 2008), the stacked (soft) outputs of the multiple classifiers as a new feature space subsequently training a new classifier on this new space (Huang et al., 2009) or the simple weighted averaging schemes to conclude the results (Huang and Zhang, 2010, 2012; Dalla Mura et al., 2011). With this understanding, many studies have been conducted to explore a variety of combinations of features or results from different classifiers for data fusion at feature or decision level. The final combination may yield results similar to those of object-based techniques (Blaschke et al., 2000). However, the overall performance is less sensitive to region segmentation (Du, 2009). Nevertheless, proper integration of spatial data fusion and classifier fusion strategies incorporating any parametric or non-parametric classifiers would most likely result in a more accurate classification outcome.

Varying spatial, spectral, and temporal resolutions of satellite remote sensing images has compounded the feature extraction (MIT, 2016). Different types of satellite imageries pre-processed for feature extraction impact results due to the ranges of variations of the band values, as do fused or merged images. Feature extraction and classification methods are constrained by their learning capacity. It is noticeable that several non-parametric approaches including Artificial Neural Networks (ANN) (Benediktsson et al., 1990; Serpico and Roli, 1995), self-learning Decision Trees (DT) (Friedl and Brodley, 1997; Pal and Mather, 2003), and Support Vector Machines (SVM)

(Huang et al., 2002; Foody and Mathur, 2006) are not constrained to prior assumptions on the distribution of input data as is the case when using the maximum-likelihood methods and other statistical methods. These non-parametric approaches have laid down the foundation for better learning regarding intelligent feature extraction and classification. Among the family of ANN models, Backward Propagation Neural-network (BPN) is mainly a supervised learning method to solve non-linear problems, where outputs come close to target outputs for a training set containing a pattern (Rumelhart et al., 1986; Werbos, 1990). These ANN-based models, such as the Single Layer Forward Network (SLFN), have a common architecture of interconnected hidden layers in the middle to connect the single input and output layers, providing basic learning capacity which might have a long computational time for learning and over-fitting issues (Rumelhart et al., 1986; Werbos, 1990). In response to such a long computational time, Morphological Shared Weight Neural Networks (MSNNs) (Won and Gader, 1995) were developed. The recent regime shift from "traditional feature extraction methods" to "intelligent or advanced feature extraction methods" by using deep and fast machine-learning tools has resulted in a significant impact on numerous signal/ image processing applications. Well-known examples of fast and deep learning models include, but are not limited to, Extreme Learning Machine (ELM) (Huang et al., 2006a,b), Deep Boltzmann Machines (DBMs) (Hinton et al., 2006), and Convolutional Neural Networks (CNNs) (Nogueira et al., 2015), which have not been well studied by the environmental science, geoscience, and the remote sensing community. While CNNs are more suitable for 2-dimensional image processing and pattern recognition for unsupervised or semi-supervised classification, ELM and DBM could be more applicable for earth observation and environmental monitoring.

In addition to the fast advancement of machine learning for feature extraction and classification, the process of preserving the spectral properties via specific image processing during data fusion between hyperspectral and multispectral images at different scales has also received wide attention. Hyperspectral or multispectral imaging sensors often record scenes in which numerous disparate material substances contribute to the ultimate spectrum in a single mixed pixel (Keshava, 2003). Spectral unmixing is an image processing technique used to preserve the spectral property at the subpixel level by which the measured spectrum of a mixed pixel is decomposed into a collection of endmembers (i.e., constituent spectra or thematic classes). The proportion of each constituent spectra present in the mixed pixel can then be calculated by a linear mixing model to extract the spectral mixing pattern in terms of a set of corresponding fractions for each fused pixel band by band during the spatial-temporal data fusion. For every pixel, the least square inversion criterion must be confirmed when solving the linear mixing model band by band. If the substances comprising the medium are not organized proportionally on the surface, a nonlinear spectral unmixing method incorporating multiple scattering into the expression may be applied to improve the overall classification accuracy. However, all these inversion strategies have been developed in a pixel-by-pixel context, taking no account of possible spatial correlations between the different pixels of the hyperspectral or multispectral image. Hence, enhancing image unmixing with spatial correlations can be of further importance in image and data fusion (Alsahwa et al., 2016). For example, the Gaussian Markov Random Field (GMRF) model is another representative method for extracting spatial information to aid in such spatial unmixing techniques (Dong et al., 1999; 2001). Such a spatial spectral unmixing method enhances image unmixing with spatial correlations while preserving the spectral properties at the subpixel level.

With these advancements, the primary objective of this chapter aims to demonstrate an integration of image and data fusion methods with data mining or machine learning technologies for environmental mapping applications. The newly developed Integrated Data Fusion and Machine-learning (IDFM) algorithm will be delineated stepwise for illustration (Chang et al., 2014a,b). The case study of IDFM is dedicated to linking the merits of data fusion and intelligent feature extraction, trying to utilize the learning capacity embedded in the ELM and DBM, and maximizing the synergistic potential between data fusion and machine learning to reach the goal in system of systems engineering, which is "The whole is greater than the sum of its parts."

13.2 INTEGRATED DATA FUSION AND MACHINE LEARNING APPROACH

This section aims to demonstrate the IDFM algorithm with the aid of the Spatial and Temporal Adaptive Reflectance Fusion Model (STARFM) and evaluate the influences of fast and deep learning models on feature extraction based on fused images. The process described in this section led to the improvement of multisource heterogeneous remote sensing data fusion via a collection of enriched feature extraction comparisons among ANN, ELM, and DBM. The DBM applied herein is the same as the one in Figure 7.23. The five procedural steps of the IDFM algorithm undertaken in this study are marked in Figure 13.1 in which a new machine learning or data mining scheme of regular, deep, and fast learning methods is emphasized for intelligent feature extraction based on the fused images of MODIS and Landsat.

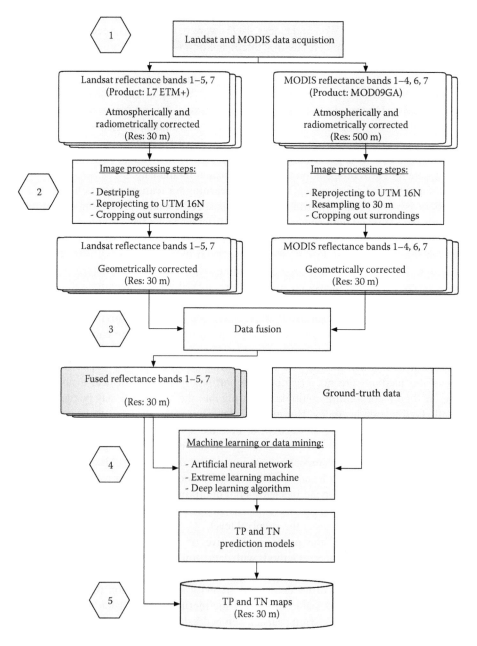

FIGURE 13.1 The flowchart of IDFM.

The illustrative example led to an enhanced water quality monitoring assessment in Lake Nicaragua (Central America) over both dry and wet seasons for change detection and environmental management. The Lake Nicaragua watershed is very extensive and is shared between Nicaragua and Costa Rica, spanning 13,707 km^2 in Nicaragua (excluding the lake itself and its islands) and 2,577 km^2 in Costa Rica (World Bank, 2013). Lake Managua has a surface area of 1,042 km^2. The lake is approximately 65 km long and 25 km wide. Lake Nicaragua and Lake Managua are two large freshwater lakes in Central America. Both lakes are located in the San Juan watershed and they are interconnected by a canal in the middle. Lake Managua connects to Lake Nicaragua through the Tipitapa River, creating a pollution plume from Lake Managua to Lake Nicaragua (World Bank, 2013). Managua, the capital city of Nicaragua, lies on the southwestern shore of the lake. Water quality in these two interconnected lakes has been deteriorated over time due to intensive agricultural activities in both watersheds. Remote sensing can help assess this impact via multitemporal change detection of water quality in these two lakes.

In Figure 13.1, step one pertains to acquiring the Landsat and MODIS swath path images containing Lake Nicaragua. The second step involves the procedures required to prepare the images for fusion. Data or image fusion procedures are encompassed in step three based on the STARFM for data fusion (Gao et al., 2006), although other data fusion algorithms can also be applied. The fourth step involves coalescing the ground-truthing data and fused spectral data into a few designated inversion models (i.e., ELM, DBM, and SLFN) for machine learning and feature extraction. A final comparison among ANN, ELM, and DBM is prepared for new knowledge discovery with respect to original, fused or merged satellite images. Within this thematic area, an interesting question in remote sensing is "Can deep learning methods outperform regular and fast learning methods in intelligent feature extraction based on the fused images?" Finally, in the machine learning steps, the dataset is fed into the three feature extraction models to carry out a comparative study of water quality patterns in the lake. The best one may be applied to generate Total Nitrogen (TN) and Total Phosphorus (TP) concentration maps in both wet and dry seasons for lake eutrophication assessment. All essential steps are included below for demonstration.

13.2.1 STEP 1—DATA ACQUISITION

Reflectance data over Lake Nicaragua and Lake Managua were collected from the Landsat 5/7 Thematic Mapper (TM)/Enhanced Thematic Mapper Plus (ETM+) and Terra MODIS (MOD09GA) satellite sensors. For Landsat 7 ETM+ data, surface reflectance data are generated from the Landsat Ecosystem Disturbance Adaptive Processing System (LEDAPS), a specialized software originally developed through National Aeronautics and Space Administration (NASA) Making Earth System Data Records for Use in Research Environments (MEaSUREs) grant by NASA Goddard Space Flight Center (GSFC) and the University of Maryland (Masek et al., 2006). The software applied Moderate Resolution Imaging Spectroradiometer (MODIS) atmospheric correction routines to Level-1 data products; MOD09GA has been atmospherically corrected by the same process. The Second Simulation of a Satellite Signal in the Solar Spectrum (6S) radiative transfer model was used and ozone, water vapor, geopotential height, aerosol optical thickness, and digital elevation were input with Landsat data to generate top of atmosphere reflectance, surface reflectance, brightness temperature, and masks for clouds, cloud shadows, adjacent clouds, land, and water (Masek et al., 2006). MODIS-Terra images were obtained from the online data pool managed by the United States Geological Survey (USGS). Relatively cloud free days of MODIS-Terra images that corresponded with available ground-truth data and cloud free Landsat images were selected for data fusion.

MODIS surface reflectance data were first processed into GEOTIFF images using the online processing system available at https://mrtweb.cr.usgs.gov/. The USGS also provided the Landsat imageries, which can be obtained from the website http://earthexplorer.usgs.gov/. All cloud free satellite data for bands 1–5 and 7 of Landsat 5 and 7 for the days near the sampling dates were downloaded. The near sampling dates were selected so that higher levels of prediction accuracy could be achieved in the later stage after fusing the images using the STARFM algorithm. A

TABLE 13.1

Satellite Products Utilized in this Study

Satellite Sensor	Product Selection	Spatial Resolution	Temporal Resolution	Bands Used
Terra MODIS	Surface Reflectance (MOD09GA)	250/500 m	Daily	1–4, 6,7
Landsat 5 TM	Surface Reflectance	30 m	16 days	1–5, 7
Landsat 7 ETM+				

comparison between these satellites in regard to the choice of bands and the spatial and temporal resolution is summarized in Table 13.1 (Chang et al., 2014a).

Supervised classification was adopted in this study and field campaigns were conducted intensively to enrich the ground truth dataset. Since the two lakes are interconnected through a canal, the two lakes can be regarded as a unified water body. This study thus grouped all ground-truth samples together over these two lakes for model training, testing, and validation toward intelligent feature extraction.

13.2.2 STEP 2—IMAGE PROCESSING AND PREPARATION

The collected MODIS data are at a level-2G basis, where the data have been radiometrically and atmospherically corrected to account for the impact of scattering and aerosols (Vermote and Kotchenova, 2008). The Landsat data are on a level-1T basis, with radiometric and geometric corrections (USGS, 2017). As denoted by Step 2 of Figure 13.1, ArcGIS, a mapping and spatial analysis software, was used to process the images in preparation for the data fusion process. It was necessary to perform the following actions on the Landsat images to: (1) conduct a gap-filling operation using ENVI software to account for data gaps resulting from sensor failure, (2) carry out reprojection to the Universal Transverse Mercator (UTM) zone 16 North, and (3) crop out land data around Lake Nicaragua. In addition, the following steps were taken to process the MODIS images to: (1) conduct reprojection to the UTM zone 16 North, (2) carry out resampling to a 30-m spatial resolution, and (3) crop out land data around Lake Nicaragua.

The image processing consists of two quintessential categories and they are: (1) modifying the images to have the same projection, pixel size, and scale in order to fuse them, and (2) preparing the images to increase fusion accuracy by cropping out the land and narrow portions of the lake. In the first image processing category, images of disparate geographic map projections cannot be accurately compared. Therefore, UTM 16 North projection was applied to all Lake Nicaragua images to ensure the same viewing angle.

Pre-processing steps may achieve radiometric correction, resampling, and reprojection, by which measurement errors were corrected and compatibility was assured between the two data streams. First, in our case, only the MODIS images were pre-processed to adjust for backscattering effects of the atmosphere. MODIS 6S radiative transfer code was applied to algorithmically correct the pixel values to produce more accurate surface reflectance values. It is known that each type of satellite image needs to have the same number of pixels, rows, and columns before data fusion. Resampling of the MODIS imagery to the resolution of the Landsat images was thus needed since the STARFM software package compares images on a pixel-by-pixel basis. Landsat and MODIS surface reflectance products store the reflectance values at different scales. The Landsat product stores the surface reflectance on a scale from 0 to 255 and the MODIS data ranges from −100 to 16,000 (Vermote and Kotchenova, 2008). In order for STARFM to compare pixel values, each of the images needs to have the same scale. The LEDAPS Processing Toolbox automatically increases the bit depth storage from unsigned 8-bit to signed 16-bit and scales the data from −100 to 16,000 when applying atmospheric correction techniques. The second element of image processing was to crop out the land and narrow portions of the lake. The rationale behind this approach is to reduce the potential for surrounding land to contaminate the fused image if the focus is the water body. The land contamination may occur during the data fusion process when the STARFM algorithm searches through neighboring pixels. Besides, the land was removed in order to limit the search to the lake only, and narrow areas of the lake were blacked out; when water channels were smaller than MODIS 250–500 m resolution, it was essential that a part of the land surface was averaged into the pixel value standing for the reflectance of the water.

13.2.3 STEP 3—DATA FUSION

Data fusion is the algorithmic fusion of the spectral, temporal or spatial properties of two or more images into a composite or synthetic image based on the characteristics of the input images. There are a set of data fusion techniques available, and selecting an algorithm to apply depends upon the type of output data required for the application, the accuracy of the fused data, and the characteristics of the input data streams that the user would like to fuse. Data fusion techniques are classified into three groups per the level at which the processing takes place (Pohl and van Genderen, 1998), including: (1) pixel level, (2) feature level, and (3) decision level.

Pixel level image fusion refers to the fusion of the measured physical attributes of the data prior to significant processing, as is the case in our study. With an appropriate pixel level fusion, the fused data streams in a single fused image has the potential to increase the reliability of the data and displays more defining attributes of a target object (Pohl and van Genderen, 1998). Such benefits can lead to more informed feature extraction. Furthermore, data fusion at the feature level takes the measured input data and extracts objects with the aid of segmentation methods (Pohl and van Genderen, 1998). The successful classification of objects may depend on their shape, location, pixel value, and extent. Classified objects from the data streams are then fused in preparation for feature extraction through various statistics or artificial intelligence methods (Pohl and van Genderen, 1998). Decision level fusion, which is also called interpretation level fusion, takes feature level fusion a step farther by processing the classified data to make a valid decision in regard to the identity of the target object based on a preselected classifier iteratively. With a set of preselected classifiers, all the predictions are then fused according to some user-defined decision rules, such as fuzzy operators, to finally identify the target object (Shen, 1990).

As introduced in section 9.2.2.3, the STARFM algorithm was used in this study to produce a set of fused images of enhanced spatial, spectral, and temporal properties, thereby allowing for accurate prediction of the selected water quality indicator in a water body. Before performing data fusion, spectral characteristics of each product should be matched to guarantee accurate fusion. Table 13.2 details the proper band matched between MODIS and Landsat reflectance. The spectral reflectance value for each pixel of the MODIS image is a conglomeration of the surface reflectance from each object in the 250 by 250 (m^2) area. Alternatively, spectral reflectance values provided by the Landsat image are an average of objects contained within a 30 by 30 (m^2) pixel. In regard to the data fusion process, the STARFM algorithm uses the original Landsat image as a base data set and fuses the

TABLE 13.2
Landsat 7 ETM+ and Terra MODIS Band Comparisons

Landsat 7 ETM+ Band	Landsat Bandwidth (nm)	Terra MODIS Band	MODIS Bandwidth (nm)
1	450–520	3	459–479
2	520–600	4	545–565
3	630–690	1	620–670
4	760–900	2	841–876
5	1550–1750	6	1628–1652
7	2080–2350	7	2105–2155

MODIS and Landsat images of the same data to produce the predicted image (Gao et al., 2006). The CPU time is about 25 minutes for fusing Landsat and MODIS images for four days including all the corresponding bands for Lake Managua and about an hour for Lake Nicaragua. The above computational time was obtained on a PC (CPU: Intel Core i7-4790U (four physical cores, 3.60 GHz), memory: 16 GB) using STARFM V1.2.1. After the inverse modeling for feature extraction is developed, the fused image should correctly conform to the ground-truthing data as well as the original high resolution of the Landsat image. Then a near-real-time water quality monitoring system is constructed for use on a daily basis. One caution in the method is noted in fusing the data streams from Landsat and MODIS for high-resolution water quality prediction toward early warning.

As the STARFM program translates through the matrix of pixels in the Landsat and MODIS images, it may choose a central pixel every few steps and reassign the pixel's value based upon a set of candidate pixels that are located near the central pixel and spectrally similar. The candidate pixels are then filtered out if they show more change over time than the central pixel, or if the spectral features of the candidate pixel are greater than the difference between the spectral features of the central pixel in the Landsat and MODIS images (Gao et al., 2006). Thus, the surface reflectance of the central pixel will be predicted from the group of selected candidate pixels. However, the surface reflectance of the central pixels is not simply the average value of all surface reflectance values of candidate pixels involved. A weighted average of all surface reflectance values of candidate pixels involved is applied based on how likely it is that each of the selected candidate pixels could represent the central pixel. Higher weighting factors are assigned if the candidate pixel is spatially, spectrally, and temporally similar to the central pixel (Gao et al., 2006). Through this predictive process, the synthetic Landsat image is generated based on the inputs from all candidate pixels involved within the MODIS image taken for prediction of the central pixel. The STARFM algorithm is proven effective at filling in data gaps temporally and spatially when both pre- and post-conditions of the target satellite images are available.

There is a need to verify the effectiveness of the data fusion based on some statistical metrics. This study chose Shannon entropy as a representative index for verification. Entropy is a statistical measure of randomness or richness of information in an image, and hence can be used to characterize the texture of an image (Shi et al., 2005). The expression of the entropy of an image is defined as:

$$H = -\sum_{i=1}^{n} p_i \log_2 p_i \tag{13.1}$$

where n denotes the number value of gray level, and p_i is the probability mass function which contains the histogram counts of valid grid data values. The higher the value of entropy, the richer the information content.

The results of data fusion are thus validated by using the Shannon entropy index (H). A statistical example of the original Landsat image, original MODIS image, and fused image at Lake Nicaragua on March 22, 2015 based on the Landsat images collected on March 10 and March 25, 2015 and the daily MODIS images between March 11 and March 24, 2015, is shown in Figure 13.2. The estimated Probability Density Function (PDF) of the fused image is close to that of the Landsat image, while it is titled by the effect from the MODIS image (Figure 13.3). The estimated Shannon entropy for group 1 is 4.21, 6.68, and 5.21 for Landsat, MODIS, and fused images in group 1, respectively, whereas the estimated Shannon entropy for group 2 is 4.29, 6.77, and 5.34 for Landsat, MODIS, and fused images, respectively. The Shannon entropy of the fused image is between that of the two source images. Both PDF and entropy properties imply that the data fusion can maintain the important properties from the input images across different bands.

13.2.4   Step 4—Machine Learning for Intelligent Feature Extraction

Whereas machine learning is a good tool for a supervised classification process with the aid of ground-truth data, data mining is also suitable for an unsupervised classification process with no

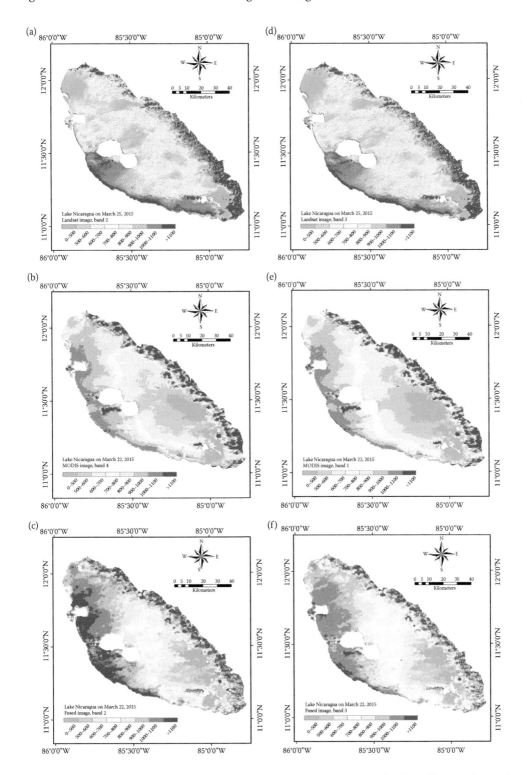

FIGURE 13.2 The two groups of source images at Lake Nicaragua on March 25, 2015. Group 1: (a) Landsat image at band 2, (b) MODIS image at band 4, and (c) fused image at band 2 on March 22, 2015, as well as Group 2: (d) Landsat image at band 3, (e) MODIS image at band 1, and (f) fused image at band 3 on March 22, 2015. The legend is classified homogeneously based on the surface reflectance values in order to show the gradients.

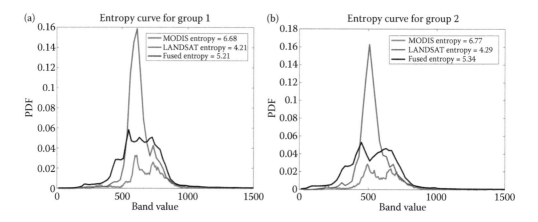

FIGURE 13.3 The estimated PDF for (a) group 1: the Landsat image (band 2), MODIS image (band 4) and fused image (band 2), as well as (b) group 2: Landsat image (band 3), MODIS image (band 1) and fused image (band 3) at Lake Nicaragua on March 22, 2015.

ground-truth data. With the aid of STARFM, fused images for the time period of the ground-level sampling dates were created in this case study. The ground-truth data were collected by the National Autonomous University of Nicaragua during 2011–2016 (see Appendixes 1 and 2). For each ground-truth observation, a MODIS image as well as a temporally similar Landsat image were acquired. Several different inverse models with regular, fast, and deep learning algorithms for feature extraction were applied to identify water quality patterns and perform classification within a data set for content-based mapping. Empirical models such as regression models are also suited for determining the statistical relationships between measured water quality parameters at the ground level and fused spectral values from space-borne satellite images although the prediction accuracy is not always satisfactory (Dekker et al., 1996; Yu, 2005). SLFN's learning capacity is governed by the lower bound of the number of hidden nodes (Narasimha et al., 2008). In this case study, it is $N/(1 + n/m) = 28/(1 + 3/1) = 7$ and the upper bound is 28.

Deep learning models, such as DBM, inspired by the structure of the brain, might be a new solution to provide better feature extraction capacity (Bengio, 2009; Goodfellow et al., 2016; Penatti et al., 2015). Deep learning models are normally formulated as multiple-layer models to learn representations of input/output data with abstractions from lower to higher levels. Specifically, DBM is a probabilistic generative model consisting of multiple layers of stochastic latent variables, whose building blocks are restricted Boltzmann machines (RBMs) for deep learning (Hinton et al., 2006). On the other hand, ELM fills a gap between different algorithms (such as support vector machine, polynomial networks, etc.) by considering universal approximation and classification capabilities to reduce the computational time (Huang et al., 2006a,b). In addition to the input and output layers an ANN model has, the ELM model also has a single hidden layer with multiple hidden nodes requiring much less computational time relative to ANN. The merit is that the random assignment of weights to the hidden nodes and the learning of weights between hidden nodes and outputs are done in a single step in an ELM model to save computational time (Kasun et al., 2013; Ding et al., 2015; Zhang et al., 2016). In this study, a single hidden layer structure of ANN, ELM, and DBN is chosen on such a common basis for a fair comparison.

When multiple thematic information retrieval techniques (i.e., feature extraction) are developed, there must be a set of statistical indexes to assess the performance of these inversion models (i.e., classifiers). In general, a model is deemed accurate if the predicted values closely match the observed values. In this study, model performance for different algorithms/inversion methods was analyzed by using five statistical indexes, including root mean squared error (rmse), ratio of the standard deviations (CO), mean percentage of error (PE), Pearson product moment correlation coefficient (RSQ), and coefficient of determination (R^2). The model performs well if the value of rmse is

TABLE 13.3

statistical Indices of Feature Extraction for TP Based on the Random Grouping Scheme

	Training			Testing			Validation		
	SLFN (20)	ELM (20)	DBN (20)	SLFN	ELM	DBN	SLFN	ELM	DBN
RMSE	0.41	0.25	0.59	1.03	2.63	0.67	0.69	0.72	0.67
CO	0.67	0.82	0.30	0.51	3.26	0.12	0.78	1.14	0.13
PE	−42.54	−20.09	−68.83	−123.07	−419.43	−92.92	−22.38	−20.25	−96.80
RSQ	0.70	0.90	0.31	0.49	0.38	0.87	0.70	0.63	0.18
R^2	0.49	0.81	0.10	0.24	0.14	0.75	0.49	0.40	0.03

Note: The values marked in red represent the best value in each category. Numbers in brackets are the number of hidden nodes used to create the neural network structure.

close to zero, CO is close to one, PE is close to zero, or R^2 is close to one. 28 (70%), 6(15%), and 6 (15%) samples out of 40 samples were chosen by the MATLAB® program via a random grouping scheme for training, testing, and validation based on their predicted values of training, testing, and validation output from 50 numerical experiments in computer simulation. Overall, a model is deemed accurate if the predicted values closely match the observed values. Model performance is analyzed using four statistical indexes for this study.

To evaluate the performance among three algorithms (SLFN, DBM, ELM) using the normalized training, testing and validation data, four statistical indexes (rmse, CO, PE, and R^2), as defined by Equations 7.19 through 7.23, were calculated for different numbers of hidden nodes to determine which may lead to the best performance associated with each water quality constituent of concern. The results of the performance evaluation are summarized in Tables 13.3 and 13.4 for TP and TN, respectively. The common number of hidden neurons in the hidden layer of all of the three selected machine learning models is 20 for training, testing, and validation consistently. However, the DBN is stacked by multiples of RMB. Inside the multiple layer structure of the DBN, the higher level RBM's input (via visible units) is from the lower level RBM's output (via hidden units) after the lower level RBM has been learned (see Chapter 7). After layer-by-layer pre-training in the DBN, back-propagation techniques fine tune all the parameters in the DBN, which include the input weights connecting the network outputs with the top level RBM outputs (Keyvanrad and Homayounpour, 2014). Thus, the 20 nodes represent the sum of all nodes in the RBM.

ELM is known as a fast learning machine, making it stand out in the training stage, and which implies an overfitting issue; ELM lacks the persistency in both testing and validation stages due to

TABLE 13.4

Statistical Indices of Feature Extraction for TN Based on the Random Grouping Scheme

	Training			Testing			Validation		
	SLFN (20)	ELM (20)	DBN (20)	SLFN	ELM	DBN	SLFN	ELM	DBN
RMSE	0.32	0.18	0.54	0.57	60.98	0.34	0.47	0.69	0.30
CO	0.70	0.86	0.22	0.30	103.25	0.10	0.56	1.37	0.20
PE	29.51	113.43	181.08	948.21	5.15e3	190.68	−30.24	145.92	−40.48
RSQ	0.78	0.93	0.43	0.00	0.22	0.21	0.66	0.64	0.71
R^2	0.61	0.86	0.19	0.00	0.05	0.05	0.43	0.40	0.51

Note: The values marked in red represent the best value in each category. Numbers in brackets are the number of hidden nodes used to create the neural network structure.

the overfitting issue. DBN exhibits a strong advantage in deep learning. Even though DBN cannot stand out in the training stage, it evolves quickly and excels in the testing and/or validation stage. This is evidenced by the fact that DBN has strength in terms of at least two statistical indexes in Tables 13.3 and 13.4 whereas ELM cannot exhibit widespread strength in this regard during the testing and validation stages, as shown in these two tables.

13.2.5 STEP 5—WATER QUALITY MAPPING

The preselected water quality constituents (i.e., TN, TP) at each pixel can be obtained by using a suite of feature extraction techniques which are functions in terms of surface reflectance values. Each pixel of the fused images represents a 30-m by 30-m square of the lake. Within the pixel there are 3 bands available, characterizing the surface reflectance at the wavelengths. This estimation process is then repeated for different water quality constituents for each of the pixels making up the lake map.

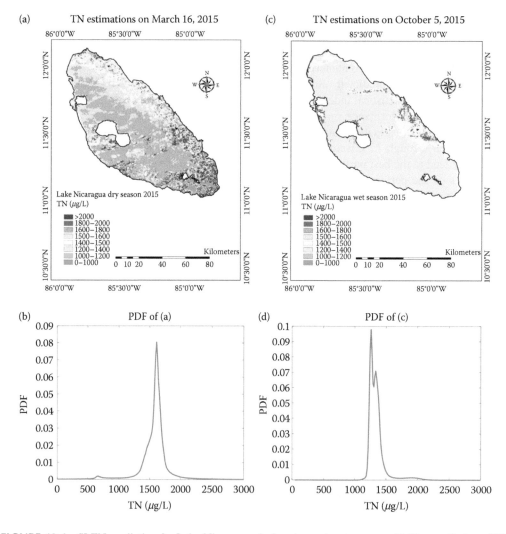

FIGURE 13.4 SLFN prediction for Lake Nicaragua during dry and wet seasons. (a) The prediction of TN value on March 16, 2015. (b) The estimated probability density function of TN value within [0, 3000] on March 16, 2015. (c) The prediction of TN value on October 5, 2015. (d) The estimated probability density function of TN value within [0, 3000] on October 5, 2015.

The validated models associated with the three intelligent feature extraction practices in Tables 13.3 and 13.4 can be used for content-based mapping to generate the spatial water quality patterns for wet and dry seasons. The best DBN model may translate the surface reflectance values relatively correctly to a map of water quality constituent throughout Lake Nicaragua. As determined by the validated model, certain wavelengths may have a stronger explanatory power in the determination of the water quality concentration. To deepen the understanding of the feature extraction outcome, the PDF associated with each feature extraction effort was also summarized for examination along with the derived water quality patterns in both wet and dry seasons. In general, normal distribution is anticipated for the pixels within each water quality map based on the probability and statistical theory (i.e., the law of large numbers).

The higher the TN values in a water body, the worse the water quality. In this regard, the TN maps in Figures 13.4 through 13.6 associated with the SLFN, ELM, and DBM models,

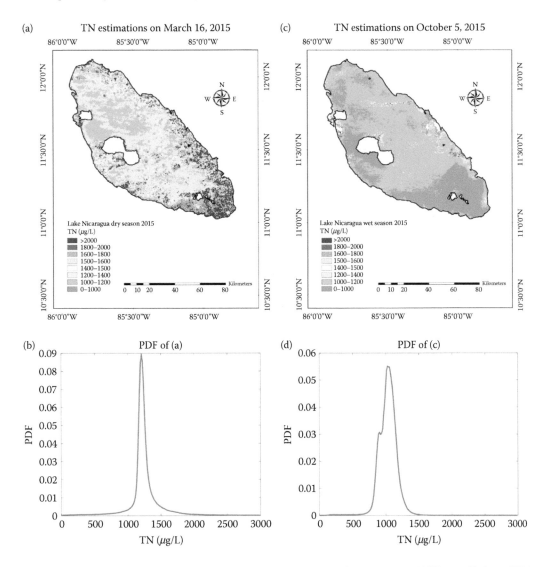

FIGURE 13.5   ELM prediction for Lake Nicaragua during dry and wet seasons. (a) The prediction of TN value on March 16, 2015. (b) The estimated probability density function of TN value within [0, 3000] on March 16, 2015. (c) The prediction of TN value on October 5, 2015. (d) The estimated probability density function of TN value within [0, 3000] on October 5, 2015.

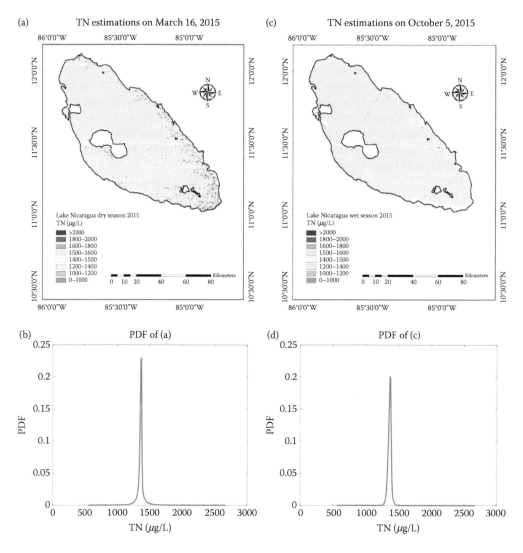

FIGURE 13.6 DBN prediction for Lake Nicaragua during dry and wet seasons. (a) The prediction of TN value on March 16, 2015. (b) The estimated probability density function of TN value within [0, 3000] on March 16, 2015. (c) The prediction of TN value on October 5, 2015. (d) The estimated probability density function of TN value within [0, 3000] on October 5, 2015.

respectively, reveal a suite of common water quality patterns with seasonality effect over the three modeling scenarios. First, the water quality in terms of TP and TN in the dry season is generally worse than that in the wet season. This is because sunlight in the dry season is not strong enough to trigger the fast growth of algal blooms to consume those nutrients and less rainfall cannot dilute the contamination quickly. Second, there are a few hot spots in the southern part of the lake in the dry season in 2015 no matter which machine learning model was applied for feature extraction. Third, water quality in the central part of the lake is generally better than that of the lake shore area, indicating human impacts on water quality regardless of the season. When considering the three machine-learning models for comparison, SLFN and ELM predictions for Lake Nicaragua during both dry and wet seasons have slight overestimation issues of the TN values, especially in the dry season of 2015. SLFN prediction

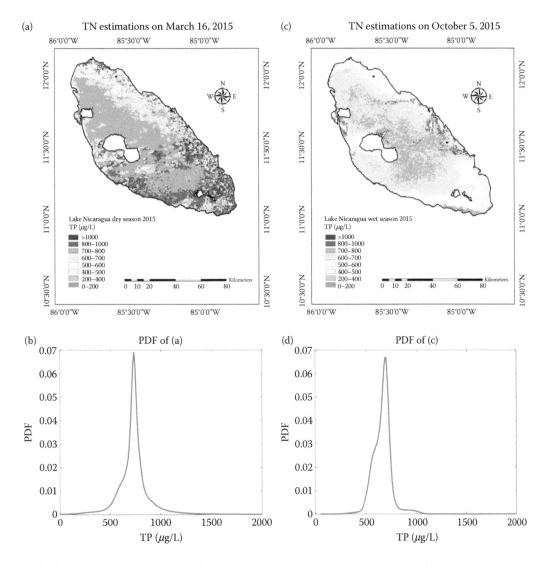

FIGURE 13.7 SLFN prediction for Lake Nicaragua during dry and wet seasons. (a) The prediction of TP value on March 16, 2015. (b) The estimated probability density function of TP value within [0, 2000] on March 16, 2015. (c) The prediction of TP value on October 5, 2015. (d) The estimated probability density function of TP value within [0, 2000] on October 5, 2015.

exhibits the worst scenario in terms of TN, however, as evidenced by the color distribution in Figure 13.4 relative to the other two figures (i.e., Figures 13.5 and 13.6). So does the case in TP mapping. Taking the PDF into account, the one associated with DBN has a relatively narrow distribution, albeit no impact on prediction accuracy. The higher the TP values in a water body, the worse the water quality. It is noticeable that the TP maps in Figures 13.7 through 13.9 share a few very similar water quality patterns with seasonality effect with the TN maps. DBM prediction for Lake Nicaragua exhibits the best water quality scenario with no overestimation issues, unlike the ANN and ELM redictions. Besides, the TP maps generated by the SLFN prediction show the worst scenario in terms of TP. The TP map generated by the ELM prediction is somewhat moderate toward the water quality condition when compared to the other two types

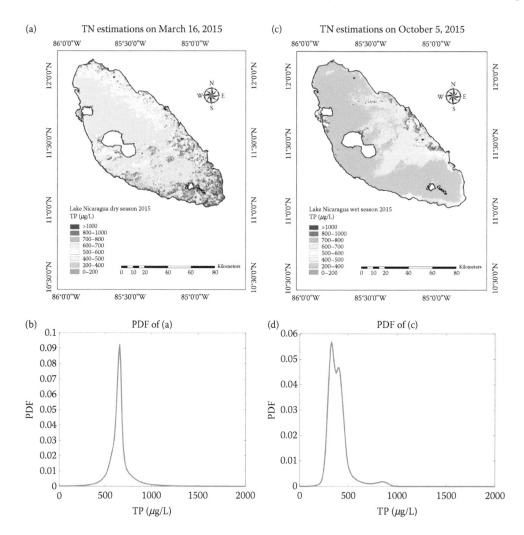

FIGURE 13.8 ELM prediction for Lake Nicaragua during dry and wet seasons. (a) The prediction of TP value on March 16, 2015. (b) The estimated probability density function of TP value within [0, 2000] on March 16, 2015. (c) The prediction of TP value on October 5, 2015. (d) The estimated probability density function of TP value within [0, 2000] on October 5, 2015.

of predictions. Again, the one associated with DBM has a relatively narrow distribution when taking the PDF into account.

The derived water quality patterns in both TN and TP give rise to some insightful linkages to the nutrient sources in the lake watershed. The environmental problems in the Lake Nicaragua watershed are mainly related to inadequate land use, such as the farming of hillside areas, unregulated lake fishing, lake aquaculture, agricultural and livestock development, the construction of poorly designed roads, deforestation of tropical forest species, and the destruction of wetlands and plant cover in ecologically sensitive areas (World Bank, 2013). These human activities have resulted in multiple sources of pollution and caused erosion and land degradation leading to transporting more nutrients, fertilizers, and pesticides into the lake (World Bank, 2013). In addition, the water quality hot spots identified in the southern part of the lake in the TN and TP maps are very close to an unregulated lake aquaculture site that further contributed to more nutrient input and contaminated the lake water body.

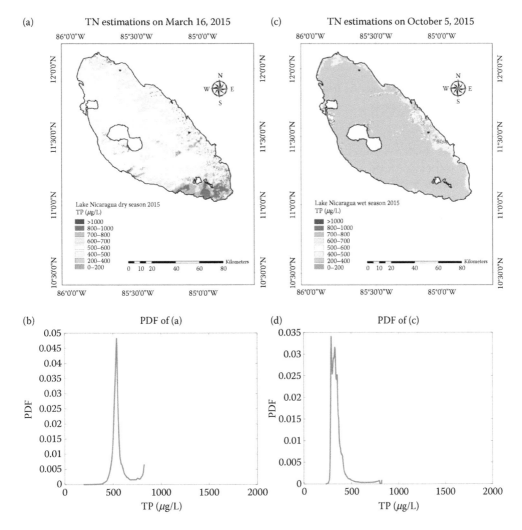

FIGURE 13.9 DBM prediction for Lake Nicaragua during dry and wet seasons. (a) The prediction of TP value on March 16, 2015. (b) The estimated probability density function of TP value within [0, 2000] on March 16, 2015. (c) The prediction of TP value on October 5, 2015. (d) The estimated probability density function of TP value within [0, 2000] on October 5, 2015.

13.3 SUMMARY

In this chapter, an integrated data fusion and mining application is demonstrated by application for water quality mapping. Intelligent feature extraction methods for fused images are discussed as well as suggestions for further case studies. In the context of IDFM, these intelligent feature extraction methods, including SLFN, ELM, and DBN, are compared with each other in terms of prediction accuracy based on a wealth of ground-truth data after supervised classification. It shows the deep learning model outperformed others with a profound learning capacity when considering the water quality indicators of TP and TN both of which have no direct optical property to address changing concentrations. In addition, findings gathered from the water quality maps are intimately tied to human land use and agricultural production in the lake watershed. Ensemble learning algorithms are needed to aggregate the advantage over different machine leaning tools for better prediction of water quality. In the next chapter, concepts of complex large-scale systems analysis in remote sensing will be introduced for further extension of IDFM.

APPENDIX 1: GROUND-TRUTH DATA

Sta. No.	Date	Band 1	Band 2	Band 3	TP	TN	Training	Testing	Validation
(a) 29 Samples from Lake Managua for Training, Testing, and Validation									
2	08/01/2014	541	616	560	738	1492	x		
2	04/10/2015	824	819	819	718	2451	x		
3	08/01/2014	943	999	864	994	1554	x		
3	04/10/2015	617	727	727	717	2120	x		
4	08/01/2014	490	721	596	812	1600	x		
4	04/10/2015	557	648	648	727	1070			x
5	04/09/2015	451	806	868	754	1343	x		
6	07/31/2014	712	906	797	901	1555	x		
6	04/09/2015	4876	4759	5145	711	1248		x	
7	07/31/2014	730	943	789	779	1609		x	
7	04/09/2015	1954	1950	1820	728	934	x		
8	07/31/2014	704	981	841	765	1633		x	
8	04/09/2015	4561	4532	5103	684	1053	x		
9	07/30/2014	404	355	328	808	1650	x		
9	04/08/2015	1247	1335	1340	776	2601	x		
9	03/15/2016	1218	1131	941	787	1543	x		
10	04/09/2015	4696	3851	4745	728	2672	x		
10	03/15/2016	774	810	676	876	1200		x	
11	04/08/2015	1358	1566	1550	780	1076	x		
12	07/30/2014	516	814	1081	1038	2171	x		
12	04/08/2015	1210	1355	1488	790	1208	x		
13	07/30/2014	952	1612	1826	835	1638	x		
13	04/08/2015	1152	1399	1283	749	2399	x		
14	04/08/2015	1278	1401	1262	790	1214	x		
15	04/08/2015	1174	1418	1252	788	1183			x
16	03/17/2016	768	779	570	647	972			x
17	07/29/2014	1118	1041	933	826	1783	x		
XOL-02	08/30/2016	869	924	650	647	1585			x
XOL-16	09/01/2016	2301	2432	2145	580	837	x		
(b) 11 Samples from Lake Nicaragua for Training, Testing, and Validation									
22	03/16/2015	712	954	1034	198	1471	x		
25	03/16/2015	789	928	896	87	801		x	
26	03/18/2015	368	590	440	53	870	x		
29	03/18/2015	497	592	507	65	898		x	
31	03/16/2015	843	950	892	58	788	x		
34	03/20/2015	151	281	162	43	555	x		
40	03/20/2015	845	1055	818	44	727	x		
43	03/22/2015	161	227	110	46	555			x
44	03/21/2015	980	1254	860	51	620	x		
TOM-4400	03/08/2016	656	756	595	113	748			x
COC-05	09/06/2016	935	818	770	40	1460	x		

Note: The checked marks in the training/testing/validation column indicate the assignment scheme for fixed samples.

APPENDIX 2

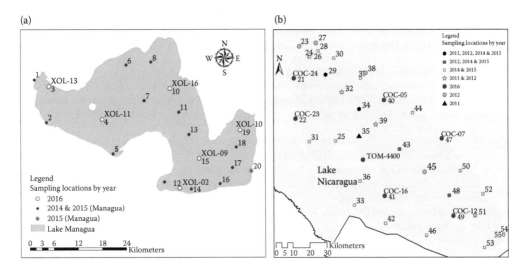

FIGURE 13.A2   Locations of water quality monitoring stations in (a) Lake Managua and (b) Lake Nicaragua.

REFERENCES

Abdikan, S., Sanli, F. B., Sunar, F., and Ehlers, M., 2014. A comparative data-fusion analysis of multi-sensor satellite images. *International Journal of Digital Earth*, 7, 671–687.

Alsahwa, B., Solaiman, B., Bossé, É., Almouahed, S., and Guériot, D., 2016. A method of spatial unmixing based on possibilistic similarity in soft pattern classification. *Fuzzy Information and Engineering*, 8, 295–314.

Ardeshir Goshtasby, A. and Nikolov, S., 2007. Image fusion: Advances in the state of the art. *Information Fusion*, 8, 114–118.

Benediktsson, J. A., Swain, P. H., and Ersoy, O. K., 1990. Neural network approaches versus statistical methods in classification of multisource remote sensing data. *IEEE Transactions on Geoscience and Remote Sensing*, 28, 540–552.

Bengio, Y., 2009. Learning deep architectures for AI. *Foundations and trends® in Machine Learning*, 2, 1–127.

Blaschke, T., Lang, S., Lorup, E., Strobl, J., and Zeil, P., 2000. Object-oriented image processing in an integrated GIS/remote sensing environment and perspectives for environmental applications. In: *Environmental Information for Planning, Politics, and the Public*, Marburg, Germany, Metropolis, II, 555–570.

Castanedo, F., 2013. A review of data fusion techniques. *The Scientific World Journal*, 2013, Article ID 704504, 19 pages.

Chang, N. B., Vannah, B., and Yang, Y., 2014a. Comparative sensor fusion between hyperspectral and multispectral satellite sensors for monitoring microcystin distribution in Lake Erie. *IEEE Journal of Selected Topics in Applied Earth Observations and Remote Sensing*, 7, 2426–2442.

Chang, N. B., Vannah, B., Yang, Y., and Elovitz, M., 2014b. Integrated data fusion and mining techniques for monitoring total organic carbon concentrations in a lake. *International Journal of Remote Sensing*, 35, 1064–1093.

Dalla Mura, M., Villa, A., Benediktsson, J. A., Chanussot, J., and Bruzzone, L., 2011. Classification of hyperspectral images by using extended morphological attribute profiles and independent component analysis. *IEEE Geoscience and Remote Sensing Letters*, 8, 542–546.

Dekker, A. G., Zamurović-Nenad, Ž., Hoogenboom, H. J., and Peters, S. W. M., 1996. Remote sensing, ecological water quality modelling and in situ measurements: A case study in shallow lakes. *Hydrological Sciences Journal*, 41, 531–547.

Ding, S., Zhang, N., Xu, X., Guo, L., and Zhang, J., 2015. Deep extreme learning machine and its application in EEG classification. *Mathematical Problems in Engineering*, 2015, Article ID 129021, 11 pages.

Dong, Y., Forster, B. C., and Milne, A. K., 1999. Segmentation of radar imagery using Gaussian Markov random field model. *International Journal of Remote Sensing*, 20, 1617–1639.

Dong, Y., Milne, A. K., and Forster, B. C. 2001. Segmentation and classification of vegetated areas using polarimetric SAR image data. *IEEE Transactions on Geoscience and Remote Sensing*, 39, 321–329.

Du, Q., 2009. Decision fusion for classifying hyperspectral imagery with high spatial resolution. SPIE Newsroom.

Fok, H.S., 2015. Data fusion of multisatellite altimetry for ocean tides modelling: A spatio-temporal approach with potential oceanographic applications. *International Journal of Image and Data Fusion*, 6, 232–248.

Foody, G. M. and Mathur, A., 2006. The use of small training sets containing mixed pixels for accurate hard image classification: Training on mixed spectral responses for classification by a SVM. *Remote Sensing of Environment*, 103, 179–189.

Friedl, M. A. and Brodley, C. E., 1997. Decision tree classification of land cover from remotely sensed data. *Remote Sensing of Environment*, 61, 399–409.

Gao, F., Masek, J., Schwaller, M., and Hall, F., 2006. On the blending of the Landsat and MODIS surface reflectance: Predicting daily Landsat surface reflectance. *IEEE Transactions on Geoscience and Remote Sensing*, 44, 2207–2218.

Goodfellow, I., Bengio, Y., and Courville, A., 2016. Deep learning, book in preparation for MIT Press. URL http://goodfeli.github.io/dlbook.

Hinton, G. E., Osindero, S., and Teh, Y. W., 2006. A fast learning algorithm for deep belief nets. *Journal Neural Computation*, 18, 1527–1554.

Huang, C., Davis, L. S., and Townshend, J. R., 2002. An assessment of support vector machines for land cover classification. *International Journal of Remote Sensing*, 23, 725–749.

Huang, X. and Zhang, L., 2010. Comparison of vector stacking, multi-SVMs fuzzy output, and multi-SVMs voting methods for multiscale VHR urban mapping. *IEEE Geoscience and Remote Sensing Letters*, 7, 261–265.

Huang, X. and Zhang, L., 2012. A multilevel decision fusion approach for urban mapping using very high-resolution multi/hyperspectral imagery. *International Journal of Remote Sensing*, 33, 3354–3372.

Huang, G. B., Zhu, Q. Y., and Siew, C. K., 2006a. Real-time learning capability of neural networks. *IEEE Transactions on Neural Networks*, 17, 863–878.

Huang, G. B., Zhu, Q. Y., and Siew, C. K., 2006b. Extreme learning machine: Theory and applications. *Neurocomputing*, 70, 489–501.

Kasun, L. L. C., Zhou, H. M., Huang, G. B., and Vong, C. M., 2013. Representational learning with extreme learning machine for big data. *IEEE Intelligent System*, 28, 31–34.

Keshava, N. 2003. A survey of spectral unmixing algorithms. *Lincoln Laboratory Journal*, 14, 55–78.

Keyvanrad, M. A. and Homayounpour, M. M., 2014. A brief survey on deep belief networks and introducing a new object-oriented toolbox (DeeBNet). arXiv preprint arXiv:1408.3264.

Luo, R. C., Yih, C. C., and Su, K. L., 2002. Multisensor fusion and integration: Approaches, applications, and future research directions. *IEEE Systems Journal*, 2, 107–119.

Masek, J. G., Vermote, E. F., Saleous, N. E., Wolfe, R., Hall, F. G., Huemmrich, K. F., Gao, F., Kutler, J., and T.-K. Lim, 2006. A landsat surface reflectance dataset for North America, 1990–2000. *IEEE Geoscience and Remote Sensing Letters*, 3, 68–72.

MIT Technology Review. URL https://www.technologyreview.com/s/513696/\deep-learning accessed by January 2016.

Mitrakis, N. E., Topaloglou, C. A., Alexandridis, T. K., Theocharis, J. B., and Zalidis, G. C., 2008. Decision fusion of GA self-organizing neuro-fuzzy multilayered classifiers for land cover classification using textural and spectral features. *IEEE Transactions on Geoscience and Remote Sensing*, 46, 2137–2152.

Narasimha, P. L., Delashmit, W. H., Manry, M. T., Li, J., and Maldonado, F., 2008. An integrated growing-pruning method for feedforward network training. *Neurocomputing*, 71, 2831–2847.

Nguyen, H., 2009. Spatial statistical data fusion for remote sensing applications, Ph.D. dissertation, University of California-Los Angeles, USA.

Nikolakopoulos, K. G. K., 2008. Comparison of nine fusion techniques for very high resolution data. *Photogrammetric Engineering & Remote Sensing*, 74, 647–659.

Nogueira, K., Miranda, W. O. and Dos Santos, J. A., 2015. Improving spatial feature representation from aerial scenes by using convolutional networks. In: *28th IEEE SIBGRAPI Conference on Graphics, Patterns and Images*, 289–296.

Pal, M. and Mather, P. M., 2003. An assessment of the effectiveness of decision tree methods for land cover classifications. *Remote Sensing of Environment*, 86, 554–565.

Penatti, O., Nogueira, K., and Santos, J., 2015. Do deep features generalize from everyday objects to remote sensing and aerial scenes domains? In: *Proceedings of the IEEE Conference on Computer Vision and Pattern Recognition Workshops*, 44–51.

Petrakos, M., Atli Benediktsson, J., and Kanellopoulos, I., 2001. The effect of classifier agreement on the accuracy of the combined classifier in decision level fusion. *IEEE Transactions on Geoscience and Remote Sensing*, 39, 2539–2546.

Pohl, C. and van Genderen, J. L., 1998. Multisensor image fusion in remote sensing: Concepts, methods and applications. *International Journal of Remote Sensing*, 19, 823–854.

Reiche, J., Souza, C. M., Hoekman, D. H., Verbesselt, J., Persaud, H., and Herold, M., 2013. Feature level fusion of multi-temporal ALOS PALSAR and Landsat data for mapping and monitoring of tropical deforestation and forest degradation. *IEEE Journal of Selected Topics in Applied Earth Observations and Remote Sensing*, 6, 2159–2173.

Rumelhart, D., Hinton, G., and Williams, R., 1986. Learning internal representations by error propagation. In: Rumelhart, D. E., McClelland, J. L., and PDP Research Group (Eds.), *Parallel Distributed Processing* (Vol. 1). MIT Press, Cambridge, MA, USA.

Serpico, S. and Roli, F., 1995. Classification of multisensory remote-sensing images by structured neural networks. *IEEE Transactions on Geoscience and Remote Sensing*, 33, 562–578.

Shen, S., 1990. Summary of types of data fusion methods utilized in workshop papers. Multisource Data Integration in Remote Sensing, Proceedings of Workshop, Maryland, U.S.A., 14–15 June 1990, NASA Conference Publication 3099 (Greenbelt, MD: NASA).

Shi, W., Zhu, C., Tian, Y., and Nichol, J., 2005. Wavelet-based image fusion and assessment. *International Journal of Applied Earth Observation and Geoinformation*, 6, 241–251.

United States Geological Survey (USGS). Landsat Processing Details [Online]. Available: http://landsat.usgs.gov/Landsat_Processing_Details.php

Vermote, E. F. and Kotchenova, S., 2008. Atmospheric correction for the monitoring of land surfaces. *Journal of Geophysical Research-Atmospheres*, 113, D23S90.

Waltz, E., 2001. Principles and practice of image and spatial data fusion. In: David Hall and James Llinas (Eds.), *Multisensor Data Fusion*. CRC Press, Boca Raton, Florida, USA, 50–63.

Wang, Z., Ziou, D., Armenakis, C., Li, D., and Li, Q., 2005. A comparative analysis of image fusion methods. *IEEE Transactions on Geoscience and Remote Sensing*, 43, 1391–1402.

Waske, B. and Benediktsson, J. A., 2007. Fusion of support vector machines for classification of multisensor data. *IEEE Transactions on Geoscience and Remote Sensing*, 45, 3858–3866.

Werbos, P. J., 1990. Back propagation through time: What it does and how to do it. *The IEEE Proceedings*, 78, 1550–1560.

Won, Y. and Gader, P., 1995. Morphological Shared-Weight Neural Network for Pattern Classification and Automatic Target Detection, University of Missouri-Columbia.

World Bank, 2013. Policy and investment priorities to reduce environmental degradation of the lake Nicaragua watershed (Cocibolca). Report No. 76886. Washington, DC.

Yu, T., 2005. Utility of remote sensing data in retrieval of water quality constituents concentrations in coastal water of New Jersey. Ph.D. dissertation, New Jersey Institute of Technology, available at http://archives.njit.edu/vol01/etd/2000s/2005/njit-etd2005-091/njit-etd2005-091.pdf, accessed by August 2016.

Zhang, N., Ding, S., and Zhang, J., 2016. Multilayer ELM-RBF for multi-label learning. *Applied Soft Computing*, 43, 535–545.

14 Integrated Cross-Mission Data Merging, Fusion, and Machine Learning Algorithms Toward Better Environmental Surveillance

14.1 INTRODUCTION

Image fusion, which has been extensively discussed in previous chapters, relies on satellite imageries and data collected from different sensors for synergistically improving spatial, spectral, and temporal resolution of image data simultaneously. The integration of data from several sources to monitor a single target in the environment is also known as data fusion. Possible spatial data gaps may still exist after data fusion due to either internal or external factors, limiting the remote sensing capacity for environmental monitoring. Whereas the internal factors may be due to mechanical failure of the instrument, such as Landsat ETM+ SLC-off images, the external factors may be caused by cloud contamination, such as a heavy cloud event. Hence, gap filling or image reconstruction is often required as one of the essential pre-processing techniques.

In Chapters 11 and 12, the introduction of the Spectral Information Adaptation and Synthesis Scheme (SIASS) (Bai et al., 2015) and SMart Information Reconstruction (SMIR) (Chang et al., 2015a) in sequence has laid down the mathematical foundation for some big data analytics via image enhancement, reconstruction, pattern recognition, and classification in support of various feature extraction schemes based on data integration, localization, aggregation, fusion, merging, and scaling. Following the logical sequence of alignment, association, merging, and fusion, the prior knowledge gained by developing integrated algorithms and heuristics for creating simulated or synthetic data after data verification and validation has also triggered a new direction for a possible complex large-scale systems analysis. It will lead to the systems analysis for situation awareness by linking the knowledge base to an inference engine for continuous situation refinement. With such a goal, this chapter will focus on conducting a complex large-scale systems analysis with a system of systems engineering approach. While there are a myriad of ways to explore this complex large-scale systems analysis, let us start with the following sequence of options based on the core modules of SIASS, SMIR, and Integrated Data Fusion and Mining (IDFM).

To deal with data gaps, data merging is one option for concurrently merging several satellites' imageries to improve the spatial coverage (i.e., fill in data gaps) after removing inherent bias among sensors. In the case study of SIASS, as we recalled, three cross-mission sensors, including Visible Infrared Imaging Radiometer Suite (VIIRS) onboard Suomi-NPP, MODIS-Terra, and MODIS-Aqua, provide a synergistic opportunity to enhance spatial coverage. This can be completed by merging their complementary remote sensing data collected in almost the same time window for the same target area with slightly different overpassing times and view angles. Nevertheless, SIASS can be applied to enhance temporal coverage as well to extend the utility of data products from a set of satellites (i.e., could be more than three) with similar functionalities (Bai et al., 2016) (see Chapter 15 for a real world application). If data merging is not able to attain a spatially complete imagery, other algorithms like SMIR may be used as a supplementary tool to fill in the spatial data gaps. SMIR is a model-based approach applied for simulation among cloudy and clear pixels based on the relationships of plausible network topologies among pixels that the model can identify. With the long-term memory effect over spatial, temporal, and spectral domains, SMIR may be implemented for image classification, pattern recognition, and image reconstruction in sequence for cloudy pixels via the time–space–spectrum continuum after SIASS is fully carried out. If there is a need for data fusion to improve the spatial and temporal resolution at this stage, the Integrated Data Fusion and Machine Learning (IDFM) may be included later to achieve the goal (Chang et al., 2014a,b). Yet this is not the only sequence of options in such a complex large-scale systems analysis.

In fact, SMIR can be applied either before or after the data merging step in support of IDFM (either follow route 1 or 2 in Figure 14.1). In other words, reversing the order from route 2 to 1 would not be preventive if one satellite's image data need to have gap filling as priority before data merging. Or SMIR may work with IDFM alone from route 8 to 9 in Figure 14.1 to reconstruct the cloudy images before data fusion without involving data merging. On the other hand, data merging can become a step prior to IDFM from route 7 to 9 without involving SMIR in a case of few data gaps as long as the image reconstruction can be successful using SIASS alone. Yet challenges in feature extraction require even more attention due to the acute need to support more effective and efficient data merging, data fusion, and gap filling, as well as to retrieve more complicated environmental indicators. In this context, different machine learning or data mining models may perform an independent role in support of SIASS, SMIR, or IDFM in due course on an as-needed basis (i.e., route 4 and 5 in Figure 14.1). For example, feature extraction may be performed right after SIASS if data merging successfully removes most of the cloudy pixels (i.e., route 6 in Figure 14.1). The process of weaving all these essential components together seamlessly to provide a new software platform is deemed a contemporary system of systems engineering mission in remote sensing science, for which a total solution has been pending in the last few decades (Figure 14.1). Following the thorough literature review in the previous chapters, this chapter aims to demonstrate the total solution via possible automation of satellite-based environmental monitoring through the introduction of a new software platform called "Cross-mission Data merging with Image Reconstruction and fusion in support of Machine learning" (CDIRM).

CDIRM aims to continuously monitor and decipher environmental conditions from a suite of analytical procedures for different types of environmental systems, such as a lake watershed, an urban airshed or an ecosystem of concern. At the intersection of data merging, gap filling, data fusion, and intelligent feature extraction, the CDIRM that integrates SIASS, SMIR, and IDFM in

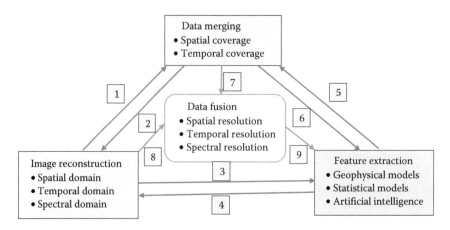

FIGURE 14.1 The structure of the CDIRM software package from a system of systems engineering perspective.

support of intelligent feature extraction has the primary objective of removing the cross-mission biases among sensors, merging the relevant satellite imageries to improve spatial and/or temporal coverage, and fusing the complementary satellite imageries to elevate the spatial and spectral resolutions for better feature extraction and environmental monitoring. There are multiple ways that CDIRM can reach its goal, as described in the multiple cross-linkages embedded within the nexus of Figure 14.1.

14.2 ARCHITECTURE OF CDIRM

CDIRM is a multiscale modeling system in a common software platform that incorporates a variety of remote sensing image processing and spatial analysis techniques to better monitor and understand the Earth's environment. The software package enables users to manipulate remote sensing images with the aid of data merging/fusion, image reconstruction, and machine learning/data mining techniques in a flexible way (Figure 14.2). The following delineation follows a fixed streamline (i.e., option) from the route 2 to the route 3 in Figure 14.1 to demonstrate the theoretical foundations (Figure 14.3). Readers may extend the principles for various applications to meet their own study's goal individually. As shown in Figure 14.2, one of the workflows in CDIRM generally follows even main steps to: (1) collect radiometric data from satellite observations, (2) pre-process the raw images in preparation for further manipulation, (3) perform data merging by using SIASS to generate synthetic images, (4) apply image reconstruction methods (i.e., SMIR) to create spatially

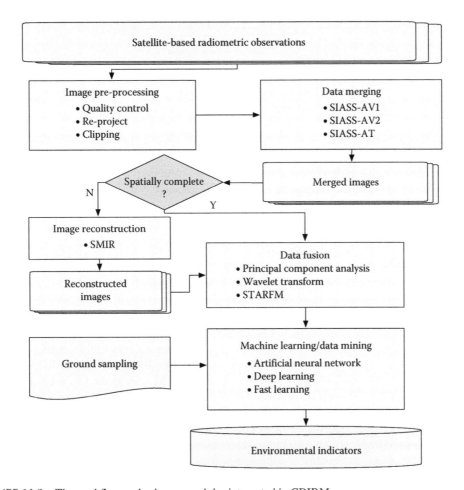

FIGURE 14.2 The workflow and primary modules integrated in CDIRM.

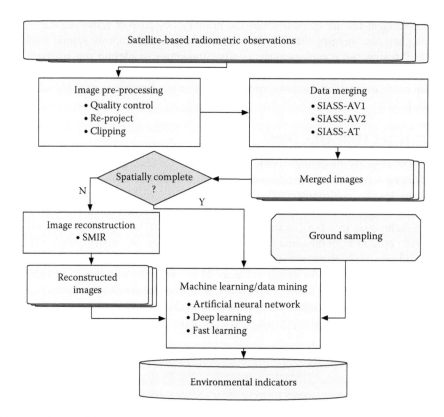

FIGURE 14.3   The workflow and primary modules in CDMIM (a subset of CDIRM). (From Chang, N. B., Bai, K., and Chen C. F., 2017. *Journal of Environmental Management*, 201, 227–240.)

simulated images with no data gap, (5) fuse the reconstructed image with other complementary images to enhance spatial, temporal, or spectral resolution if necessary, (6) carry out a selective feature extraction to assimilate ground truth data in an inverse modeling procedure via the use of merged and reconstructed images, and (7) generate content-based environmental quality maps based on featured ocean color reflectance data.

For more details, let us start with step 1, the collection of common observations over three satellite sensors (i.e., MODIS-Aqua, MODIS-Terra, and VIIRS) during the overlapped time period which may provide a synergistic opportunity for sensor inter-comparisons and data merging. Then, a long-term memory of the baseline satellite images can be used to reconstruct the value-missing pixels based on the merged and/or fused datasets. Data fusion occasionally contributes to the continuous elevation of high quality images with better spatial and temporal resolutions. The endeavor of IDFM supports the subsequent feature extraction with different fast, regular, and deep learning algorithms toward content-based mapping. With this multiscale modeling system, the proposed software platform and cyberinfrastructure can automatically retrieve embedded environmental indicators or patterns of interest over weeks, months, or seasons via different feature extraction models. For demonstration in this chapter, CDIRM was applied to predict the water quality in terms of Total Phosphorus (TP), Total Nitrogen (TN), chlorophyll-a, and water clarity in Lake Nicaragua (Central America), enabling the automatic multitemporal change detection. The following subsections elucidate the essence of the implementation of a brand new software platform for "Cross-mission Data Merging with Image reconstruction and Machine learning" (abbreviated as CDMIM) to decipher water quality conditions based on a suite of satellite observations for aiding in watershed management (Chang et al., 2017). CDMIM is a simplified version of CDIRM and is depicted in the flowchart shown in Figure 14.3, in which the data fusion was excluded tentatively for demonstration purposes.

To evaluate the quality of data merging, data reconstruction, and data mining, statistics such as the ensemble mean of daily clear pixels Coverage Ratio (*CR*) (Equation 11.27), root mean squared error (*rmse*) (Equation 7.19), Pearson product moment correlation coefficient (*r*-value) (Equation 7.22), and the coefficient of determination (R^2) (Equation 7.23), were used for performance evaluation. In addition, the information entropy (*H*), a statistical measure of randomness that can be used to characterize the texture of the target image, was also used to assess the performance of data merging stepwise. Generally, entropy can be defined as Equation 13.1 for the assessment of data fusion.

14.2.1 IMAGE PRE-PROCESSING

MODIS-Aqua, MODIS-Terra, and VIIRS-NPP sensors orbiting Earth may collect remote sensing imageries on a daily basis at a moderate resolution. Evaluation of the literature of the ocean color products collected from these sensors indicates comparable accuracies among these three sensors at most wavelengths. Detailed band settings and properties of these two types of sensors can be found in Table 11.3. These radiometric products have previously been atmospherically and radiometrically corrected by NASA to remove effects of atmospheric attenuation and solar orientation caused by aerosols and scattering. Pre-processing is an essential step in image collection and preparation, as many important manipulations, such as quality control, need to be conducted first to remove suspicious data/information from a separate analysis. For instance, processing flags of MODIS-Aqua and MODIS-Terra images indicate certain conditions (Table 14.1); a flag is applied to a pixel that meets a certain condition to mark the pixel as essential. However, a pixel can have more than one flag applied to it. A comparative example can be made possible by showing the Figure 14.4 with multiple flagging or without flagging of a level 2 ocean color product collected on January 5, 2012.

Conventional manual sampling for water quality monitoring is labor-intensive, costly, and unable to capture spatial variability simultaneously. To provide an all-inclusive insight with a large spatial coverage in a timely manner, satellite ocean color observations should be used on a daily basis. At present, the satellite sensors orbiting Earth with the ability to monitor ocean color and inland waters mainly include: (1) MODIS aboard the Terra (1999–present, MODIS-Terra hereafter), (2) MODIS aboard the Aqua (2002–present, MODIS-Aqua hereafter), (3) the VIIRS aboard the Suomi-NPP (2011–present, VIIRS-NPP hereafter), and many others.

Because the raw NASA Earth Observing System (EOS) images are always archived as granule files rather than a gridded raster dataset, the raw granule files must be reprojected into gridded datasets with a unified spatial reference to collocate pixels at different temporal phases at the same grid cell. Toward this goal, several remote sensing image processing softwares can be of help. For example, the SeaWiFS Data Analysis System (SeaDAS) software provided by NASA's OceanColor science team was applied to pre-process the Level-2 MODIS and VIIRS ocean color images in this study. To manipulate enormous raw image files, Windows batch scripts were created to execute the reproject module embedded in the SeaDAS package to process the input raw images sequentially.

As shown in Figure 14.5, the images were processed in a similar manner to: (1) filter out pixels with poor quality flags, (2) reproject to UTM zone 16 North with a spatial resolution of 1000 m, (3) crop out land data surrounding the study region in an aquatic environment, such as a lake, with the corresponding boundary vector file in ArcMap, (4) remove duplicated images for the same date and keep the image with the largest spatial coverage through a MATLAB program, and (5) remove images with partial coverage of the study region to produce the final product. In terms of quality control, a smaller set of flags that are often applied to real world applications are highlighted in bold for demonstration in Table 14.2. A sample binary map of Lake Nicaragua with multiple selected flagging on March 22, 2015, was produced (shown in Figure 14.6). The green pixels thus denote those questionable pixels that need to be removed.

TABLE 14.1

NASA Quality Assurance Flags of MODIS-Aqua and MODIS-Terra Level 2 Ocean Color Products with Descriptions

Bit	Name	Description	L2 Mask Default	L3 Mask Default
00	ATMFAIL	Atmospheric correction failure		ON
01	LAND	Pixel is over land	ON	ON
02	PRODWARN	One or more product algorithms generated a warning		
03	HIGLINT	Sunlight: reflectance exceeds threshold		ON
04	HILT	Observed radiance very high or saturated	ON	ON
05	HISATZEN	Sensor view zenith angle exceeds threshold		ON
06	COASTZ	Pixel is in shallow water		
07	spare			
08	STRAYLIGHT	Probable stray light contamination	ON	ON
09	CLDICE	Probable cloud or ice contamination	ON	ON
10	COCCOLITH	Coccolithophores detected		ON
11	TURBIDW	Turbid water detected		
12	HISOLZEN	Solar zenith exceeds threshold		ON
13	spare			
14	LOWLW	Very low water-leaving radiance		ON
15	CHLFAIL	Chlorophyll algorithm failure		ON
16	NAVWARN	Navigation quality is suspect		ON
17	ABSAER	Absorbing Aerosols determined (disabled?)		ON
18	spare			
19	MAXAERITER	Maximum iterations reached for NIR iteration		ON
20	MODGLINT	Moderate sun glint contamination		
21	CHLWARN	Chlorophyll out-of-bounds		
22	ATMWARN	Atmospheric correction is suspect		ON
23	spare			
24	SEAICE	Probable sea ice contamination		
25	NAVFAIL	Navigation failure		ON
26	FILTER	Pixel rejected by user-defined filter OR Insufficient data for smoothing filter?		
27	spare	(used only for SST)		
28	spare	(used only for SST)		
29	HIPOL	High degree of polarization determined		

Source: NASA, 2017. *Quality of Level 2 Ocean Color products.* https://oceancolor.gsfc.nasa.gov/atbd/ocl2flags/ accessed by January 2017.

To avoid the propagation of large biases in the following data merging, cloudy pixel reconstruction, and machine learning steps, quality control was conducted in this case study by excluding all problematic pixels with negative reflectance at any wavelength and any of the Level-2 processing flags in this study (Table 14.3). Note that they are the same flags selected by NASA's Ocean Biology Processing Group (OBPG) in order to remove any questionable data for processing the global Level-3 data products.

14.2.2   Data Merging via SIASS

Cross-mission ocean color reflectance products collected from the three satellite sensors after essential pre-processing steps were merged to increase the spatial coverage at this stage. Biases between cross-sensors' observations can be a big issue and need to be handled first. A schematic flowchart of the cross-mission ocean color reflectance merging scheme with SIASS was shown in Figure 11.1 (Bai et al., 2015). By taking advantage of SIASS, biases across missions can be identified,

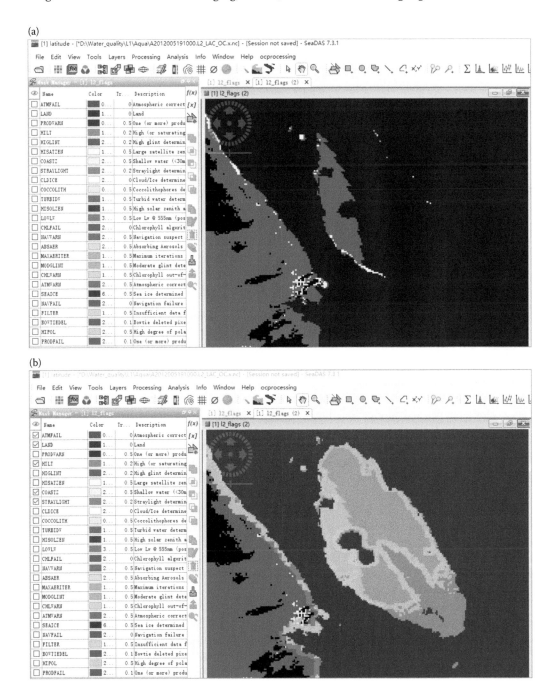

FIGURE 14.4 ENVI snapshots of Lake Nicaragua level 2 product with multiple flagging or without flagging on January 5 2012. (a) Image with flags turned off (January 5, 2012) and (b) images with multiple flags turned on (January 5, 2012).

quantified, and removed with integrated signal processing, statistical inference, and machine learning techniques for cross-mission sensor observations via the proper integration of singular value decomposition (SVD) (Businger and Golub, 1969; Nash, 1990), Quantile-Quantile (Q-Q) adjustment (Wilk and Gnanadesikan, 1968), Empirical Mode Decomposition (EMD) (Drazin, 1992), and Extreme Learning Machine (ELM) (Huang et al., 2006). Both systematic bias and associated

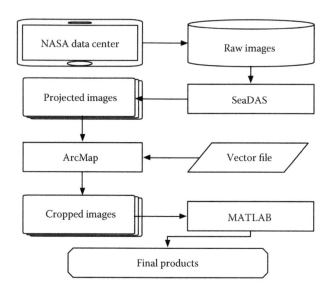

FIGURE 14.5   Data pre-processing steps and the software platform.

location-dependent bias can be removed for each cross-mission sensor observation (i.e., pixel value) when monitoring dynamic environments such as an aquatic environment (Bai et al., 2015). Within the systematic bias correction, spectral feature extraction via SVD was proposed to retrieve the major spectral information (i.e., the first principal component) leading to the generation of both referenced datasets for reference and baseline satellite sensors (i.e., MODIS-Terra and MODIS-Aqua or VIIRS-NPP and MODIS-Aqua given that MODIS-Aqua was selected as the baseline in this case).

Then, the systematic bias embedded in two relevant referenced datasets (i.e., MODIS-Terra and VIIRS-NPP in this case) can be mostly removed by the Q-Q adjustment. Following the SIASS algorithm, the location-dependent bias can be removed through inverse modeling by using EMD and ELM in sequence. Whereas EMD may remove the high frequency oscillations embedded in the two reference datasets after the Q-Q adjustment, ELM may help predict the relationship between the target pixel value and the candidate pixel values for any of the two reference datasets with respect to a nonlinear and non-stationary process afterwards. With such a common procedure, the SIASS algorithm was designed to not only eliminate biases among cross-mission sensor observations for wavelengths of the synchronized bands, but also reconstruct spectral information for wavelengths of mismatched bands (Bai et al., 2015).

The effectiveness in terms of spatial coverage ratio (i.e., the *CR* value) when dealing with a heavy cloud event is shown in Figure 14.7. Considering the discrepancy arising from the local overpassing time differences and error propagations, ocean color reflectance data from VIIRS-NPP were first merged to MODIS-Aqua because both sensors have a similar local overpassing time (i.e., 1:30 PM) (Figure 14.7a,b). With the aid of SIASS, the spatial coverage ratio (i.e., the *CR* value) of the merged dataset (i.e., A + V in Figure 14.7b) was improved significantly from 18.60% to 24.64%. To further improve the spatial coverage ratio, ocean color reflectance data collected from MODIS-Terra were merged with the merged image from the previous step (i.e., A + V) to finally generate a fully merged dataset (A + V + T in Figure 14.7c). With the aid of the SIASS, the *CR* increased from the original 18.60% to 24.92% finally.

14.2.3   Data Reconstruction via SMIR

Although SIASS can be of help in improving the spatial coverage ratio by merging cross-mission sensors' observations to some extent, data gaps may still exist due to a lack of sufficient observations. In some heavy cloud events, the cloudy satellite imageries can never be reconstructed wholly by

TABLE 14.2

NASA Quality Assurance Flags of MODIS-Aqua and MODIS-Terra Level 2 Ocean Color Products with Selected Flags of Interest (in Bold)

Name	Value	Bit	Description
ATMFAIL	**1**	**0**	**Atmospheric correction failure**
LAND	2	1	Land
PRODWARN	4	2	One (or more) product algorithms generated a warning
HIGLINT	8	3	High glint determined
HILT	**16**	**4**	**High (or saturating) TOA radiance**
HISATZEN	**32**	**5**	**Large satellite zenith angle**
COASTZ	64	6	Shallow water (<30 m)
SPARE8	128	7	Unused
STRAYLIGHT	256	8	Straylight determined
CLDICE	512	9	Cloud/Ice determined
COCCOLITH	**1024**	**10**	**Coccolithophores detected**
TURBIDW	2048	11	Turbid water determined
HISOLZEN	**4096**	**12**	**High solar zenith angle**
SPARE14	8192	13	Unused
LOWLW	**16384**	**14**	**Low Lw @ 555 nm (possible cloud shadow)**
CHLFAIL	32768	15	Chlorophyll algorithms failure
NAVWARN	**65536**	**16**	**Navigation suspect**
ABSAER	**131072**	**17**	**Absorbing Aerosols determined**
SPARE19	262144	18	Unused
MAXAERITER	524288	19	Maximum iterations reached for NIR iteration
MODGLINT	1048576	20	Moderate glint determined
CHLWARN	2097152	21	Chlorophyll out-of-bounds (<0.01 or >100 mg m^{-3})
ATMWARN	**4194304**	**22**	**Atmospheric correction warning; Epsilon out-of-bounds**
SPARE24	8388608	23	Unused
SEAICE	16777216	24	Sea ice determined
NAVFAIL	**33554432**	**25**	**Navigation failure**
FILTER	67108865	26	Insufficient data for smoothing filter
SPARE28	134217728	27	Unused
SPARE29	268435456	28	Unused
HIPOL	536970912	29	High degree of polarization determined
PRODFAIL	1073741824	30	One (or more) product algorithms produced a warning
SPARE32	2147483648	31	Unused

using SIASS alone with the inclusion of the three selected satellite sensors. SMIR was, therefore, proposed to reconstruct spectral information (e.g., radiometric values) of value-missing pixels to fill in data gaps in remotely sensed imagery caused by a variety of deficits such as cloud contaminations.

The assumption of the SMIR method lies in the existence of interrelationships between objects present in nature, either spatially or temporally. For example, in one partially cloudy ocean color image, a certain relationship exists between cloud free and cloudy pixels nearby in the scalable time-space-spectrum domain. This hidden relationship can be considered as a kind of lost memory in the historical time scales although values of those cloudy pixels are missing. This assumption supports following the cognitive science concept to recover the forgotten pixel values. Based on this concept, it is entirely possible to recall the missing values via a fast synergistic and universal scanning over the scalable time-space-spectrum continuum to dig up the relationships between cloud free and cloudy pixels embedded in the relevant image via simulation (Chang et al., 2015a). SMIR thus takes advantage of a floating pattern recognition process with the aid of a machine

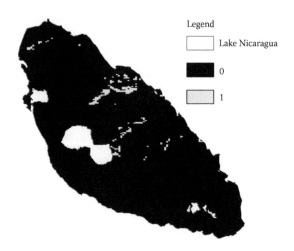

FIGURE 14.6   A sample binary map of Lake Nicaragua level 2 product with multiple selected flagging on March 22, 2015.

learning tool, such as ELM, to characterize and quantify the complex relationships between cloud free and cloudy pixels over highly heterogeneous and dynamic environments where other cloud removal approaches might fail (Chang et al., 2015a). In fact, SMIR may be extended to choose an artificial neural network (ANN), conditional restricted Boltzmann machines, deep believe network or genetic programming for image reconstruction if the neural network-based learning capacity associated with ELM is not strong enough.

As stated in Chapter 12, the current SMIR method mainly consists of four basic steps to recover the missing values of cloudy pixels: (1) creation of a historical time series image via acquisition and

TABLE 14.3

Level-2 Processing Flags Applied to Ocean Color Data

Bit Position (0 Based)	Flag Name	Description
0	ATMFAIL	Atmospheric correction failure
1	LAND	Pixel is over land
3	HIGLINT	High sun glint detected
4	HILT	Very high or saturated radiance
5	HISATZEN	High sensor view zenith angle ($>60°$)
8	STRAYLIGHT	Likely straylight contamination
9	CLDICE	Probable cloud or ice contamination
12	HISOLZEN	High solar zenith angle
14	LOWLW	Low water leaving radiance
15	CHLFAIL	Failure to derive chlor-a product
16	NAVWARN	Reduced navigation quality
19	MAXAERITER	Aerosol iterations exceed maximum allowable
21	CHLWARN	Derived chlor-a product quality is reduced
22	ATMWARN	Atmospheric correction is suspect
25	NAVFAIL	Navigation failure
26	FILTER	User-defined

Note: These are the NASA standard flags that have been used to produce Level-3 data products by NASA OBPG (Barnes and Hu, 2015). Pixels with these flags are removed for quality control purposes.

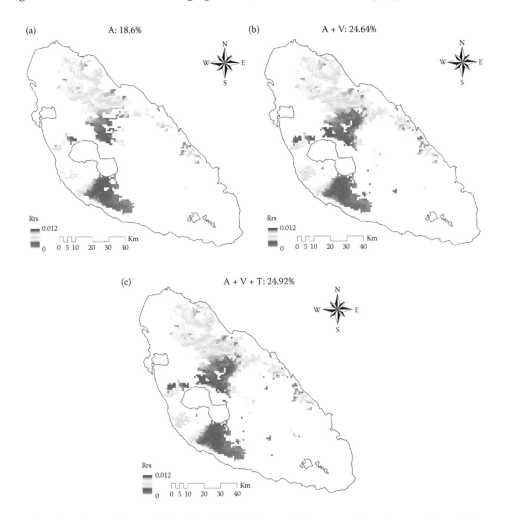

FIGURE 14.7 Comparisons of (a) original MODIS-Aqua (A1) ocean color reflectance image before and after fusing at 678 nm with (b) VIIRS-NPP (V) and (c) MODIS-Terra (T) in a heavy cloud event on June 17, 2014. The percent coverage was given as the coverage ratio (i.e., *CR* as defined in Chapter 11) between the number of pixels having valid data values and the total number of pixels covered in the lake, and a value of 100% means full coverage.

pre-processing, (2) cloud detection, (3) selection of candidate images and target cloudy pixels for model inputs, and (4) prediction model building for cloudy pixel information recovery. Therefore, the image reconstruction based on the merged data file with the largest data coverage (e.g., the merged image of MODIS-Aqua, VIIRS-NPP, and MODIS-Terra; A + V + T) would have some advantages relative to that of MODIS-Aqua and VIIRS-NPP (A + V) for a day. If none of the merged image is available, the historic image of MODIS-Aqua may be directly used once an adequate number of valid pixels is guaranteed. Following the definition of the ensemble mean of daily clear pixels coverage ratio (*CR*) in Chapter 11, the reconstructed image increases the information content of the observed scene dramatically (Figure 14.8).

The data gaps were particularly salient over Lake Nicaragua, which is evidenced by the fact that the *CR* values of the original products associated with these three sensors were all lower than 60% (Table 14.4). Due to the heavy cloud impact, improving the data coverage is essential to advancing our understanding of the aquatic environment. With the aid of data merging algorithms, the spatial coverage of merged images increases gradually, and the spatial coverage improves notably after the

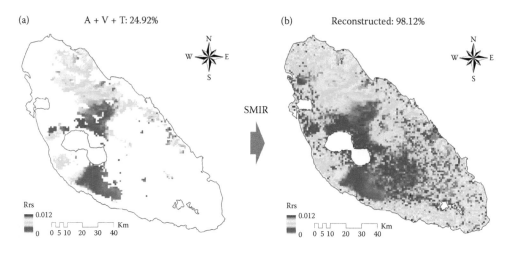

FIGURE 14.8 Comparisons of (a) the merged ocean color reflectance image and (b) the reconstructed one on June 17, 2014.

data reconstruction method based on those merged images, with the largest improvement by 74% (i.e., from 24.92% to 98.12% on 2014168). In principle, the reconstructed image should be spatially complete, having a *CR* value of 100%, if the memory effect is strong, which in turn advances the environmental monitoring capacity and improves understanding of the aquatic environment, allowing us to retrieve some specified environmental parameters or indicators. In this study, however, no value was predicted for a few pixels with too few inputs to avoid the generation of large outliers arising from poorly characterized relationships. This trade-off made the *CR* values of reconstructed images lower than 100% (Table 14.4).

In addition to *CR*, the information entropy of every available image in the data merging and reconstruction processes was calculated (Table 14.5). As information entropy is a statistical measure of randomness characterizing the texture of an image, the larger the value of *H*, the more inhomogeneity of the image expected. The *H* value of the merged image depends highly on the image having a larger *CR* value. In principle, the reconstructed images should have larger *H* values than that of the merged images. In practice, images with limited spatial coverage might have larger *H* values than the reconstructed images (e.g., June 17, 2014). This unexpected effect is mainly caused by the fact that data values in images with limited spatial coverage may show larger variance than those in the reconstructed images, that is, are more heterogeneous.

Overall, SMIR relies mainly on the historical time series to characterize and quantify possible relationships between candidate images and target cloudy pixels via model building and simulation. The final prediction accuracy of simulations is highly related to the screening and selection of adequate training inputs via classification and pattern recognition. Thus, some additional criteria with better learning capacity in machine learning may be proposed based on the correlation and overlap ratio between candidate images and target cloudy pixels to further improve the overall performance of SMIR.

14.2.4 Feature Extraction and Content-Based Mapping

Chapters 6 and 7 introduced a wealth of feature extraction techniques. Although a variety of remote sensing data products are provided by different agencies, they may still be insufficient for certain environment monitoring applications because of the lack of important features specified for addressing unique environmental quality status and ecosystem state. To better understand the changing environment, certain informative parameters or indicators should be derived, and techniques like machine learning or data mining can help retrieve complex features in lakes, estuaries, coastal

TABLE 14.4

Calculated Percent Coverage Ratio for Images at Different Processing Levels

DOY	A	V	T	A + V	A + T	A + V + T	Reconstructed
2011335	37.35	0.00	0.00	/	/	/	98.41
2014161	2.87	0.00	0.18	/	/	/	/
2014163	1.78	0.00	27.01	/	28.47	/	98.30
2014165	0.00	0.00	0.00	/	/	/	/
2014168	18.60	12.62	0.37	24.64	/	24.92	98.12
2014211	0.01	0.00	0.00	/	/	/	/
2015070	0.19	0.00	0.44	/	/	/	/
2015071	0.00	13.18	0.00	/	/	/	/
2015072	0.37	59.05	0.37	59.57	/	/	98.00
2015073	4.56	2.61	0.00	/	/	/	/
2015075	55.08	42.37	0.15	60.08	/	60.13	98.41
2015076	17.83	50.77	0.00	54.91	/	/	98.30
2015099	20.06	40.18	0.00	45.24	/	/	97.81
2016060	43.12	18.10	0.01	46.76	/	46.77	98.58
2016063	0.00	36.99	0.00	/	/	/	98.14
2016124	0.00	0.00	0.00	/	/	/	/
2016155	0.10	0.44	0.00	/	/	/	/

Note: DOY: day-of-year; A: MODIS-Aqua; V: VIIRS-NPP; T: MODIS-Terra; A + V: fused images between MODIS-Aqua and VIIRS-NPP; A + T: fused images between MODIS-Aqua and MODIS-Terra; A + V + T: fused images between MODIS-Aqua, VIIRS-NPP, and MODIS-Terra; Reconstructed: reconstructed images. If the CR of the original file is too low, then no data merging and data reconstruction is conducted and hence the CR is given as blank (i.e., denote as "/" therein) (Chang et al., 2017).

bays, forests, and atmospheric environments. In other words, the cognitive extraction of the existing relations between the ground truth data and the merged and/or fused reflectance data via supervised classification appears to be the research frontier recently. Notwithstanding, CDIRM does not rule out the use of traditional feature extraction techniques such as statistical, bio-optical, or geophysical models and does not intend to overemphasize the use of machine learning or data mining models. Rather, CDIRM would count on a model base and a data base for screening and selection based on a set of statistical indexes (Chang et al., 2015b, 2016).

Ocean color bands have narrower wavelengths than those of land bands, making them more sensitive to the aquatic environment. Given that the eutrophication level of Lake Nicaragua is significant, water quality parameters, such as chlorophyll-a, TN, TP, and Secchi Disk Depth (SDD), should be derived to monitor the water pollution issues. Because no remote sensing products are currently available for TN, TP, and SDD, machine learning approaches can be applied to derive those parameters by establishing relationships between the ground truth data of those parameters and the relevant ocean color reflectance at 10 different bands.

By using ANN tools, relationships between the observed dataset (i.e., the ground truth dataset listed in Appendix) and ocean color reflectance at 10 different bands were established based on the supervised classification introduced in Chapter 5. To avoid randomness, 20 independent models were established for this study. For each training process, 70% of the data pairs were randomly assigned for training and the remaining 30% were used for validation. The training process continued until a correlation coefficient of 0.8 was attained between the observed dataset and the predicted dataset for validation. Once a model is well trained and validated, it can be saved for further prediction analysis. The ensemble means of predicted results from the 20 trials agreed well

TABLE 14.5

Calculated Entropy for Images at Different Processing Levels

DOY	A	V	T	A + V	A + T	A + V + T	Reconstructed
2011335	0.94	/	/	/	/	/	1.07
2014161	0.70	/	2.45	/	/	/	/
2014163	2.14	/	0.77	/	0.56	/	0.94
2014165	/	/	/	/	/	/	/
2014168	0.26	0.45	1.21	0.34	/	0.35	0.70
2014211	/	/	/	/	/	/	/
2015070	2.26	/	0.00	/	/	/	/
2015071	/	0.03	/	/	/	/	/
2015072	0.96	1.91	0.56	1.75	/	/	1.88
2015073	0.68	0.55	/	/	/	/	/
2015075	0.99	0.33	1.72	0.99	/	0.99	1.05
2015076	1.63	0.89	/	1.38	/	/	1.65
2015099	0.34	2.01	/	1.65	/	/	1.94
2016060	1.06	1.49	0.00	1.20	/	1.20	1.15
2016063	/	0.96	/	/	/	/	0.99
2016124	/	/	/	/	/	/	/
2016155	0.72	2.00	/	/	/	/	/

Source: The entropy value shown here is calculated based on the first ocean color band of MODIS with a wavelength of 412 nm (Chang et al., 2017).

with the sampled chlorophyll-a concentration, with an R^2 of 0.93 and an *RMSE* of 2.43 µg/L (Figure 14.9a). The predicted values not only capture the spatial pattern of chlorophyll-a concentrations at those sampled points, but also the extreme values, for example the local minimum value at the 12th grid cell and the local maximum value at the 30th grid cell. The results strongly demonstrate the effectiveness and reliability of the established models. Similarly, 20 independent models were established for SDD, TN, and TP. The established model can characterize the relationships between the sampled parameters and ocean color reflectance at 10 different bands, with R^2 ranging from 0.7 to 0.88 (Figure 14.9b–d).

The concentration maps of the four parameters derived on June 17, 2014 (DOY: 2014168) and on March 16, 2015 (DOY: 2015075) are shown in Figures 14.10 through 14.13, respectively. Findings indicate that the chlorophyll-a concentrations appeared better in the wet season spatially; SDD conditions were also relatively better in the wet season than those in the dry season. It is significant that higher SDD were always present throughout the central lake area where the lower chlorophyll-a concentrations are present. This evidence demonstrates the prediction accuracy of CDIRM. By contrast, the TN and TP concentrations are relatively higher in the dry season than those in the wet season, partially due to an amount of sunlight too insufficient to trigger more growth of algal bloom cells. Such a phenomenon strongly echoes the seasonality effect, that is, the concentrations of chlorophyll-a are higher in the wet season due to abundant sunlight in the summer, whereas the TN and TP concentrations are higher in the dry season due to the cumulative effect of farming and municipal wastewater discharge without sufficient consumption driven by the algal blooms.

14.3 SUMMARY

This chapter provides a holistic discussion of data merging/fusion, image reconstruction, and machine learning/data mining with a case study for water quality monitoring in Lake Nicaragua.

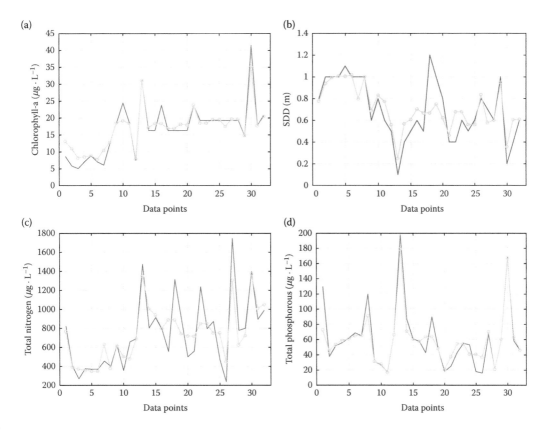

FIGURE 14.9 Comparisons of observed (red lines) and simulated (green lines) water quality parameters in Lake Nicaragua. (a) Chlorophyll-a concentration, (b) SDD, (c) TN, and (d) TP. (From Chang, N. B., Bai, K., and Chen C. F., 2017. *Journal of Environmental Management*, 201, 227–240.)

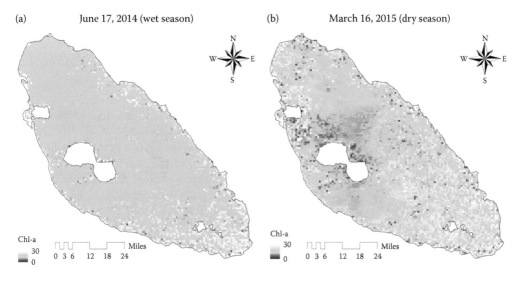

FIGURE 14.10 Predicted chlorophyll-a (Chl-a) concentrations in Lake Nicaragua on (a) June 17, 2014 (DOY: 2014168) and (b) March 16, 2015 (DOY: 2015075) using the ANN approach.

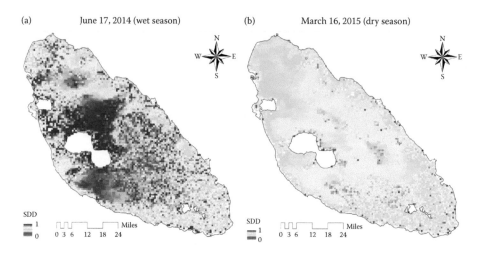

FIGURE 14.11   Predicted SDD concentrations in Lake Nicaragua on (a) June 17, 2014 (DOY: 2014168) and (b) March 16, 2015 (DOY: 2015075) using the ANN approach.

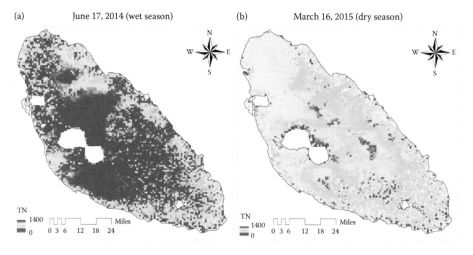

FIGURE 14.12   Predicted TN concentrations in Lake Nicaragua on (a) June 17, 2014 (DOY: 2014168) and (b) March 16, 2015 (DOY: 2015075) using data mining approaches.

The CDIRM incorporates a multiscale modeling system over a scalable time-space-spectrum continuum to advance environmental monitoring and better understand the Earth's dynamic environment. It also demonstrates the ability of the CDIRM software platform to flexibly weave a wealth of options for system analysis based on integrated signal processing, statistical analysis, and machine learning/data mining technologies. In the case study, a final generation of a series of concentration maps of chlorophyll-a, TN, TP, and SDD in Lake Nicaragua aid in lake watershed management. This well-structured software package with a suite of well-connected algorithms, models, and theories lays down the application foundations and enables end-users to pre-process interdisciplinary knowledge, manipulate remote sensing images, explore possible streaming capacity, and mine the spatiotemporal patterns via image merging, fusion, reconstruction, and data mining/machine learning. The advancement is significant in satellite remote sensing science in the nexus of environmental modeling and software development, elevating the exploitation level of the current remote sensing products with the most recent machine learning algorithms.

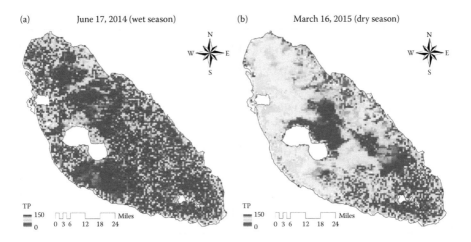

FIGURE 14.13 Predicted TP concentrations in Lake Nicaragua on (a) June 17, 2014 (DOY: 2014168) and (b) March 16, 2015 (DOY: 2015075) using data mining approaches.

APPENDIX: FIELD DATA COLLECTION FOR GROUND TRUTHING

To carry out the supervised classification, a series of field sampling campaigns were carried out collectively by National Central University, National Automous University of Nicaragua, Centro para la Investigación en Recursos Acuáticos de Nicaragua, and the Wisconsin Department of

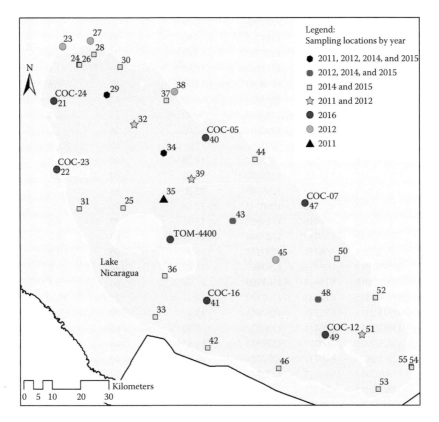

FIGURE A.1 Sampling locations by the year for Lake Nicaragua. (Chang, N. B., Bai, K., and Chen, C. F., 2017. *Journal of Environmental Management*, 201, 227–240.)

TABLE A.1

Water Quality Parameters Sampled in Lake Nicaragua

Station No.	Date	UTM-X	UTM-Y	Chl-a ($\mu g \cdot L^{-1}$)	TP ($\mu g \cdot L^{-1}$)	SDD (m)	TN ($\mu g \cdot L^{-1}$)
21	17/03/2015	619718	1319936	19.24	43	0.4	1235
21	17/03/2015	619718	1319936	19.24	55	0.6	794
22	16/03/2015	620781	1296609	31.08	198	0.1	1471
22	14/06/2014	620781	1296609	39.96	173	0.3	1162
23	17/03/2015	622433	1338152	19.24	650	0.3	2060
24	01/12/2011	628274	1332239	8.59	130	0.8	821
24	01/03/2012	628274	1332239	11.54	68	0.4	460
24	12/03/2015	628274	1332239	10.36	115	1.0	1326
25	16/03/2015	628334	1283269	16.28	87	0.4	801
26	17/03/2015	628337	1331989	19.24	53	0.5	870
26	17/03/2015	628337	1331989	19.24	18	0.6	472
27	16/03/2015	632204	1340248	23.68	168	0.4	1854
28	11/03/2015	633383	1335621	14.80	58	0.4	769
28	17/06/2014	633383	1335621	18.50	31	0.6	618
29	14/03/2015	637608	1321833	17.76	65	0.4	898
29	01/12/2011	637608	1321833	5.80	38	1.0	402
29	01/03/2012	637608	1321833	11.96	70	0.5	613
29	01/03/2012	637608	1321833	9.62	81	1.0	1397
30	16/03/2015	642276	1331405	16.28	61	0.5	911
31	16/03/2015	643442	1283708	23.68	58	0.6	788
31	17/03/2015	643442	1283708	19.24	16	0.8	235
32	01/12/2011	647076	1311788	5.03	52	1.0	269
32	01/03/2012	647076	1311788	13.44	76	0.4	520
32	11/03/2015	647076	1311788	14.80	102	1.1	1126
33	14/03/2015	654493	1246764	17.76	53	0.6	834
34	16/03/2015	657023	1302101	16.28	43	0.5	555
34	17/06/2014	657023	1302101	24.42	27	0.8	356
34	01/12/2011	657023	1302101	7.05	56	1.0	375
34	01/03/2012	657023	1302101	12.91	65	0.5	563
34	16/03/2015	657023	1302101	16.28	90	1.2	1313
35	01/12/2011	657105	1286481	8.82	62	1.1	370
36	11/03/2015	657662	1260747	22.20	65	0.6	760
36	12/06/2014	657662	1260747	12.58	120	1.0	400
37	14/03/2015	658011	1319992	13.32	32	0.6	758
38	17/03/2015	660730	1323040	19.24	67	0.7	1744
39	01/12/2011	666716	1293407	6.99	69	1.0	370
39	16/03/2015	666716	1293407	16.28	48	1.0	907
40	10/03/2015	671843	1307700	11.84	44	0.5	727
40	17/03/2015	671843	1307700	19.24	21	0.6	777
41	14/03/2015	672275	1252484	13.32	39	0.8	631
42	13/03/2015	672414	1236372	7.40	66	0.5	689
43	10/03/2015	680717	1279374	11.84	46	0.6	555
43	12/03/2015	680717	1279374	10.36	42	0.6	608
43	11/06/2014	680717	1279374	17.02	74	1.0	1596
44	14/03/2015	688451	1300279	17.76	51	0.6	620
44	17/06/2014	688451	1300279	18.50	17	0.6	657
45	01/03/2012	695801	1266210	12.08	55	0.8	777

(Continued)

TABLE A.1 (*Continued*)

Water Quality Parameters Sampled in Lake Nicaragua

Station No.	Date	UTM-X	UTM-Y	Chl-a ($\mu g \cdot L^{-1}$)	TP ($\mu g \cdot L^{-1}$)	SDD (m)	TN ($\mu g \cdot L^{-1}$)
45	14/03/2015	695801	1266210	17.76	72	1.2	830
46	11/03/2015	696842	1229468	14.80	86	0.4	1152
46	10/03/2015	696842	1229468	11.84	31	0.4	601
47	11/03/2015	706066	1285770	14.80	53	0.4	694
47	16/06/2014	706066	1285770	37.00	53	0.4	1308
48	12/03/2015	710672	1252821	8.88	48	0.6	683
48	16/03/2015	710672	1252821	16.28	18	0.8	500
48	01/03/2012	710672	1252821	10.54	59	0.8	872
48	09/04/2015	710672	1252821	14.65	60	1.0	798
49	12/03/2015	713001	1241001	10.36	41	0.6	633
50	11/03/2015	716912	1266951	14.80	50	0.6	739
50	10/03/2015	716912	1266951	20.72	29	0.6	377
51	01/12/2011	725610	1241195	6.04	65	1.0	455
51	01/03/2012	725610	1241195	10.26	70	0.6	1137
51	30/07/2014	725610	1241195	15.39	55	1.2	1158
52	11/03/2015	730124	1253706	22.20	69	0.5	1051
52	16/03/2015	730124	1253706	23.68	25	0.4	559
53	10/03/2015	731306	1222787	11.84	69	0.5	845
53	11/03/2015	731306	1222787	14.80	40	0.4	642
54	10/03/2015	742762	1230025	20.72	92	0.4	856
54	10/06/2014	742762	1230025	6.51	20	0.4	331
55	01/12/2011	742772	1230198	8.41	56	1.1	439
55	29/07/2014	742772	1230198	17.09	55	0.8	1416
COC-05	03/03/2016	671843	1307700	20.72	46	0.6	988
COC-07	03/04/2016	706066	1285770	23.68	59	0.6	807
COC-12	03/05/2016	713001	1241001	17.76	59	0.6	877
COC-16	03/06/2016	672275	1252484	20.72	51	0.6	777
COC-23	02/29/2016	620781	1296609	41.44	168	0.2	1397
COC-24	02/29/2016	619718	1319936	17.76	58	0.4	894
TOM-4400	03/08/2016	659725	1273072	20.72	113	0.6	748

Natural Resources to monitor water quality at various points in Lake Nicaragua from 2011 to 2016 to come up with a database (Figure A.1 and Table A.1). The selected water quality parameters should be representative and must comply with the surrounding anthropogenic activities and main polluting sources near the lake.

REFERENCES

Bai, K., Chang, N. B., and Chen, C. F., 2015. Spectral information adaptation and synthesis scheme for merging cross-mission consistent ocean color reflectance observations from MODIS and VIIRS. *IEEE Transactions on Geoscience and Remote Sensing*, 54, 311–329.

Bai, K., Chang, N. B., Yu, H., and Gao, W., 2016. Statistical bias corrections for creating coherent total ozone records with OMI and OMPS observations. *Remote Sensing of Environment*, 182, 150–168.

Barnes, B. B. and Hu, C., 2015. Cross-sensor continuity of satellite-derived water clarity in the Gulf of Mexico: Insights into temporal aliasing and implications for long-term water clarity assessment. *IEEE Transactions on Geoscience and Remote Sensing*, 53, 1761–1772.

Businger, P. A. and Golub, G. H., 1969. Algorithm 358: Singular value decomposition of a complex matrix. *Communications of the ACM*, 12, 564–565.

Chang, N. B., Bai, K., and Chen, C. F., 2015a. Smart information reconstruction via time-space-spectrum continuum for cloud removal in satellite images. *IEEE Journal of Selected Topics in Applied Earth Observations and Remote Sensing*, 8, 1898–1912.

Chang, N. B., Bai, K., and Chen, C. F., 2017. Integrating multisensor satellite data merging and image reconstruction in support of machine learning for better water quality management. *Journal of Environmental Management*, 201, 227–240.

Chang, N. B., Bai, K. X., Imen, S., Chen, C. F., and Gao, W., 2016. Multi-sensor satellite image fusion, networking, and cloud removal for all-weather environmental monitoring. *IEEE Systems Journal*, 1–17.

Chang, N. B., Imen, S., and Vannah, B., 2015b. Remote sensing for monitoring surface water quality status and ecosystem state in relation to the nutrient cycle: a 40-year perspective. *Critical Reviews of Environmental Science and Technology*, 45, 101–166.

Chang, N. B., Vannah, B., and Yang, Y., 2014a. Comparative sensor fusion between hyperspectral and multispectral satellite sensors for monitoring microcystin distribution in Lake Erie. *IEEE Journal of Selected Topics in Applied Earth Observations and Remote Sensing*, 7, 2426–2442.

Chang, N. B., Vannah, B. W., Yang, Y. J., and Elovitz, M., 2014b. Integrated data fusion and mining techniques for monitoring total organic carbon concentrations in a lake. *International Journal of Remote Sensing*, 35, 1064–1093.

Drazin, P. G., 1992. *Nonlinear Systems*. Cambridge University Press, Cambridge, USA.

Huang, G.-B., Zhu, Q.-Y., and Siew, C.-K., 2006. Extreme learning machine: Theory and applications. *Neurocomputing*, 70, 489–501.

NASA, 2017. *Quality of Level 2 Ocean Color products*. https://oceancolor.gsfc.nasa.gov/atbd/ocl2flags/, accessed by January 2017.

Nash, J. C., 1990. The singular-value decomposition and its use to solve least-squares problems. In: *Compact Numerical Methods for Computers: Linear Algebra and Function Minimisation*, Adam Hilger (Ed.), Bristol, England, CRC Press, Boca Raton, Florida, USA, 30–48.

Wilk, M. B., and Gnanadesikan, R., 1968. Probability plotting methods for the analysis for the analysis of data. *Biometrika*, 55, 1–17.

Part V

Remote Sensing for Environmental
Decision Analysis

15 Data Merging for Creating Long-Term Coherent Multisensor Total Ozone Record

15.1 INTRODUCTION

In Earth system science, ozone in the atmosphere can be routinely divided into two typical groups simply by considering their locations residing in the atmosphere. Specifically, these groups are: stratospheric ozone residing in the upper atmosphere (i.e., stratosphere) and tropospheric ozone residing in the lower atmosphere (troposphere) (Figure 15.1). As the second major layer of the Earth's atmosphere, the stratosphere resides above the troposphere while being below the mesosphere, with a varying vertical height at different latitudes. Investigations indicate that the tropospheric ozone poses adverse impacts on living beings on Earth by affecting human health, natural vegetation growth, crop yield, and so on, whereas the stratospheric ozone enables to protect life on Earth by absorbing the harmful ultraviolet (UV) radiation from the Sun. When stratospheric ozone absorbs high frequency radiation, such as the UV light, it often involves a photochemical reaction process, which affects and modifies the background chemistry and energy budget of the stratosphere, in turn resulting in variations of atmospheric dynamics (McPeters et al., 2013).

Since the late 1970s, dramatic ozone depletion over Antarctica and its resultant impacts on regional climate have been observed. The ozone in the stratosphere that is deemed "good ozone" to the biosphere on Earth's surface has received extensive scientific attention worldwide. As driven by the unconstrained emissions of anthropogenic chlorofluorocarbons (CFCs) into the atmosphere, in the past few decades, the depletion of the ozone layer has resulted in an extended ozone hole over Antarctica that was observed in each austral spring. For this reason, the Montreal Protocol and its amendments have been in place since 1989 to limit the release of anthropogenic Ozone Depleting Substances (ODS) into the atmosphere. Further investigations have revealed that the dramatic ozone depletion over Antarctica has resulted in significant climatic changes in the Southern Hemisphere (SH), and such effects are observed extending over much of the SH, even to the tropics (e.g., Reason, 2005; Son et al., 2009; Feldstein, 2011; Thompson et al., 2011; Fyfe et al., 2012; Gillett et al., 2013; Gonzalez et al., 2013; Manatsa et al., 2013; Bai et al., 2016a). Consequently, monitoring the abundance of ozone in the atmosphere is of great importance to the Earth's environment.

Essentially, a long-term coherent Total Column Ozone (TCO) record is critical for linking regional climate changes and Earth system processes to the observed ozone variability. Since the late 1970s, satellite-based TCO observations have been available as collected by different satellite instruments and platforms, and some popular instruments and platforms are presented in Figure 15.2. TCO observations from these sensors over different periods have been widely used for ozone variability assessment at different time scales. Nevertheless, due to the inherent differences in the system design of platforms, orbits, and sensors, as well as calibration processes and retrieval algorithms (Table 15.1), apparent discrepancies are observed among TCO observations from different missions.

One way to harmonize the discrepancies across those sensors and platforms is to reprocess the low-level radiometric record from each instrument by performing inter-instrumental calibration and then retrieve TCO with the same retrieval algorithm (McPeters et al., 2013). Such an approach

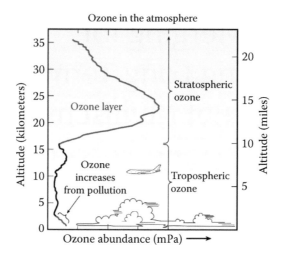

FIGURE 15.1   The vertical structure of ozone in the atmosphere. (From World Meteorological Organization, 2007. Scientific Assessment of Ozone Depletion: 2006. *Global Ozone Research and Monitoring Project Report No. 50*, Geneva, Switzerland.)

can guarantee a long-term coherent TCO record because nonlinearly propagated instrumental errors can be significantly removed, thus preventing the spatiotemporal dependent errors from the relevant ozone retrievals (McPeters et al., 2013). However, such a collective calibration effort is by no means an easy task since it requires collaborative efforts from the original instrument teams of each sensor. In addition, data assimilation provides another capable way to create a coherent TCO record by assimilating satellite- and ground-based TCO observations. A representative example is the multisensor reanalysis ozone described in Van Der A et al. (2010, 2015).

Aside from the aforementioned two complicated methods, bias correction through statistical modeling can be used as an alternative to create a coherent TCO record, and such approaches are generally termed as statistical bias correction methods. In practice, there exist many methods with such capabilities, such as the simple approaches like linear scaling and delta-change (e.g., Vila et al., 2009; Teutschbein and Seibert, 2012), more advanced methods like Kalman filtering (e.g., Delle Monache et al., 2006; Sicardi et al., 2012), distribution mapping (e.g., Li et al., 2010;

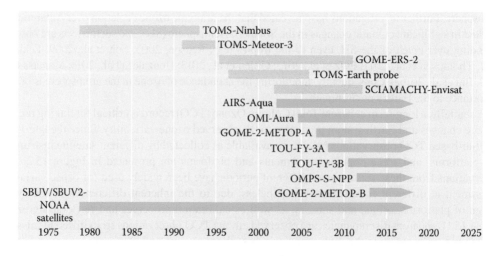

FIGURE 15.2   The timeline of popular satellite-based total ozone mapping missions. Shapes with arrows indicate the ongoing missions.

TABLE 15.1

Several Current and Retired TCO Sensors Onboard Satellites

Instrument	Platforms	Operation Time	Temporal Resolution	Spatial Resolution	Retrieval Algorithms	Operating Agency
TOMS	Nimbus-7	1978.11–1993.6	Daily	50×50 km^2	TOMS	NASA
	Meteor-3	1991.8–1994.12				
	Earth Probe	1996.7–2004.12				
GOME	ERS-2	1995.5–2011.9	3 days	320×40 km^2	GDP	ESA
SCIAMACHY	ENVISAT	2002.3–2012.4	6 days	30×60 km^2	TOSOMI	ESA
GOME-2	METOP-A	2006.10–	1.5 days	80×40 km^2	GDP	ESA
	METOP-B	2012.9–				
OMI	Aura	2004.10–	Daily	24×13 km^2	TOMS DOAS	NASA
TOU	FY-3A	2008.5–	Daily	50×50 km^2	Similar to TOMS	CMA
	FY-3B	2010.11–				
	FY-3C	2013.9–				
OMPS	Suomi-NPP	2011.10–	Daily	50×50 km^2	SBUV	NOAA
SBUV/SBUV-2	NOAA series	1978–	Daily	170×170 km^2	SBUV	NOAA

GDP: GOME Data Processor operational retrieval algorithm; TOSOMI: Total Ozone retrieval scheme for SCIAMACHY based on the OMI DOAS algorithm; CMA: China Meteorological Administration; DOAS: Differential Optical Absorption Spectroscopy.

Piani et al., 2010a,b; Haerter et al., 2011; Hagemann et al., 2011; Ahmed et al., 2013; Argüeso et al., 2013; Bai et al., 2015a), and nonlinear regression (e.g., Loyola and Coldewey-Egbers, 2012; Bordoy and Burlando, 2013). Such methods have been successfully employed to correct the climate model biases in order to better predict future scenarios; however, the inherent limitations associated with each method are still observed. For example, simple methods like delta-change can only remove a mean bias level between observations and modeling results. In other words, such methods are incapable of handling nonlinear biases. On the other hand, nonlinear methods, such as the Artificial Neural Network (ANN) models, are prone to be constrained by the training inputs, also having the chance of overfitting issues. More specifically, the outputs from ANN models could be biased with a limited number of training samples. Kalman filtering, one of the effective methods that are commonly used to optimize modeling predictions by addressing modeling biases, is incapable of removing biases from a discrete time series due to its cascade biases modeling scheme (i.e., bias at the current state is estimated from previous observations) (Bai et al., 2016b). Nevertheless, bias correction methods with more advanced modeling schemes still outperform those simple methods due to their higher-level bias estimation schemes (Lafon et al., 2013; Teutschbein and Seibert, 2013).

Commonly, cross-mission sensors, especially for satellite missions, are designed to allow for a certain overlapped time period. In other words, there are common observations between them, providing the opportunity to quantify discrepancies between sensors through inter-instruments comparisons (Barnes and Hu, 2015). Typically, modeling with matchups between cross-mission sensors' observations through either linear or nonlinear regressions, is the most straightforward way (Bai et al., 2016b). For instance, with the aid of ANN models, spatiotemporal drifts of TCO between three European satellite sensors were estimated and then applied sequentially to calibrate cross-mission TCO observations, in turn yielding a long-term homogeneous TCO record at the global scale (Loyola and Coldewey-Egbers, 2012). Moreover, cross-mission sensors' bias can also be addressed by calibrating satellite observations with colocated ground-based measurements (Bai et al., 2016b).

For instance, the *in situ* measurements were used as a baseline to remove the inconsistency of ocean color data derived from the Sea-viewing Wide Field-of-view Sensor (SeaWiFS) and the Moderate Resolution Imaging Spectroradiometer (MODIS), as both sensors' observations were mapped onto such a baseline for possible bias correction (Gregg and Casey, 2010). Such a process is also called vicarious calibration, and its applications can be widely found in the literature (e.g., D'Alimonte et al., 2008; Hoyer et al., 2013). According to Bai et al. (2016b), however, such bias correction schemes are also associated with salient limitations. Specifically, cross-mission biases cannot be well addressed simply with *in-situ* observations alone if they are spatially heterogeneous due to the lack of abundant *in-situ* measurements. Moreover, it is difficult to address latitudinal and seasonal dependent biases.

This chapter aims to demonstrate a unique data merging scheme with the aid of a statistical bias correction algorithm, namely the modified quantile-quantile adjustment, to remove cross mission TCO biases between Ozone Mapping and Profiler Suite (OMPS) on board the Suomi National Polar-orbiting Partnership (Suomi-NPP) satellite and Ozone Monitoring Instrument (OMI) on board the Aura satellite. The reason to choose OMPS and OMI TCO data is mainly due to the fact that their data are generally compatible with each other in the common time-space domain. Despite this, salient cross-mission biases are still observed.

15.2 DATA COLLECTION AND ANALYSIS

Both Level-3 TCO data sets from OMPS and OMI were used toward the creation of a long-term coherent TCO record. Descriptions of both data sets were presented in Table 15.2. In addition, ground-based TCO measurements were also used as ground truth to evaluate the improvement of consistency before and after bias correction with respect to the merged TCO record. The ground-based TCO data were obtained from the World Ozone and Ultraviolet radiation Data Centre (WOUDC). Figure 15.3 shows the distributions of ground stations providing TCO measurements from Brewer and Dobson instruments worldwide.

15.2.1 OMPS TCO Data

The OMPS sensor was deployed on board the Suomi-NPP satellite that was successfully launched into the orbit on October 28, 2011. As a space-borne ozone mapping instrument, OMPS was designed to monitor ozone by measuring backscattered UV and visible (VIS) radiations from the Earth's surface. In general, OMPS is an instrumental suite with three hyperspectral instruments, dedicated to measuring total column ozone and ozone profile (vertical distribution of ozone) on a daily basis. The goal is to extend the historical total ozone records measured by previous backscatter UV sensors on board a variety of satellites since 1978. TCO observations used in this study were derived from the OMPS Total Column Nadir Mapper's measurements with the latest SBUV V8.6 algorithm. As opposed to previous algorithms where TCO were derived based on a separate set

TABLE 15.2

Features of TCO Data Sets from OMPS and OMI

Instrument/ Platform	Data Set	Spatial Resolution	Temporal Resolution	Retrieval Algorithm	Time Period	Data Provider
OMPS/Suomi-NPP	EDR_TO3_L3	1°	Daily	SBUV V8.6	2012.1–2015.3	OMPS science team[a]
OMI/Aura	OMTO3d	1°	Daily	TOMS V8.5	2004.10–2015.3	OMI science team[b]

[a] NASA's Goddard Space Flight Center/OMPS science team. https://ozoneaq.gsfc.nasa.gov.

[b] NASA's Goddard Space Flight Center/OMI science team. http://disc.gsfc.nasa.gov/datacollection/OMTO3d_V003.html.

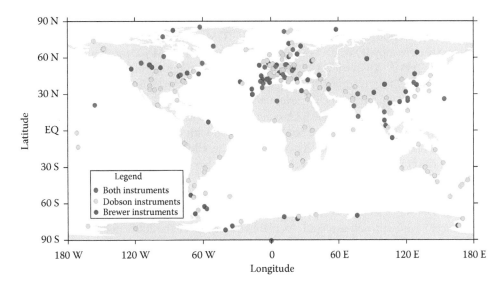

FIGURE 15.3   The global distribution of the ground-based stations for TCO measurements in WOUDC.

of wavelengths, TCO from the SBUV-type instruments were derived by simply integrating the retrieved profile (Bai et al., 2016b). In fact, the latest SBUV V8.6 algorithm is an analog to the V8 algorithm, which is operationally used by NOAA for producing their total ozone products. In the V8.6 algorithm, the ozone absorption cross-sections developed by Malicet et al. (1995) were applied instead of those used in V8 developed by Bass and Paur (1985). In addition, a monthly latitudinal temperature climatology derived from NOAA temperature datasets was applied in the V8.6 algorithm, aiming to minimize temperature dependent ozone cross-section related biases. Generally, major improvements in the V8.6 algorithm include the updating of instrument calibration process, the application of new ozone absorption cross-sections in the forward model, and the implementation of new ozone and cloud height climatology (Bai et al., 2016b). More details related to improvements in the SBUV 8.6 algorithm can be found in Bhartia et al. (2013).

Based on the early 14-month V8.6 TCO products from OMPS, the quality of OMPS TCO products was well assessed by comparing with TCO measurements from 39 Dobson and 34 Brewer spectrophotometers (Bai et al, 2015b). The results indicated a mean positive bias around 1% relative to the colocated ground-based TCO measurements. In addition, no apparent bias was observed varying with time, latitude, Solar Zenith Angle (SZA), viewing geometry, and cloud fraction. In other words, OMPS TCO data are of good quality.

15.2.2   OMI TCO Data

The mission of OMI is to continue the TCO record established by the TOMS series of instruments since 1978, providing daily global TCO measurements. As a nadir viewing and wide swath UV-visible hyperspectral spectrometer, OMI was deployed on board the NASA EOS-Aura satellite that was successfully launched into its orbit on July 15, 2004 (Levelt et al., 2006). Essentially, OMI measures backscattered and reflected solar radiation from both the atmosphere and ground surface at UV (270–370 nm) and VIS (350–500 nm) wavelength ranges, with a spectral resolution of 450 nm in the UV and 630 nm in the VIS, respectively. Specifically, the UV channel can be further divided into UV-1 and UV-2 sub-channels at 310 nm, with a ground swath of 2,600 km, and each swath has 60 and 30 cross-track pixels for UV-2/VIS and UV-1 spectra, respectively. At nadir, each ground pixel has a size of 13 km in the flight direction and 24 km (UV-2/VIS) and 48 km (UV-1) in the across-track direction.

Regarding the OMI total ozone data sets, there are two distinct TCO products with respect to two different retrieval algorithms, that is, the OMI-TOMS algorithm and the Differential Optical Absorption Spectroscopy (OMI-DOAS) algorithm. The OMI-TOMS algorithm is associated with the traditional TOMS V8 retrieval algorithm, which uses sun-normalized radiances at paired wavelengths for TCO derivation (Bhartia and Wellemeyer, 2002). In most cases, 317.5 nm and 331.2 nm are used in pair for ozone retrieval under most conditions, whereas 331.2 nm and 360 nm are used under high ozone and high SZA conditions. The rationale is intimately tied to different levels of absorption of ozone at these two wavelengths, with one significantly absorbed by ozone whereas the other is less absorbed. As opposed to the OMI-TOMS algorithm, the OMI-DOAS retrieval algorithm was specifically developed for OMI. For more details, please refer to Veefkind et al. (2006).

By comparing with ground-based TCO measurements collected from globally distributed Brewer and Dobson spectrophotometers, performances of both OMI-TOMS and OMI-DOAS TCO products have been well assessed. As claimed in Balis et al. (2007), a bias level of ~1% was observed for OMI-TOMS and about 2% for OMI-DOAS TCO data, respectively. In summary, the TCO product derived from the OMI-TOMS algorithm has higher overall accuracy. In addition, there is no significant latitude or SZA dependent bias observed. In contrast, the OMI-DOAS TCO product overestimated the colocated ground-based TCO measurements at the high latitudes in the SH. Moreover, a significant SZA dependent bias was observed in the OMI-DOAS TCO product when comparing with ground-based TCO measurements. Additionally, a seasonal dependent bias (an average of ~1.65% underestimation) was observed when comparing both TCO products with ground-based Brewer measurements, which was further confirmed by Bak et al. (2015). Similar results were also reported in Antón et al. (2009), with an approximation of 2% underestimation by OMI-TOMS and 1.4% underestimation by OMI-DOAS.

In this study, the daily Level-3 gridded OMI-TOMS TCO data set, that is, OMTO3d, generated by the NASA OMI science team for the period 2004 to 2015, was employed. This product was created by gridding quality-assured Level-2 TCO swath data on a $1° \times 1°$ grid as a weighted average under a variety of quality control criteria. They include, but are not limited to, removing Level-2 observations flagged by the solar eclipse possibility, row anomaly, and others.

15.2.3 WOUDC TCO DATA

Currently, the WOUDC has been considered as the most critical data archive providing ground-based ozone measurements, collected mainly from the well-established World Meteorological Organization/Global Atmosphere Watch (WMO/GAW) network (Balis et al., 2007). Brewer and Dobson spectrophotometers are two dominant instruments in providing TCO measurements that are widely used to evaluate satellite-based TCO observations. Essentially, TCO is derived from the observed spectral irradiance of solar radiation at specific wavelengths by both instruments, under the principle of differential absorption in the Huggins band in the UV spectrum in which ozone absorbs significantly (Dobson, 1968; Brewer, 1973). Brewer and Dobson instruments have been functional with high accuracies. According to Basher (1982), TCO derived from a well maintained Dobson instrument has an accuracy level of 1% with direct sun observations, and 2%–3% under zenith sky or SZA lower than 75°. A comparable error level (1%) was observed for a well-calibrated Brewer instrument with direct sun observations (Antón et al., 2009). Nevertheless, a minute TCO difference (i.e., within $\pm 0.6\%$) was still observed between two instruments, which can be largely ascribed to wavelengths difference and varying temperature dependences embedded in the ozone absorption coefficients (Van Roozendael et al., 1998). In this study, daily TCO products collected by 38 Dobson spectrophotometers during 2004–2015 were used as the ground truth (Figure 15.4). Meanwhile, only those TCO measurements recorded under direct-sun and blue-sky observing mode were used in order to avoid larger biased TCO measurements.

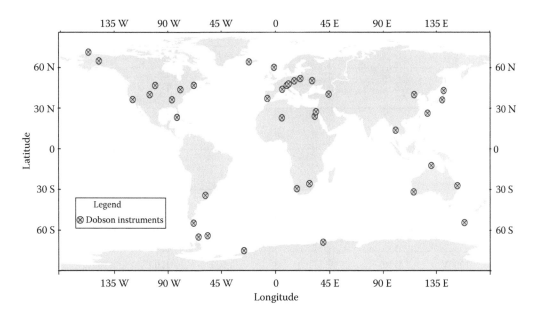

FIGURE 15.4   Spatial distribution of 38 ground-based Dobson instruments collecting TCO records.

15.2.4   COMPARATIVE ANALYSIS OF TCO DATA

As indicated by Figures 15.5 and 15.6, apparent cross-mission biases were observed between overlapped observations of OMPS and OMI TCO during 2012–2015 at the global scale. The Mean Absolute Relative Difference (MARD) (Figure 15.5a) and the associated standard deviation (Figure 15.5b) indicate significant latitudinal dependent biases. The relative bias was defined as TCO difference between OMPS and OMI, denoted as OMPS minus OMI mathematically (Bai et al., 2016b). In addition, apparent seasonal dependent biases were observed from the daily zonal mean difference (Figure 15.6). It is clear that large standard deviations were observed at the mid-to-high latitudes in the Northern Hemisphere (NH), in particular over the Arctic. Temporally, larger standard deviations were observed mainly in the NH during the boreal spring (i.e., March–April–May), at which large TCO values were observed (Bai et al., 2016b). At the same time, significant biases were observed near the polar nights in the SH. Such an effect, to a larger extent, could be related to the low signal-to-noise ratio during those time periods because of poor viewing conditions (i.e., large SZA), which in turn may result in large uncertainties in the TCO retrievals (Bai et al., 2015b). This effect can be also reflected by large standard deviations (large uncertainty) of biases over those regions (Figure 15.5b). It is noticeable that the TCO biases between OMPS and OMI are spatiotemporal heterogeneous, varying nonlinear and non-stationary in nature, which is a big challenge in bias correction.

The TCO observations collected by the TSUKUBA station (36.1°N/140.1°E) must be taken into account (Figure 15.7), for instance, OMI slightly underestimated the ground-based TCO measurements; in contrast, OMPS overestimated TCO values during the overlapped time period. In this context, larger differences are observed between OMPS and OMI, with a mean positive bias larger than 2% observed (Figure 15.8). Meanwhile, apparent seasonal dependence is observed as well in both the monthly and daily time-scales.

15.3   STATISTICAL BIAS CORRECTION SCHEME

In Figure 15.9, TCO data from OMPS and OMI were processed at first for bias removal through a statistical bias correction scheme to generate the OMPS-like coherent TCO time series record

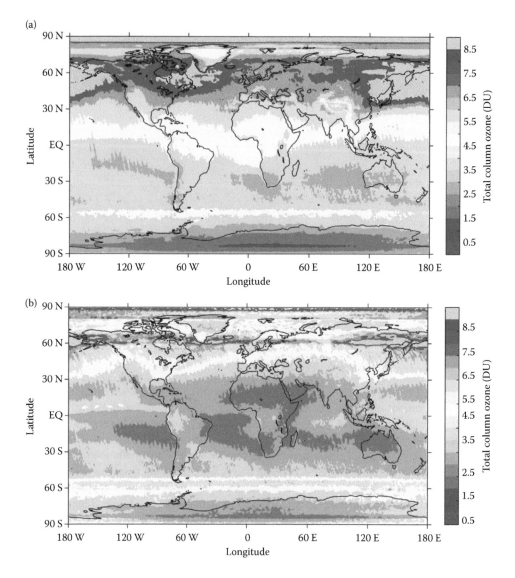

FIGURE 15.5 (a) Mean absolute relative bias and (b) associated standard deviation of the relative bias between OMPS and OMI TCO observations during 2012–2015. The *MARD* was calculated as TCO from OMPS minus those from OMI (abbreviated as OMPS-OMI in this context). (From Bai, K. et al. 2016b. *Remote Sensing of Environment*, 182, 150–168.)

based on the modified Q-Q adjustment method. The traditional Q-Q adjustment method was originally developed to remove the modeling biases from the Regional Climate Model (RCM) outputs by comparing with the observed observations (Amengual et al., 2011). Then data merging was conducted after removing a myriad of biases systematically, denoted as phase I bias correction in this context. Finally, the merged TCO time series data were vicariously calibrated by the ground-based measurements collected at 38 stations with ground-based Dobson instruments worldwide, denoted as phase II bias correction in this context (Figure 15.9).

By applying the modified statistical bias correction method to each OMI TCO record, a global long-term, consistent TCO data set can be attained by merging the bias corrected OMI observations with those of OMPS observations. Essentially, cross-mission biases can be removed from each OMI record individually by referring to the common observations of OMI and OMPS during the overlapped

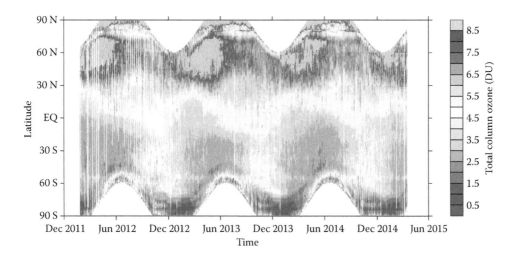

FIGURE 15.6   Daily zonal mean absolute TCO bias between OMPS and OMI during 2012–2015. (From Bai, K. et al. 2016b. *Remote Sensing of Environment*, 182, 150–168.)

time period (i.e., 2012–2015). In this study, the OMPS TCO data were used as the baseline while the OMI TCO data were used as complementary, although OMI is still active in the orbit to provide TCO observations simultaneously. Such a selection can be ascribed to the fact that OMPS has a longer expected life period in the orbit than OMI, because OMI has been operated for more than 10 years in the orbit and will finish its mission soon in the upcoming years. Thus, selecting TCO observations from OMPS as the baseline information is advisable. Through such a merging scheme, the merged TCO record can be extended easily by adding the latest OMPS TCO observations to the historical merged TCO record, and no further bias correction is required with respect to the OMPS TCO record. With this setting, a long-term coherent TCO record can be eventually obtained by simply merging bias-corrected OMI TCO observations during 2004–2012 with those from OMPS thereafter.

As prominent seasonal variability was observed within cross-mission TCO biases (e.g., Bak et al., 2015), in practice, monthly TCO records were used as control in the operational bias correction processes, rather than TCO observations through the year. In addition, control observations were created individually over each geographic grid for the bias modeling scheme because significant spatial heterogeneity of TCO biases is observed as well. For instance, to correct one OMI TCO

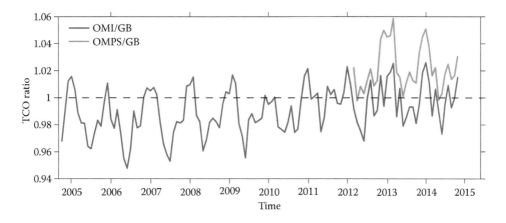

FIGURE 15.7   Comparison of monthly TCO ratios between OMI and OMPS relative to colocated ground-based measurements at the TSUKUBA station.

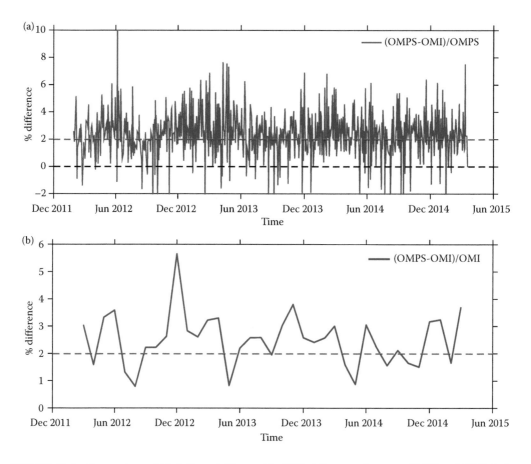

FIGURE 15.8 Comparison of (a) daily and (b) monthly TCO differences between OMPS and OMI during 2012–2015.

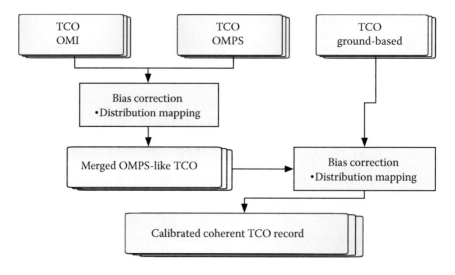

FIGURE 15.9 Flowchart of the data merging for generating bias corrected coherent TCO time series record.

value observed in January 2010, only those common TCO observations observed in January during the overlapped time period over the same geographic grid will be used.

15.3.1 BASICS OF THE Q-Q ADJUSTMENT METHOD IN THIS STUDY

As described in Amengual et al. (2011), the traditional Q-Q adjustment in the literature was designed to calibrate future climate model projections by removing the inherent modeling biases through a matching scheme between Cumulative Distribution Functions (CDFs) of the recent past observations and RCM simulation outputs. Such a process is also known as distribution mapping. Specifically, differences between the recent past observations and model simulated outputs are first detected by matching the CDFs of both data sets. Next, the inherent bias associated with each prediction will be removed from the projected regional climate scenario. In this step, the relevant bias for each prediction is estimated as the relative difference between the recent past observation and model simulations at the same percentile in CDFs of the projected time series. Such a scheme has been demonstrated to work effectively in calibrating many climate model outputs, such as precipitation (Osca et al., 2013) and air temperature (Gerelchuluun and Ahn, 2014).

15.3.1.1 Traditional Bias Correction Method

In Figure 15.10, the CDFs clearly depict the roles of projected, simulated, and observed datasets in the traditional Q-Q adjustment scheme for bias correction. To create a long-term coherent TCO record simply based on OMPS and OMI TCO observations, TCO observations from one of these two sensors should be mapped onto another sensor's level, and the mapping process is to remove the inherent biases between these two datasets. As stated earlier, TCO observations from OMI should be projected onto the OMPS level, and thus OMPS TCO data are used as the baseline information and those from OMI are applied as the complementary.

Nevertheless, in order to guarantee an accurate distribution mapping, the traditional Q-Q adjustment bias correction scheme requires common observations having the same number of

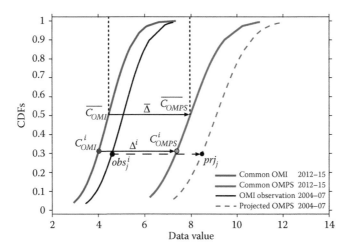

FIGURE 15.10 Schematic plot of the traditional Q-Q adjustment method for removing cross-mission TCO biases between OMPS and OMI. Note that the raw observations (OMI observation: thick black line) have a time scale (3 years: 2004–2007) the same as that (3 years: 2012–2015) of the two control observations (common observations of OMI and OMPS: blue and red lines), meaning that calibrating a length of 3 years OMI observations (i.e., 2004–2007) to OMPS level requires the same length of control observations during the overlapping time period between OMI and OMPS (i.e., 2012–2015). The projected OMPS denotes the anticipated data at the OMPS level that should be corrected from the OMI observations. (From Bai, K. et al. 2016b. *Remote Sensing of Environment*, 182, 150–168.)

samples (or same time scales) as the input time series for adjustment. In other words, if we want to calibrate one time series with 3-year observations, the common observations should also have a length of 3-year samples. For example, in order to calibrate OMI observations for the period of 2004–2007, a 3-year long (e.g., 2012–2015 in this study) common observations should be provided as references to fulfill this calibration (Figure 15.10). Given a TCO value obs_j in one OMI time series (defined as raw observations and depicted as the black line in Figure 15.10), it can be mapped onto the OMPS level by removing the inherent cross-mission bias. Mathematically, this bias correction process can be formulated as:

$$prj_j = obs_j + \Delta TCO_j \tag{15.1}$$

where prj_j is the projected TCO value at the OMPS level, and the related time series is thus defined as the projected OMPS (dashed line in Figure 15.10). obs_j denotes the jth data value in the raw observation time series, and ΔTCO_j represents the associated bias between OMPS and OMI for the given obs_j that needs to be addressed.

Based on the rationale of traditional Q-Q adjustment in this study, ΔTCO_j is to be estimated via a distribution mapping scheme, by referring to common TCO observations between OMPS and OMI during the overlapped time period (2012–2015). Two common observations, denoted as C_{OMPS} and C_{OMI}, are defined as control observations for bias modeling. The bias estimation scheme can be formulated as (Bai et al., 2016b):

$$\Delta TCO_j = g\bar{\Delta} + f\Delta_i' \tag{15.2}$$

where

$$i = p_1 \mid_{obs_{OMI} \cong obs_j} \tag{15.3}$$

$$\Delta_i = C_{OMPS}^i - C_{OMI}^i \tag{15.4}$$

$$\bar{\Delta} = \frac{1}{N}\sum \Delta_i = \overline{C_{OMPS}} - \overline{C_{OMI}} \tag{15.5}$$

$$\Delta_i' = \Delta_i - \bar{\Delta} \tag{15.6}$$

$$g = \frac{\overline{obs_{OMI}}}{\overline{C_{OMI}}} \tag{15.7}$$

$$f = \frac{IQR_{obs_{OMI}}}{IQR_{C_{OMI}}} \tag{15.8}$$

$$IQR_{obs_{OMI}} = obs_{OMI}\mid_{p=75\%} - obs_{OMI}\mid_{p=25\%} \tag{15.9}$$

$$IQR_{C_{OMI}} = C_{OMI}\mid_{p=75\%} - C_{OMI}\mid_{p=25\%} \tag{15.10}$$

In Equations 15.2–15.6, p_1 denotes the associated CDFs of obs_{OMI} (i.e., raw observations of OMI); i is the related percentile of obs_j in p_1; N is the number of samples in the raw observation time series; $\overline{C_{OMPS}}$ and $\overline{C_{OMI}}$ denote the mean values of C_{OMPS} and C_{OMI}, respectively; g and f are two modulation parameters. $IQR_{obs_{OMI}}$ and $IQR_{C_{OMI}}$ are two inter-quantile ranges between 25% and 75% percentile for raw and control OMI TCO observations, respectively.

It is clear that the bias estimation scheme is highly dependent on the CDFs mapping between raw observation and the two control observations, which in turn suggests the need to guarantee the same number of samples in raw observation and the two control time series. In this context, for instance, if we want to calibrate a 10-year TCO time series, two 10-year overlapped TCO time series must be applied as control observations; otherwise, cross-mission biases may not be fully removed, especially for those with dependence effects. However, such a requirement raises a big constraint to the original method, in particular in dealing with cross-mission biases between satellite observations, because the overlapped time periods between cross-mission sensors are often limited.

15.3.1.2 Modified Bias Correction Method

In order to address the aforementioned temporal constraint, some essential modifications should be conducted to optimize the bias estimation scheme. In general, the modified bias estimation scheme can be summarized as follows, taking OMPS and OMI as an example:

1. The common observations between two sensors during the overlapped time period are used as control observations. In practice, adequate samples, covering at least a full period cycle (e.g., one year for TCO), should be guaranteed in order to address seasonal dependent cross-mission biases. There is no doubt that better accuracy will be reached with more samples in control observations. In this study, therefore, all common observations between these two sensors during 2012–2015 were used to create two control observations.

2. As opposed to the original bias estimation scheme which requires that the number of samples in each time series should be identical, the new bias estimation scheme is designed to be highly adaptive regardless of the number of samples in the input time series. In other words, the bias correction method should be independent of the input data, and it should work even with an input data set with only one data value. Given any observed OMI TCO time series, no matter how many data samples are included, the modified bias correction method should be able to map OMI observations onto the OMPS level by removing the relevant bias for each sample. Therefore, given an obs_j, the ΔTCO_j can be estimated from the following modeling scheme (Bai et al., 2016b):

$$\Delta TCO_j = g'\overline{\overline{\Delta}} + f'\Delta'_{ii} \tag{15.11}$$

where

$$ii = p_2 \big|_{C_{OMI} \cong obs_j} \tag{15.12}$$

$$\Delta_{ii} = C^{ii}_{OMPS} - C^{ii}_{OMI} \tag{15.13}$$

$$\overline{\overline{\Delta}} = \overline{C_{OMPS}} - \overline{C_{OMI}} \tag{15.14}$$

$$\Delta'_{ii} = \Delta_{ii} - \overline{\overline{\Delta}} \tag{15.15}$$

$$g' = \frac{\overline{C_{OMPS}}}{\overline{C_{OMI}}} \tag{15.16}$$

$$f' = \frac{IQR_{C_{OMPS}}}{IQR_{C_{OMI}}} \tag{15.17}$$

$$IQR_{C_{OMPS}} = C_{OMPS}\mid_{p=75\%} - C_{OMPS}\mid_{p=25\%} \tag{15.18}$$

where p_2 denotes the associated CDFs of C_{OMI} (i.e., control observations of OMI), and ii is the ith percentile in p_2 and is estimated by finding the data value C_{OMI} equal (or close) to the given observation obs_j. $\overline{C_{OMPS}}$ and $\overline{C_{OMI}}$ are median values of the two control observations; g' and f' are two modulation parameters, which are applied to adjust the distribution of the projected time series. $IQR_{C_{OMPS}}$ denotes the inter-quantile range of control TCO observations from OMPS at 75% and 25% percentile.

As stated above, the modified bias correction scheme is to model the cross-mission bias for each sample in an adaptive manner, without referring to other external samples of the raw observation time series. More specifically, the principle can be summarized as follows: for each observed OMI TCO value (i.e., obs_j), the relevant bias associated with this observation is only related to the TCO value itself and the distributions of two control observations. In practice, the percentile with respect to the given obs_j in the OMI control observation time series was first determined simply by searching the pixel with the same or closest data value to obs_j (Equation 15.12). The next step is to find the data value in the OMPS control observation time series at the same percentile (i.e., C_{OMPS}^{ii}) that is determined in the first step. Once these two data values are obtained, the raw bias Δ_j with respect to the given obs_j can be easily estimated from Equation 15.13. Finally, the cross-mission bias ΔTCO_j can be attained by following Equation 15.11. During this process, all other modulation parameters, such as g' and f' as well as $\overline{C_{OMPS}}$ and $\overline{C_{OMI}}$, can be easily calculated simply based on the two control observation time series (Figure 15.11).

Compared to the traditional Q-Q adjustment, the primary difference relative to the modified bias correction method lies in the bias estimation scheme. In the modified scheme, the median value, rather than mean value, was used so as to avoid uncertainties driven by any outliers. Additionally, the distribution mapping was performed simply between the two control observations C_{OMPS} and C_{OMI}, regardless of the raw observations.

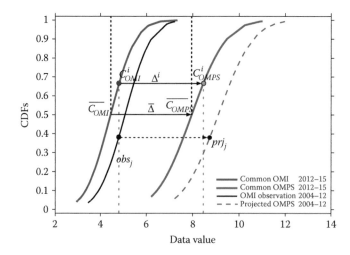

FIGURE 15.11 Distribution mapping for the modified method. Note that the raw observations (thick black line) have a time scale (8 years: 2004–2012) different from that (3 years: 2012–2015) of the two control observations (blue and red lines). (From Bai, K. et al. 2016b. *Remote Sensing of Environment*, 182, 150–168.)

Analogously, the two modulation parameters g' and f', were also estimated simply from the two control observations. Due to its independence to the raw observations, the modified bias correction method is more adaptive relative to the traditional one, which renders this modified bias correction method the capability to handle various observations at different time scales.

15.3.2 OVERALL INCONSISTENCY INDEX FOR PERFORMANCE EVALUATION

Statistical metrics play a critical role in performance evaluation of the bias correction method. In order to better assess the consistency improvement after bias correction, one index called Overall Inconsistency (OI) was developed here based on the Mahalanobis distance (D_{Mah}), aiming to use this as an indicator to check the consistency (or similarity) between time series. The D_{Mah} is one popular time series similarity measure, which not only quantifies the difference between two time series, but also accounts for non-stationarity of variance and temporal cross-correlation (Mahalanobis, 1936; Lhermitte et al., 2011). Nevertheless, the D_{Mah} cannot indicate the effect of overestimation and/or underestimation due to the root squared mathematical limitation. Moreover, it is more likely to emphasize the mean squared root rather than the relative difference. In this study, the OI was developed as a more comprehensive time series similarity measure, in which the Mean Relative Difference (*MRD*), *MARD*, Root Mean Squared Error (*rmse*), and D_{Mah} were all integrated. Generally, the OI emphasizes the overall difference between two time series data sets, which is also capable of indicating the overestimation and/or underestimation effect because it can be either positive or negative. These statistical indexes are formulated as (Bai et al., 2016b):

$$MRD = 100 \times \frac{1}{N} \sum_{i=1}^{N} \frac{SAT_i - GB_i}{GB_i} \tag{15.19}$$

$$MARD = 100 \times \frac{1}{N} \sum_{i=1}^{N} \left| \frac{SAT_i - GB_i}{GB_i} \right| \tag{15.20}$$

$$rmse = 100 \times \sqrt{\frac{1}{N} \sum_{i=1}^{N} \left(\frac{SAT_i - GB_i}{GB_i} \right)^2} \tag{15.21}$$

$$D_{Mah} = \sqrt{\epsilon' \Sigma^{-1} \epsilon} \tag{15.22}$$

$$OI = MRD \times \frac{RMSE}{MARD} \times \frac{D_{Mah}^2}{N} \tag{15.23}$$

where *SAT* and *GB* denote satellite- and ground-based TCO observations, respectively. In Equation 15.22, ϵ denotes the difference between *SAT* and *GB* while ϵ' is the transposition of ϵ. Σ is the error covariance matrix; and N is the number of samples in ϵ. It is clear that *OI* varies between $-\infty$ and $+\infty$, with a perfect value of zero. A negative value of *OI* thus indicates the underestimation, whereas a positive value implies overestimation.

To assess the effectiveness of the bias correction method, the relative difference of *OI* before and after bias correction can be calculated by

$$\Delta OI\% = 100 \times \frac{\frac{1}{N} \sum_{i=1}^{N} |OI_i^{before}| - \frac{1}{N} \sum_{i=1}^{N} |OI_i^{after}|}{\frac{1}{N} \sum_{i=1}^{N} |OI_i^{before}|} \tag{15.24}$$

15.4 PERFORMANCE OF MODIFIED BIAS CORRECTION METHOD

Figure 15.12 shows comparisons of relative differences between OMPS and OMI before and after bias correction, which indicates that cross-mission biases have been significantly reduced by the modified bias correction method, with a mean bias level around 0.1% after the bias correction (Figure 15.12b) (Bai et al., 2016b). The estimated biases by the modified bias correction method are presented in Figure 15.12c. It is clear that the estimated biases show significant seasonal variability,

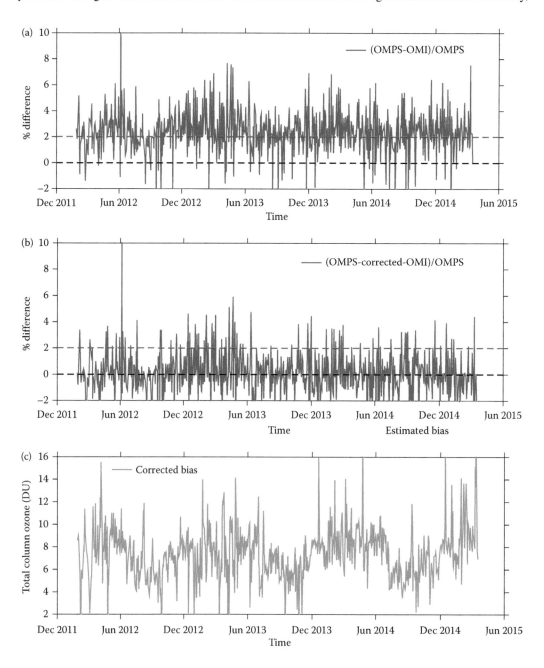

FIGURE 15.12 Comparisons of cross mission TCO bias level between OMPS and OMI (a) before and (b) after bias correction. Both TCO records were collected at the TSUKUBA station. Panel (c) shows the removed biases estimated by the modified bias correction method.

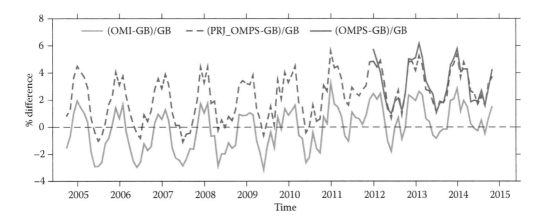

FIGURE 15.13 Monthly mean relative differences between satellite- and ground-based TCO observations at the TSUKUBA station. PRJ_OMPS denotes the bias corrected OMI TCO observations.

which means that the removed bias level is not a simple mean difference (e.g., delta-change) or a linear fitted value (e.g., linear transformation). Such an effect indicates that the modified bias correction method is capable of handling seasonal dependent biases, which are nonlinear and nonstationary in nature.

Performance of the modified bias correction method was further assessed with different comparisons over the prescribed temporal and spatial domains (Figures 15.13 through 15.15). Figure 15.13 shows the vicarious comparisons of monthly mean relative differences of satellite observations with respect to the ground truth at the TSUKUBA station before and after performing the bias correction. The results show that the original OMI TCO observations underestimated ground-based TCO by 1%, whereas OMPS overestimated by 3%. In other words, biases between OMPS and OMI TCO observations are significant (about 4%). However, the bias level was reduced to 0.1% after performing bias correction to the OMI TCO observations. In addition, probability density functions of TCO differences between OMPS and OMI before and after bias correction

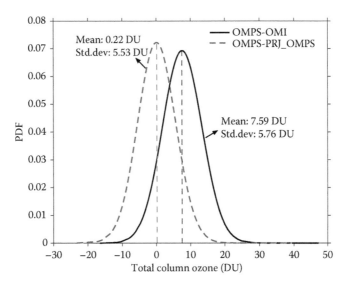

FIGURE 15.14 Probability density functions of TCO differences (OMPS-OMI) between OMPS and OMI at the TSUKUBA station. The thick black curve shows the one before correction and the dashed blue curve represents the one after bias correction.

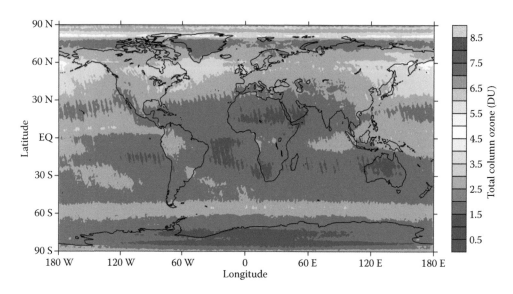

FIGURE 15.15   *MARD* of TCO between OMPS and bias corrected OMI during 2012–2015.

were compared to show the possible consistency improvements (Figure 15.14). It shows that the bias level was reduced from a mean positive bias of 7.6 DU before bias correction to 0.2 DU after correction. Moreover, the standard deviation of biases was also reduced after bias correction. Such improvements can be also observed from the *MARD*, its standard deviation, and the associated zonal mean of TCO between OMPS and bias corrected OMI during 2012–2015 spatially and temporally (Figures 15.15 through 15.17). All these evidences reveal the efficacy of the modified bias correction method in removing spatially and temporally heterogeneous cross-mission biases.

The monthly cascade differences between OMPS and OMI were also calculated to further depict the significance of cross-mission bias correction as well as the associated performance of

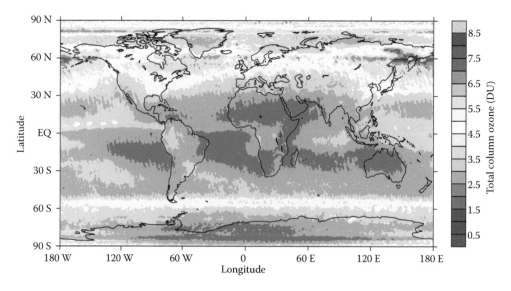

FIGURE 15.16   Standard deviation of *MARD* of TCO between OMPS and bias corrected OMI during 2012–2015.

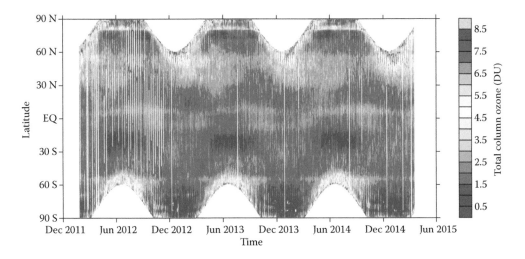

FIGURE 15.17   Zonal mean TCO deviation between OMPS and bias corrected OMI during 2012–2015.

the modified bias correction method. As shown in Figure 15.18, the monthly cascade differences were plotted in a 2-dimensional matrix. Note that the monthly cascade differences are expected to be positive and negative every three months, which can be explained by the fact that TCO values vary in a sinusoid manner throughout the year because of seasonality. However, significant positive cross-mission biases were observed between OMPS and the original OMI observations, and these biases overwhelm the seasonal variations of TCO (Figure 15.18a). Yet these cross-mission biases were significantly reduced after bias correction, revealing an obvious seasonal variability of TCO difference (Figure 15.18b). Compared to the case in Figure 15.18a before bias correction, the bias corrected OMI data were much more consistent with OMPS. This is evidenced by not only the TCO differences in the same month (the diagonal values of the 2-dimensional matrix), but also the more salient symmetric pattern of positive and negative values alternately of the monthly TCO differences after bias correction (Figure 15.18b). Note that the diagonal values of the 2-dimensional matrix in Figure 15.18b always turn out to be around zero.

In addition, the OI was also calculated to assess the consistency improvements between OMPS and OMI TCO observations after bias correction. Before bias correction, OMPS overestimated OMI observations globally, with an OI greater than 5 in the NH, in particular over the Arctic and tropic regions (Figure 15.19a). This effect can be ascribed to large biases observed therein (Figures 15.5 and 15.6). Moreover, significant inconsistencies (OI > 7) were also observed in the tropics, although small cross-mission TCO biases were observed therein. Further investigations indicate poor correlations between OMPS and OMI TCO observations, despite small cross-mission biases in magnitude observed between them. Such an effect reveals the advantage of OI for performance evaluation, in contrast to other simple statistical indexes like *MRD* and *rmse* that can only show deviations between two time series, being incapable of reflecting variance and correlation simultaneously. Within such a context, OI is deemed to be a more comprehensive and reliable statistical index for time series similarity assessment as variances and correlations are considered as well. The derived OI after bias correction is observed varying within ±1 over each geographic grid at the global scale (Figure 15.19b), which in turn means that the apparent inconsistency has been significantly removed, yielding an overall consistency improvement (i.e., $\Delta OI\%$) by 90% (Bai et al., 2016b). After bias correction, OMPS slightly overestimated the bias-corrected OMI TCO in the NH with a negligible underestimation in the SH. In general, the modified bias correction method showed promising efficacy in removing those nonlinear latitudinal and seasonal dependent biases between OMPS and OMI TCO observations in this study.

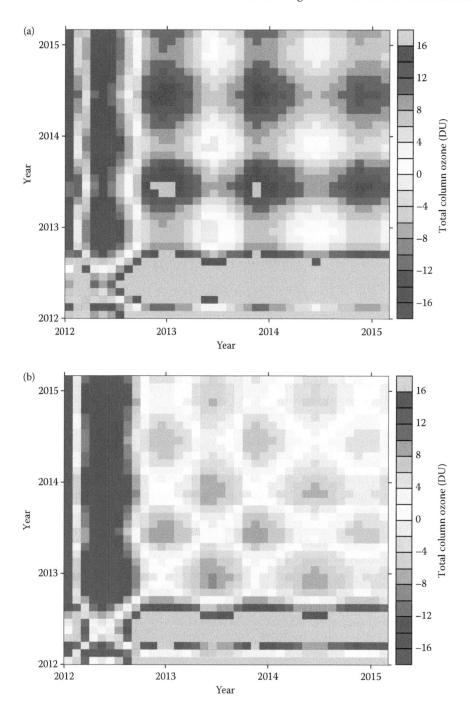

FIGURE 15.18 Monthly cascade TCO differences (OMPS-OMI) (a) before and (b) after bias correction at the TSUKUBA station. The monthly cascade TCO differences indicate differences between TCO from OMPS and OMI over different months, and the diagonal elements thus represent TCO differences over the same months. The pixel value at the lower left corner represents the TCO difference between OMPS and OMI during January, whereas the second pixel in the same row represents the difference between TCO from OMPS during February and that of OMI during January. (From Bai, K. et al. 2016b. *Remote Sensing of Environment*, 182, 150–168.)

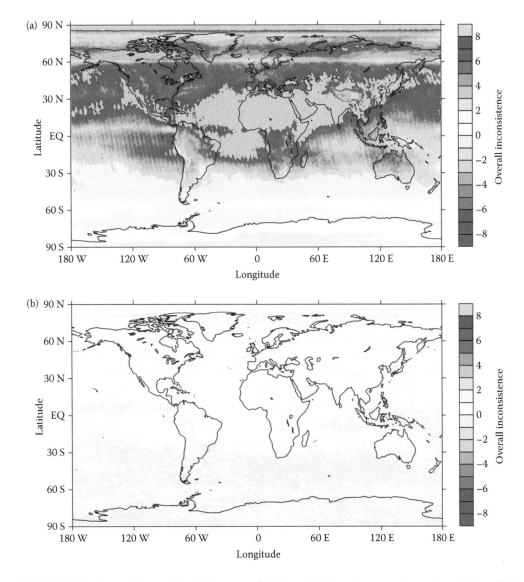

FIGURE 15.19 Comparisons of the OI between OMPS and OMI total ozone observations during 2012–2015. (a) OI between OMPS and OMI; (b) OI between OMPS and bias-corrected OMI. (From Bai, K. et al. 2016b. *Remote Sensing of Environment*, 182, 150–168.)

15.5 DETECTION OF OZONE RECOVERY BASED ON THE MERGED TCO DATA

With the bias correction efforts in phase I, the coherent TCO data gave rise to an opportunity to conduct a long-term assessment of the trend of ozone recovery at the global scale. As reported in Kuttippurath et al. (2013), an ozone recovery rate by 1–2.6 Dobson Units (DU) per year was observed during 2000–2010 over Antarctica. These small trend values raise the concern that a small drift in TCO time series could introduce large bias into the estimated ozone recovery speed. More specifically, untreated cross-mission biases in the long-term TCO time series could result in large uncertainties in trend estimations, which may yield an earlier expectation of ozone layer recovery at the global scale in turn (Bai et al., 2016b). Therefore, cross-mission biases should be well addressed

TABLE 15.3

Linear Trends Estimated from Different TCO Records at 38 Ground-Based Stations

WMO ID	Latitude	Longitude	N	OMI DU/month	OMI+OMPS DU/month	PRJ_OMPS+OMPS DU/month	Ground-Based DU/month
2	22.8	5.5	120	0.035	0.089	0.037	0.068
12	43.1	141.3	122	−0.085	0.005	−0.081	−0.094
14	36.1	140.1	122	0.023	0.097	0.023	−0.032
19	46.77	−100.75	113	0.170	0.240	0.177	0.199
27	−27.42	153.12	109	0.048	0.077	0.054	0.032
29	−54.5	158.97	108	0.077	0.096	0.077	−0.008
40	43.93	5.7	102	0.154	0.235	0.166	0.054
43	60.1	−1.2	103	0.133	0.208	0.149	0.113
51	64.1	−21.9	91	0.213	0.321	0.226	0.244
67	40.03	−105.25	117	0.081	0.128	0.082	0.077
68	51.84	20.79	111	−0.035	0.025	−0.032	−0.054
84	−12.42	130.88	109	0.043	0.073	0.049	0.032
91	−34.58	−58.48	110	0.046	0.076	0.050	−0.003
96	50.2	15.8	116	0.027	0.089	0.032	0.031
99	47.8	11	122	0.002	0.055	0.003	−0.014
101	−69	39.6	92	0.257	0.278	0.247	0.206
105	64.82	−147.87	78	−0.058	0.063	−0.044	0.183
106	36.25	−86.57	114	0.073	0.135	0.076	−0.001
107	37.93	−75.48	106	0.070	0.152	0.073	−0.110
111	−89.98	−24.8	52	0.627	0.606	0.651	0.690
152	30.1	31.3	105	0.051	0.125	0.055	0.076
159	−31.92	115.95	98	−0.018	0.014	−0.017	−0.066
190	26.2	127.7	120	0.032	0.084	0.032	0.033
191	−14.25	−170.56	83	0.126	0.163	0.129	0.064
199	71.32	−156.6	57	0.066	0.234	0.075	0.239
208	40	116.4	119	−0.049	0.030	−0.047	−0.132
209	25.03	102.68	83	0.057	0.140	0.058	0.077
214	1.33	103.88	70	0.001	0.021	0.000	0.087
216	13.67	100.61	119	0.033	0.081	0.036	0.034
219	−5.84	−35.21	24	0.439	0.439	0.477	−0.053
233	−64.23	−56.62	79	0.192	0.228	0.188	0.232
245	23.97	32.78	118	0.035	0.094	0.037	0.105
256	−45.03	169.68	104	0.014	0.043	0.011	−0.039
265	−25.91	28.21	95	0.122	0.141	0.127	0.137
268	−77.83	166.67	69	0.306	0.344	0.309	0.310
284	64.2	19.8	79	−0.269	−0.162	−0.266	−0.186
293	38	23.7	120	−0.034	0.033	−0.030	−0.058
339	−54.85	−68.28	109	0.064	0.087	0.069	0.097
Average				0.081	0.136	0.086	0.068

Source: Bai, K. et al. 2016b. *Remote Sensing of Environment*, 182, 150–168.

WMO ID: Identifying Number in WMO Ground-Based Station Network. Latitude: North Positive. Longitude: East Positive. N: Number of Months. OMI: Untreated Monthly TCO Time Series from OMI Observations During 2004–2015. OMI+OMPS: Merged Monthly TCO Time Series Between Untreated OMI (2004–2012) and OMPS (2012–2015). PRJ_OMPS+OMPS: Merged Monthly TCO Time Series Between Bias-Corrected OMI (i.e., PRJ_OMPS, 2004–2012) and OMPS (2012–2015). Ground-Based: Monthly TCO Time Series from Ground-Based Measurements Coincident with that of Satellite Observations.

prior to the merging scheme toward the creation of a long-term TCO record for the assessment of the long-term ozone variability.

In this study, the ozone layer variability was depicted by the linear trends derived from the merged TCO time series, represented by the slope of the fitted straight line to each time series. To examine the impacts of cross-mission bias on long-term ozone trends, linear trends derived from four different TCO data sets, including the original OMI record, the merged TCO time series without bias correction (i.e., OMI+OMPS), the merged TCO time series after bias correction (i.e., PRJ_OMPS+OMPS), and the ground-based TCO measurements, were compared (Table 15.3). Both the original monthly OMI TCO time series and colocated ground-based TCO measurements during 2004–2015 were used as references. The results from the inter-comparison analyses clearly indicate that trends derived from the merged TCO time series without bias correction (i.e., OMI+OMPS) significantly overestimate those from the other three datasets, by a speed of 0.136 DU/month on average (Bai et al., 2016b). This increasing speed is almost twice the original OMI (0.081 DU/month), the bias-corrected OMI (0.086 DU/month), and the ground-based TCO measurements (0.068 DU/month) (Bai et al., 2016b). This effect strongly suggests the need to perform bias correction before data merging (Bai et al., 2016b). Moreover, it can be observed that cross-mission biases yielded opposite signs of trends, for instance, the one at BELSK station (WMO/GAW ID: 68) (Table 15.3). Specifically, a positive trend by 0.025 DU/month was derived from the merged TCO time series without bias correction. However, negative trends were calculated based on the untreated OMI (-0.035 DU/month), bias-corrected (-0.032 DU/month), and ground-based (-0.054 DU/month) TCO time series.

15.6 CALIBRATION OF THE MERGED TCO RECORD WITH GROUND-BASED MEASUREMENTS

Significant improvements confirm the efficacy of the proposed approach for removing residual biases from phase I. In fact, to substantiate the effort of bias correction further, comparisons of OI between satellite and colocated ground-based TCO observations over 38 geographic grids during 2004–2015 may shed light on some more measurement errors from the space-borne sensors based on the results of bias correction in phase I with the aid of ground-based TCO observations (Figure 15.20). The further bias correction effort is denoted as phase II in this study.

In Table 15.4, statistics of corrected TCO in phase I (i.e., TCO from OMI corrected to OMPS level with systematic bias removal) was shown inside brackets while that of phase II (further

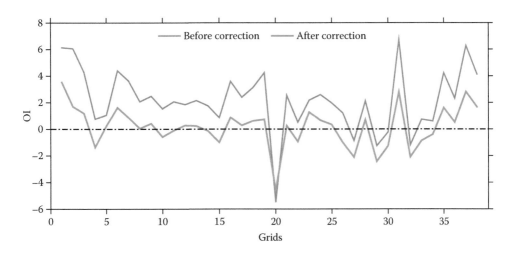

FIGURE 15.20 Comparisons of OI between satellite- and colocated ground-based TCO observations over 38 geographic grids during 2004–2015 before and after further bias correction in Phases I and II.

TABLE 15.4

Comparison of Statistics between Corrected Record and Ground-Based Measurements from 38 Dobson Instruments

WMO ID	Latitude	Longitude	N	R²	rmse DU	rmse %	MRD (%)
2	22.80	5.50	3068	0.92 (0.92)	**5.71 (6.73)**	**2.17 (2.57)**	**1.51 (2.01)**
12	43.10	141.30	2456	**0.97 (0.96)**	10.51 (14.91)	**2.89 (4.08)**	1.11 (2.76)
14	36.10	140.10	2632	**0.94 (0.93)**	9.10 (12.04)	**2.81 (3.75)**	0.83 (2.24)
19	46.77	−100.75	1567	0.94 (0.94)	10.47 (10.11)	3.01 (2.97)	−0.83 (0.50)
27	−27.42	153.12	2223	0.96 (0.96)	**4.18 (4.49)**	**1.51 (1.62)**	**0.18 (0.68)**
29	−54.50	158.97	991	0.94 (0.94)	**11.69 (13.73)**	**3.69 (4.30)**	1.09 (2.31)
40	43.93	5.70	1072	0.91 (0.91)	**12.21 (13.34)**	**3.90 (4.32)**	0.54 (1.91)
43	60.10	−1.20	1161	0.95 (0.95)	**10.92 (12.31)**	**3.100 (3.48)**	0.04 (1.25)
51	64.10	−21.90	743	0.94 (0.94)	**13.42 (15.02)**	**3.83 (4.40)**	0.28 (1.60)
67	40.03	−105.25	2085	**0.93 (0.92)**	**8.68 (9.41)**	**2.85 (3.10)**	−0.37 (0.92)
68	51.84	20.79	2095	0.96 (0.96)	**8.41 (10.04)**	**2.72 (3.29)**	−0.07 (1.20)
84	−12.42	130.88	2365	**0.84 (0.83)**	**4.04 (4.97)**	1.60 (1.98)	0.21 (1.07)
91	−34.58	−58.48	2312	0.95 (0.95)	**5.54 (6.61)**	1.95 (2.33)	0.18 (1.26)
96	50.20	15.80	1405	0.97 (0.97)	**6.85 (7.90)**	2.06 (2.39)	−0.09 (1.08)
99	47.80	11.00	1587	0.98 (0.98)	**6.46 (6.80)**	1.96 (2.06)	−0.67 (0.61)
101	−69.00	39.60	1579	**0.99 (0.98)**	**6.98 (8.41)**	3.00 (3.60)	0.58 (1.84)
105	64.82	−147.87	815	**0.94 (0.93)**	**13.51 (16.16)**	3.72 (4.40)	0.17 (1.36)
106	36.25	−86.57	1693	0.93 (0.93)	**7.75 (9.45)**	2.50 (3.06)	0.45 (1.69)
107	37.93	−75.48	736	0.89 (0.89)	**11.66 (12.40)**	4.86 (5.15)	0.36 (1.70)
111	−89.98	−24.80	969	**0.98 (0.97)**	**9.90 (10.56)**	4.14 (4.53)	**−2.34 (−2.72)**
152	30.10	31.30	846	**0.86 (0.83)**	**7.74 (9.35)**	2.66 (3.24)	0.19 (1.51)
159	−31.92	115.95	1431	0.91 (0.91)	7.08 (6.88)	2.39 (2.36)	−0.63 (0.37)
190	26.20	127.70	2317	0.93 (0.93)	**5.93 (6.31)**	2.16 (2.33)	**0.86 (1.27)**
191	−14.25	−170.56	873	**0.77 (0.75)**	**4.96 (6.04)**	2.00 (2.45)	**0.48 (1.38)**
199	71.32	−156.60	463	0.98 (0.98)	**8.66 (10.14)**	2.18 (2.54)	**0.26 (1.21)**
208	40.00	116.40	2062	**0.91 (0.90)**	11.43 (13.02)	3.30 (3.78)	**−0.74 (0.91)**
209	25.03	102.68	1300	0.91 (0.91)	6.68 (6.11)	2.45 (2.25)	−1.23 (−0.62)
214	1.33	103.88	1381	**0.50 (0.48)**	**9.92 (10.50)**	4.39 (4.65)	**0.46 (1.31)**
216	13.67	100.61	2304	0.84 (0.84)	7.66 (6.98)	2.85 (2.61)	−1.42 (−0.85)
219	−5.84	−35.21	372	**0.80 (0.78)**	6.12 (5.92)	2.24 (2.20)	−0.83 (−0.18)
233	−64.23	−56.62	1111	0.97 (0.97)	**11.11 (12.80)**	4.79 (5.50)	**1.77 (3.23)**
245	23.97	32.78	1663	**0.87 (0.86)**	6.88 (6.46)	2.44 (2.28)	−1.22 (−0.81)
256	−45.03	169.68	1633	0.96 (0.96)	**8.79 (8.94)**	**2.83 (2.88)**	−0.58 (0.51)
265	−25.91	28.21	1629	0.80 (0.80)	**7.37 (7.39)**	**2.80 (2.83)**	**−0.28 (0.42)**
268	−77.83	166.67	586	0.97 (0.97)	**11.14 (12.15)**	**5.83 (6.11)**	0.76 (1.84)
284	64.20	19.80	503	0.98 (0.98)	**7.34 (8.55)**	**2.16 (2.54)**	**0.36 (1.31)**
293	38.00	23.70	1617	0.87 (0.88)	**12.34 (14.29)**	**4.13 (4.78)**	1.50 (2.81)
339	−54.85	−68.28	1872	0.92 (0.92)	**10.80 (12.38)**	**3.94 (4.49)**	1.08 (2.23)
All			57517	0.97 (0.97)	**8.65 (9.87)**	**2.96 (3.35)**	**0.15 (1.17)**

Statistics of Corrected TCO in Phase I (TCO from OMI corrected to OMPS Level) was shown inside brackets while that of Phase II (further correction to TCO from Phase I) was shown outside. Improvements were highlighted in bold.

correction to TCO from phase I) was shown outside brackets. Improvements that indicate the marginal contribution of 2%–4% in terms of *rmse* and 0%–2.5% in terms of *MRD* from the 38 ground-based stations in phase II were highlighted in bold (Table 15.4). The visualization effect via spatial comparisons of OI between satellite and colocated ground-based TCO observations before and after bias correction in phase II indicates that location-specific residual biases still impact the bias removal in phase II, especially in the equatorial area (Figure 15.21). Taking Figure 15.13 into account again, corrected OMI denotes corrected TCO from OMI to OMPS level and calibrated OMI denotes corrected TCO from previously corrected OMI to the ground-based measurements level. The improvement from phase I is salient. Similarly, the improvements from phase II can be summarized by Figure 15.22, in which the average relative difference was reduced from 2.24% to 0.83%.

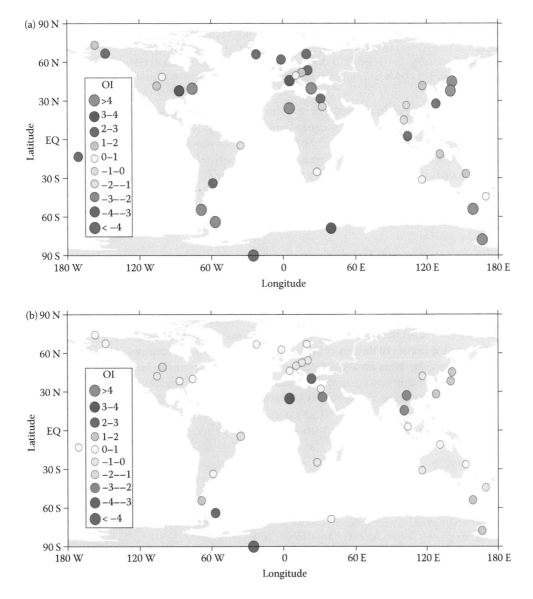

FIGURE 15.21 Spatial comparisons of OI between satellite and colocated ground-based TCO observations: (a) before bias correction; (b) after bias correction.

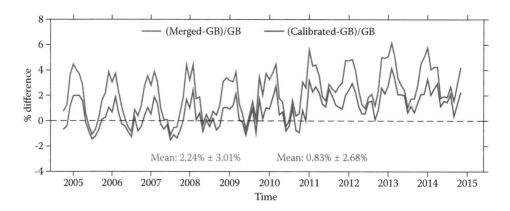

FIGURE 15.22 Comparisons of monthly mean difference based on TCO ratios at the TSUKUBA station before and after bias correction with respect to merged OMPS data and ground-based measurements as references, respectively. Corrected OMI denotes corrected TCO from OMI to OMPS level. Calibrated OMI denotes corrected TCO from previous corrected OMI to the ground-based measurements level.

In contrast to changes in magnitudes in estimated trends, trends with opposite phases (positive or negative) are more outrageous, because opposite projections could be misleading in scientific investigation. Although there exists inconsistency between trends derived from estimated satellite- and ground-based measurements, differences between derived trends from the bias-corrected time series and untreated OMI time series are negligible. These findings strongly demonstrate that bias correction is necessary in time series analysis, in particular for trend analysis. Hence, future long-term trend analyses of the ozone recovery should be performed based on the bias corrected TCO data in phase II for advanced verification to confirm or refute whether the ozone layer is on a healing path.

15.7 SUMMARY

In this chapter, a statistical bias correction method for data merging between OMI and OMPS observations was proposed to create a coherent total column ozone record globally. This leads to an opportunity to conduct the detection of early signs of ozone hole recovery after years of depletion after showing the impact of data merging on the trend analysis of ozone hole recovery. Such a bias correction scheme has been proven effective and useful to generate the merged total column ozone record that may be further vicariously calibrated by ground-based measurements. Future long-term trend analyses of the ozone recovery may be further performed based on bias corrected TCO data with the aid of not only the merged TCO record between OMI and OMPS but also ground-based TCO observations for advanced verification to confirm or refute whether the ozone layer is on a healing path in the future.

REFERENCES

Ahmed, K. F., Wang, G., Silander, J., Wilson, A. M., Allen, J. M., Horton, R., and Anyah, R., 2013. Statistical downscaling and bias correction of climate model outputs for climate change impact assessment in the U.S. northeast. *Global and Planetary Change*, 100, 320–332.

Amengual, A., Homar, V., Romero, R., Alonso, S., and Ramis, C., 2011. A statistical adjustment of regional climate model outputs to local scales: Application to platja de palma. *Spain. Journal of Climate*, 25, 939–957.

Antón, M., López, M., Vilaplana, J. M., Kroon, M., McPeters, R., Bañón, M., and Serrano, A., 2009. Validation of OMI-TOMS and OMI-DOAS total ozone column using five brewer spectroradiometers at the iberian peninsula. *Journal of Geophysical Research: Atmospheres*, 114, 1–10.

Argüeso, D., Evans, J. P., and Fita, L., 2013. Precipitation bias correction of very high resolution regional climate models. *Hydrology and Earth System Sciences*, 17, 4379–4388.

Bai, K., Chang, N. B., and Chen, C. F., 2015a. Spectral information adaptation and synthesis scheme for merging cross-mission ocean color reflectance observations from MODIS and VIIRS. *IEEE Transactions on Geoscience and Remote Sensing*, 99, 1–19.

Bai, K., Chang, N. B., and Gao, W., 2016a. Quantification of relative contribution of antarctic ozone depletion to increased austral extratropical precipitation during 1979–2013 period. *Journal of Geophysical Research: Atmospheres*, 121, 1459–1474.

Bai, K., Chang, N. B., Yu, H., and Gao, W., 2016b. Statistical bias corrections for creating coherent total ozone records with OMI and OMPS observations. *Remote Sensing of Environment*, 182, 150–168.

Bai, K., Liu, C., Shi, R., and Gao, W., 2015b. Comparison of suomi-NPP OMPS total column ozone with brewer and dobson spectrophotometers measurements. *Frontiers of Earth Science*, 9, 369–380.

Bak, J., Liu, X., Kim, J. H., Chance, K., Haffner, D. P., and Systems, S., 2015. Validation of OMI Total Ozone Retrievals from the SAO Ozone Profile Algorithm and three operational algorithms with brewer measurements. *Atmospheric Chemistry and Physics*, 15, 667–683.

Balis, D., Kroon, M., Koukouli, M. E., Brinksma, E. J., Labow, G., Veefkind, J. P., and McPeters, R. D., 2007. Validation of ozone monitoring instrument total ozone column measurements using brewer and dobson spectrophotometer ground-based observations. *Journal of Geophysical Research: Atmospheres*, 112, D24S46.

Barnes, B. B. and Hu, C., 2015. Cross-sensor continuity of satellite-derived water clarity in the gulf of mexico: Insights into temporal aliasing and implications for long-term water clarity assessment. *IEEE Transactions on Geoscience and Remote Sensing*, 53, 1761–1772.

Basher, R. E., 1982. Review of the Dobson spectrophotometer and its accuracy. *Global Ozone Research and Monitoring Project- Report 13*. Geneva, Switzerland.

Bass, A. M. and Paur, R. J., 1985. The Ultraviolet Cross Section of Ozone, I. The Measurements. In: *Proceedings of the Quadrennial Ozone Symposium*, Halkidiki, Greece, 606–616.

Bhartia, P. K., McPeters, R. D., Flynn, L. E., Taylor, S., Kramarova, N. A., Frith, S., Fisher, B., and DeLand, M., 2013. Solar Backscatter UV (SBUV) total ozone and profile algorithm. *Atmospheric Measurement Techniques*, 6, 2533–2548.

Bhartia, P. K. and Wellemeyer, C. W., 2002. *TOMS-V8 Total O3 Algorithm. OMI algorithm theoretical basis document, volume II, OMI ozone products*. NASA-OMI, Washington, DC.

Bordoy, R. and Burlando, P., 2013. Bias correction of regional climate model simulations in a region of complex orography. *Journal of Applied Meteorology and Climatology*, 52, 82–101.

Brewer, A. W., 1973. A replacement for the dobson spectrophotometer? *Pure and Applied Geophysics*, 106–108, 919–927.

D'Alimonte, D., Zibordi, G., and Mélin, F., 2008. A statistical method for generating cross-mission consistent normalized water-leaving radiances. *IEEE Transactions on Geoscience and Remote Sensing*, 46, 4075–4093.

Delle Monache, L., Nipen, T., Deng, X., Zhou, Y., and Stull, R., 2006. Ozone ensemble forecasts: 2. A kalman filter predictor bias correction. *Journal of Geophysical Research: Atmospheres*, 111, 1–15.

Dobson, G. M. B., 1968. Forty years' research on atmospheric ozone at Oxford: A History. *Applied Optics*, 7, 387–405.

Feldstein, S. B., 2011. Subtropical rainfall and the Antarctic ozone hole. *Science*, 332, 925–926.

Fyfe, J. C., Gillett, N. P., and Marshall, G. J., 2012. Human influence on extratropical Southern Hemisphere summer precipitation. *Geophysical Research Letters*, 39, L23711.

Gerelchuluun, B. and Ahn, J. B., 2014. Air temperature distribution over mongolia using dynamical downscaling and statistical correction. *International Journal of Climatology*, 34, 2464–2476.

Gillett, N. P., Fyfe, J. C., and Parker, D. E., 2013. Attribution of observed sea level pressure trends to greenhouse gas, aerosol, and ozone changes. *Geophysical Research Letters*, 40, 2302–2306.

Gonzalez, P. L. M., Polvani, L. M., Seager, R., and Correa, G. J. P., 2013. Stratospheric ozone depletion: A key driver of recent precipitation trends in South Eastern South America. *Climate Dynamics*, 42, 1775–1792.

Gregg, W. W. and Casey, N. W., 2010. Improving the consistency of ocean color data: A step toward climate data records. *Geophysical Research Letters*, 37, L04605.

Haerter, J. O., Hagemann, S., Moseley, C., and Piani, C., 2011. Climate model bias correction and the role of timescales. *Hydrology and Earth System Sciences*, 15, 1065–1079.

Hagemann, S., Chen, C., Haerter, J. O., Heinke, J., Gerten, D., and Piani, C., 2011. Impact of a statistical bias correction on the projected hydrological changes obtained from three GCMs and two hydrology models. *Journal of Hydrometeorology*, 12, 556–578.

Hoyer, J. L., Le Borgne, P., and Eastwood, S., 2013. A bias correction method for arctic satellite sea surface temperature observations. *Remote Sensing of Environment*, 146, 201–213.

Kuttippurath, J., Lefèvre, F., Pommereau, J.-P., Roscoe, H. K., Goutail, F., Pazmiño, A., and Shanklin, J. D., 2013. Antarctic ozone loss in 1979–2010: First sign of ozone recovery. *Atmospheric Chemistry and Physics*, 13, 1625–1635.

Lafon, T., Dadson, S., Buys, G., and Prudhomme, C., 2013. Bias correction of daily precipitation simulated by a regional climate model: A comparison of methods. *International Journal of Climatology*, 33, 1367–1381.

Levelt, P. F., van den Oord, G. H. J., Dobber, M. R., Malkki, A., Stammes, P., Lundell, J. O. V., and Saari, H., 2006. The ozone monitoring instrument. *IEEE Transactions on Geoscience and Remote Sensing*, 44, 1093–1101.

Lhermitte, S., Verbesselt, J., Verstraeten, W. W., and Coppin, P., 2011. A comparison of time series similarity measures for classification and change detection of ecosystem dynamics. *Remote Sensing of Environment*, 115, 3129–3152.

Li, H., Sheffield, J., and Wood, E. F., 2010. Bias correction of monthly precipitation and temperature fields from Intergovernmental Panel on Climate Change AR4 models using equidistant quantile matching. *Journal of Geophysical Research*, 115, D10101.

Loyola, D. G. and Coldewey-Egbers, M., 2012. Multi-sensor data merging with stacked neural networks for the creation of satellite long-term climate data records. *EURASIP Journal on Advances in Signal Processing*, 2012, 91.

Mahalanobis, P. C., 1936. On the generalised distance in statistics. *Proceedings of the National Institute of Sciences of India*, 2, 49–55.

Malicet, J., Daumont, D., Charbonnier, J., Parisse, C., Chakir, A., and Brion, J., 1995. Ozone UV spectroscopy. II. Absorption cross-sections and temperature dependence. *Journal of Atmospheric Chemistry*, 21, 263–273.

Manatsa, D., Morioka, Y., Behera, S. K., Yamagata, T., and Matarira, C. H., 2013. Link between Antarctic ozone depletion and summer warming over southern Africa. *Nature Geoscience*, 6, 934–939.

McPeters, R. D., Bhartia, P. K., Haffner, D., Labow, G. J., and Flynn, L., 2013. The version 8.6 SBUV ozone data record: An overview. *Journal of Geophysical Research: Atmospheres*, 118, 8032–8039.

Osca, J., Romero, R., and Alonso, S., 2013. Precipitation projections for Spain by means of a weather typing statistical method. *Global and Planetary Change*, 109, 46–63.

Piani, C., Haerter, J. O., and Coppola, E., 2010a. Statistical bias correction for daily precipitation in regional climate models over Europe. *Theoretical and Applied Climatology*, 99, 187–192.

Piani, C., Weedon, G. P., Best, M., Gomes, S. M., Viterbo, P., Hagemann, S., and Haerter, J. O., 2010b. Statistical bias correction of global simulated daily precipitation and temperature for the application of hydrological models. *Journal of Hydrology*, 395, 199–215.

Reason, C. J. C., 2005. Links between the Antarctic Oscillation and winter rainfall over western South Africa. *Geophysical Research Letters*, 32, L07705.

Sicardi, V., Ortiz, J., Rincón, A., Jorba, O., Pay, M. T., Gassó, S., and Baldasano, J. M., 2012. Assessment of Kalman filter bias-adjustment technique to improve the simulation of ground-level ozone over Spain. *Science of the Total Environment*, 416, 329–342.

Son, S.-W., Tandon, N. F., Polvani, L. M., and Waugh, D. W., 2009. Ozone hole and Southern Hemisphere climate change. *Geophysical Research Letters*, 36, L15705.

Teutschbein, C., and Seibert, J., 2012. Bias correction of regional climate model simulations for hydrological climate-change impact studies: Review and evaluation of different methods. *Journal of Hydrology*, 456–457, 12–29.

Teutschbein, C., and Seibert, J., 2013. Is bias correction of regional climate model (RCM) simulations possible for non-stationary conditions. *Hydrology and Earth System Sciences*, 17, 5061–5077.

Thompson, D. W. J., Solomon, S., Kushner, P. J., England, M. H., Grise, K. M., and Karoly, D. J., 2011. Signatures of the antarctic ozone hole in Southern Hemisphere surface climate change. *Nature Geoscience*, 4, 741–749.

Van Der A, R. J., Allaart, M. A. F., and Eskes, H. J., 2010. Multi sensor reanalysis of total ozone. *Atmospheric Chemistry and Physics*, 10, 11277–11294.

Van der A, R. J., Allaart, M. A. F., and Eskes, H. J., 2015. Extended and refined multi sensor reanalysis of total ozone for the period 1970–2012. *Atmospheric Measurement Techniques*, 8, 3021–3035.

Van Roozendael, M., Peeters, P., Roscoe, H. K., De Backer, H., Jones, A. E., Bartlett, L., Vaughan, G. et al., 1998. Validation of ground-based visible measurements of total ozone by comparison with dobson and brewer spectrophotometers. *Journal of Atmospheric Chemistry*, 29, 55–83.

Veefkind, J. P., De Haan, J. F., Brinksma, E. J., Kroon, M., and Levelt, P. F., 2006. Total ozone from the Ozone Monitoring Instrument (OMI) using the DOAS technique. *IEEE Transactions on Geoscience and Remote Sensing*, 44, 1239–1244.

Vila, D. A., de Goncalves, L. G. G., Toll, D. L., and Rozante, J. R., 2009. Statistical evaluation of combined daily gauge observations and rainfall satellite estimates over continental South America. *Journal of Hydrometeorology*, 10, 533–543.

World Meteorological Organization, 2007. Scientific Assessment of Ozone Depletion: 2006. *Global Ozone Research and Monitoring Project Report No. 50,* Geneva, Switzerland.

16 Water Quality Monitoring in a Lake for Improving a Drinking Water Treatment Process

16.1 INTRODUCTION

The Total Organic Carbon (TOC) content is the measure of organic molecules of carbon in water or sediment or both. TOC is the sum of Dissolved Organic Carbon (DOC), Particulate Organic Carbon (POC) or Suspended Organic Carbon (SOC), and colloids, which serves as a key water quality parameter in lakes, rivers, and reservoirs that the water supply industry is required to monitor. TOC is introduced to surface waters from both natural and anthropogenic sources. The headwater lakes in peat-rich areas have high organic matter concentration commonly. The level of TOC concentrations in surface waters also affects pH, redox reactions, bioavailability of metals, and the sorption capacity of suspended solids with regards to hydrophobic organic chemicals (Thurman, 1985; Parks and Baker, 1997). Whereas naturally occurring sources of TOC include humic substances, as well as degraded vegetation and animal matter flushed out by stormwater runoff in a watershed (Thurman, 1985; GEAS, 1994; Bayram et al., 2011), anthropogenic sources include fertilizers captured by stormwater runoff and irrigation return flows, a release of contaminants from a spill or improper usage in the watershed, as well as pesticides, surfactants, and solvents from sewage treatment plants (Visco et al., 2005).

The surface water treatment rules promulgated by the United States Environmental Protection Agency (USEPA) require water disinfection in treatment plants to protect consumers from microbiological contaminants. Disinfection ByProducts (DBPs) are formed when organic matter (i.e., TOC) reacts with oxidants, such as free chlorine residuals in drinking water disinfection networks. The USEPA also regulates and restricts the concentration of the DBPs below the levels that are harmful to human health (USEPA, 1998). DBPs in water after the drinking water treatment include TriHaloMethantes (THMs) and HaloAcetic Acids (HAAs) and the sum of them is called Total Trihalomethanes (TTHMs). TTHMs is a family of suspected carcinogens and has been related to possible birth defects (Aschengrau et al., 1993; Bove et al., 1995; Dodds et al., 1999; Klotz and Pyrch, 1999). Knowledge of TOC concentrations spatially and temporally in the source water body enables treatment plant operators to alter the treatment strategy to limit DBP generation. This can include adjustment of conventional water treatment processes or activation of advanced treatment process units. As described by TCEQ (2002), one common example of adjustment of conventional water treatment processes regards changing coagulant feed rates or altering the pH for improved TOC removal through coagulation. The inclusion of advanced treatment process units, however, such as Granular Activated Carbon (GAC) adsorption, membrane filtration, and ion exchange can be very costly. Hence, it is necessary to know in real time the source water TOC concentration and variations for smooth implementation of all possible operational adjustments in conventional water treatment processes. This can be done via integrated sensing, monitoring, networking, and control toward decision making.

The National Source Water Assessment and Protection Program was established based on the Safe Drinking Water Act Amendments of 1996, targeting drinking water sources for all public water systems in the United States. Monitoring TOC for drinking water source protection reports can be an expensive and time consuming process. Manual water sampling and laboratory TOC measurement of source water has been widely used by the water supply industry, which has a substantial cost and a significant time delay (O'Connor, 2004). Purchasing equipment such as TOC analyzers for on-site measurement of TOC in the water body around the intake costs between $20,000 and $30,000, and using an external laboratory certified for TOC analysis takes 2 to 4 weeks to obtain results (TCEQ, 2002). For large TOC changes in source water driven by natural disaster events or anthropogenic activities, the time delay and lack of information on TOC spatial distribution can unexpectedly delay or even impede the preparation of operational adjustments in drinking water treatment during emergency response. The use of air-borne or space-borne remote sensing for an accurate and timely TOC estimation is, therefore, favored (Chang et al., 2014).

The Integrated Data Fusion and Mining (IDFM) techniques introduced in Chapter 13 may support daily monitoring capacity for providing timely water quality conditions in space-time domain via a series of near-real-time TOC concentration maps around the water intake and associated water body. Instead of having to sample 3 weeks ahead of time to collect TOC data available for water quality reports in drinking water treatment, IDFM may be regarded as an early warning system that enables the treatment plant operators to account for instantaneous TOC variations for all clear days throughout the year. Niche applications of IDFM include predicting whether TOC levels have quickly increased right after heavy rains and aiding in spatial analyses for adaptive management when water treatment plants have multiple intakes located at reservoirs, rivers, or lakes simultaneously from which to pump source water. The latter case might trigger an advanced spatial analysis to generate the optimal pumping strategies at each intake location over time.

Optical sensors onboard satellites can detect the surface reflectance emissions of selected light spectrums with medium to high resolution images. Since reflectance is a function of wavelength, the measured reflectance will vary throughout the wavelength of the electromagnetic spectrum for any given substance. Yet every dissolved substance gives off a different spectral signature or pattern, some of which can overlap, for instance, particulate organic carbon influenced remotely sensed chlorophyll-a readings from phytoplankton (Smith and Baker, 1978). In more recent decades, the use of remote sensing for organic carbon detection has been proven effective by several studies (Stramski et al., 1999; Stedmon et al., 2000; Ohmori et al., 2010). TOC cannot be easily identified based on its unique spectral response and differentiated from other compounds in the water column, and the POC portion is not measurable by remote sensing. It is noticeable that a peak or valley in a reference curve will be incurred on the spectral signature graph when a substance is especially reflective or absorptive to a specific wavelength of light. For this study, the spectral reflectance peaks of interest for TOC are based on published experiments measuring the peaks of chromophoric or Colored Dissolved Organic Matter (CDOM) (Vertucci, 1989; Arenz et al., 1995; Menken et al., 2005). CDOM is the light absorbing fraction of dissolved organic carbon. In the three different case studies cited above, the spectral reflectance associated with varying levels of CDOM was measured in over 38 different lakes and reservoirs. The observed spectral peaks are detailed in Table 16.1. This information can be very instrumental for space-borne remote sensing.

To provide a daily monitoring capacity from space, the individual Landsat and MODIS sensors have their own stand-alone advantages and disadvantages for CDOM detection in water in terms of revisit times, spatial resolution, and the ability to sense most of the reflected light at the CDOM peaks. One of the possible solutions for overcoming this technological challenge is to incorporate the best spatial and temporal features of MODIS and Landsat sensors so that a composite image may be generated with both high temporal and spatial resolutions when spectrally sensitive enough to capture the presence of TOC. This is feasible via the use of data fusion algorithms followed by effective feature extraction algorithms, which is exactly the essence of IDFM, as discussed in Chapter 13.

Within the context of IDFM, it is imperative that satellites with sensors that detect light at bandwidths corresponding to reflectance peaks be chosen when determining TOC concentrations

TABLE 16.1

CDOM Spectral Peaks as Determined from Case Study Analysis

CDOM Spectral Peaks (nm)	Sampling Locations	Sampling Instrument	Source
570	25	Spectron Engineering SE-590 spectrometer	Menken et al. (2005)
550, 571, 670, 710	8	Scanning spectrophotometer	Arenz et al. (1995)
560, 650–700	5+	Spectron Instrument CE395 spectroradiometer	Vertucci (1989)

Source: Chang, N. B. et al. 2014. *International Journal of Remote Sensing*, 35, 1064–1093.

based upon surface reflectance (Table 16.2) (Chang et al., 2014). Matching the bands for data fusion is based upon corresponding band centers instead of the same band number. For example, Landsat band 1 is fused with MODIS band 3 since they represent similar portions of the electromagnetic spectrum.

The ability for MODIS and Thematic Mapper (TM) onboard Landsat to detect the main spectral peaks of CDOM is shown in Figure 16.1, in which two main spectral peaks for CDOM are marked in gray for band selection. To ensure success in data fusion, the bands for these two instruments must overlap with the two CDOM spectral peaks for the MODIS and TM to properly determine TOC concentrations. For the first CDOM peak between 550 and 571 nm, whereas the TM can detect the entirety of the peak, the MODIS can detect most of the reflected light caused by CDOM. By the same token, for the second CDOM peak between 650 and 710 nm, TM can detect the first two-thirds of the CDOM peak, whereas MODIS detects only the first half. Thus, TM not only enhances spatial resolution of the fused image, but also provides sufficient spectrally-relevant data for TOC estimation (Chang et al., 2014).

This study conducted TOC prediction based on remotely-fused images of Landsat 5 TM, Landsat 7 ETM+, and Terra MODIS (Table 16.3), producing daily fused images with 30-m resolution. The emphasis of feature extraction was placed on using Genetic Programming (GP) as an alternative feature extraction algorithm to Artificial Neural Networks (ANNs) (Chang et al., 2013, 2014). In addition, a performance comparison between the GP model and a traditional two-band ratio model (i.e., a geophysical model) was carried out as a baseline to further justify the use of more complex and computationally intensive machine learning techniques.

TABLE 16.2

Band Selection between Landsat 5/7 and MODIS in IDFM

Landsat 5/7 TM/ETM+ Band	Landsat Bandwidth (nm)	Terra MODIS Band	MODIS Bandwidth (nm)
1	450–520	3	459–479
2	520–600	4	545–565
3	630–690	1	620–670
4	760–900	2	841–876
5	1550–1750	6	1628–1652
7	2080–2350	7	2105–2155

Source: Chang, N. B. et al. 2014. *International Journal of Remote Sensing*, 35, 1064–1093.

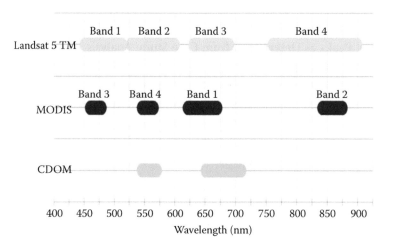

FIGURE 16.1   Band selection in support of data fusion in IDFM. (From Chang, N. B. et al. 2014. *International Journal of Remote Sensing*, 35, 1064–1093.)

TABLE 16.3
Satellite Products Utilized in This IDFM Study

Satellite Sensor	Product Selection	Spatial Resolution	Temporal Resolution	Bands Used
Terra MODIS	Surface Reflectance (MOD09GA)	250/500 m	Daily	1–4,6,7
Landsat 5 TM Landsat 7 ETM+	Surface Reflectance	30 m	16 Days	1–5,7

Source: Chang, N. B. et al. 2014. *International Journal of Remote Sensing*, 35, 1064–1093.

16.2 STUDY REGION

The William H. Harsha Lake is located in Clermont County, Ohio, roughly 40 km (25 miles) east of Cincinnati, and covers an area of 8,739,360 m² (2,160 acres) (Figure 16.2a). Located near Batavia and serving 29,948 people with 11,664 service connections, the Bob McEwen surface water treatment plant has a design capacity of 37,600 m³/day (10 MGD) (marked in red in Figure 16.2b). Surface water is withdrawn from Harsha Lake, which was created in 1973 by constructing a 62-m (205-foot) dam across the East Fork Little Miami River. There are three 300 HP pumps used at the remote intake structure on Harsha Lake, and each pump is rated at 99.12 m³/min (3,500 gallons per minute) (Ohio EPA, 2004). Since its impoundment, the lake has prevented over $77 million in flood damages, and it has generated $32.7 million in visitor expenditures (USACE, n.d.).

The raw water characterization record is shown in Table 16.4, in which the arithmetic average of TOC is 5.7 mg/L; that is deemed quite high in general. According to the Ohio Environmental Protection Agency, TOC values over 6 mg/L are considered high, and DBPs form in water with a TOC value greater than 6 mg/L (Ohio EPA, 2004). In 2012, to deal with the TOC issue and the fast growing population, the nominal capacity of the existing Bob McEwen Water Treatment Plant was expanded from 37,600 m³/day (10 MGD) to 75,200 m³/day (20 MGD) and the expansion project also included the addition of a GAC facility to enhance the level of treatment for TOC removal at the water treatment plant.

(a)

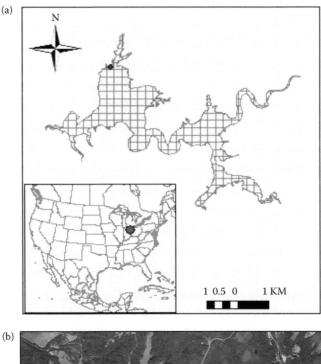

(b)

FIGURE 16.2   Geographical location of the study region. (a) Location of the William H. Harsha lake (Chang et al., 2014) and (b) location of the drinking water treatment plant and the water intake (Google, map data ©2012 Google).

16.3   STUDY FRAMEWORK

The IDFM procedural steps undertaken in this study are depicted in Figure 16.3. As in Chapter 13, the chart is split into five main steps: (1) data acquisition, (2) image pre-processing, (3) data fusion, (4) training and validating the GP models, and (5) TOC concentration mapping. Step one pertains to obtaining the Landsat and MODIS swath path images covering Harsha Lake. The second step involves the preparation of the relevant images for data fusion, including screening, gap filling, image corrections, and so on. The data fusion step needs to be performed prior to performing feature extraction and classification toward generating TOC predictions. The data fusion step in this study is tied to the use of the Spatial and Temporal Adaptive Reflectance Fusion Model (STARFM) at the pixel level, although other data fusion algorithms can also be applicable (Table 16.5). The computer

TABLE 16.4

Harsha Lake Water Quality Characterization

Raw Water Characterization		
Parameter/Dose	Range	Average
Turbidity (ntu)	3.00–38.70	9.10
UV-254 (cm^{-1})	0.15–0.23	0.18
pH	7.10–7.86	7.60
Temperature (°C)	11.50–17.00	14.25
Alkalinity (mg/L as CaCO$_3$)	98.00–108.00	103.00
Total Organic Carbon (mg/L)	5.60–5.90	5.70
Total Manganese (µg/L)	17.00–618.00	120.00

Source: Chang, N. B. et al. 2014. *International Journal of Remote Sensing*, 35, 1064–1093.

processing time of STARFM is of importance in such an early warning system. The fourth step involves relating the ground-truth data to fused image band data into a machine learning algorithm, such as a GP algorithm or a geophysical model (e.g., a two-band model) for comparison. In this study, Discipulus® (Francone, 1998) was used to solve the GP model for the estimation of TOC concentrations via a highly nonlinear equation. Last, step five specifically applies the selected GP model to generate the TOC concentration maps based on all the required band values associated with the fused images. The five essential steps are delineated in greater detail below.

- *Data Acquisition* (Figure 16.3; Step 1): The ground-truth data for TOC in Harsha Lake were collected by the United States Army Corps Engineers (USACE) during 2008 and 2009 and the USEPA from 2010 to 2012. In addition, MODIS and Landsat images were acquired from the online Data Pool managed by the Land Processes Distributed Active Archive Center (LP DAAC) of the National Aeronautics and Space Administration (NASA), and the Earth Resources Observation and Science (EROS) Center of the United States Geological Survey (USGS). Maintained by the LP DAAC and USGS EROS Center, these images were handled by the Global Visualization Viewer. Dates for downloading MODIS and Landsat imageries were based on two criteria: (1) each date for downloading MODIS imageries must be cloud free so as to use the ground-truth data fully in inversion models and (2) a Landsat image is required before and/or after each of the ground-truth dates. Otherwise, the ground-truth date was given up due to insufficient information for data fusion.
- *Image Pre-processing* (Figure 16.3; Step 2): The acquired MODIS data are at a level-2G basis, and the data have been radiometrically calibrated and atmospherically corrected to account for scattering and aerosols (Vermote et al., 2011). The Landsat data is on a level-1T basis, with radiometric and geometric corrections (USGS, n.d.). As denoted by step 2 of Figure 16.3, ArcGIS, a mapping and spatial analysis software was used to process the images for the subsequent data fusion. It was necessary to perform the following steps on the Landsat images: (1) perform atmospheric correction by applying MODIS 6S radiative transfer code using the LEDAPS toolbox supplied by NASA for this operation; (2) carry out reprojection to the Universal Transverse Mercator (UTM) zone 16 North; and (3) crop out land data around Lake Harsha (Figure 16.4). In addition, the following steps were taken to process the MODIS images: (1) perform reprojection to the UTM zone 16

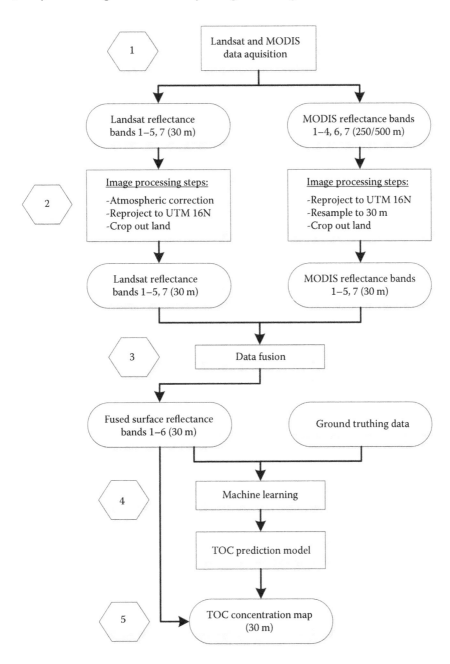

FIGURE 16.3 Methodological flowchart in the TOC concentration retrieval procedures. (From Chang, N. B. et al. 2014. *International Journal of Remote Sensing*, 35, 1064–1093.)

North; (2) resample to a 30-m spatial resolution, and (3) crop out land data around Lake Harsha (Figure 16.4). The rationale for cropping out land data is to reduce the potential for the surrounding land to contaminate the fused image. Narrow areas of the lake were blacked out when water channels were smaller than MODIS' 250/500 m resolution to avoid contamination. The contamination may impact the data fusion process when the STARFM algorithm searches through neighboring pixels along the lake shorelines.

TABLE 16.5

Fusion Techniques and Their Associated Fusion Level

Fusion Technique	Fusion Level	Enhanced Properties	Source
Color Composites (RGB)	Pixel	Spectral	Pohl and van Genderen (1998)
Intensity-Hue-Saturation (HIS)	Pixel	Spectral	Pohl and van Genderen (1998)
Principal Component Analysis	Pixel	Spatial	Pohl and van Genderen (1998)
Wavelets	Pixel	Spatial	Pohl and van Genderen (1998)
High Pass Filtering (HPF)	Pixel	Spatial	Pohl and van Genderen (1998)
STARFM	Pixel	Spatial/temporal	Gao et al. (2006)
Bayesian Inference	Decision/feature	Spatial/temporal	Robin et al. (2005)
Dempster-Shafer	Decision/feature	Spatial/temporal	Zeng et al. (2006)

Source: Chang, N. B. et al. 2014. *International Journal of Remote Sensing*, 35, 1064–1093.

- *Data Fusion* (Figure 16.3; Step 3): Fused images with both near and off-nadir viewing angles of the lake corresponding to the ground-truth dates were created for every clear day once a ground-truth sample was available. A MODIS image was acquired for each ground-truth observation. Both Landsat and MODIS images taken before and after the ground-truth date were grouped for data fusion. The STARFM algorithm performs data fusion for Landsat and MODIS images at the pixel level based upon spatial, temporal, and spectral similarities (Tables 16.2 and 16.3). It allows for a series of daily predictions of TOC concentrations in the water body. The spectral reflectance value for each pixel of the MODIS image is a conglomeration of the surface reflectance from each object in the 250 by 250 m^2 area for MOD09GQ and 500 by 500 m^2 for MOD09GA (Chang et al., 2014). In data fusion, spectral reflectance values provided by the Landsat image are an average of objects contained within a 30 by 30 m^2 pixel which may complement the extremely coarse spatial resolution of MODIS spatially. As the STARFM program translates through the matrix of pixels in the Landsat and MODIS images to generate the fused images, it may select a central pixel every few steps and reassign the pixel's value based upon candidate pixels that are both near the central pixel and spectrally similar (Gao et al., 2006). The candidate pixels are then filtered out if they exhibit more change over time than the central pixel or if the spectral features of the candidate pixel are greater than the difference between the spectral features of the central pixel in the Landsat and MODIS images (Gao et al., 2006). However, one must be cautious in fusing the data streams from Landsat and MODIS as Landsat band 1 does not have the same spectral range as MODIS band 1. Instead, band 1 of Landsat corresponds to Band 3 of MODIS and so on. Table 16.2 details the proper pairing of those bands for the data fusion of MODIS and Landsat. During the data fusion, the surface reflectance of the central pixel will be generated from the group of selected candidate pixels

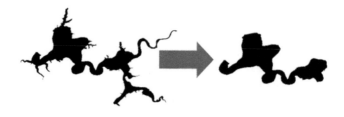

FIGURE 16.4   Cropping out narrow channels of the lake to reduce land surface reflectance contamination. (From Chang, N. B. et al. 2014. *International Journal of Remote Sensing*, 35, 1064–1093.)

via a weighted average approach that is applied based upon how likely it is that each of the selected candidate pixels could represent the central pixel (Gao et al., 2006). Higher weighting factors are assigned if the candidate pixel is spectrally and temporally similar to the central pixel; in addition to spectral information, weighting factors are also influenced by its geometric distance between the candidate pixels and the central pixel (Gao et al., 2006). An integrated weighting scheme will be produced in the end according to all factors. Through this process, the synthetic Landsat image is generated based on the input candidate pixels in the MODIS image taken during the desired prediction date. The computational time requirement for fusing Landsat and MODIS images for a lake of this size was under 5 seconds for the computer used in this experiment (computer specifications: Intel® Core™ i7-3720QM CPU at 2.6 GHz, 8 192 MB RAM, and 500 GB hard drive).

- *Machine Learning or Data Mining* (Figure 16.3; Step 4): There are many feature extraction methods, as introduced in Chapters 6 and 7. The GP model was selected in this study for nonlinear feature extraction since CDOM has complex reflective patterns. The GP model is presented as a series of mathematic or logic operations that must be applied in sequence to the fused surface reflectance band data in order to generate a TOC concentration value via a nonlinear equation. The GP model used in this study was specifically developed to relate the band values (i.e., surface reflectance values) associated with the fused images to the ground truth data (i.e., TOC values) using the Discipulus® software package, created by Francone (1998). Discipulus® is designed to sort through GP models using supervised classification techniques and determine the 30 best models based on the fitness of the training and validation data (Francone, 1998). The arithmetic operations selected for training the GP model include addition, subtraction, division, multiplication, absolute value, square root, and trigonometric functions. The suggested values of 95% and 50% were used for mutation and crossover frequency (Francone, 1998). Last, the initial program size was set to 80 Mb with a max program size of 256 Mb. The max program size is gradually increased if an accurate GP model cannot be developed based on the available training and validation data sets. This improves the explanatory power of the GP model by increasing the number of mathematical operations to relate surface reflectance to TOC. After the Discipulus® program has finished creating models, the 30 models with the best overall performance are saved and analyzed. An advantage of Discipulus® is its capability to adapt current models with new ground truth data (Francone, 1998). The ground truth data set was divided into two smaller data sets for model training and calibration in advance. Since Discipulus® ranks the models based on the average fitness between the training and calibration data sets, the model that yields high fitness values for both training and calibration was selected for the GP model in this study. As an early warning system for water treatment plant operators is necessary, it is imperative that the selected GP model be capable of predicting peak TOC concentrations in the lake. This ensures that the plant operators are able to observe and track plumes of TOC in the lake that are in the vicinity of the source water intake of the treatment plant. As additional TOC samples are collected, the model can be updated to reflect hydrological and anthropogenic changes over time. With this knowledge, the treatment operations can be adjusted to minimize the production of disinfection by-products. In this study, the GP model was further compared against a traditional two-band model (Vincent et al., 2004) which was solved through a linear regression model in MATLAB® using band ratios instead of individual bands as explanatory variables. The generic form of the two-band ratio model (i.e., a bio-optical model) is shown in Equation 16.1:

$$C_{TOC} = A\left(\lambda_1/\lambda_2\right) + B \tag{16.1}$$

where C_{TOC} is the concentration of TOC, A is the slope, λ_1 is the wavelength of the first band sensitive to CDOM, λ_2 is the wavelength of the second band sensitive to CDOM in the

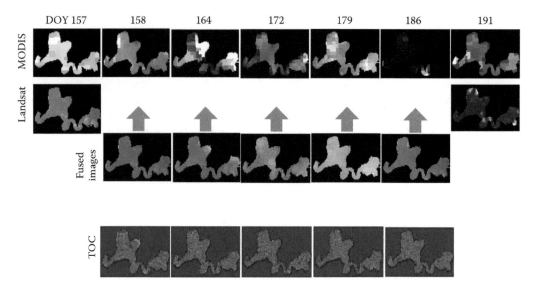

FIGURE 16.5 The final TOC concentration mapping in the IDFM method. (From Chang, N. B. et al. 2014. *International Journal of Remote Sensing*, 35, 1064–1093.)

ratio, and *B* is the intercept. The same training and calibration data sets used for creating the GP model were employed again to train and calibrate the two-band model toward a fair comparison.

- *Concentration Map Generation* (Figure 16.3; Step 5): The GP model translates the surface reflectance values to TOC concentrations and maps the TOC concentrations based on the derived nonlinear GP equation for those new dates when the newly fused reflectance images become available. The TOC concentration can be obtained pixel by pixel via the derived nonlinear GP equation that is in a functional form (i.e., a convoluted form) in terms of surface reflectance values. This estimation process is then repeated for each of the pixels making up the lake map. Certain bandwidths may have a stronger explanatory power in the determination of the TOC concentration because not all bandwidths were used in the determination of the TOC concentration in some iterations of the GP model. Statistical indexes for model assessment and selection between the two-band model and the GP model are listed below. Model performance was analyzed using four statistical indexes in this study. The indices include the root mean square error (rmse), ratio of the standard deviations (CO), mean percent of error (PE), and the square of the Pearson product moment correlation coefficient (RSQ). These equations have been defined in Chapter 7 (see Equations 7.19–7.23). Model performance ranks well when the PE approaches zero, and the RSQ should ideally equal 1 for a good model. The final selected GP model can then be used for mapping TOC concentration daily (Figure 16.5).

16.4 TOC MAPPING IN LAKE HARSHA USING IDFM

16.4.1 FIELD CAMPAIGN

The sampling campaign in this study was carried out from 2008 to 2012 by the USACE and the National Risk Management Research Laboratory of the USEPA (Table 16.6). A statistical analysis

TABLE 16.6

Ground-Truth Acquisition Record

Date	TOC (mg/L)	Date	TOC (mg/L)	Date	TOC (mg/L)
August 20, 2008	4.50	June 16, 2010	6.90	July 27, 2011	6.20
May 18, 2009	8.36	June 16, 2010	7.20	July 27, 2011	7.70
June 29, 2009	9.12	June 16, 2010	7.20	July 28, 2011	8.50
July 6, 2009	8.27	September 7, 2010	5.92	July 28, 2011	11.00
July 13, 2009	8.66	September 8, 2010	5.50	July 28, 2011	10.00
July 21, 2009	7.77	September 8, 2010	5.70	August 1, 2011	5.92
August 3, 2009	7.75	September 8, 2010	6.70	August 22, 2011	5.37
August 17, 2009	7.65	September 8, 2010	5.60	August 23, 2011	14.00
August 31, 2009	7.33	September 9, 2010	5.60	August 23, 2011	12.00
September 14, 2009	7.73	September 9, 2010	5.80	August 24, 2011	11.00
September 28, 2009	6.45	March 2, 2011	6.14	August 24, 2011	12.00
October 5, 2009	6.33	March 17, 2011	6.91	August 24, 2011	6.30
October 26, 2009	5.82	April 13, 2011	7.35	August 29, 2011	5.49
April 12, 2010	8.23	May 23, 2011	7.01	October 5, 2011	5.38
April 19, 2010	8.96	June 1, 2011	7.05	November 2, 2011	5.45
May 3, 2010	7.26	June 7, 2011	7.57	November 17, 2011	5.99
May 24, 2010	7.88	June 13, 2011	6.98	May 24, 2012	5.70
June 14, 2010	6.72	June 21, 2011	5.98	June 13, 2012	4.30
June 15, 2010	7.20	July 5, 2011	6.34		
June 15, 2010	6.90	July 11, 2011	7.03		

Source: Chang, N. B. et al. 2014. *International Journal of Remote Sensing*, 35, 1064–1093.

Note: Grey images correspond to data used GP model training and the data in the blue cells were designated for GP model validation.

of the ground-truth data is provided in Table 16.7 to provide a level of general understanding of the TOC characteristics and fluctuations in the lake (Change et al., 2014). It is observed that the TOC level was as high as 10 mg/L or above in many sampling days from which the arithmetic average of TOC is 7.2 mg/L; this is deemed very high in general, requiring further attention in water quality management.

The training set was allotted 67% of the ground-truth data, which corresponds to 39 of the grey colored cells in Table 16.6, to aid the training and calibration procedure. The remaining 33% or

TABLE 16.7

Ground-Truth Data Analysis

Parameter	Value
Ground-truth time period	2008–2012
Number of ground-truth samples	58
Average TOC value (mg/L)	7.20
Maximum TOC value (mg/L)	14.00
Minimum TOC value (mg/L)	4.30
Sample standard deviation (mg/L)	1.90

Source: Chang, N. B. et al. 2014. *International Journal of Remote Sensing*, 35, 1064–1093.

19 ground-truth observations of the blue colored cells in Table 16.6 were used for calibration. It is necessary to ensure that both the training and calibration data sets are exposed to the widest range of TOC values available to increase the accuracy of the model's prediction at extremes. Determining which observations were selected for training and calibration data sets is based purely on the measured concentration without regard to the temporal aspects. First, the data were sorted from low to high values, and then 67% of the low, medium, and high TOC concentrations were allotted into the training data set. The remainder data were employed for calibration. After training, the calibration stage confirms whether the model that is well suited for calculating TOC concentrations may aptly predict a peak TOC concentration when using the calibration data set. The final selection of a model for feature extraction must be based on the prediction capabilities when both the training and calibration data sets exemplify fitness (Francone, 1998).

16.4.2 Impact of Data Fusion

It is interesting to know whether the MODIS-based or fused MODIS-Landsat will perform better for predicting TOC. Both the fused and MODIS images were thus separately processed for 75 runs in Discipulus®. A single run is characterized by the Discipulus® building a model until the maximum size of 256 Mb has been reached. Thus, both the MODIS-based and fused GP models were appropriated equal amounts of computational effort. The correlation between the predicted and observed TOC concentration for the MODIS surface reflectance training and calibration data sets are presented in Figure 16.6. The coefficients of determination for the training data and calibration data sets are 0.6836 and 0.4570, respectively. These results indicate a strong correlation for the training data set, and a moderate correlation for the calibration data set. Examination of Figure 16.6 shows that the model performs well when predicting concentrations within one standard deviation of the average TOC value. Yet, the model underestimates higher TOC concentration values, since several predicted TOC values fall below the 45-degree line past 9 mg/L. The corresponding correlation plot for the GP model using the fused data as inputs is shown in Figure 16.7. The GP model using the fused data yields better coefficients of determination of 0.8745 and 0.5635 for the training and calibration data sets, respectively. These values reveal that a strong relationship exists between the predicted and observed TOC values. Unlike the MODIS-based GP model that underestimated TOC values past 9 mg/L, the GP model using the fused data stands out for predicting peak concentrations without a bias toward under or overestimation. A more lucid performance comparison of the MODIS-based GP model (Figure 16.6) and the GP model using the fused data (Figure 16.7) is

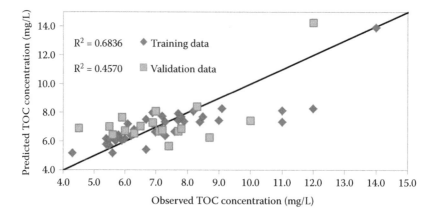

FIGURE 16.6 Correlation between predicted versus observed TOC concentrations formulated using the MODIS surface reflectance as inputs to the GP model. (From Chang, N. B. et al. 2014. *International Journal of Remote Sensing*, 35, 1064–1093.)

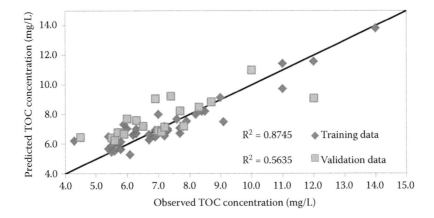

FIGURE 16.7 Correlation between estimated versus observed TOC concentrations formulated using the fused image surface reflectance as inputs to the fusion-based GP model. (From Chang, N. B. et al. 2014. *International Journal of Remote Sensing*, 35, 1064–1093.)

presented by plotting the predicted and observed results as a time series in Figure 16.8 collectively. Through visual examination of Figure 16.8, it can be observed that both GP models exhibit moderate to strong performance, since the predicted values (dotted lines) replicate the temporal trends of observed TOC values in the lake (Chang et al., 2014).

A comparison between Figures 16.6 and 16.7 indicated that both MODIS-based and the fusion-based GP models had little error when predicting TOC concentrations below 9 mg/L.

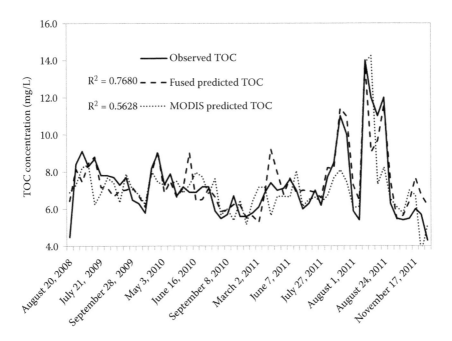

FIGURE 16.8 Time series plots comparing the predicted TOC values from the MODIS and fused GP models. The predicted TOC values based on the MODIS surface reflectance share similar accuracy to the TOC values with predicted ones using the fused surface reflectance; however, the GP model using the fused image surface reflectance as inputs excelled at predicting the peak TOC values. (From Chang, N. B. et al. 2014. *International Journal of Remote Sensing*, 35, 1064–1093.)

The MODIS-based GP model exhibited an underestimation bias for TOC concentrations at and above 9 mg/L, but this was not the case in the fusion-based GP model. Possible explanations for this phenomenon are associated with sensor limitations and possible constraints embedded in the GP modeling procedure. First, band 1 of the MODIS sensor that is centered at 645 nm does not capture the entire CDOM spectral feature between 650 and 710 nm, given that its upper range is 670 nm only. However, the fused image benefits from additional spectral data of Landsat band 3, which is centered at 660 nm with an upper range of 690 nm. The MODIS-based GP model cannot distinguish itself in the comparison due to the constraints imposed while training the GP model. The best GP model was selected after 75 runs, and this limitation permitted insufficient time for the MODIS-based GP model to build the nonlinear equation for cases at higher TOC concentrations. Since both models experienced the same training time, it is indicative that the fusion-based GP model stood out due to a better input with fused information given the same number of runs.

16.4.3 IMPACT OF FEATURE EXTRACTION ALGORITHMS

- *GP modeling – identifying important spectral bands*: One of the GP models depicts the nonlinear relationship between TOC and the fused image band data, shown below in Equation (16.2) for demonstration.

$$C_{TOC} \ (mg/L) = \left| 0.19907 * X_3 - X_4 \right| - X_2 - \frac{4914.3}{v_3^2} + v_1 \right|^{\frac{1}{4}} \tag{16.2}$$

Variables X_1, X_2, X_3, and X_4 are defined as follows for simplification purposes:

$$X_1 = \sin\left(|\sin(|1.119 * \sin\left(|v_3 - 0.36978|\right) + v_3|) - v_3|\right)$$

$$X_2 = \cos\left(X_1 + v_2 - v_1 - 0.21070 + v_5\right) * v_1$$

$$X_3 = \left[\left(\left| 1.709 * \left| -\frac{4914.3}{v_3^2} + X_1 + v_2 - v_1 - 0.21070 + v_5 - 28.990 * \frac{X_2}{v_3} \right|^{\frac{1}{2}} + v_0 \right|^2 + 1.7099 * v_4 \right)^2 \right]^{\frac{1}{2}} + v_0$$

$$X_4 = 0.19907 * X_2 - \frac{978.30}{v_3^2} - 0.19907 * v_2 - X_1 + v_1 + 0.21070 - v_5 + 28.990 * \frac{X_2}{v_3}$$

Variables v_0, v_1, v_2, v_3, v_4, and v_5 characterize the band data of the fused images (Table 16.2). As previously noted, the Landsat and MODIS images were fused in accordance with their bandwidths rather than their band numbers. The frequency of use explains which bands the GP model found most useful in explaining the relationship between surface reflectance and TOC concentration. The frequency of use of band data that describes how often a variable (a specific bandwidth) was used among the 30 best GP models is beneficial for determining which satellites have the most useful bands for monitoring the lake, as well as limiting the amount of band data to be downloaded and stored for feature extraction. 100% frequency of use means that the variable was used to compute TOC in all 30 models (Chang et al., 2014). The GP model using fused images developed to fit the observed data curve in Figure 16.9 is explicitly depicted in Appendix A of Vannah (2013). For the 30 best candidate models being generated, the frequency of use for each bandwidth is shown in Table 16.8 as well as each variable corresponding to the surface reflectance of a specific bandwidth. v_1 was used in all of the models regardless of whether the model utilized the fused images, while

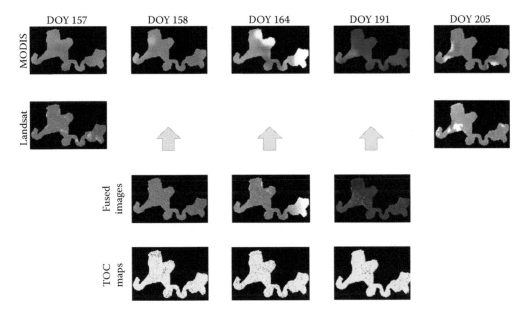

FIGURE 16.9   Data fusion and TOC map generation for June–July 2011. This represents both the gap-filling capabilities of the IDFM technique and the ability to predict TOC concentrations for cloud free regions of the water body. The top row is the daily MODIS images, and the second row is comprised of 2 Landsat images. STARFM is used to create high resolution synthetic images for ground-truth dates. Surface reflectance values at sampling locations in the fused images are used to train the genetic programming model for TOC map generation featuring the lake. (From Chang, N. B. et al. 2014. *International Journal of Remote Sensing*, 35, 1064–1093.)

v_0 and v_2 were determined to be of varying significance from 67% to 97% and from 80% to 97% of the MODIS-based GP and fusion-based GP models, respectively. It is noticeable that although v_1 was heavily used by both models, the similarities between band usage across these two types of GP models deviate over the subsequent bands.

Overall, v_1 and v_2 would be commonly used in most occasions for TOC estimation, since these bands correspond with the spectral reflectance peaks for TOC occurring at 550 and 675 nm as seen in Figure 16.1. During the inversion process, the MODIS-based model still prioritized the use of the v_2 band in 80% of the best 30 GP models, while the v_4 band in 87% of the best 30 GP

TABLE 16.8

Frequency of Use for Bands in the TOC GP Model

Variable	Band Number	Bandwidth (nm)	MODIS-Based GP Frequency of Use (%)	Fusion-Based GP Frequency of Use (%)
v_0	MODIS Band 3 Landsat Band 1	459–479 450–520	67	97
v_1	MODIS Band 4 Landsat Band 2	545–565 520–600	100	100
v_2	MODIS Band 1 Landsat Band 3	620–670 630–690	80	97
v_3	MODIS Band 2 Landsat Band 4	841–876 760–900	70	93
v_4	MODIS Band 6 Landsat Band 5	1628–1652 1550–1750	87	80
v_5	MODIS Band 7 Landsat Band 7	2105–2155 2080–2350	63	70

Source: Chang, N. B. et al. 2014. *International Journal of Remote Sensing*, 35, 1064–1093.

models was given a slightly higher significance. By analyzing Figure 16.1 once more, a possible explanation can be observed for why the fusion-based model uses v_2 in 97% of the best models, while it is only used 80% of the time in the MODIS-based models. Landsat is capable of detecting more of the CDOM peak occurring at v_2 (Band 3 in Figure 16.1) than MODIS. Since the key spectral information from Landsat band 3 has been smoothly fused with MODIS band 1 to form a single synthetic image, it is indicative that such an advantage associated with using data fusion largely helps identify the CDOM peak. Lastly, the variables v_4 and v_5 were used in the least priority in the fusion-based GP model, which gives rise to an insight that there is not a strong relation between the TOC concentration in the water and TOC's reflectance at these wavelengths. Overall data streams from the 6 fused bands are collectively used to generate daily TOC maps of the lake with varying levels of contribution.

- *GP model versus two-band model*: The four statistical indexes presented in the previous section are used to compare the performance between the MODIS-based and fusion-based GP models, as well as a traditional two-band ratio model. The resulting two-band model is shown in Equation (16.3).

$$C_{TOC} = -0.04630 v_3/v_5 + 7.2087 \tag{16.3}$$

where C_{TOC} is the concentration of TOC. A comparison using the four statistical indexes is presented in Table 16.9.

As shown in Table 16.9, the mean of predicted values of the two-band model is equal to the mean of the observed values. However, a more detailed analysis of the predicted values indicates poor performance. The *rmse* of 1.716 mg/L and the *CO* of 0.394 are far from the ideal values of 0 mg/L and 1. Yet the *PE* of 5.046% is relatively low, which is likely due to the fact that the linear two-band model yielded predicted values close to the observed average of 7.276 mg/L, and the range of TOC values in the lake were typically between 6 and 9 mg/L. Last, the R^2 value for this model was 0.1974, which means that only 19.74% of the variation is explained by the two-band model. Such finding sheds light on the necessity of using multiple methods to assess prediction accuracy. Given that the traditional two-band model exhibited poor TOC prediction capabilities in terms of all statistical indexes relative to GP models, there is an acute need to employ a more powerful modeling technique.

TABLE 16.9

Observed vs. Predicted TOC Values and Indices of Accuracy for the Traditional Two-Band Model and the GP Models Created from MODIS and Fused Images

	2-Band Model	MODIS-Based GP Model	Fusion-Based GP Model
TOC observation mean (mg/L)	7.276	7.276	7.276
TOC prediction mean (mg/L)	7.276	7.085	7.433
Percent difference of the means (%)	0.000	2.629	−2.166
Root mean square error (mg/L) (*rmse*)	1.716	1.248	0.900
Ratio of st. dev. (*CO*)	0.394	0.851	0.855
Mean percent error (%) (*PE*)	5.046	−0.448	3.921
Square of the Pearson product moment correlation coefficient (R^2)	0.1974	0.5628	0.7680

Source: Chang, N. B. et al. 2014. *International Journal of Remote Sensing*, 35, 1064–1093.

In regard to the performance of GP models, the average values of estimated TOC concentrations based on the MODIS-based and fusion-based GP models are within the range of 2.629% and −2.166% of measured TOC concentrations in the lake, respectively. Taking the relative performance of the GP model into account based on the MODIS-based GP model versus the fusion-based GP model, *rmse* values of 1.248 and 0.900 mg/L for the MODIS-based and fusion-based GP models are reasonably close to the minimum error of zero. The fusion-based GP model exhibited higher accuracy. Both GP models performed quite well with the ratios of 0.851 and 0.855, respectively. This performance is much better than the two-band model. When taking the *PE* into account, both models are under 5%, yet the MODIS-based GP model outperforms the fusion-based GP. In principle, the higher the levels of accuracy, the closer the *PE* value is to zero. The negative value of −0.448 indicates that the estimated TOC concentrations are lower than the observed values. It is known that accurate results are depicted as the R^2 value approaches 1. In this case, the R^2 values achieved by the MODIS-based GP model and the fusion-based GP model are 0.5628 and 0.7680, respectively. Therefore, the fusion-based GP model successfully outperforms its counterpart. Hence, the statistical analysis presented in Table 16.9 suggests that the fusion-based GP model noticeably outperformed the MODIS-based GP model in terms of *rmse*, *CO*, and R^2.

One of the aims in this study is to compare a traditional two-band model to a GP model. Two-band models are generally analytically derived based on knowledge of which band contains a salient spectral feature for TOC and then divided by another band to reduce systematic noise, backscattering or reflectance contamination from other water quality parameters. This can be effective in case I waters, where the surface reflectance is the sum of clean water and a low number of water quality constituents (Chang et al., 2014). Yet this method is less effective in case II waters, in which the spectral reflectance is a product of numerous water quality constituents (Chang et al., 2014). Per the results in Table 16.9, both GP models yielded more accurate *rmse*, *CO*, *PE*, and R^2 values relative to the proposed two-band model. The two GP models are well suited for decomposing complex relationships without prior knowledge or input relative to two two-band models.

There are two reasons that the fusion-based GP model stands out in the end. On one hand, this outcome is due to the use of machine learning tools that are especially handy for developing an empirical or semi-empirical model specifically tuned to the unique water quality characteristics in case II waters for exploring water quality trends of the lake under investigation. On the other hand, this achievement is partially because the enhanced spatial resolution exhibits a salient advantage when monitoring water quality in such a small lake. This is because data fusion limits the amount of surface reflectance from the land contaminating shoreline pixels when the fused image offers delineation between the land, the shoreline, and open water clearly. The finer spatial resolution, which is only 30 m in size, can detect plumes in water pollution events, whereas MODIS would require the same plume to be as large as 250 to 500 m for detection. This is crucial for detecting plumes coming toward source water intakes of water treatment plants.

16.4.4 Spatiotemporal TOC Mapping for Daily Monitoring

Model predictability for TOC concentrations can be further assessed by reconstruction of the TOC distributions in Harsha Lake, leading to more insightful spatial analyses. First, the fusion-based GP model with the solution procedure, listed as a series of equations in Appendix A of Vannah (2013), may yield the TOC concentration map for the dates of missing high-resolution Landsat imaginaries. As an example, the fusion-based GP model developed above was applied based on the proposed data fusion process using surface reflectance values collected from mostly cloud free days in June and July of 2011 (Chang et al., 2014). Fused images and derived TOC concentration maps are shown in Figure 16.9 for the 158th, 164th, and 191st days of the year in 2011, all of which were missing high-resolution Landsat imageries (Chang et al., 2014).

In this data fusion context, when the Landsat images serve as a pre- and post-reference for conditions on the lake, MODIS images are available for each of these days. The high spatial and temporal fused images fill in the gaps left by Landsat on the 158th, 164th, and 191st days of the year via using STARFM. The fourth row in Figure 16.9 shows the TOC concentration maps that are produced to describe the spatiotemporal variations of TOC on the lake. A closer review of the TOC concentration maps indicates no significant spatial patterns in TOC concentration levels during individual dates of the 2011 summer season given the fact that the predicted concentrations fall into a narrow range of 5 and 9 mg/L with few extreme values. Closer evaluation of a TOC concentration map is provided for the case on May 24, 2010, in Figure 16.10. The TOC concentrations identified are as low as 5 mg/L and as high as 15 mg/L. Most of the TOC concentrations range between 5.5 mg/L and 9 mg/L. While the TOC concentration is mostly uniform throughout the lake, the fusion-based GP model has identified some patches of extreme TOC levels in the lake. Streams and tributaries feeding into the lake do not exhibit an influx or dilution of TOC at these conjunction points. In Figure 16.11, there is an obvious spatial variation of TOC in the lake with a possible TOC plume. Yet the accuracy of these plumes cannot be completely realized due to the narrow range and limited distribution of the available ground-truth TOC concentrations that were used to train the model.

Seasonal TOC maps provide a visual realization into the dynamics of TOC variations throughout the year in a two-dimensional sense. As shown in Table 16.10, the seasonal TOC maps in Figure 16.11 were generated by grouping the images into seasons and averaging the predicted TOC values during the season. The average TOC values in the winter, spring, summer, and fall were 7.2 mg/L, 8.0 mg/L, 7.2 mg/L, and 7.4 mg/L. In the spring, TOC concentrations that are homogenous throughout the lake remained high at 7 to 9 mg/L. The TOC level in summer and fall seasons decreased to a range of 6 to 8 mg/L, and the TOC turned out to be the highest toward the western side of the lake close to the water intake. In winter, there appeared to be a transition in which higher TOC concentrations began to occur in the middle portion of the lake. These maps would provide a firm basis in regard to the optimal location to install the new water intake location, which should be close to the western shoreline of the lake.

Previous studies have reported that TOC values were lowest during the winter (Agren et al., 2008; Bayram et al., 2011). Likewise, our study found that winter had one of the lowest TOC values, due to low precipitation and temperatures reducing microbial activity. Spring yielded the highest TOC

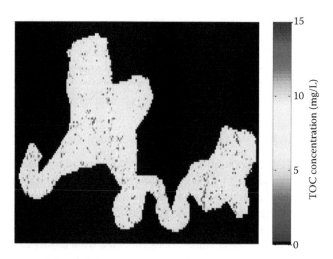

FIGURE 16.10   TOC concentration map for Harsha Lake on May 24, 2010. (From Chang, N. B., Vannah, B., Yang, Y. J., and Elovitz, M., 2014. *International Journal of Remote Sensing*, 35, 1064–1093.)

TABLE 16.10
Separation of the Sampling Data by Season

Spring		Summer			Fall	Winter
May 18, 2009	April 13, 2011	August 20, 2008	September 14, 2009	July 28, 2011	September 28, 2009	March 2, 2011
April 12, 2010	May 23, 2011	June 29, 2009	September 7, 2010	August 1, 2011	October 5, 2009	March 17, 2011
April 19, 2010	June 1, 2011	July 6, 2009	September 8, 2010	August 22, 2011	October 26, 2009	
May 3, 2010	June 7, 2011	July 13, 2009	September 9, 2010	August 23, 2011	October 5, 2011	
May 24, 2010	June 13, 2011	July 21, 2009	June 21, 2011	August 24, 2011	November 2, 2011	
June 14, 2010	May 24, 2012	August 3, 2009	July 5, 2011	August 29, 2011	November 17, 2011	
June 15, 2010	June 13, 2012	August 17, 2009	July 11, 2011			
June 16, 2010		August 31, 2009	July 27, 2011			

Source: Chang, N. B. et al. 2014. *International Journal of Remote Sensing*, 35, 1064–1093.

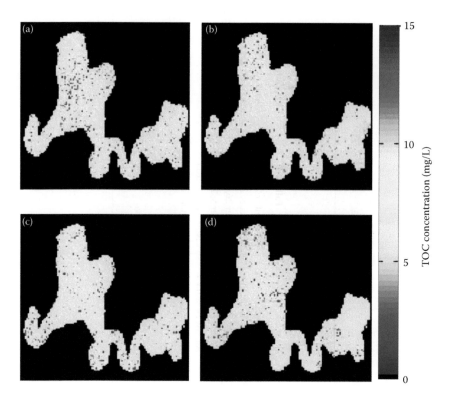

FIGURE 16.11 Seasonal average TOC concentration maps predicted by IDFM. (a) Spring, (b) Summer, (c) Fall, and (d) Winter. (From Chang, N. B. et al. 2014. *International Journal of Remote Sensing*, 35, 1064–1093.)

value in the study, whereas summer yielded one of the lowest. This was largely beyond expectation, since Bayram et al. (2011) reported that spring TOC values were lower than fall and late summer. At least, the fall TOC value was higher than those of the summer and winter, which coincides with the findings of Bayram et al. (2011). Agren et al. (2008) found that spring TOC values peaked slightly, yet late summer and fall TOC values were still higher due to increased temperature and precipitation. Possible explanations can be attributed to differences in the number of variables that factor into TOC generation and export, such as regional climate influences, municipal discharge, land usage, and soil types (Chang et al., 2014).

16.5 SUMMARY

In this chapter, IDFM techniques have been proven effective to provide near-real-time monitoring capability for mapping TOC concentrations spatially and temporally with better feature extraction results. The majority of the *in situ* data used in training were from 6 to 9 mg/L. The fusion-based GP model was able to accurately predict peak TOC values at 11, 12, and 14 mg/L, although it does not showcase the potential for plume tracking. In other words, the current fusion-based GP model may not be thoroughly trained to detect events that are significantly higher than the narrow range of the sample data. Ideally, a wider range of TOC *in situ* data would have been provided for the study, as well as samples and the location of a major TOC plume detected in the lake. However, this could still be explored in a future study due to the flexibility of the IDFM technique and the Discipulus® GP software. Discipulus allows for GP models to be updated as new *in situ* data are obtained, and IDFM can use the new GP model and previously derived fused images to generate updated TOC concentration maps right away. Thus, periodic sampling of source water allows a water treatment plant to update their fusion-based GP model to more accurately predict TOC fluctuations and track the plume movement due to external forces such as anthropogenic influences, climate change, and weather. This ensures that the early warning capabilities of the IDFM stay relevant as water quality conditions change on a long-term scale.

REFERENCES

Agren, A., Jansson, M., Ivarsson, H., Biship, K., and Seibert, J., 2008. Seasonal and runoff-related changes in total organic carbon concentrations in the River Ore, Northern Sweden. *Aquatic Sciences*, 70, 21–29.

Arenz, R., Lewis, W., and Saunders, J., III, 1995. Determination of chlorophyll and dissolved organic carbon from reflectance data for colorado reservoirs. *International Journal of Remote Sensing*, 17, 1547–1566.

Aschengrau, A., Zierler, S., and Cohen, A., 1993. Quality of community drinking water and the occurrence of late adverse pregnancy outcomes. *Arch Environ Health*, 48, 105–113.

Bayram, A., Onsoy, H., Akinci, G., and Bulut, V., 2011. Variation of total organic carbon content along the stream harsit, eastern black sea basin, Turkey. *Environmental Monitoring and Assessment*, 182, 85–95.

Bove, F. J., Fulcomer, M. C., Klotz, J. B., Esmart, J., Dufficy, E. M., and Savrin, J. E., 1995. Public drinking water contamination and birth outcomes. *Am J Epidemiol*, 141, 850–862.

Chang, N. B., Vannah, B., Yang, Y. J., and Elovitz, M., 2014. Integrated data fusion and mining techniques for monitoring total organic carbon concentrations in a lake. *International Journal of Remote Sensing*, 35, 1064–1093.

Chang, N., Xuan, Z., and Yang, Y., 2013. Exploring spatiotemporal patterns of phosphorus concentrations in a coastal bay with MODIS images and machine learning models. *Remote Sensing of Environment*, 134, 100–110.

Dodds, L., King, W., Woolcott, C., and Pole, J., 1999. Trihalomethanes in public water supplies and adverse birth outcomes. *Epidemiology*, 10, 233–237.

Francone, D. F., 1998. *Discipulus Software Owner's Manual, version 3.0 DRAFT*. Machine Learning Technologies, Inc., Colorado.

Gao, F., Masek, J., Schwaller, M., and Hall, F., 2006. On the blending of Landsat and MODIS surface reflectance: Predicting daily Landsat surface reflectance. *IEEE Transactions on Geoscience and Remote Sensing*, 44, 2207–2218.

General Electric Analytical Systems (GEAS), 1994. *USEPA FAQs for TOC Monitoring*. Water & Process Technologies Analytical Instruments, rev. B, Boulder, Colorado.

Klotz, J. B. and Pyrch, L. A., 1999. Neural tube defects and drinking water disinfection by-products. *Epidemiology*, 10, 383–90.

Menken, K., Brozonik, P., and Bauer, M., 2005. *Influence of Chlorophyll and Colored Dissolved Organic Matter (CDOM) on Lake Reflectance Spectra: Implications for Measuring Lake Properties by Remote Sensing*. University of Minnesota, MN.

O'Connor, J., 2004. *Removal of Total Organic Carbon*. H2O'C Engineering.

Ohio Environmental Protection Agency, 2004. *Drinking Water Source Assessment for Clermont County's Bob McEwen Water Treatment Plant. Public Water System # 1302212*, http://wwwapp.epa.ohio.gov/gis/swpa/OH1302212.pdf, accessed by April, 2017.

Ohmori, Y., Kozu, T., Shimomai, T., Seto, K., and Sampei, Y., 2010. Feasibility study of TOC and C/N ratio estimation from multispectral remote sensing data. *Remote Sensing of Spatial Information Science*, 38, 8.

Parks, S., and Baker, L., 1997. Sources and transport of organic carbon in an Arizona river-reservoir system. *Water Research*, 31, 1751–1759.

Pohl, C. and van Genderen, J., 1998. Multisensor image fusion in remote sensing: Concepts, methods, and applications. *International Journal of Remote Sensing*, 19, 823–854.

Robin, A., Hegarat, S., and Moisan, L., 2005. A multiscale multitemporal land cover classification method using a bayesian ppproach. In: *Proceedings of SPIE 5982, Image and Signal Processing for Remote Sensing XI, 598204* (October 18, 2005).

Smith, R. and Baker, K., 1978. The bio-optical state of ocean waters and remote sensing. *Limnology and Oceanography*, 23, 247–259.

Stedmon, C., Markager, S., and Kass, H., 2000. Optical properties and signatures of chrolophoric dissovled organic matter in danish coastal waters. *Estuarine, Coastal and Shelf Science*, 51, 267–278.

Stramski, D., Reynolds, R., Kahru, M., and Mitchell, B., 1999. Estimation of particulate organic carbon in the ocean from satellite remote sensing. *Science*, 285, 239–242.

Texas Commission on Environmental Quality (TCEQ), 2002. *Total Organic Carbon (TOC) Guidance Manual*. Water Supply Division, Austin, Texas.

Thurman, E., 1985. *Organic Geochemistry of Organic Waters*. Martinus Nijhoff/Dr. W. Junk Publishers, Dordrecht, The Netherlands.

United States Army Corps of Engineers (USACE). William H. Harsha Lake: Benefits [Online]. Available: http://www.lrl.usace.army.mil/whl/

United States Environmental Protection Agency (USEPA), 1998. National Primary Drinking Water Regulations: Disinfectants and Disinfection Byproducts. In: *Federal Register: Rules and Regulations, 68*, Boulder, Colorado.

United States Geological Survey (USGS). Landsat Processing Details [Online]. Available: http://landsat.usgs.gov/Landsat_Processing_Details.php

Vannah, B., 2013. Integrated Data Fusion and Mining (IDFM) Technique for Monitoring Water Quality in Large and Small Lakes. M.S. Thesis, University of Central Florida: U.S.A.

Vermote, E., Kotchenova, S., and Ray, J., 2011. MODIS Surface Reflectance User's Guide. v1.3, MODIS Landsat Surface Reflectance Science Computing Facility.

Vertucci, F., 1989. Spectral reflectance and water quality of adirondack mountain region lakes. *Limnology and Oceanography*, 34, 1656–1672.

Vincent, R., Xiaoming, Q., McKay, R., Miner, J., Czajkowski, K., Savino, J., and Bridgeman, T., 2004. Phycocyanin detection from landsat TM data for mapping cyanobacterial blooms in Lake Erie. *Remote Sensing of the Environment*, 89, 381–392.

Visco, G., Campanella, L., and Nobili, V., 2005. Organic carbons and TOC in waters: an overview of the international norm for its measurements. *Microchemical Journal*, 79, 185–191.

Zeng, Y., Zhang, J., and Genderen, J., 2006. Comparison and Analysis of Remote Sensing Data Fusion Techniques at Feature and Decision Levels. In: *ISPRS 2006: ISPRS mid-term symposium 2006 remote sensing: from pixels to processes*, 8–11 May 2006, Enschede, the Netherlands. Enschede: ITC, 5 p.

17 Monitoring Ecosystem Toxins in a Water Body for Sustainable Development of a Lake Watershed

17.1 INTRODUCTION

Continuous population growth, urban development, and agricultural production have inevitably yielded the increase of nutrients-laden stormwater runoff, making the surface waters eutrophic. Such eutrophic conditions have fueled the mass production of cyanobacteria-dominated algal blooms in polluted waters, which are commonly referred to as Harmful Algal Blooms (HABs). The HABs contain toxins that can negatively impact the ecosystem and human health. The predominant species of cyanobacteria that produce cyanotoxins are *Microcystis aeruginsa*, *Microcystis viridis*, *Aphanizomenon flos-aquqe*, and *Anabaena,* all of which are a form of blue-green algae that produces liver toxins causing numbness, vomiting, nausea, and occasional liver failure (WHO, 1999; Lekki et al., 2009; NASA, 2013). The dynamic occurrence and movement of the HABs driven simultaneously by hydrodynamic currents, ocean circulation, and enhanced wind fetching over the surface waters requires constant monitoring and appropriate control measures.

Zebra mussels (*Dreissena polymorpha*), named for the striped pattern of their shells, are filter-feeding organisms that normally attach to solid surfaces in water. The introduction of zebra mussels in some shallow experimental lakes in Europe has proven effective for improving water clarity and macrophyte growth after changing the primary production (i.e., food chain) in water bodies (Vanderploeg, 2003). Each adult zebra mussel can filter about one liter of water per day, thereby removing particles from the water column and increasing water clarity (Ohio Sea Grant, 1994). The zebra mussels eat mostly algae, selecting primarily the 15–40 μm size range for consumption, thereby providing possible improvement of water quality (Ohio Sea Grant, 1994). In fact, the application of zebra mussels was deemed an effective ecological control measure to clean up the HABs as far back as the early 1980s. However, by the fall of 1989, zebra mussels had colonized the surfaces of nearly every firm object in Lake Erie and become an invasive species (Ohio Sea Grant, 1994). Sometimes zebra mussels could even block out the water intakes of drinking water treatment plants, causing failure in pumping raw water for treatment.

When the zebra mussel had become established in Lake Erie in the 1990s, water clarity increased from 0.15 m up to 0.9 m in some areas (Vanderploeg, 2003). Due to the rapid increase in zebra mussel populations that reduced chlorophyll concentration in the water column via rapid filtering, zebra mussels can selectively reject certain toxic strains and promote toxic blooms of *Microcystis* in nature (Vanderploeg et al., 2001). As a result, there were marked decreases in water clarity in Lake Erie due to the presence of massive blooms of *Microcystis*, a potentially toxic colonial cyanobacterium (i.e., blue-green alga), in 1992, 1994, and 1995 (Vanderploeg, 2003). Severe blooms of *Microcystis* occurred in western Lake Erie in 1995 and 1998 (Vanderploeg, 2003). While there are a variety of cyanotoxins, microcystin is the main toxin produced (WHO, 1999; Lekki et al., 2009). Microcystin is a potent hepatotoxin produced by *Microcystis* bacteria during intensive

HABs in eutrophic lakes (WHO, 1999; Chang et al., 2014). Most of the microcystin is contained within healthy *Microcystic* cells, and the toxin remains inside the cell until death or induced rupture of the cell wall of healthy *Microcystic* cells by other organisms such as zebra mussels.

Freshwater algae that quickly spread out in a water body do not accumulate to form dense surface scums or blooms as do some cyanobacteria. Since *Microcystis* is a bacterium that uses photosynthesis for energy production, high concentrations of *Microcystis* can be correlated with elevated chlorophyll-a levels. Chlorophyll-a levels in *Microcystis* blooms are thus related to the amount of microcystin in a water body (WHO, 1999; Rogalus and Watzin, 2008; Rinta-Kanto et al., 2009). Budd et al. (2001) used the Advanced Very High Resolution Radiometer (AVHRR) and Landsat Thematic Mapper (TM) images to determine chlorophyll-a concentrations in a lake, leading to the detection and tracking of the pathways of HABs. Wynne et al. (2008) also employed the surface reflectance of chlorophyll-a values to specifically predict *Microcystis* blooms. Mole et al. (1997) and Ha et al. (2009) had similar findings in regard to using chlorophyll-a as an indicator for identifying and quantifying microcystin in algae blooms which had reached the late exponential growth and stationary phase. Their studies proved that surface reflectance data may be used to detect and track HABs based upon chlorophyll-a levels. In addition, it was discovered that *Microcystis* blooms can be distinguished from other cyanobacteria blooms through a spectral analysis of the detected surface reflectance at 681 nm if there is a satellite band covering this wavelength (Ganf et al., 1989). As the surface reflectance at 681 nm is closely related to phycocyanin, phycocyanin may be regarded as an alternative indicator of *Microcystis*. In fact, phycocyanin is a pigment-protein complex from the light-harvesting, water-soluble phycobiliprotein family and is an accessory pigment to chlorophyll that all cyanobacteria own. Phycocyanin concentrations also share a positive correlation with microcystin levels (Rinta-Kanto et al., 2009).

The surface reflectance curves for chlorophyll-a and phycocyanin in surface waters peak at 525 nm, 625 nm, 680 nm, and 720 nm (Figure 17.1a). MEdium Resolution Imaging Spectrometer (MERIS) was one of the main instruments on board the European Space Agency (ESA)'s ENVISAT platform, and could have the capability to detect phycocyanin. Bands 5 and 8 of MERIS help distinguish the phycocyanin and chlorophyll that offer a powerful opportunity to quantify the microcystin concentration (Chang et al., 2014). Additional resources include the Moderate Resolution Imaging Spectroradiometer (MODIS), which is a payload scientific instrument that was launched into Earth orbit by NASA in 1999 on board the Terra (EOS AM) Satellite and in 2002 on board the Aqua (EOS PM) satellite as a paired operation. Vincent et al. (2004) used Landsat TM images in the visible and infrared spectral bands to predict phycocyanin concentrations with prediction accuracies from 73.8% to 77.4%. Shi et al. (2015) adopted a regression model that was validated and then applied to an 11-year series of MODIS-Aqua data to investigate the spatial and temporal distributions of microcystin levels in Lake Taihu, China. They concluded that, in addition to the existing spectral information, cyanobacterial bloom scums, temperature, wind, and light conditions could probably affect the temporal and spatial distribution of microcystin levels in Lake Taihu. Thus, the surface reflectance of phycocyanin, chlorophyll-a, and *Microcystis* are suitable indicators for the prediction of microcystin levels in a water body. However, in Figure 17.1a, the defining peaks and troughs for Landsat are smoothed out as the total reflectance is averaged for the bands. For this reason, due to the lack of such information associated with band 681 nm, the resulting Landsat band is most likely unable to detect embedded reflectance information essential for discerning some water quality constituents sensitive to optical properties and characterizing the species within a phytoplankton bloom (Chang et al., 2014). However, hyperspectral sensors like MERIS have more bands with much thinner bandwidths. These hyperspectral bands with more narrow bandwidths may accurately depict the spectral reflectance curve of the target constituents in a water body, as evidenced by Figure 17.1b. Hyperspectral information has two advantages due to the ability to introduce key hyperspectral information to include more degrees of freedom in species identification. One advantage is that it allows for optical models of higher explanatory

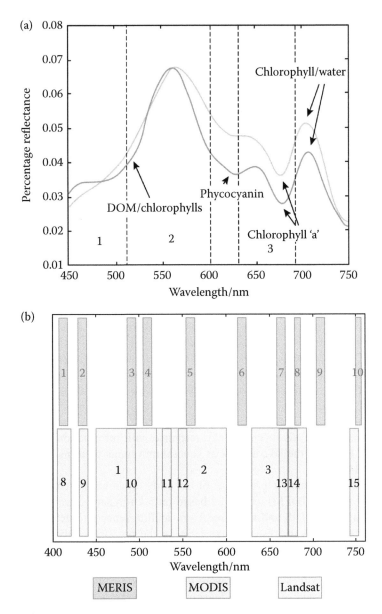

FIGURE 17.1 Multisensor data fusion potential in regard to (a) the reflectance features of lake water containing chlorophyll-a and phycocyanin (Adapted from Vincent, R. et al. 2004. *Remote Sensing of Environment*, 89, 381–392.) and (b) bandwidths between multispectral (Landsat and MODIS) bands and hyperspectral (MERIS) bands.

power to quantify the nonlinear relationships between surface reflectance and water quality constituents, and the other is that it enhances the determination of inherent optical properties that vary with water depth for water column surveillance (Chang et al., 2004; Torrecilla et al., 2009).

Use of multispectral and hyperspectral sensor imageries for ocean color remote sensing has been reported by Lubac et al. (2008). Multispectral sensors may collect the surface reflectance, which is the combination of a few water quality parameters at the water's surface (O'Reilly et al., 1998). Case 2 water bodies (such as Lake Erie) exhibit significantly more complex optical properties than the open ocean. The feature extraction algorithms utilizing multispectral data sensor products demonstrate reduced performance in these Case 2 waters (Hu et al., 2000; Lee and Carder, 2002). The multispectral data fusion pair of Landsat and MODIS may jointly capture half of the peak at 625 nm and half of the peak at 680 nm, while the hyperspectral data fusion pair of MERIS and MODIS may jointly detect the full spectral feature at 680 nm, as shown in Figure 17.1b (Chang et al., 2014). The multispectral fusion pair also has bands in the shortwave infrared range with the band centers at 1,650 nm and 2,090 nm, which shed insight on chlorophyll concentrations in the water (Chang et al., 2014). The hyperspectral data fusion pair can detect electromagnetic radiation centered at 412 nm and 443 nm. It is known that 412 nm is associated with the degradation products of once-living organisms (ESA, 2006). This is important for toxic microcystin prediction because the toxin is only released from the *Microcystis* bacteria once the organism dies (Chang et al., 2014). In addition, 443 nm is associated with chlorophyll maximum absorption (ESA, 2006). While Landsat and MERIS can detect spectral features for chlorophyll-a and phycocyanin, a significant drawback is their longer revisit cycles of 16 and 3 days, respectively (Chang et al., 2014). The daily revisit time of the MODIS sensor can fill in the data gaps using the data fusion technique. The niche for our data fusion rests on the fact that MODIS alone cannot be used as a substitute because of its poor spatial resolution (250/500 m) for the land bands, which is outclassed by Landsat with 30 m spatial resolution, and its 1,000 m spatial resolution for the ocean bands, which is enhanced by MERIS with 300 m spatial resolution (Chang et al., 2014). Therefore, data or image fusion between hyperspectral and multispectral images is recognized as a synergistic image processing technique that helps improve the monitoring capability of surface water body and water column surveillance. Thus, an ultimate solution may be to fuse Landsat and MODIS (MODIS' land bands) or MERIS and MODIS (MODIS' ocean color bands) pairwise to generate a series of synthetic images with enhanced spatial and temporal resolutions. Such a series of synthetic images can enable near real-time monitoring of microcystin concentrations, creating seasonal maps for regular trend analyses and populating an environmental informatics database for decision analysis of emergency response with information on spatial occurrence and its timing of HABs in the lake as well as general movement patterns of *Microcystin* blooms (Chang et al., 2014).

This chapter demonstrates a prototype of an early warning system using the Integrated Data Fusion and Mining (IDFM) technique based on both hyperspectral images (MERIS) and multispectral images (MODIS and Landsat) from satellite sensors to determine spatiotemporal microcystin concentrations in western Lake Erie. In the proposed IDFM, the MODIS images with temporal resolution were fused with the MERIS and Landsat images with higher spatial resolution to create daily synthetic images (Chang et al., 2014). The spatiotemporal distributions of microcystin within western Lake Erie were then reconstructed using the band data from the fused products with the aid of machine learning or data mining techniques, such as Genetic Programming (GP), for feature extraction. The performance of these feature extraction models using fused hyperspectral and fused multispectral sensor data were then assessed by four statistical indexes to prove their effectiveness. These machine learning or data mining models were further compared with traditional two-band bio-optical models in terms of microcystin prediction accuracy. This study hypothesized that GP models could outperform traditional two-band models and additional spectral reflectance data offered by the hyperspectral sensor (i.e., MERIS) might noticeably increase the prediction accuracy, especially in the range of low microcystin concentrations. Overall, this study aimed to explore (1) the feasibility of predicting microcystin concentrations in a lake using the IDFM technique with the aid of computational intelligence methods, (2) which of the fused band combinations are most useful in determining microcystin concentrations in a Case 2 inland water body, and (3) whether hyperspectral sensor products may provide a significant advantage over products from multispectral sensors for microcystin prediction (Chang et al., 2014).

17.2   STUDY REGION AND POLLUTION EPISODES

Lake Erie is one of the five Great Lakes located in North America. These lakes provide drinking water for over 40 million Americans, in addition to 207 million tons (56 billion gallons) per day withdrawn from the lakes for industrial, agricultural, and municipal use (Lekki et al., 2009). The Maumee River Basin is considered an area of concern due to its impact on western Lake Erie's eutrophic condition (Figure 17.2). The Maumee River is a primary tributary in Ohio that flows into western Lake Erie. The basin contributes a significant portion of soluble reactive phosphorus (a bioavailable form of phosphorus), which has been increasing since the 1990s (Ohio EPA, 2010). Nutrients from urban areas and agricultural operations are being conveyed straight into the Maumee River due to poor soil infiltration and the increase in impervious surface area in urban regions (Chang et al., 2014). Each summer, the Great Lakes are threatened by *Microcystis* blooms, and the blooms in western Lake Erie are the most severe (Chang et al., 2014).

Throughout the 2000s, *Microcystis* blooms increased in frequency and severity (Bridgeman 2005; Ouellette et al., 2006) and the situation has been getting worse in the years following 2010. In Figure 17.3, a snapshot of the recent worst case taken by the Landsat images on October 5, 2011, can be seen. In Figure 17.4, the waters of Lake Erie seem to grow green in the image taken by the MODIS-Aqua satellite in the summer 2013; the false color comes from an algae bloom that has been growing in the lake since July 2013. The bloom is comprised of *Microcystis* according to the archived "Harmful Algal Bloom Operational Forecast System Bulletins" managed by the National Oceanic and Atmospheric Administration (NOAA).

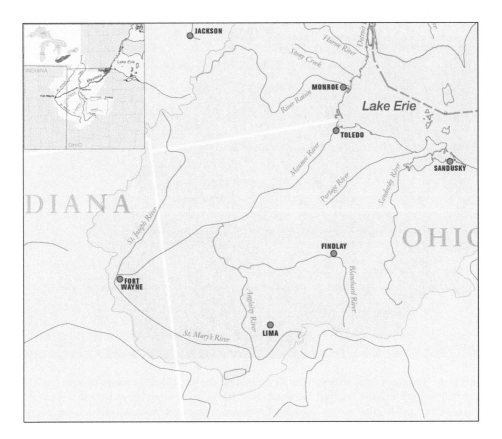

FIGURE 17.2   The Maumee River basin and western Lake Erie. (United States Environmental Protection Agency (US EPA), 2012. https://www.epa.gov/greatlakes/lake-erie. Accessed on April 2017.)

(a)

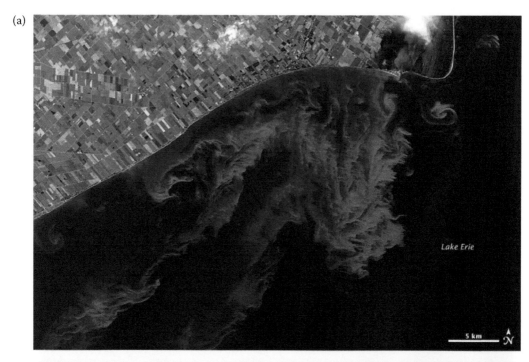

(b)

FIGURE 17.3 The green scum shown in the image above is the worst algae bloom in western Lake Erie and was acquired by the Landsat-5 satellite; the bloom is primarily microcystis aeruginosa shown in (a) the top image on October 5, 2011, that zooms in on a hot spot with vibrant green filaments extending out from the northern shore of Lake Erie and (b) the entire bloom in western Lake Erie. (National Aeronautical and Space Agency (NASA), 2011. https://earthobservatory.nasa.gov/IOTD/view.php?id=76127.)

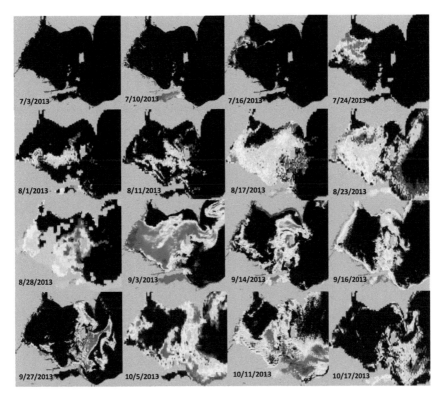

FIGURE 17.4 The HABs that occurred in August, September, and October 2013. However, the current bloom is smaller than the record-setting bloom on October 5, 2011 shown in Figure 17.3. (National Oceanic and Atmospheric Administration (NOAA), 2013. https://tidesandcurrents.noaa.gov/hab/bulletins.html. Accessed on May 2017.)

17.3 SPACE-BORNE AND *IN SITU* DATA COLLECTION

The space-borne data of the surface reflectance values utilized in this study were obtained from Landsat TM, MERIS, and MODIS sensors. MERIS is a hyperspectral sensor with a moderate 300-m resolution which has a 3-day revisit time for sites near the equator. While the MERIS sensor itself has the capability to acquire hyperspectral data from 390 nm to 1,040 nm at a 1.8 nm spectral resolution for a total of 520 bands, the processing onboard the satellite yields a multispectral product set (ESA, 2006). Because the band widths and locations are programmable, the remaining spectral values are averaged into 15 discrete bands while unwanted spectral data are disposed (ESA, 2006). Nevertheless, the 15 product bands of MERIS still offer greater detail about the visible, near-infrared, and infrared frequencies than many other satellite sensor products available in this range (ESA, 2006). Landsat offers superior spatial resolution at 30-m, but only seven spectral bands are available compared to the 15 of MERIS. The revisit time of Landsat is 16 days, which is significantly longer than MERIS and MODIS. In developing a near real time monitoring system, daily satellite images of the area of interest are required; data fusion techniques are used to fill in the data gaps in MERIS and Landsat by using the MODIS sensor with its daily revisit time (Chang et al., 2014). The spatial, temporal, and spectral resolutions of the two paired satellites central to this data fusion approach are compared in Table 17.1. Note that in this study, MERIS bands were fused with the ocean color bands of MODIS and Landsat bands were fused with the land bands of MODIS. Data fusion at the pixel level using the Spatial and Temporal Adaptive Reflectance Fusion Model (STARFM) introduced in Chapter 9 requires input images to be spectrally similar (Gao et al., 2006). Accordingly, Landsat TM bands were fused with the land bands of MODIS.

TABLE 17.1

Spatial, Temporal, and Spectral Properties of the Satellite Sensors Used in This Study

Parameters	Hyperspectral Sensor Pair		Multispectral Sensor Pair	
	MERIS	**MODIS TERRA** (Ocean Bands)	**Landsat TM**	**MODIS TERRA** (Land Bands)
Product	MER_FR_2P	MODOCL2	LT5	MODO9
Spatial resolution	300-m	1,000-m	30-m	250/500 m
Temporal resolution	1–3 days	1 day	16 days	1 day
Band number: Band center ± band width (nm)	1: 412 ± 10	8: 413 ± 15	1: 485 ± 35	3: 469 ± 10
	2: 443 ± 10	9: 443 ± 10	2: 570 ± 40	4: 555 ± 10
	3: 490 ± 10	10: 488 ± 10	3: 660 ± 30	1: 645 ± 25
	4: 510 ± 10		4: 840 ± 60	2: 859 ± 18
		11: 531 ± 10	5: 1650 ± 100	6: 1640 ± 12
	5: 560 ± 10	12: 551 ± 10	7: 2090 ± 130	7: 2130 ± 25
	6: 620 ± 10			
	7: 665 ± 10	13: 667 ± 10		
	8: 681 ± 10	14: 678 ± 10		
	9: 708 ± 10			
	10: 753 ± 10	15: 748 ± 10		
	11: 760 ± 10			
	12: 779 ± 10			
	13: 865 ± 10	16: 869 ± 15		

Source: Chang, N. B. and Vannah, B., and Yang, J., 2014. *IEEE Journal of Selected Topics in Applied Earth Observations and Remote Sensing*, 7, 2426–2442.

Note: The band centers shared between the satellites have been aligned in the table. Band combinations that occur on the same row are suitable candidates for spectral fusion.

According to NOAA, a common form of microcystin is Microcystin-LR, one of the most toxic strains of microcystin in the Great Lakes (Leshkevich and Lekki, 2017). NOAA is a long-term provider of the *in situ* data for microcystin concentration and the monitoring stations located at western Lake Erie are summarized in Figure 17.5. NOAA collects these surface water samples at stations in western Lake Erie periodically to provide surface microcystin concentrations that coincide with the surface reflectance observed in satellite or air-borne remote sensing data products. Total microcystin concentration of each sample was quantified by using ELISA kits (Abraxis; 520011) after cell lysis (Abraxis; Quik-lyse kit 529911QL) of unfiltered samples (Michalak et al., 2013). A total of 44 microcystin measurements were screened out from 2009 to 2011 and are suitable for ground-truth usage in this study (Table 17.2). The data set used in Table 17.2 for the final feature extraction of the corresponding satellite images only includes those with sampling locations free from cloud cover, aerosol contamination, and significant suspended sediment levels in the study region (Chang et al., 2014).

17.4 STUDY FRAMEWORK

The IDFM technique for the prediction of microcystin that is designed to fuse satellite data streams and apply machine learning or data mining algorithms to derive a working algorithm is shown in Figure 17.6. Such a working algorithm enables us to relate the fused data streams of remote sensing images to the desired outputs of a specified parameter by a tailored approach. In this study, Landsat,

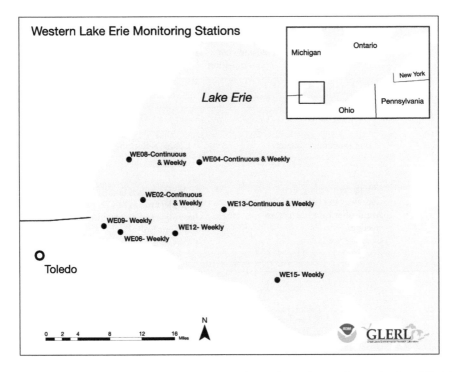

FIGURE 17.5 Western Lake Erie monitoring stations for surface microcystin monitoring. (NOAA Great Lakes Environmental Research Laboratory, 2015. Lake Erie Microcystin Sampling Data, https://www.glerl. noaa.gov/res/HABs_and_Hypoxia/WLEMicrocystin.html.)

MERIS, and MODIS surface reflectance imagery serve as the data streams, and the estimated concentrations of the microcystin at each ground-truth sampling location are the desired outputs of a pre-specified parameter. Machine learning or data mining techniques were applied to incorporate data into a single image for feature extraction, which led to the creation of a prediction model for monitoring microcystin.

The IDFM technique in such a context-aware intelligent system consists of five main steps, as shown in Figure 17.6. For example, Step One is the acquisition of the surface reflectance data from MERIS and MODIS. Step Two formats images from different sources into a unified condition. Step Three formalizes the data fusion with the aid of STARFM algorithms. Again, this study employed the STARFM algorithm to fuse the pair of MODIS (ocean color bands) and MERIS as well as the pair of MODIS (land bands) and Landsat. A fused image is created by the algorithmic fusion of

TABLE 17.2

Ground-Truth Samples Were Taken at Various Sites in Western Lake Erie on These Days

	June	July	August	September
2009		7, 14		
2010	28	26	2, 16, 30	2
2011		12	11	14

Source: Chang, N. B. and Vannah, B., and Yang, J., 2014. *IEEE Journal of Selected Topics in Applied Earth Observations and Remote Sensing*, 7, 2426–2442.

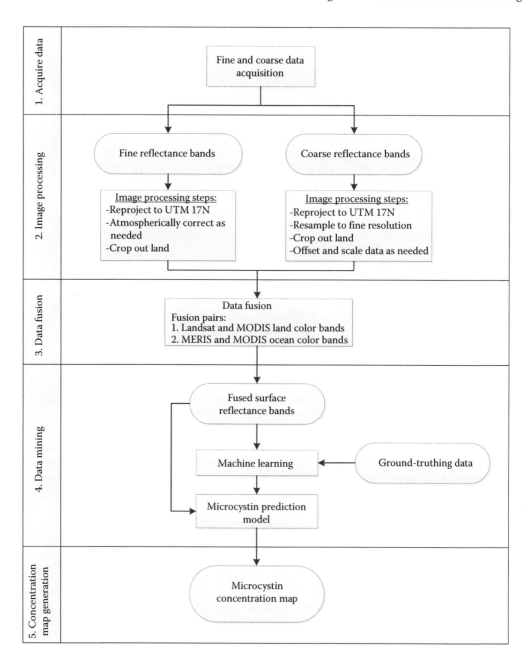

FIGURE 17.6 Methodological flowchart for the IDFM procedure using hyperspectral or multispectral data. (Chang, N. B., Vannah, B., and Yang, J., 2014. *IEEE Journal of Selected Topics in Applied Earth Observations and Remote Sensing*, 7, 2426–2442.)

the spectral, temporal, and spatial properties of two or more images (Van Genderen and Pohl, 1994). The resulting synthetic or fused image has all the characteristics of the input images and incorporates objects' defining attributes into a single image with the potential to increase the reliability of the data (Pohl and Van Genderen, 1998). In Step Four, the synthetic images and ground-truth data are used as inputs for feature extraction via machine learning or data mining to retrieve a specific parameter of interest. An GP model is trained to create an explicit, nonlinear equation relating the fused band data to the ground-truth data by using Discipulus®. Finally, Step Five uses the GP model created in Step Four to compute microcystin concentration maps using the fused band data generated in Step Three (Figure 17.6). The following five subsections delineate the details of these five steps (Chang et al., 2014).

17.4.1 DATA ACQUISITION [FIGURE 17.6; STEP 1]

In this study, ENVISAT MERIS, Terra MODIS, and Landsat TM sensors helped collect surface reflectance data over Lake Erie. Ocean color band images for 2009–2011 from the MODIS-Terra satellite were pre-processed at a level 2 basis and formatted as HDF-EOS images by a team with the National Aeronautics and Space Administration (NASA), and were available from the online repository through the NASA Ocean Color Web. The MODIS images were selected for the cloud free condition at each of the sampling locations since multiple ground-truth samples were taken at different sampling locations over Lake Erie on the same day. The level 2 image was downloaded only when at least one ground-based sampling location was not hindered by cloud. The same criterion was applied to the rest of the acquired satellite data images one by one. In addition, land band images for 2009–2011 from MODIS-Terra pre-processed at a level 2 basis were downloaded from the online repository managed by the NASA Land Processes Distributed Active Archive Center (LP DAAC) and Earth Resources Observation and Science Center (EROS) in the United States Geological Survey (USGS). MERIS data pre-processed at a level 2 basis were obtained through the ESA. Note that the MERIS sensor in 2012 did not work well, so ground-truth data during this time period cannot be utilized. Finally, the Landsat TM data were collected from the EROS (USGS) by way of the Global Visualization Viewer managed by LP DAAC, USGS.

17.4.2 IMAGE PROCESSING [FIGURE 17.6; STEP 2]

The MODIS images were pre-processed at a level 2 basis, which includes the radiometrically calibrated data that were atmospherically corrected for aerosols and scattering (Vermote et al., 2011). MERIS data were also processed at a level 2 basis with radiometric, geometric, and atmospheric corrections (ESA 2006). Landsat data were processed at a level-1T basis with radiometric and geometric corrections. Because the data fusion process requires input image pairs to have the same bit-depth and spatial resolution, the input images were processed in ArcGIS, a mapping and spatial analysis software. Specifically, MERIS images were processed to (1) re-project to the Universal Transverse Mercator (UTM) zone 17 North, and (2) crop out land data surrounding Lake Erie. The MODIS ocean band images were processed in a similar manner to (1) re-project to UTM zone 17 North, (2) resample to the 300-m spatial resolution to match those of MERIS, (3) crop out land from around Lake Erie, and (4) recalculate surface reflectance values using the same offset and scaling applied to MERIS data. The MODIS land-band images were processed in a similar manner to (1) re-project to UTM zone 17 North, (2) resample to 30 m spatial resolution to match Landsat, (3) crop out land from around Lake Erie, and (4) recalculate surface reflectance values using the same offset and scaling applied to MERIS data. In a similar way, Landsat images were processed to (1) atmospherically correct Landsat images using the LEDAPS Processing software available through NASA, (2) re-project to UTM zone 17 North, and (3) crop out land around Lake Erie.

Overall, the common image processing for all three satellite products consists of three categories of actions and they include (Chang et al., 2014) (1) modification of the geometric projections, pixel size, bit-depth, and scale to fuse them properly, (2) atmospheric correction, and (3) preparation of the image to increase the accuracy of the fused image by removing land parcels from the original images. Geographic projection and scaling prior to fusion are required to the make images consistent within the same geographic system. Otherwise data representing different swaths of land with differently scaled values for the same pixels between the satellite pairs might become incomparable. Note that all images were projected to the UTM 17N in this study to avoid the incompatibility issue. In addition, the MODIS ocean band data were resampled to match the spatial resolution of

MERIS and Landsat for the subsequent data fusion. In color modification, the MODIS ocean color band surface reflectance values were scaled upward to match those of MERIS. This ensures that input images collected from both satellite sensors have the same number of pixels, enabling the pixel-based fusion process chosen by STARFM. It is known that the Landsat and MODIS data are not the same bit-depth. Hence, atmospheric correction using the LEDAPS Processing tool scales the Landsat values from 0–255 to −100–16,000 (Vermote et al., 2002; Masek et al., 2005). This is due to the fact that the same MODIS 6S radiative transfer technique has to be applied to correct the Landsat data.

For the other two categories, atmospheric correction is required to remove the scattering effects of the atmosphere from the raw data, thereby producing surface reflectance instead of top of atmosphere radiance. The last category of processing for cropping out the land surrounding Lake Erie was performed on all relevant images. This step, with the purpose of masking the land pixel values, is required to prevent fusing land pixel values with surface water values during the data fusion process with the STARFM algorithm.

17.4.3 Data Fusion [Figure 17.6; Step 3]

Pixel-based fusion of spatial and temporal properties was conducted at this stage based on the STARFM algorithm developed by NASA. The Landsat and MERIS images are of higher quality than MODIS in terms of spatial resolution, but they are sparse in time. Consequently, MODIS data are used as the baseline to capture temporal changes during the periods of data gaps. For the first pair, the STARFM algorithm was used to fill in data gaps caused by the 3-day revisit time of MERIS and the daily MODIS ocean color bands; for the second pair, the STARFM algorithm was used to fill in data gaps caused by the 16-day revisit time of Landsat and the daily MODIS land color bands. The overall workflow of the STARFM algorithm is detailed in Chapter 9.

17.4.4 Machine Learning or Data Mining [Figure 17.6; Step 4]

The IDFM technique permits the use of a suite of machine learning or data mining techniques, in addition to traditional bio-optical or geophysical models, to retrieve features by deriving an explicit equation or black box model in each case relating the fused surface reflectance data to the ground-truth observations. Notable machine learning or data mining algorithms include GP, Artificial Neural Networks (ANN), Adaptive Resonance Theory, Constrained Optimization Techniques, Adaptive Dynamic K-means, Principal Component Analysis, and Support Vector Machines (SVM). The GP model was created for this study using the Discipulus® software package. The user provides the software with inputs (e.g., surface reflectance values) and outputs (e.g., target values at the ground level), both of which are used to train and calibrate the prediction model (e.g., a GP model). In artificial intelligence, a fitness function is defined as a particular type of objective function that is used to summarize how close the predicted values are to a given set of targets. GP is one of the artificial intelligence techniques designed to encode a computer program in association with a set of artificial chromosomes and to evaluate its fitness with respect to the predefined targets. During training, Discipulus® identifies 30 of the best programs and finally selects the model exhibiting the highest fitness (e.g., minimal discrepancies between the inputs and outputs) (Francone, 1998).

There are six surface reflectance values, one corresponding to each band from MODIS and MERIS, to pair with Landsat for each pixel of the fused image. These six band values were used as input variables of the GP model to help generate the explicit nonlinear equation created for content-based mapping. The 44 ground-truth samples listed in Table 17.2 were used to train and calibrate the GP models. In this study, 60% of the input data was used to train the models, and the remaining 40% was used to validate the performance of the model. The method for splitting up the data into the training and calibration sets follows a unique procedure to improve the learning capacity of the proposed inversion models. In the beginning, we ordered the ground-truth values from low to high and then assigned the ordered data to the training and validation data sets in an alternating manner to guarantee that both the training and calibration process can entertain the same range of microcystin concentration values during training and validation. When generating the GP model, certain band values played a strong role in the determination of the microcystin concentration, while others just offered weak explanatory power.

To achieve the goal of academic rigor, the selected GP model was finally compared against a traditional two-band inversion model. The traditional two-band inversion model was generated through a linear regression model using band ratios as explanatory variables instead of individual bands (Vincent et al., 2004). The generic setup for a two-band model is shown in Equation 17.1:

$$C_{MS} = a * \frac{Rrs(\lambda_1)}{Rrs(\lambda_2)} + b \qquad (17.1)$$

where $Rrs(\lambda)$ is the atmospherically corrected surface reflectance at the band center λ. The coefficients a and b denote the slope and intercept obtained through regression. Additionally, a spectral slope two-band inversion model was included in the analysis (Dash et al., 2011). The spectral slope is calculated using Equation 17.2:

$$Slope = \frac{Rrs(\lambda_1) - Rrs(\lambda_2)}{|\lambda_1 - \lambda_2|} \qquad (17.2)$$

A non-linear exponential fit was carried out to determine the spectral slope coefficients relating the exponential increase of absorption with wavelength for phycocyanin and chlorophyll. For both types of two-band inversion models, band combinations were compared to determine the two bands exhibiting the highest correlation with microcystin and phycocyanin estimation, both of which are ecological indicators of *Microcystis*. On a comparative basis, the same training and calibration data sets used for creating the GP models were employed to train and calibrate the two-band inversion models.

17.4.5   CONCENTRATION MAP GENERATION [FIGURE 17.6; STEP 5]

Microcystin concentration maps were produced for western Lake Erie by using the GP model on the fused data image generated in Step Three. The well-calibrated GP model enabled us to apply the fused surface reflectance values of the pixel to predict the microcystin concentration at each pixel location to map the entire western Lake Erie. Then a clear depiction of microcystin blooms is available. These maps can support the knowledge discovery of probable pathways of the HABs, yearly hot spots, and seasonal or short-term factors that might trigger the incipience of microcystin generation.

17.5   MODEL COMPARISON FOR FEATURE EXTRACTION

17.5.1   RELIABILITY ANALYSIS

In an attempt to create such an early warning system, an IDFM-based tool is available for quantifying toxin levels in HABs using satellite remote sensing data. Such an advancement depends upon two primary factors. One is the accurate fused remote sensing surface reflectance data of the water body, and the other is a reliable algorithm for predicting microcystin concentrations in a water body. The advantages of using enhanced spatial data (MERIS and MODIS ocean color bands) over less detailed products (Landsat and MODIS land bands) to support the derivation of computationally-intensive GP inversion models versus traditional two-band inversion models need to be elucidated further by a comparative approach.

For the 44 ground-truth samples listed in Table 17.2, the traditional two-band inversion models will always yield the same coefficients when solved using linear regression techniques. However, this is not the case for GP model generation. As mentioned, Discipulus® identifies 30 of the best programs, in which we find the model exhibiting the highest fitness during training. Because the equation and performance of each GP model will vary during different runs, the final pool of the 30 best models is subject to an evolutionary computation process. To lucidly depict the variation and average performance of the GP models, five separate training runs were carried out using Discipulus® for both multispectral and hyperspectral inputs of the paired fusion exercises. Only the single best model from each training run in each exercise was selected for comparative analysis in terms of computational time and level of fitness. Of course, computational time is also dependent upon the computer hardware used to run Discipulus®. During each training run, Discipulus® was the only active program on the computer. The computer specifications are as follows: Intel® Core™ i7-3720QM CPU at 2.6 GHz, 24 574 MB RAM, and 500 GB hard drive. Finally, a thorough comparison between the fused data sets derived from the multispectral and hyperspectral sensors was conducted to derive the GP inversion models. The coefficient of determination, computer time required for computing each GP model, and the run number of each GP model are detailed in Table 17.3.

The GP models derived by using the fused hyperspectral sensor data products took 322 seconds longer to solve on average, yet the resulting square of Pearson product moment correlation coefficient (i.e., coefficient of determination) was 0.8883, which is 0.0717 greater than that derived from the counterpart by using fused multispectral sensor data products (Chang et al., 2014). The multispectral sensor GP model yielded a fitness of 0.1652 which is larger than the fitness of 0.1425 produced by the hyperspectral sensor GP model. This is due to the positive and negative deviations being balanced out more uniformly in the multispectral sensor GP model. The multispectral sensor GP models had shorter run times since they had less information to process. In contrast, the greater coefficients of determination for the hyperspectral sensor GP models are attributed to the finer band widths (refer to Figure 17.1a and b), which allow for telltale peaks and troughs of chlorophyll-a and phycocyanin (indicators of microcystin) to be more readily identified (Chang et al., 2014). An exceptional observation is made regarding the multispectral sensor GP models by analyzing the run times. The fifth multispectral sensor GP model was derived in a mere 5 seconds, while the next closest run time was obtained in 92 seconds. The order of magnitude difference is unusual and is probably due to the randomly generated starting weights and the randomly selected input data that are used to initiate model formulation (Chang et al., 2014). In this case, the program was initialized by an excellent combination of parameters for determining the relationship between surface reflectance and microcystin concentration (Chang et al., 2014).

TABLE 17.3

Statistical Comparison between GP Models Created Using Fused Data from Multispectral and Hyperspectral Sensors

	Model Number	1	2	3	4	5	AVG
Fused multispectral GP models	R^2	0.8425	0.7931	0.7683	0.8449	0.8344	0.8166
	Run time (s)	194	272	246	92	5	162
	Run number	34	43	40	26	4	29
Fused hyperspectral GP models	R^2	0.8243	0.9270	0.8847	0.9177	0.8879	0.8883
	Run time (s)	211	450	437	932	389	484
	Run number	30	50	48	71	43	48

Source: Chang, N. B., Vannah, B., and Yang, J., 2014. *IEEE Journal of Selected Topics in Applied Earth Observations and Remote Sensing*, 7, 2426–2442.

Identification of high microcystin values is imperative for emergency response. The best GP model is selected based on the coefficient of determination, fitness level achieved, and a visual confirmation that the model can accurately identify peak microcystin values. Based on these criteria, the fourth fused multispectral sensor GP model and the second fused hyperspectral sensor GP model from Table 17.3 were selected for advanced analysis (Chang et al., 2014). The predictive capabilities of these three GP models developed from pure MERIS (a & d), fused multispectral sensor data (b & e), and fused hyperspectral sensor data (c & f) are presented in Figure 17.7. As can be seen in the time series plot of a, b, and c, the models aptly predict peak microcystin values; however, the ability to predict low microcystin values varies among these models. In the timer series plot of a, the predicted values at low concentrations show mediocre correlation with the observed values. In the timer series plot of b, the model has a horizontal line for predicting observed values

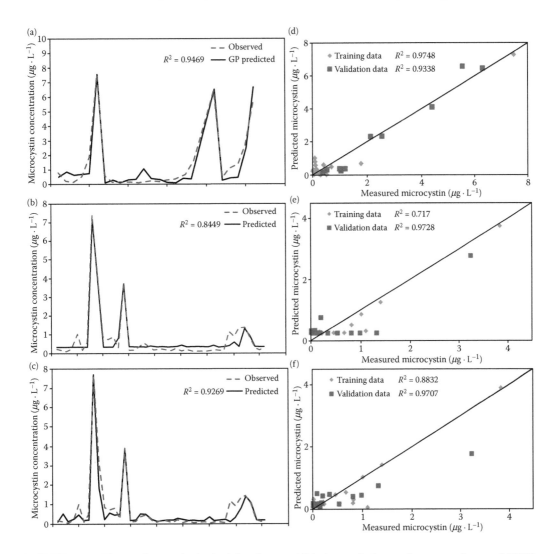

FIGURE 17.7 Time series graphs in the left column exhibit the predictive performance of a pure MERIS GP model (a), a fused multispectral sensor GP model (Landsat and MODIS land bands) (b), and a fused hyperspectral sensor GP model (MERIS and MODIS ocean color bands) (c). For images in the right column (d, e, and f), the predicted microcystin values have been plotted against the observed values to accentuate any biases that are present in the predicted values. (Chang, N. B. and Vannah, B., and Yang, J., 2014. *IEEE Journal of Selected Topics in Applied Earth Observations and Remote Sensing*, 7, 2426–2442.)

below 0.3 μg/L. Overall, the hyperspectral sensor GP model (c) has the best success estimating the microcystin at low concentrations due to some bands (such as bands 5 and 8 of MERIS in Figure 17.1) that are sensitive to even the minute levels of microcystin concentrations. In the right column of Figure 17.7, the predicted microcystin values have been plotted against the observed values to accentuate any biases that are present in the predicted values.

The MERIS GP model (Figure 17.7a–d) served as a baseline to compare the two fused GP models derived using the fused spectral data as inputs. Limited by the single source of satellite images, only twenty-six data points were available to generate the MERIS model as opposed to the case for the fused models with forty-one data points. However, the pure MERIS GP model exhibits the best Pearson product moment correlation coefficient in both calibration (i.e., 0.9748) and validation (i.e., 0.9338) and admirable capacity at predicting peak values, although the pure MERIS GP model cannot capture the time series trend well over the regime of the low microcystin values. More data points were available for the fused models due to the additional information provided by fusion with MODIS. The MERIS GP model obviously demands less computational power and data storage capacity to develop, since data from only one satellite sensor are necessary.

When analyzing the multispectral fusion (Figure 17.7b–e) and hyperspectral fusion (Figure 17.7c–f) performance, the GP model created from the hyperspectral sensor product yields a better coefficient of determination of 0.8832 compared to 0.7170 in model training, although the coefficients of determination of both GP models in model validation look similar. Both of the GP models are capable of predicting peak microcystin values, enabling the system to differentiate between a HAB laced with microcystin and an algal bloom comprised of nontoxic algal species. The predictive capabilities of the fused hyperspectral sensor GP model excelled at predicting microcystin concentrations less than 1 μg/L because the multispectral sensor GP model simply plateaus in this region, as shown in Figure 17.7b. This discovery is evidenced by the difference in predictive power in this region when comparing Figure 17.7e and Figure 17.7f. In fact, the claim can be supported by the horizontal set of data points in Figure 17.7e in the fused multispectral sensor GP model, which consistently underestimates low microcystin values. In contrast, the fused hyperspectral sensor GP model possesses the advantage of more accurately predicting low microcystin values, allowing for the identification of HAB formation at an incipient stage and the evolutionary pathways of HAB later on. Such a monitoring capability is not only useful for assessing water quality changes in ecologically sensitive areas but also the provision of a near real time early warning system with more accurate microcystin prediction at all concentrations. Overall, the GP model formulated with inputs from the hyperspectral sensor successfully achieved this goal. In this context, the exemplary GP model from the hyperspectral sensor depicting the nonlinear relationship between microcystin and the hyperspectral fused band data is convoluted as below:

$$C_{MS} = abs(-1*(X8+X6)*(X6)/X9)^{\wedge}(1/2)$$
$$X1 = cos(v0 + 1.9076)$$
$$X2 = cos(X1*v5-1.*v2 + v0 + 1.5308)$$
$$X3 = cos(2.*cos(X2)^{\wedge}2)$$
$$X4 = abs((2.1683*cos(2.*X3*v2) + 1.2392)/(X1 + X2 + X3)^{\wedge}2) \qquad (17.3)$$
$$X5 = cos((X4 + .13876)^{\wedge}2-1.*v2 + v0 + 1.9876)$$
$$X6 = X1 + X2 + X3 + X5$$
$$X7 = 1.0842*cos(X5*v4/v1) + 1.0842*X1 + 1.0842*X2$$
$$X8 = cos((X7 + 1.0842*X3 + 1.0842*X5 + .86796)/(X6)^{\wedge}2)$$
$$X9 = v5*v3 + .57862$$

where C_{MS} is the concentration of microcystin (μg/L), and variables include v_0, v_1, v_2, v_3, v_4, and v_5 corresponding to the fused MODIS and Landsat bands, respectively. Each X represents an intermediate variable being used by Discipulus®.

17.5.2 Prediction Accuracy over Different Models

17.5.2.1 Comparison between Bio-Optical Model and Machine Learning Model for Feature Extraction

Prediction accuracy between the GP models and traditional inversion methods, including a two-band ratio model and a spectral slope model (Vincent et al., 2004; Dash et al., 2011), were compared in terms of five criteria in Table 17.4. The ideal bands for the two-band models were found by testing all possible band combinations and choosing the one that yielded the highest coefficient of determination and fitness (Chang et al., 2014). For the GP models, as mentioned before, all of the bands were supplied as inputs to Discipulus®, and the program generated a nonlinear relationship between all of the bands (input variables) and the microcystin concentration (output variable). A comparison of the two-band models along with the GP models was finalized using four statistical indexes and the computational time to come up with the five criteria. The four statistical indexes include the root mean square error (rmse), ratio of standard deviations (CO), mean percent error (PE), and the square of the Pearson product moment correlation coefficient (RSQ = R^2). The results are presented in Table 17.4 with special attention to the computational time required to solve the models. The bold value in each category represents the best one compared to the counterparts in the same category.

In the first row of the data matrix in Table 17.4, the observed microcystin mean values are the same for each of the models since they share the same set of ground-truth data. In general, the traditional two-band models performed worse than the spectral slope and the GP models. The spectral slope models performed poorly when using multispectral input values ($R^2 = 0.09625$) given that

TABLE 17.4

GP and Two-Band Models Using Multispectral and Hyperspectral Sensor Surface Reflectance Input Data are Evaluated Using Four Indices of Accuracy

	Fused Multispectral Input[a]			Fused Hyperspectral Input[b]		
	Two-Band Ratio Model	Spectral Slope Model	GP Model	Two-Band Ratio Model	Spectral Slope Model	GP Model
Observed microcystin mean (µg/L)	0.6718	0.6718	0.6718	0.6718	0.6718	0.6718
Predicted microcystin mean (µg/L)	2.226	0.1792	0.6360	1.008	0.3571	0.5936
Root mean square error (µg/L) (*rmse*)	1.348	1.340	**0.3451**	1.356	0.7583	**0.3530**
Ratio of St. Dev. (*CO*)	0.8270	0.1238	**0.6787**	0.5540	0.5589	**0.6837**
Mean percent error (%) (*PE*)	87.57	**5.251**	38.07	61.87	**2.177**	25.01
Square of the Pearson product moment correlation coefficient (R^2)	0.02393	0.09625	**0.8449**	0.2710	0.7062	**0.9269**
Computational time (seconds)	<1	<1	92	<1	<1	450

Source: Chang, N. B., Vannah, B., and Yang, J., 2014. *IEEE Journal of Selected Topics in Applied Earth Observations and Remote Sensing*, 7, 2426–2442.

Note: Bolded values indicate the two models exhibiting the highest performance in seconds required to generate the model. As expected, machine learning methods took longer to solve than regression techniques. The fused hyperspectral input provided the most accurate results overall (N = 44).

[a] Fused Multispectral Sensor Input Pair: Landsat and MODIS land bands.

[b] Fused Hyperspectral Sensor Input Pair: MERIS and MODIS ocean color bands.

multispectral bands tend to average out spectral peaks and troughs within the neighborhood, thereby losing its shape and detail. In contrast, this model performed significantly better for the hyperspectral sensor surface reflectance inputs due to the inclusion of the correct hyperspectral bands with the defining features in the spectral reflectance curves for chlorophyll-a and phycocyanin ($R^2 = 0.7062$).

Yet the hyperspectral sensor GP model underestimates the mean observed microcystin value by 0.0782 µg/L, while the multispectral sensor GP model is much closer to the mean with an average underestimation of 0.0338 µg/L (Chang et al., 2014). Recall from Figure 17.7b that the multispectral model often over predicted the microcystin values at low concentrations, while it underestimated the microcystin values when they were close to 1 µg/L. The hyperspectral model with fused images matched the observed microcystin values more closely in both regimes of microcystin values at low and high concentrations; yet the hyperspectral model with the fused model underestimated concentrations more often than it overestimated concentrations. The hyperspectral GP model yielded an RMSE value of 0.3530 µg/L, which is slightly worse than the 0.3451 µg/L obtained by the multispectral fused GP model. On the other hand, the fused multispectral and hyperspectral GP models had CO values of 0.6787 and 0.6837, both of which are close to the ideal value of 1. The hyperspectral GP model yielded a PE of 25%, while the multispectral sensor GP model had 13.06% higher error. According to the R^2 values, both models are statistically significant based on a positive correlation with values of 0.8449 for the multispectral sensor GP sensor model and 0.9269 for the hyperspectral sensor GP model. Overall, the GP models are better suited for determining the complex, nonlinear relationship between microcystin and surface reflectance. Hyperspectral surface reflectance inputs yielded better results than multispectral sensor surface reflectance inputs with regard to the CO, PE, and R^2 values.

In terms of the computational time required to derive the feature extraction models, machine learning techniques require much more time to retrieve and formulate the feature extraction models. In detail, computational time required for solving the multispectral and hyperspectral sensor GP models are 92 and 450 seconds, respectively. The hyperspectral spectral slope model presented reasonable predictions for microcystin within less than a second for model training. The short training time for the hyperspectral spectral slope model is a minor benefit given the current computational power available in a small laptop.

17.5.2.2   Impact of Data Fusion on Final Feature Extraction

Given that the spectral integrity of the data is altered during the data fusion, a suite of statistical indexes is proposed for comparison among those GP models derived from the single sensor data or the fused sensor data. The outcome is shown in Table 17.5. Each of the GP models using single sensor data had either similar or better accuracy than those trained from fused sensor data. However, fused images provide better insight into HAB identification and delineation due to shorter revisit time (daily) and higher spatial resolution (300-m) (Table 17.5). Therefore, the fused GP models offer a superior early warning capability because they can provide microcystin concentration maps daily with higher spatial resolution. Overall, the drawbacks of additional computing power and storage capacity required to formulate the fused hyperspectral sensor GP models is outweighed by the benefits of daily 300 m concentration maps with comparable performance (Chang et al., 2014).

17.5.2.3   Influences of Special Bands on Model Development

Valuable insights can be gained by investigating the spectral bands used to generate each of the GP models. The bands used to train the two-band models fall in the range of 560–570 nm and 660–681 nm (Table 17.6). Chlorophyll produces a distinct reflectance peak or trough in the range of 660–681 nm (Figure 17.1), which is why this wavelength exhibits a strong relationship with microcystin prediction. Furthermore, hundreds of millions of programs were generated by Discipulus® while solving for the thirty best models and "frequency of use" information generated by Discipulus® stands for how often a specific band was used in the thirty best programs or models created by Discipulus®. For example, if a band was used in each of the thirty programs it would have 100% frequency of use and it would likely share a high correlation between surface reflectance and microcystin. As a result, only the 30

TABLE 17.5

Comparison between GP Model Performance from Single Sensor Inputs and Fused Sensor Inputs

	Single Sensor		Fused Sensor Pair	
	MODIS GP Model	**MERIS GP Model**	**Multispectral GP Model**	**Hyperspectral GP Model**
Root mean square error (μg/L) (*rmse*)	0.14	0.50	0.35	0.35
Ratio of St. Dev. (*CO*)	0.69	0.71	0.68	0.68
Mean percent error (%) (*PE*)	23.80	5.81	38.07	25.01
Square of the Pearson product moment correlation coefficient (R^2)	0.92	0.95	0.85	0.94
Spatial resolution (m)	1000	300	30	300
Temporal resolution (days)	1	3	1	1

Note: On average, the GP models created using fused sensor pairs exhibited slightly lower accuracy, yet the temporal and spatial resolution of the fused sensor GP model's outputs are superior to the single sensor GP model's outputs.

TABLE 17.6

Spectral Band Centers with the Highest Performance for the Traditional Two-Band Models

Model Type	Band Centers (nm)	R^2
Multispectral two-band ratio	570 & 840	0.02393
Multispectral spectral slope	570 & 660	0.09625
Hyperspectral two-band ratio	560 & 681	0.2710
Hyperspectral spectral slope	665 & 681	0.7062

Source: Chang, N. B., Vannah, B., and Yang, J., 2014. *IEEE Journal of Selected Topics in Applied Earth Observations and Remote Sensing*, 7, 2426–2442.

best models are used to infer the relative importance of each band. The GP models were analyzed for the "frequency of use" of the 6 bands comparatively (Table 17.7).

"Frequency of use" in Table 17.7 indicates that the fused multispectral and hyperspectral sensor inputs do not share many of the same band centers. Nevertheless, an independent analysis of "frequency of use" can be organized for fused multispectral or hyperspectral sensor input. According to Table 17.7, the fused multispectral sensor GP model favored band center at 660, 1650, and 2,090 nm, which corresponds to bands 3, 5, and 7 of Landsat and the MODIS land bands 1, 6, and 7, respectively. Figure 17.1 shows that the wide band center at 660 nm averages spectral features unique to phycocyanin and chlorophyll-a (both are indicators of microcystin). The strong emphasis placed on the shortwave infrared bands for predicting microcystin can be attributed to the decrease in absorption at this band as the chlorophyll content in the water increases (Chang et al., 2014). The fused hyperspectral sensor GP model frequency used bands centered at 412, 443, and 681 nm. As 412 nm plays a critical role in recognizing decaying matter from once living organisms, this is directly linked to microcystin concentrations in the water (Chang et al., 2014). This is because *Microcystis* produce the toxin microcystin at the end of their exponential growth phase, and the toxin is retained within the bacteria until death causes the cell walls to rupture and release the toxin into the water (Chang

TABLE 17.7

Frequency of Use for the Band Centers Used as Spectral Inputs for the Multispectral and Hyperspectral Sensor GP Models

Fused Multispectral Input		Fused Hyperspectral Input	
Band Center (nm)	Frequency of Use (%)	Band Center (nm)	Frequency of Use (%)
477	67	**412.5**	**83**
562.5	53	**443**	**80**
652.5	**97**	489	57
849.5	57	555.5	27
1645	**80**	666	7
2010	**70**	**689.5**	**100**

Source: Chang, N. B., Vannah, B., and Yang, J., 2014. *IEEE Journal of Selected Topics in Applied Earth Observations and Remote Sensing*, 7, 2426–2442.

Note: The top 3 bands for each sensor type have been bolded.

et al., 2014). The band center at 443 nm is the chlorophyll absorbance maximum. The band center at 681 nm was also used in the traditional two-band models, as it directly corresponds with a strong reflectance trough caused by chlorophyll-a in the water (Chang et al., 2014).

17.6 MAPPING FOR MICROCYSTIN CONCENTRATIONS

Using Equation 17.3 (i.e., the GP model derived from the fused band data), maps of the microcystin concentration throughout Lake Erie can be reconstructed to allow for retrospective environmental assessment during the sensitive period of HABs. For comparison during a unique episodic period, microcystin maps generated from a fused hyperspectral sensor GP model and a fused multispectral sensor GP model are shown in Figure 17.8, both of which may be compared to a false color image of the algal bloom occurring on the same day. The concentration map derived using the fused data with hyperspectral sensor inputs is much less detailed than that derived using the fused data with multispectral sensor inputs due to the 300-m versus 30-m spatial resolution (Figure 17.8a,b). The apparent advantage associated with using the fused data with hyperspectral sensor inputs is that the predicted medium (light yellow spots in Figure 17.8a) and high (dark red spots in Figure 17.8a) concentrations of microcystin align with the hydrodynamic patterns and the distribution of green HABs observed in the true color image (Figure 17.8c). Figure 17.8a exhibits the secondary hot spot along the western shoreline of Lake Erie northward.

The less accurate concentration map derived from multispectral sensor data (Figure 17.8b) appears to exaggerate microcystin concentrations throughout the lake due to the lack of contribution from band 5 of the MERIS images (Figure 17.1a). The high likelihood for false positives to occur greatly diminishes the usefulness of the early warning system based on the fused multispectral sensor data (Chang et al., 2014). Given this drawback, it is obvious that the bands from the hyperspectral sensor fusion pair are more powerful for differentiating between microcystin and HABs. Nevertheless, the enhanced detail provided by the multispectral sensor fusion pair with 30-m resolution provides insight into the potential benefits. Therefore, enhancing the spatial resolution of the MERIS sensor could theoretically be achieved by fusing the images obtained from MERIS with those of another sensor that has a higher spatial resolution, such as Landsat. Should this task be carried out, there are a number of factors that must be carefully considered. For instance, the satellite pair must share common spectral bands. Second, if an image of MERIS is fused with that of a satellite of higher spatial resolution and the resulting image is then fused with MODIS to improve the temporal resolution, the spectral errors and fusion approximations made during the first fusion step will

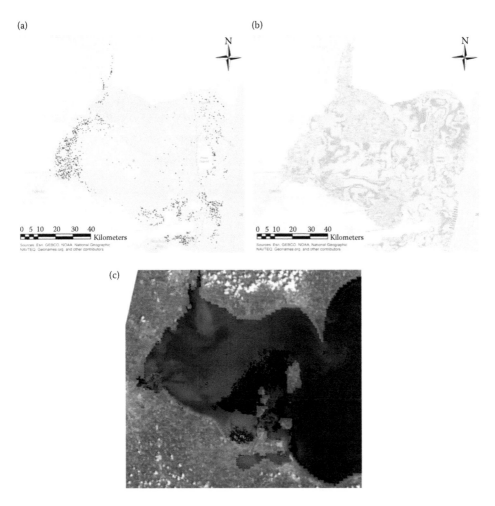

FIGURE 17.8 The concentration maps were generated using (a) the fused hyperspectral sensor GP model and (b) the multispectral GP model. The false color image of western Lake Erie is presented on the bottom (c). Large algal blooms spawning out of the Maumee and Sandusky Bays on July 26, 2010 are seen as dark green, while the sediment is a pale white in (c). Dark red spots in (a) and (b) denote areas of high microcystin concentration that pose a health threat, while yellow spots indicate low to medium concentrations. The 30 m spatial resolution of the multispectral image provides more detailed outlines, while the coarser (300 m) hyperspectral resolution predicts microcystin concentrations in locations that more closely align with HAB presence. (Chang, N. B., Vannah, B., and Yang, J., 2014. *IEEE Journal of Selected Topics in Applied Earth Observations and Remote Sensing*, 7, 2426–2442.)

propagate when fusing the resulting image the second time to improve the temporal resolution. Systematic bias must also be addressed carefully. Last, the mission of the ENVISAT satellite housing the MERIS sensor ended in April 2012 after a loss of contact; yet a suitable replacement is the Sentinel-3 mission by the ESA. Sentinel-3 similarly shares a 300-m spatial resolution. However, its spectral capabilities are higher at 21 bands compared to the 15 bands of MERIS. This could bring up more research niches for data fusion in the future.

With this advancement, spatial and temporal distributions of toxin bloom of microcystin can be further identified and assessed with respect to varying levels of proliferation and pathways of movement. It may pinpoint probable hot spots that require closer monitoring by field campaigns during the episodic period. For comparison, a single source multispectral sensor GP model based on MODIS images was used to generate the microcystin map for events on October 2, 2011 (Figure 17.1). These two maps show the hot spots of microcystin blooms along the south shoreline of western Lake

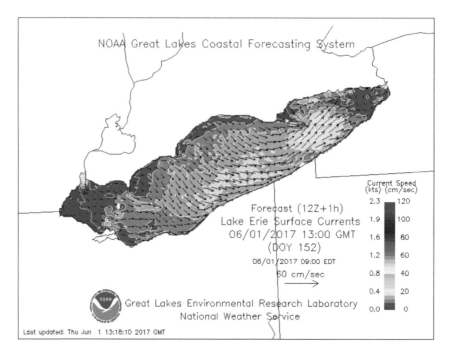

FIGURE 17.9   The average forecasted surface currents in Lake Erie and hydrodynamic patterns generated by the numerical simulations with a snapshot of forecasted surface currents at 9:00 EDT, 06/01/2017. (National Oceanic and Atmospheric Administration (NOAA), 2017. The Great Lakes Coastal Forecasting System, GLCFS, NOAA, https://www.glerl.noaa.gov//res/glcfs/. Accessed on June 2017.)

Erie driven mainly by an eastward water velocity in the hydrodynamic movement. This eastward and southward movement of surface water currents pushed the microcystin blooms to the US side; this can be justified by the simulated surface water currents for the early morning period in Figure 17.9 which allows the inflow from the mouth of Maumee River to keep moving north for quite a distance (i.e., northward along the western shoreline of Lake Erie).

17.7   SUMMARY

The City of Toledo issued a "Do Not Drink" advisory for local residents served by Toledo Water on August 4, 2014, due to the presence of unsafe levels of the algal toxin microcystin in the finished water flowing out from the drinking water plant into the water distribution networks. The affected area spanned three counties in Ohio and one in Michigan, leaving more than 400,000 people in the Toledo area without drinking water for 42 days. Avoiding this type of emergency event is deemed critical. In this chapter, IDFM techniques have been proven effective to provide the essential monitoring capability for mapping microcystin concentrations spatially, temporally, and spectrally by involving more satellites. The fused hyperspectral GP model was able to accurately predict peaks and troughs of microcystin concentrations, thus reducing the chances of false positive alarms. Discipulus® allows for GP models to be updated as new *in situ* data is obtained, and IDFM can use the new GP model and newly derived fused images to generate updated concentration maps of microcystin right away. Thus, periodic sampling of source water allows a better emergency response plan for water treatment as well as a deepened understanding of an ecosystem in transition for sustainable development. IDFM provides a basic framework of an early warning system for coastal water infrastructure on one hand, and offers insightful information about the ecosystem status and water quality state in a lake on the other hand.

Proper evaluation of influences of land development and management in the Maumee River Basin on the ecosystem changes in western Lake Erie can be made possible with the aid of such remote sensing technology and numerical simulation of the lake water quality to address the land development impact. Integrated assessment with respect to external forcing, such as anthropogenic impact, climate change, invasive species, and extreme weather, may be incorporated into the sustainable development plan in this region. This will certainly avoid some emergency events, such as the case in August 2014 at Toledo, Ohio, the United States of America.

REFERENCES

Bridgeman, T., 2005. Water quality monitoring in western Lake Erie and maumee bay, University of Toledo Lake Erie Center.

Budd, J., Beeton, A., Stumpf, R., Culver, D., and Kerfoot, W. 2001. Satellite observations of Microcystis blooms in western Lake Erie. *Verhandlungen des Internationalen Verein Limnologie*, 27, 3787–3793.

Chang, G., Mahoney, K., Briggs-Whitemire, B., Kohler, D., Mobley, C., Lewies, M., Moline, M. et al., 2004. The New Age of Hyperspectral Oceanography, Coastal Ocean Optics and Dynamics. http://digitalcommons.calpoly.edu/cgi/viewcontent.cgi?article=1146&context=bio_fac, accessed by May 2017.

Chang, N. B., Vannah, B., and Yang, J., 2014. Comparative sensor fusion between hyperspectral and multispectral remote sensing data for monitoring microcystin distribution in Lake Erie. *IEEE Journal of Selected Topics in Applied Earth Observations and Remote Sensing*, 7, 2426–2442.

Dash, P., Walker, N., Mishra, D., Hu, C., Pinckney, J., and D'Sa, E., 2011. Estimation of cyanobacterial pigments in a freshwater lake using OCM satellite data. *Remote Sensing of Environment*, 115, 3409–3423.

European Space Agency (ESA). 2006. MERIS Product Handbook Issue 2.1.

Francone, D., 1998. *Discipulus Software Owner's Manual, version 3.0 DRAFT*, Machine Learning Technologies, Inc., Colorado.

Ganf, G., Oliver, R., and Walsby, A., 1989. Optical properties of gas-vacuolate cells and colonies of *Microcystis* in relation to light attenuation in a turbid, stratified reservoir (Mount Bold Reservoir, South Australia). *Marine and Freshwater Research*, 40, 595–611.

Gao, F., Masek, J., Schwaller, M., and Hall, F., 2006. On the blending of Landsat and MODIS surface reflectance: Predicting daily Landsat surface reflectance. *IEEE Transactions on Geoscience and Remote Sensing*, 44, 2207–2218.

Ha, J., Hidaki, T., and Tsuno, H., 2009. Analysis of factors affecting the ratio of microcystin to chlorophyll-a in cyanobacterial blooms using real-time polymerase chain reaction. *Environmental Toxicology*, 26, 21–28.

Hu, C., Carder, K., and Muller-Karger, F., 2000. Atmospheric correction of SeaWIFS imagery over turbid coastal waters: A practical method. *Remote Sensing of Environment*, 74, 195–206.

Lee, Z. and Carder, K., 2002. Effect of Spectral band numbers on the retrieval of water column and bottom peroperties from ocean color data. *Applied Optics*, 41, 2191–2201.

Lekki, J., Anderson, R., Nguyen, Q., and Demers, J., 2009. Development of hyperspectral remote sensing capability for early detection and monitoring of Harmful Algal Blooms (HABs) in the Great Lakes. In: *AIAA Aerospace Conference*, Seattle, Washington, April 6–9, 2009.

Leshkevich, G. and Lekki, J., 2017. *Environmental Monitoring with Airborne Hyperspectral Imagery (HAB Prediction)*, https://www.glerl.noaa.gov/res/Task_rpts/2008/eosleshkevich09-4.html accessed by January 2017.

Lubac, B., Loisel, H., Guiselin, N., Astoreca, R., Artigas, L., and Meriax, X., 2008. Hyperspectral and multispectral ocean color inversions to detect Phaeocystis globosa blooms in coastal waters. *Journal of Geophysical Research: Oceans*, 113, C06026.

Masek, J., Vermote, E., Saleous, N., Wolfe, R., Hall, F., Huemmrich, F., Gao, F., Kutler, J., and Lim, T., 2005. A Landsat surface reflectance data set for North America, 1990–2000. *IEEE Geoscience and Remote Sensing Letters*, 3, 69–72.

Michalak, A., Anderson, E., Beletsky, D., Boland, S., Bosch, N., Bridgeman, T., Chaffin, J. et al., 2013. Record-setting algal bloom in Lake Erie caused by agricultural and meteorological trends consistent with expected future conditions. *Proceedings of the National Academy of Sciences*, 110, 6448–6452.

Mole, J., Chow, C., Drikas, M., and Burch, M., 1997. The influence of cultural media on growth and toxin production of the cyanobacterium *Microcystis aeruginosa* Kutz Emend Elenkin. Paper presented at the *13th annual conference of the Australian Society for Psychology and Aquatic Botany*, Hobart, January 1997.

National Aeronautical and Space Agency (NASA), 2011. https://earthobservatory.nasa.gov/IOTD/view.php?id=76127

National Aeronautical and Space Agency (NASA), 2013. Visiable Earth, https://visibleearth.nasa.gov/view.php?id=82165

National Oceanic and Atmospheric Administration (NOAA), 2013. https://tidesandcurrents.noaa.gov/hab/bulletins.html. Accessed on May 2017.

National Oceanic and Atmospheric Administration (NOAA), 2015. https://www.glerl.noaa.gov/res/HABs_and_Hypoxia/WLEMicrocystin.html

National Oceanic and Atmospheric Administration (NOAA), 2017. The Great Lakes Coastal Forecasting System, GLCFS, NOAA, https://www.glerl.noaa.gov/res/glcfs/. Accessed on June, 2017.

O'Reilly, J., Maritorena, S., Mitchell, B., Siegel, D., Carder, D., Garver, S., Kahru, M., and McClain, C., 1998. Ocean color chlorophyll algorithms for SeaWiFS. *Journal of Geophysical Research*, 103, 937–953.

Ohio Environmental Protection Agency (Ohio EPA), 2010. *Ohio Lake Erie Phosphorus Task Force Final Report,* Columbus, Ohio, USA.

Ohio Sea Grant 1994. Zebra mussels in Lake Erie: The invasion and its implications by Fred L. Snyder, District Extension Specialist, Ohio Sea Grant College Program. http://www.agri.ohio.gov/Public_Docs/Pest_Study_Material/3b%20Zebra%20Mussels%20in%20Lake%20Erie.pdf. Accessed on May 2017.

Ouellette, A. J. A., Handy, S. M., and Wilhelm, S. W., 2006. Toxic Microcystis is widespread in Lake Erie: PCR detection of toxin genes and molecular characterization of associated cyanobacterial communities. *Microbial Ecology*, 51, 154–165.

Pohl, C. and Van Genderen, J., 1998. Multisensor image fusion in remote sensing: Concepts, methods, and applications. *International Journal of Remote Sensing*, 19, 823–854.

Rinta-Kanto, J., Konopko, E., DeBruyn, J., Bourbonniere, R., Boyer, G., and Wilelm, S., 2009. Lake Erie microcystis: Relationship between microcystin production, dynamics of genotypes and environmental parameters in a large lake. *Harmful Algae*, 8, 665–673.

Rogalus, M. and Watzin, M., 2008. Evaluation of sampling and screening techniques for tiered monitoring of toxic cyanobacteria in lakes. *Harmful Algae*, 7, 504–514.

Shi, K., Zhang, Y., Xu, H., Zhu, G., Qin, B., Huang, C., Liu, X., Zhou, Y., and Lv, H., 2015. Long-term satellite observations of Microcystin concentrations in Lake Taihu during cyanobacterial bloom periods. *Environmental Science & Technology*, 49, 6448–6456.

Torrecilla, E., Piera, J., and Vilaseca, M. 2009. Derivative analystis of hyperspectral oceanographic data. In: Jedlovec G. (Ed.), *Advances of Geoscience and Remote sensing*, ISBN: 978-953-307-005-6, InTech, London, UK.

United States Environmental Protection Agency (US EPA), 2012. https://www.epa.gov/greatlakes/lake-erie. Accessed on April 2017.

United States Geological Survey (USGS). Landsat Processing Details [Online]. Available: http://landsat.usgs.gov/Landsat_Processing_Details.php

Van Genderen, J. and Pohl, C., 1994. Image fusion: issues, techniques, and applications. Intelligent image fusion. In: *Proc. EARSel Workshop*, Strasbourg, France, 18–26.

Vanderploeg, H. A. 2003. Ecological forecasting of impacts of Ponto-Caspian species in the Great Lakes: Describing, understanding, and predicting a system in transition. In: Valette-Silver N. and Scavia D. (Eds.) *Ecological Forecasting: New Tools for Coastal and Marine Ecosystem Management.* NOAA Technical Memorandum NOS NCCOS 1, Ann Arbor, Michigan, USA, 81–84. http://www.glerl.noaa.gov/pubs/fulltext/2003/20030015.pdf

Vanderploeg, H. A., Liebig, J. R., Carmichael, W. W., Agy, M. A., Johengen, T. H., Fahnenstiel, G. L., and Nalepa, T. F., 2001. Zebra mussel (*Dreissena polymorpha*) selective filtration promoted toxic *Microcystis* blooms in Saginaw Bay (Lake Huron) and Lake Erie. *Canadian Journal of Fisheries and Aquatic Sciences*, 58, 1208–1221.

Vermote, E., Kotchenova, S. and Ray, J., 2011. *MODIS Surface Reflectance User's Guide. v1.3*, MODIS Landsat Surface Reflectance Science Computing Facility, Greenbelt, MD, USA.

Vermote, E., Saleous, N., and Justice, C., 2002. Atmospheric correction of MODIS data in the visible to middle infrared: First results. *Remote Sensing of Environment*, 83, 97–111.

Vincent, R., Xiaoming, Q., McKay, R., Miner, J., Czajkowski, K., Savino, J., and Bridgeman, T., 2004. Phycocyanin detection from Landsat TM data for mapping cyanobacterial blooms in Lake Erie. *Remote Sensing of Environment*, 89, 381–392.

World Health Organization (WHO), 1999. *Toxic Cyanobacteria in Water: A guide to their public health consequences, monitoring and management.* E & FN Spon, London, England.

Wynne, T., Stumpf, R., Tomlinson, M., Warner, R., Tester, P., Dyble, J., and Fahnenstiel, G., 2008. Relating spectral shape to cyanobacterial blooms in the Laurentian Great Lakes. *International Journal of Remote Sensing*, 29, 3665–3672.

18 Environmental Reconstruction of Watershed Vegetation Cover to Reflect the Impact of a Hurricane Event

18.1 INTRODUCTION

Multitemporal change detection of Land-Use and Land-Cover (LULC) is an important tool for detecting the influences of human activities and natural disasters on environmental systems (Ran et al., 2009). They are also critical for understanding the complexities of environmental status and ecosystem state as well as elucidating the climate change implications surrounding human society (Jia et al., 2014). In multitemporal change detection of LULC, a suite of prescribed land use classes is normally assigned in advance for LULC classification either by unsupervised or supervised learning algorithms. Whereas unsupervised land use classification is utilized to assign land use classes on a scene-by-scene basis based on background knowledge, supervised land use classification depends on a set of ground truth data to verify the land use classification.

Some indexes in remote sensing may be applied for the holistic LULC classification or the individual assignment of a land use change class on a scene-by-scene basis. The Normalized Difference Vegetation Index (NDVI) is an index that employs the red and near-infrared bands of the electromagnetic spectrum to synthesize images and analyze whether the area of interest contains live green vegetation cover or not (Rouse et al., 1974). The NDVI data can be determined by concentrating on the satellite bands that are most associated with the vegetation data (red and near-infrared). Generally, healthy vegetation will retain a large portion of the visible light that falls on it for energy conversion and biomass growth and reflects an enormous portion of the near-infrared light back to the sky. In contrast, unhealthy vegetation reflects more of the visible light and less of the near-infrared light (Rouse et al., 1974). Uncovered soils decently reflect both the red and infrared segments of the electromagnetic spectrum (Holm et al., 1987). The greater the distinction between the near-infrared and red reflectance, the denser the vegetation must be.

There is a wide range of applicability for the aid of such a vegetation index or its extended forms. Given that the behavior of green vegetation is known across the electromagnetic spectrum, NDVI can be defined and applied for some land cover classification tasks. For example, NDVI has been used in the assessment of crop yields (Quarmby et al., 1993; Prasad et al., 2006; Mkhabela et al., 2011), rangelands conveyance capacities (Yengoh et al., 2014), and so on. It is often applied along with other ground parameters for assessment, such as ground cover percentage (Lukina et al., 1999; Scanlon et al., 2002), photosynthetic movement of the plant (Penuelas et al., 1995; Pettorelli et al., 2005), surface water (Chandrasekar et al., 2010; Fu and Burgher, 2015), leaf territory record, also known as Leaf Area Index (LAI) (Carlson and Ripley, 1997; Wang et al., 2005), and the measure of biomass (Anderson et al., 1993).

In remote sensing science, wetness stands for the moisture contents of soil/vegetation, brightness represents variations in soil background reflectance, and greenness reflects variations in the vigor of green vegetation. To discover more about the co-variations and interactions of greenness, brightness, and wetness, the Tasseled Cap Transformation (TCT) is used, which involves the conversion of original band data into composite band readings based on the weighted sum of select separate channels (Vorovencii, 2007). In fact, the TCT is a global vegetative index that separates the amount of soil brightness, vegetation, and moisture content into individual pixels (Watkins, 2005). The TCT may entail co-variations and interactions of greenness, brightness, and wetness with or without ground truth data. Such a transformation was first presented by Kauth and Thomas (1976) to describe the growth of wheat cover in an agricultural field, linking the patterns found in Landsat data from the crop lands as a function of the life cycle of the crops periodically. This transformation of the original bands of an image into a new set of bands with defined interpretations helps identify the changes of LULC directly and indirectly. Having been widely applied, the TCT enhances the spectral information content of multispectral satellite data such as Landsat TM/ETM+. Additionally, such a transformation optimizes data viewing, which helps in the study of vegetative cover changes of an area through more insightful and interactive information than the NDVI method. Yet, limited by the spatial and temporal resolution of an individual satellite, neither NDVI nor TCT are entitled to discern the daily change of LULC with high spatial resolution unless there are daily satellite images with high spatial resolution or fused images to support this initiative.

Remote sensing data, such as Landsat images, can be applied for the analyses of NDVI or TCT on a regional scale (Friedl et al., 2002; Gong et al., 2013; Adam et al., 2014). However, the Landsat satellite alone has an inherent disadvantage which might compromise efficiency when early warning or emergency response is under high concurrent demand. This is because Landsat satellite images have a high spatial resolution (i.e., 30-m by 30-m) but low temporal resolution (i.e., 16-day revisit cycle). On the other hand, MODIS has a high temporal resolution with daily revisit times. However, it has low spatial resolution (greater than 250-m). It is possible to create synthetic or fused images by using the Spatial and Temporal Adaptive Reflectance Fusion Model (STARFM) algorithm based on the high spatial resolution of Landsat and the high temporal resolution of MODIS (Gao et al., 2006). Therefore, these fused images have the high spatial resolution associated with Landsat images and the high temporal resolution associated with MODIS images. This will help generate daily Landsat imagery using MODIS images as a gap-filling technique for the missing days in between two Landsat revisits. The fused images derived from STARFM have been used in various studies for change detection of LULC, such as the generation of daily land surface temperature (Weng et al., 2014), monitoring of urban heat island (Huang et al., 2013), forest cover classification using NDVI (Jia et al., 2014; Rao et al., 2015; Zhang et al., 2015), and so on. In addition to the commonly used Landsat TM and MODIS sensor pairs, other satellite sensor pairs have also been used for data fusion to study multitemporal change detections (Table 18.1).

STARFM cannot accurately predict reflectance in heterogeneous fine-grained landscapes, however (Chang et al., 2016). The Spatial Temporal Adaptive Algorithm for mapping Reflectance CHange (STAARCH) was thus proposed to detect changes in reflectance by applying TCT of both Landsat TM/ETM and MODIS reflectance values (Hilker et al., 2009). The Enhanced Spatial and Temporal Adaptive Reflectance Fusion Model (ESTARFM) was also developed to improve the accuracy in fusing images over heterogeneous regions (Zhu et al., 2010). Along these lines, improved algorithms such as the SParse-representation-based SpatioTemporal reflectance Fusion Model (SPSTFM) (Huang and Song, 2012), the Spatial and Temporal Data Fusion Model (STDFM) (Wu et al., 2012), and Enhanced STDFM (ESTDFM) (Zhang et al., 2013) may reduce the limitation to different extents.

NDVI, which fuses two bands' information into one, is particularly useful for assessing the presence and condition of vegetation. TCT is another index which creates three band images representing brightness, greenness, and wetness for the area of interest to aid in the assessment of multitemporal change detection of LULC. With the aid of data fusion of Landsat and MODIS images based on the STARFM algorithm, finer resolution of fused images temporally and spatially becomes available for the derivation of TCT and NDVI simultaneously to allow us to see the impact much closer to the date when the event occurred. In this chapter, the dynamic mapping of NDVI and TCT for the monitoring of the selected coastal watershed was conducted to assess the impact

TABLE 18.1

Fusion of Satellite Sensors for Land Use Land Cover Change Detection

Sensors	Purpose	Reference
ERS-1 and JERS-1	Land cover classification using fuzzy concept	Solaiman et al. (1999)
IKONOS	Two classifiers are used for decision level fusion	Fauvel et al. (2006)
E-SAR (Airborne L and P Bands PolSAR and PolInSAR)	Land cover classification	Shimoni et al. (2009)
Quickbird MS & RADARSAT SAR	Analysis of urban surface temperature	Ban et al. (2010)
LiDAR & Landsat TM	Assessment of urban land cover	Singh et al. (2012)
MODIS and Landsat ETM+	Urban flood mapping; urban land cover changes via ESTARFM, SPSTFM, STDFM, and ESTDFM algorithms; LULC associated with TCT via the STAARCH algorithm	Hilker et al. (2009); Zhu et al. (2010); Wu et al. (2012); Zhang et al. (2013); Zhang et al. (2014)
Airborne LiDAR and CASI hyperspectral data	Pixel- and feature-level fusion of hyperspectral and lidar data for urban land-use classification	Man et al. (2015)
RADARSAT-2 PolSAR and HJ-1A MS	Land-cover classification	Xiao et al. (2016)
Airborne AVIRIS hyperspectral data	Analysis of spectral-spatial feature fusion for hyperspectral data classification	Liu and Li (2016)

Abbreviations: PolSAR: polarimetric synthetic aperture radar; PolInSAR: polarimetric synthetic aperture interferometric radar; MS: multispectral images; AVIRIS: the Airborne Visible Infrared Imaging Spectrometer; LIDAR: Light Detection and Ranging; CASI: Compact Airborne Spectrographic Imager; STAARCH: Spatial Temporal Adaptive Algorithm for Mapping Reflectance Change; ESTARFM: Enhanced Spatial and Temporal Adaptive Reflectance Fusion Model; SPSTFM: sparse-representation-based Spatiotemporal Reflectance Fusion Model; STDFM: Spatial and Temporal Data Fusion Model; ESTDFM: Enhanced STDFM.

of Hurricane Sandy in the northeastern United States. The present chapter aims particularly at comparing high NDVI area and greenness values given by TCT with low NDVI values and high brightness values given by TCT during the Hurricane Sandy landfall event over a coastal watershed in the northeastern United States. Data fusion empowers closer earth observation of a hurricane event right before and after its landfall. Such a side-by-side comparison signifies the environmental surveillance capability of a hurricane impact on the multitemporal changes of LULC in terms of NDVI to TCT when brightness, wetness, and greenness may be considered holistically.

18.2 STUDY REGIONS AND ENVIRONMENTAL EVENTS

The coastal watershed of interest is the Hackensack and Pascack watershed situated in New Jersey, the United States of America (USA). It is introduced below.

18.2.1 THE HACKENSACK AND PASCACK WATERSHED

This watershed is designated as watershed management area 5 by the New Jersey Department of Environmental Protection (Figure 18.1). The drainage area of this watershed is approximately 64.5 square kilometers. The watershed is comprised of three sub-watersheds. They are Hackensack River watershed, Hudson River watershed, and Pascack Brook watershed. This watershed is the most

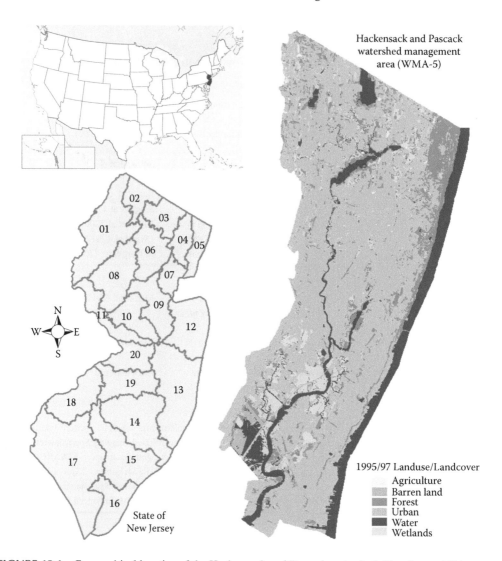

FIGURE 18.1   Geographical location of the Hackensack and Pascack watershed, New Jersey, USA.

populated of all the watersheds in the state of New Jersey. About 50% of the land is undeveloped, with more than 30% residential development. The rest of the developed land is for commercial or industrial use (Figure 18.2).

18.2.2   THE IMPACT OF HURRICANE SANDY

Hurricane Sandy is regarded as the most devastating and destructive hurricane of the 2012 Atlantic hurricane season and the second most costly hurricane in United States history (National Oceanic and Atmospheric Administration (NOAA), 2012). The hurricane is also referred to as Superstorm Sandy in unofficial terms. This was a Category 3 storm and it made landfall in Cuba when it was at its maximum strength. It became a Category 2 storm as it approached the United States. The hurricane caused damage to 24 states in the United States, including the entire eastern seaboard from Florida to Maine. The damage was particularly severe in the states of New York and New Jersey. The total damage caused by the hurricane is estimated to be 75 billion USD (2012 USD) (Blake et al., 2013).

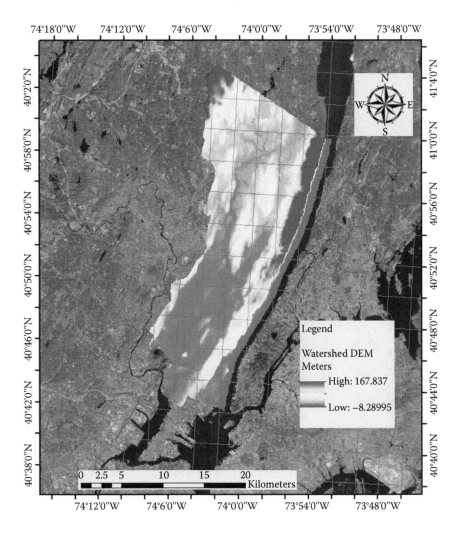

FIGURE 18.2 Regional map of the study region and digital elevation model of the Hackensack and Pascack watershed, New Jersey, USA.

Hurricane Sandy had devastating effects on the New Jersey and New York areas. The storm surge, in addition to the large and battering waves, impacted large portions of the coasts of New York and New Jersey. In New York, the hardest hit areas were Monmouth and Ocean counties (Blake et al., 2013), and the highest wave was 3.83 m (Blake et al., 2013). In New Jersey, these massive waves and tidal surges caused inundation of whole communities and many houses were destroyed. The highest wave was 2.59 m (Blake et al., 2013). Many residences lost power in these areas and power outages lasted for several weeks after the hurricane's landfall. Costs of business losses were estimated at upward of about 8.3 billion USD. The repair costs for wastewater/sewer services and gas and power lines were about 3 billion USD and 1 billion USD, respectively.

Although the hurricane caused massive damage in many areas in New York and New Jersey, the damage was relatively minimal in Massachusetts. Widespread power disruption occurred for several days. The sustained wind pressure in the Buzzards Bay and Cape Cod areas was 132.8 km/h, and the tidal surges caused by heavy wind were as high as 0.66–1.33 m in Massachusetts. (Blake et al., 2013). The total estimated damage from Hurricane Sandy in Massachusetts was about 20.8 million USD (National Climatic Data Center (NCDC), 2012). The path of the hurricane is depicted in Figure 18.3.

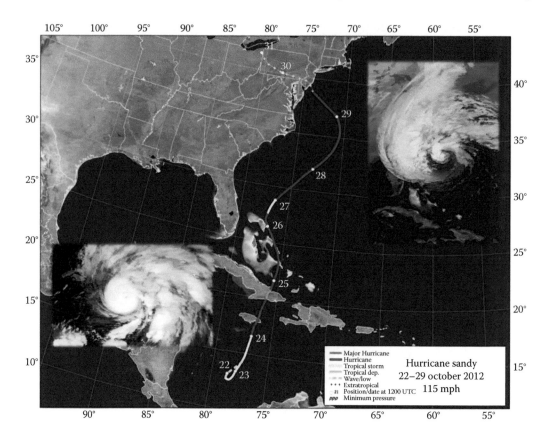

FIGURE 18.3 Trajectory of Hurricane Sandy, 2012. (National Weather Channel, NOAA; http://www. weather.gov.)

18.3 UNSUPERVISED MULTITEMPORAL CHANGE DETECTION

18.3.1 DATA FUSION

The Landsat-7 ETM+ satellite has a low temporal resolution (16 days) but high spatial resolution (30 m). However, MODIS has high temporal resolution (daily) but low spatial resolution (>250 m). In order to have both high spatial and high temporal resolutions for any of the selected days, a pair of Landsat images are used as a base with the MODIS images available for the days at which Landsat is not available. The STARFM algorithm creates a set of predicted images with high spatial and temporal resolution for the given days in between. In order to create a fused image, the Landsat and MODIS images have to be free from any type of cloud cover. Three steps are required for data fusion:

- *Data Acquisition*: The basis of data fusion involving the STARFM algorithm is the surface reflectance data. For this case, we collected surface reflectance data from Landsat-7 ETM+ and MODIS-Terra satellites. The images with few cloud covers were considered for download purposes from both satellites. For Landsat-7 ETM+, bands 1–5 and 7 were downloaded from the designated website maintained by the United States Geological Survey (USGS). For MODIS-Terra, bands 1–4, 6 and 7 were selected for download as GEOTIFF images from the online data pool managed by NASA. A comparison between these satellites is presented in Table 18.2.
- *Image Processing and Preparation*: The MODIS-Terra images are at a level-2G basis, meaning that the MODIS data are radiometrically calibrated and atmospherically corrected

TABLE 18.2

Satellite Products Utilized in this Study

Satellite Sensor	Product Selection	Spatial Resolution	Temporal Resolution	Bands Used
Landsat 5 TM Landsat 7 ETM+	Surface reflectance	30 m	16 days	1–5,7
MODIS-Terra	Surface reflectance (MOD09GA)	250–500 m	Daily	1–4,6,7

to account for scattering and aerosols (Vermote et al., 2008). The data of Landsat-7 is radiometrically and atmospherically corrected and is on a level-1T basis. ArcGIS, a spatial analysis software with mapping functionality, was employed for image processing in preparation for the data fusion process. It was essential to perform the following actions on the Landsat images (Chang et al., 2014): (1) perform a gap-filling operation using ENVI classic software to account for data loss resulting from sensor failure; (2) re-project to the Universal Transverse Mercator (UTM) zone 19 North for the watershed of Mattapoisett harbor and zone 18 North for the Hackensack and Pascack watershed; and (3) crop out unnecessary land area. In addition, following steps were taken to process the MODIS images: (1) re-project to the UTM zone 19 North for the Mattapoisett River watershed and 18 North for the Hackensack and Pascack watershed; (2) resample to a spatial resolution of 30 m, and (3) crop out unnecessary land data from both watersheds.

- *Data Fusion*: The benchmarking algorithm of STARFM is used here to fuse both data sets. This allows for creation of fused images of enhanced spatial and temporal properties. The spectral reflectance value for each pixel of the MODIS image is a conglomeration of the surface reflectance from each object within a 250 by 250 m^2 area. On the other hand, spectral reflectance values provided by the Landsat image are an average of objects contained within a 30 by 30 m^2 pixel. Thus, generating a regression model using the fused band data on a daily basis should be more accurate than using MODIS imagery alone, since the coarser MODIS data is integrated with the spectral reflectance values at Landsat's fine spatial resolution within their daily snapshots. With regard to the fusion process, the STARFM algorithm uses the original Landsat image as a base data set and fuses the MODIS and Landsat images of the same date to produce the predicted image (Gao et al., 2006). Yet, when fusing the data streams from Landsat and MODIS, Landsat band 1 does not have the same spectral range of MODIS band 1. Instead, Landsat band 1 corresponds to MODIS band 3 and so on. Table 18.3 details the proper band pairs between MODIS and Landsat.

TABLE 18.3

Landsat 7 ETM+ and Terra MODIS Band Comparisons

Landsat 7 ETM+ Band	Landsat Bandwidth (nm)	Terra MODIS Band	MODIS Bandwidth (nm)
1	450–520	3	459–479
2	520–600	4	545–565
3	630–690	1	620–670
4	760–900	2	841–876
5	1550–1750	6	1628–1652
7	2080–2350	7	2105–2155

Source: Chang, N. B. et al. 2014. *International Journal of Remote Sensing*, 35, 1064–1093.

18.3.2 NDVI Mapping Based on the Fused Images

The NDVI algorithm is calculated by subtracting the red reflectance values from the near-infrared and dividing it by the sum of the near-infrared and red reflectance values. The formula for NDVI can be written as

$$NDVI = \frac{NIR - Red}{NIR + Red} \tag{18.1}$$

The NDVI mapping effort of this study is carried out using the "Image Analysis" tool of ArcGIS software. The algorithm is used as input in the tool and the NDVI maps are generated as output based on the fused images of Landsat and MODIS. Regarding Landsat-5, the NIR corresponds to band 4 and the red corresponds to band 3, respectively. Thus, Equation 18.1 can be transformed into:

$$NDVI = \frac{Band\,4 - Band\,3}{Band\,4 + band3} \tag{18.2}$$

18.3.3 Performance Evaluation of Data Fusion

There is a need to verify the effectiveness of the data fusion based on some statistical metrics. This study chose Shannon entropy as a representative index for verification of data fusion between Landsat and MODIS in the assessment of the Hurricane Sandy event. Entropy is a statistical measure of randomness or richness of information in an image, and hence can be used to characterize the texture of an image (Shi et al., 2005). The expression of the entropy of an image is defined as

$$H = -\sum_{i=1}^{n} p_i \log_2 p_i \tag{18.3}$$

where n denotes the number value of gray level, and p_i is the probability mass function which contains the histogram counts of valid grid data values. The higher the value of entropy, the richer the information content.

18.3.4 Tasseled Cap Transformation for Hurricane Sandy Event

Generally, there exist six typical TCTs; however, only three are frequently used: brightness (measure of soil), greenness (measure of vegetation), and wetness (interrelationship of soil and canopy moisture). The guide and inspiration for this method may have been the Principal Component Analysis (PCA), which helps form new variables as weighted sums of different band readings. The first three transformations, that is, brightness, greenness, and wetness, contain most of the information. The rest of the transformations are treated as noise and rarely used.

A TCT is performed by taking linear combinations of the original spectral bands given assigned weights in the tasseled cap function. The TCT is similar to the concept of PCA. The transformation coefficients (i.e., weights) were derived statistically (Table 18.4) from remotely sensed images and empirical observations, which are always sensor-dependent. The plots of the tasseled cap for the watershed may be compared based on the rationale, as shown in Figure 18.4, in terms of landscape changes that occurred after the hurricane made landfall on the east coast of the United States.

Crist and Cicone (1984a,b, 1986) adapted the TCT to the six channels of Landsat TM data. The weights found by Crist and Cicone (1984a,b) for the Landsat TM bands are shown in Table 18.4. It is evident that brightness has all positive loadings, implying that all six bands contribute to

TABLE 18.4

Coefficients for the Tasseled Cap Functions for Landsat TM bands 1-5 and 7

Component	Band 1	Band 2	Band 3	Band 4	Band 5	Band 7
Brightness	0.3037	0.2793	0.4343	0.5585	0.5082	0.1863
Greenness	−0.2848	−0.2435	−0.5436	0.7243	0.0840	−0.1800
Wetness	0.1509	0.1793	0.3299	0.3406	−0.7112	−0.4572

its overall value, although bands 4 (NIR), 5 (SWIR1), and 7 (SWIR2) contribute the most. Study regions with high vegetation cover are found at low Red and higher NIR regions and vice versa. Since both Red and NIR bands have higher positive values of loadings associated with channels 4 and 5 (Table 18.4), any increase in vegetation density will cause less obvious changes in brightness. It is indicative that this brightness feature will only be sensitive to soil characteristics including its albedo rather than vegetation. By the same token, the loadings corresponding to greenness in Table 18.4 reveal the fact that vegetation has higher values of reflectance in NIR and lower in Red regions of the electromagnetic spectrum, since Red bands have much higher positive values of loading associated with channel 4 than that of 5. Because TCT is sensor-dependent, a new TCT based on

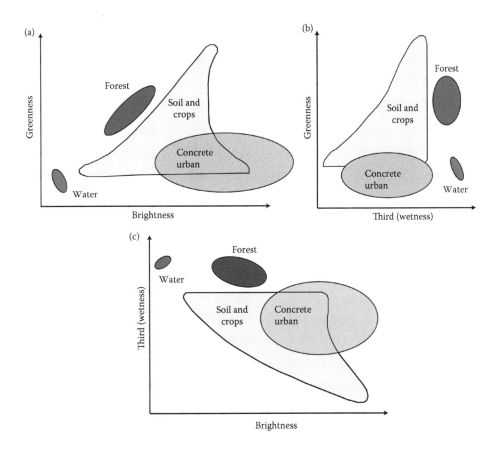

FIGURE 18.4 Tasseled cap transformation plots for landscape interpretation. (a) Greenness vs. brightness, (b) greenness vs. third (wetness), and (c) third (wetness) vs. brightness plot. (Kauth, R. J. and Thomas, G. S., 1976. *Proceedings of the Symposium on Machine Processing of Remotely Sensed Data*. Purdue University of West Lafayette, Indiana, 4B-41 to 4B-51.)

TABLE 18.5

Coefficients for the Tasseled Cap Functions for Landsat ETM+ Bands 1-5 and 7

Component	Band 1	Band 2	Band 3	Band 4	Band 5	Band 7
Brightness	0.3561	0.3972	0.3904	0.6966	0.2286	0.1596
Greenness	−0.3344	−0.3544	−0.4556	0.6966	−0.0242	−0.2630
Wetness	0.2626	0.2141	0.0926	0.0656	−0.7629	−0.5388

Landsat 7 was developed later, as listed in Table 18.5 (Huang et al., 2002). Equation 18.4 was used to accomplish the transformation.

$$u_j = R_j X_j + c \tag{18.4}$$

in which the pixel values in the four multispectral (MSS) bands are multiplied by the corresponding elements of R_i to give the position of the jth pixel in the ith Tasseled Cap axis (u). In addition, c is used to make sure that the elements of vector u are always positive.

18.4   ENTROPY ANALYSIS OF DATA FUSION

Hurricane Sandy can be closely monitored by using data fusion techniques because both Landsat and MODIS images are available. Toward such a goal, the performance of data fusion had better be evaluated by using the entropy Equation 18.3 to make sure there is no obvious information loss before and after the data fusion. The probability density functions (PDFs) associated with entropy in Figure 18.5

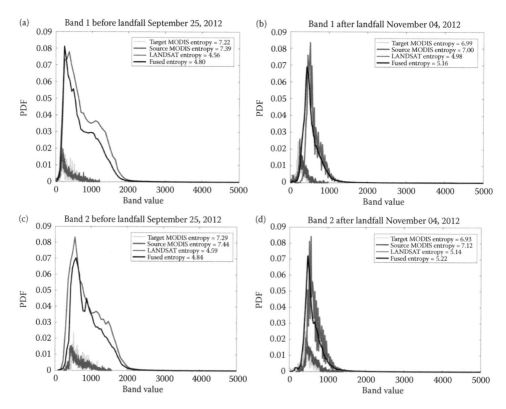

FIGURE 18.5   Entropy assessment for data fusion.   (*Continued*)

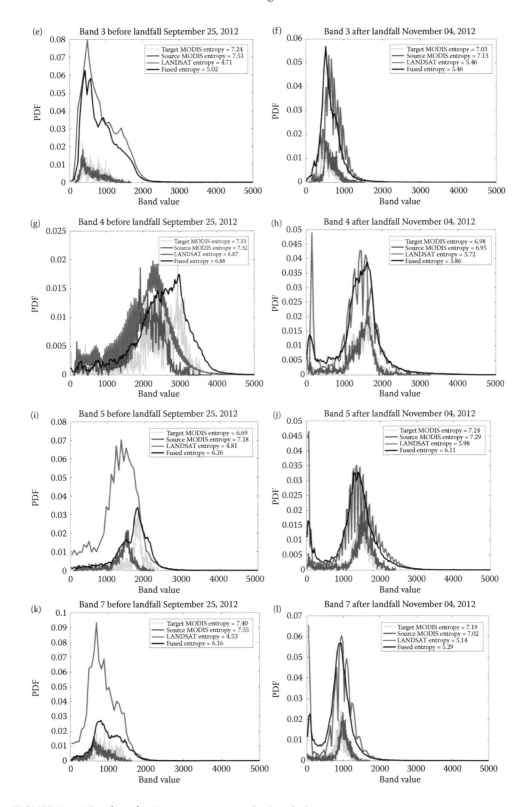

FIGURE 18.5 (Continued) Entropy assessment for data fusion.

represent the patterns of pixel values in connection to each band selected and the final entropy values of those PDFs were generated and listed in each legend.

In these subdiagrams (Figure 18.5), source MODIS entropy in the legend represents the entropy associated with the MODIS images that were used for fusing Landsat images, and target MODIS entropy in the legend represents the entropy associated with the MODIS images that were used for predicting the map at a target date. For example, the target date of Figure 18.5a is September 25, 2012, when using Landsat band 1 and corresponding MODIS band 3 to process the PDF (see Table 18.3 for pairing the two types of satellite images band by band). It shows that no obvious discrepancies are observed due to the loss of information in data fusion. Some discrepancies were observed within the range of lower band values in Figure 18.5h, j, and l although they are not significant.

18.5 COMPARISON OF THE HURRICANE SANDY IMPACT ON THE SELECTED COASTAL WATERSHED

18.5.1 NDVI Maps

Hurricane Sandy was first formed from a tropical wave in the western Caribbean Sea on October 22, 2012, and quickly turned to a tropical storm a few hours later. The hurricane wreaked havoc on Cuba, the Bahamas, and the eastern seaboard of the United States through October 29, 2012. This is also the date when it made landfall in New Jersey and New York, where the damage was the most severe. The storm spread as far as Buzzards Bay and Cape Cod, where heavy rains and storm surges caused damage to buildings and roadways in Massachusetts. Trees were uprooted and power was cut off. The extent of the damage on terrestrial vegetative cover of the coastal watershed adjacent to Mattapoisett harbor can be assessed on a day-to-day basis using the fused images created by STARFM. The NDVI maps for cloud-free days for the month before Hurricane Sandy made landfall (i.e., September) are depicted in Figure 18.6.

The NDVI maps at four days show no significant changes in the vegetation cover before the hurricane event at the Mattapoisett Bay watershed (Figure 18.6). This may be attributed to the fact that the hurricane did not make landfall in the Mattapoisett harbor area, unlike the Hurricane Bob event of 1991. The average NDVI value for the day of October 27, two days before the hurricane made landfall in New Jersey, is 0.625 (Table 18.6). The NDVI on October 30, one day after the hurricane made landfall in New Jersey, held steady with a value of 0.627 (Table 18.6).

The NDVI maps for cloud-free days for the month before and after Hurricane Sandy made landfall (i.e., September and November) are spatially depicted in Figures 18.7 and 18.8. The differences of mean values of NDVI before and after the impact of Hurricane Sandy on Hackensack and Pascack watershed confirm the impact of Hurricane Sandy on the vegetation cover (Tables 18.7 and 18.8). The low mean value of NDVI in Table 18.8 indicates that there was some damage to the vegetative cover; reports confirmed that trees were being uprooted. A comparison of the mean values of NDVI before and after the landfall is indicative that about one-third of the vegetation cover was lost in this event. In parallel with this NDVI analysis, the NDVI maps for cloud-free days at the Mattapoisett River watershed when Hurricane Sandy made landfall are quite different (Figure 18.6 and Table 18.6).

18.5.2 Tasseled Cap Transformation Plots

The TCT plots for the Mattapoisett River watershed before and after Hurricane Sandy made landfall are depicted in Figure 18.9, in which plots (a) to (c) represent October 27, 2012, (i.e., two days before the hurricane made landfall) and plots (d) to (f) represent October 30, 2012, (i.e., one day after the hurricane's landfall). It is evident that there is no significant change in the landscape conditions except more tasseled caps in the wetness versus brightness plots with a slightly different surface feature. This may be attributed to the fact that the Mattapoisett River watershed was not in the direct

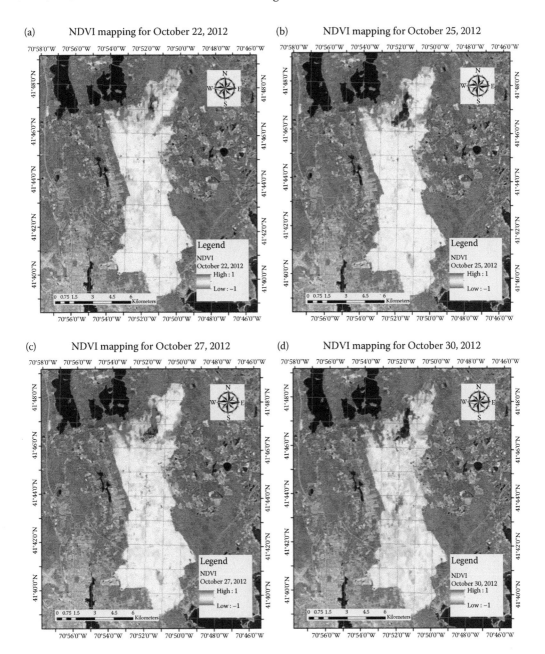

FIGURE 18.6 NDVI maps generated for observation of the impact of Hurricane Sandy on the watershed that drains into the Mattapoisett River, which in turn, flows to the Mattapoisett Bay. (a) The NDVI map generated from cloud-free Landsat 7 and (b–d) represent the NDVI maps derived from fused images for the respective dates.

path of the hurricane but simply received more rainfall with some possible soil erosion. Therefore, the damage to vegetation cover was minimal in nature. This can be deduced from the NDVI maps as well.

On the other hand, the areas that suffered the most devastating impact are in New York and New Jersey. Thus, it is essential to check the landscape conditions via the TCT plots before and after the hurricane's landfall for Hackensack and Pascack watershed (Figure 18.10). With the aid

TABLE 18.6

Mean Values of NDVI in the Mattapoisett River Watershed (Hurricane Sandy Landfall Date in New Jersey is October 29, 2012)

Date	Hurricane Condition	Mean NDVI
October 23, 2012	Before	0.556
October 25, 2012	Before	0.719
October 27, 2012	Before	0.625
October 30, 2012	After	0.627

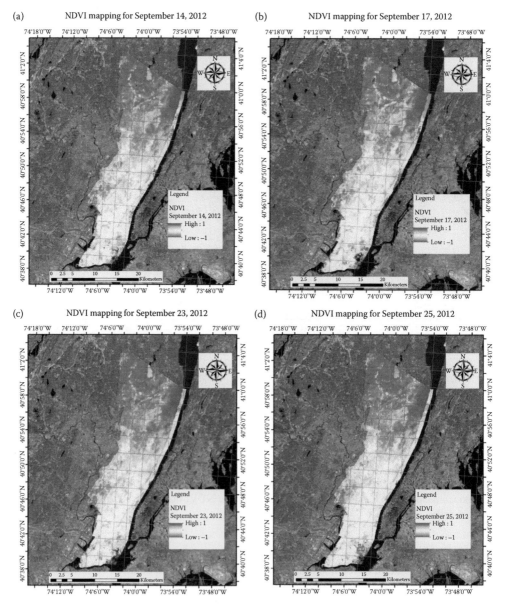

FIGURE 18.7 NDVI maps generated for observation before the impact of Hurricane Sandy on Hackensack and Pascack watershed. (a–d) represent the NDVI maps derived from fused images for the respective dates.

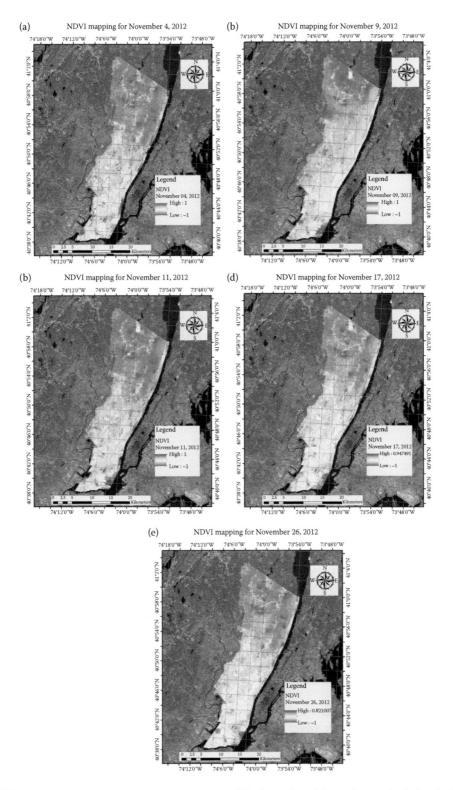

FIGURE 18.8 NDVI maps generated for observation of Hackensack and Pascack watershed after the impact of Hurricane Sandy. (a–e) represent the NDVI maps derived from fused images for the respective dates (Hurricane Sandy Landfall date in New Jersey is October 29, 2012).

TABLE 18.7

Mean Values of NDVI Before the Hurricane Sandy Impact on Hackensack and Pascack Watershed (Hurricane Sandy Landfall Date in New Jersey is October 29, 2012)

Date	Mean NDVI
September 14, 2012	0.436
September 17, 2012	0.394
September 23, 2012	0.449
September 25, 2012	0.467

TABLE 18.8

Mean Values of NDVI After the Impact of Hurricane Sandy on Hackensack and Pascack Watershed (Hurricane Sandy Landfall Date in New Jersey is October 29, 2012)

Date	Mean NDVI
November 04, 2012	0.296
November 09, 2012	0.242
November 11, 2012	0.282
November 17, 2012	0.261
November 26, 2012	0.269

of data fusion, plots for three cloud-free days for the month prior to the hurricane's landfall (i.e., for the month of September 2012) are presented. Also, plots for the month after landfall (i.e., for November 2012) are included for comparison. This setting would certainly lay down a firm foundation to help realize the implications of the comparative TCT plots. In general, dispersion phenomenon in terms of brightness-greenness, wetness-brightness, and wetness-greenness relationships is evidenced by visual examination over the cases before and after hurricane landfall with varying degrees (Figure 18.10).

The paired TCT plots in terms of the two dates before and after the hurricane's landfall in Figure 18.10 represents the time series condition of the relative level of landscape changes at the Hackensack and Pascack watershed of New Jersey, which is one of the hardest hit areas. Note that both a 3-day before and a 3-day after scenario are arranged for comparison. This watershed was somewhat directly in the path of the hurricane, resulting in vegetation cover devastation. For instance, the Figure 18.10a–i shows the landscape conditions before the hurricane's landfall in September, whereas the Figure 18.10j–r demonstrates the landscape conditions after the hurricane's landfall in November. It is clear that the differences between before and after the landfall are visible, comparable, and easily interpretable.

The TCT plot for September 17, 2012 is quite similar to the standard plot depicted in Figure 18.10a. This is also true for the plots of September 23 and September 25, 2012. After the hurricane's landfall, significant changes are noticed from the greenness versus brightness plots, especially in the clear water and turbid water sections. It is believed that huge storm surges when the hurricane hit the shoreline made the coastal area wet enough to contribute to the changes in the turbid water and the soil moisture sections. In addition, concrete urban regions were affected as well due to runoff and erosion. The

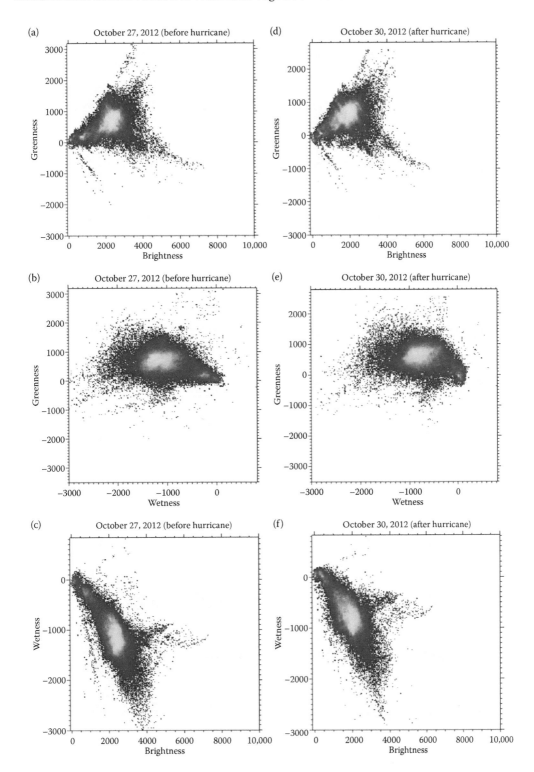

FIGURE 18.9 Tasseled cap plots (ENVI 5.3 software output) derived for the Mattapoisett River watershed area depicting land cover conditions before and after Hurricane Sandy made landfall (the landfall date is October 29, 2012). The plots (a–c) depict conditions before the hurricane's landfall and (d–f) depict the scenario after the hurricane's landfall.

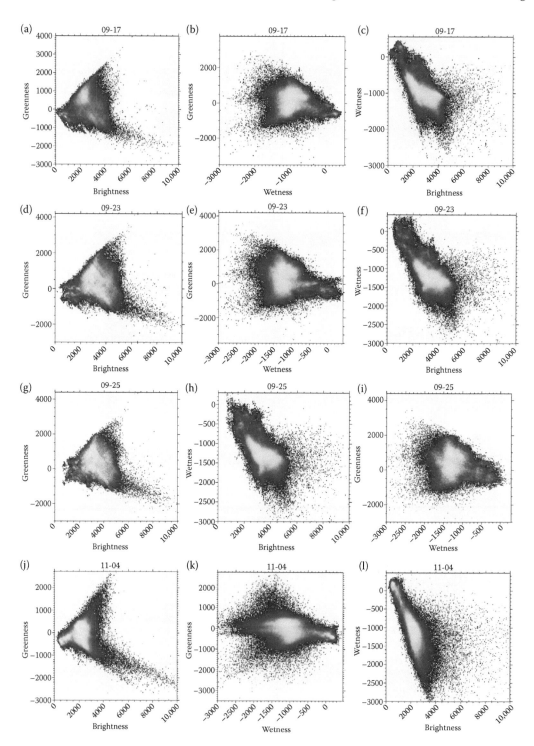

FIGURE 18.10 Tasseled cap plots (ENVI 5.3 software output) derived for the Hackensack and Pascack watershed area depicting land cover conditions before and after Hurricane Sandy made landfall (the landfall date is October 29, 2012). The plots (a–i) depict conditions before the hurricane's landfall and (g–r) depict the scenario after the hurricane's landfall and (m–r) depict the slow recovery after the hurricane's landfall.

(Continued)

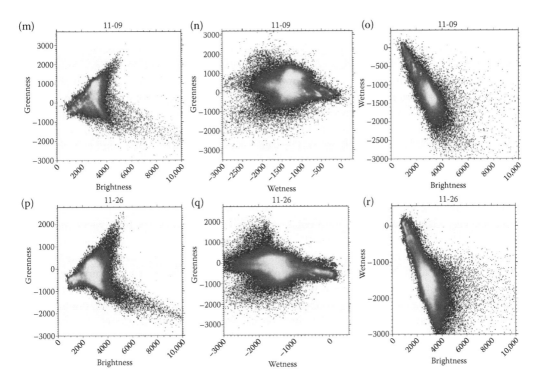

FIGURE 18.10 (Continued) Tasseled cap plots (ENVI 5.3 software output) derived for the Hackensack and Pascack watershed area depicting land cover conditions before and after Hurricane Sandy made landfall (the landfall date is October 29, 2012). The plots (a–i) depict conditions before the hurricane's landfall and (g–r) depict the scenario after the hurricane's landfall and (m–r) depict the slow recovery after the hurricane's landfall.

pixels are highly scattered in Figure 18.10a, indicating significant change in urban landscape after the hurricane's landfall. The change is noticeable for the plot of November 4, 2012 (Figure 18.10j), which is closer to the landfall date (October 29, 2012). By comparing the three plots of greenness versus brightness (Figure 18.10j, m, and p), it is observed that the scattering effects are not reduced, indicating the recovery of damages in the aftermath of the hurricane's landfall is really slow.

The greenness versus wetness plots for the days after the hurricane's landfall (Figure 18.10k, n, q) show that the shape of the plots is not consistent with that of the three days in September (Figure 18.10b, e, h). In other words, after the hurricane's landfall, the plots for the three days of November show significant distortions. The distortions are especially clear for concrete urban areas and for vegetative areas (Figure 18.10k, n, q). In the same greenness versus wetness plots, it is also observed that vegetation cover was altered quite significantly right after the hurricane's landfall. This can be evidenced in that there are a lot of tasseled caps on the right hand side with distorted shapes for the three days (Figure 18.10k, n, q). As for the wetness versus brightness plots, it is observed that in the month of September, the patterns are almost the same for the three days (Figure 18.10c, f, h) when compared to the standard plot depicted in Figure 18.10c. However, in the plots that depict the three days in November, that is, after the hurricane made landfall, there are significant changes in the areas of clear water, turbid water, and concrete urban areas (Figure 18.10i, o, r). Also, it can be seen from the plot of November 4, 2012 (Figure 18.10i) that the concrete urban areas faced significant distortions due to the hurricane. This is exemplified by the scattering of the pixels (i.e., tasseled caps) where the concrete urban area is depicted in the standard plot. Twenty-two days after landfall, the plot of November 26, 2012 (Figure 18.10r) shows that the pixels are less scattered than before the landfall occurred. This indicates the effects of slow recovery.

18.6   MULTITEMPORAL CHANGE DETECTION

A multispectral Landsat-7 ETM+ image contains a set of spectral information recorded at different wavelengths. The spectral information of each band can be displayed either individually as a gray-scale image or combined with other bands to form a composite image. This combination of bands is known as a color composite image in which the three primary colors red, green, and blue (RGB) are combined in various proportions, which enables the production of different colors in the visible spectrum. Generally, two types of combinations are widely used: (1) natural or true color composites, and (2) false color composites. The true color composite is a combination of visible red, green, and blue spectra to the red, green, and blue channels to display images in a manner similar to what is naturally seen by the human eye. The Landsat band combinations used for this purpose are Band 3, Band 2, and Band 1 for R, G, and B channels, respectively.

However, the true color images may also be low in contrast due to the scattering of blue light by the atmosphere. This is where the false color composite images can help. False color images are comprised of spectral bands other than red, green, and blue, which help in visualizing wavelengths that are not seen by the human eye. The use of bands such as near-infrared facilitates the spectral separation leading to the increase of interpretability of the images. The most commonly used false color combination for Landsat images is the combination of Band 4, Band 3, and Band 2 in the RGB channels. With such a combination, for instance, vegetation will be depicted in red, urban areas in cyan blue, soils in varying colors of dark to light brown, as well as clouds and snow in white or light cyan. The deeper the red, the more likely the vegetation is healthier or denser.

Since our study involves the observation of the impact of Hurricane Sandy on vegetation as well as other land cover aspects in the Hackensack and Pascack watersheds, the false color combination of Band 4, Band 3, and Band 2 should be used as inputs to the RGB channels. The vegetation cover changes due to the hurricane impact can be easily detected in the false color composite images (Figure 18.11). For the days before the hurricane's landfall, there was abundant healthy vegetation cover (Figure 18.11a and c). Especially in the northeastern corner of the watershed near the river (encircled in yellow), the vegetation is comparatively healthier than the rest of the watershed. After the hurricane landfall on October 29, 2012, this area was severely devastated, as evidenced by the map on November 04 (six days after the hurricane landfall) where the dense vegetation is no longer there (Figure 18.11b). Even twenty-two days after the hurricane's landfall (on the map of November 26, 2012), the vegetation at the northeastern corner of the watershed had not yet recovered (Figure 18.11d).

Following the TCT, there is an opportunity to include the brightness, greenness, and wetness maps in concert with the false color maps for comparison in the multitemporal change detection. Similarly, the impact of the hurricane's landfall can also be depicted in the brightness images of the watershed. Brightness images represent the variation in the soil background reflectance (Figure 18.12). The lower the vegetation cover, the more prominent soil reflectance is. However, the soil reflectance may co-vary due to the presence of water in the soil and urbanization. In other words, the brightness decreases as wetness increases. The impact of the hurricane's landfall on the soil of the watershed can be clearly seen as the soil reflectance is larger in both maps depicting the conditions after the hurricane's landfall (Figure 18.12b and d) than in those depicting the conditions before the hurricane's landfall (Figure 18.12a and c). This is especially true in the middle and lower watershed (encircled in yellow) which echoes the observation in Figure 18.10f and o. Obviously, heavy rains and flooding associated with the hurricane's landfall contributed to the increase in soil moisture and resulted in the decrease of soil reflectance.

The greenness index in TCT depicts the presence and density of the green vegetation. Before the hurricane's landfall, the greenness maps indicate the presence of healthy vegetation in the watershed, particularly in the northeastern side (encircled in yellow in Figure 18.13a and c). Six days after the hurricane's landfall, the density of the vegetation is lower (Figure 18.13b). The vegetation density remained lower on November 26, 2012, indicating slower recovery or no recovery at all

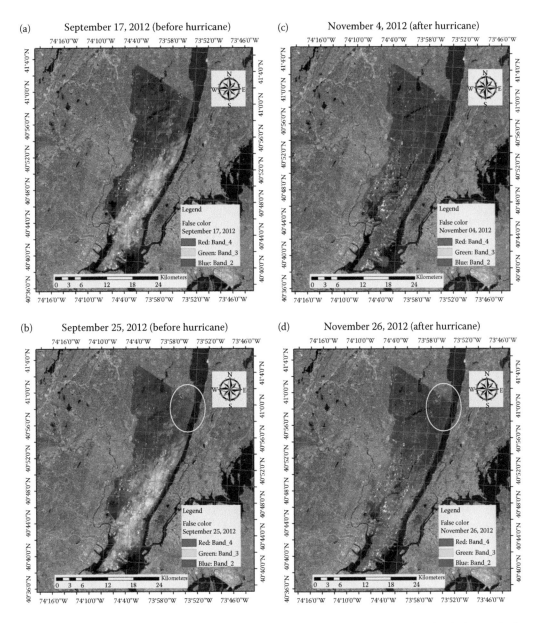

FIGURE 18.11 False color images of the Hackensack and Pascack watershed showing before and after conditions of Hurricane Sandy's landfall. (a) and (b) represent the condition before the hurricane's landfall while (c) and (d) represent the situation after the hurricane's landfall.

(Figure 18.13d). This TCT observation is consistent with the observation of the false color images depicted before.

Tasseled cap wetness is an indication of sensitivity towards soil and plant moisture. More particularly, wetness is sensitive to plant canopy moisture. Before the hurricane's landfall, the wetness maps show that there is lower wetness in the entire watershed as the reflectance is situated in the lower end of the band (Figure 18.14a and c). After the hurricane's landfall, the heavy rains and the subsequent flooding caused by tidal surges had contributed to the increase in the soil moisture and the decrease in plant canopy moisture content, as depicted in the maps (Figure 18.14b and d). The southwest part of the watershed (encircled in yellow) shows an increase in moisture content

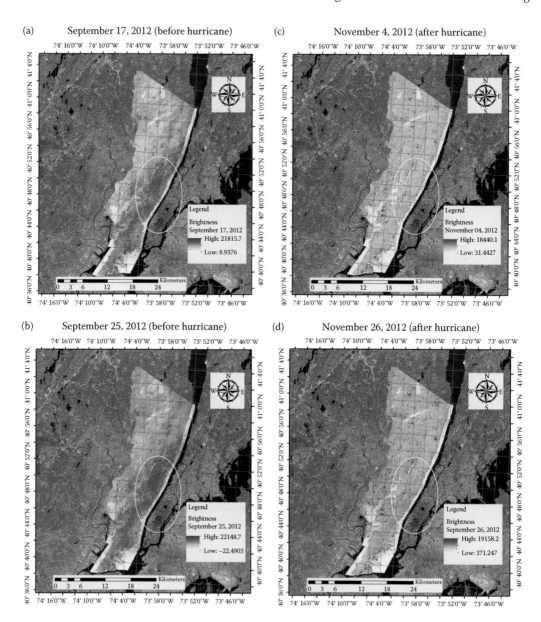

FIGURE 18.12 Brightness images of the Hackensack and Pascack watershed showing before and after conditions of Hurricane Sandy's landfall. (a) and (b) represent the conditions before landfall while (c) and (d) represent the conditions after landfall.

six days after the hurricane's landfall (encircled by yellow in Figure 18.14b). The moisture content remained high on November 26, 2012, almost one month after the hurricane's landfall (Figure 18.14d). In the northeastern corner of the watershed, where the density of vegetation is highest, as indicated in the false color and greenness maps, the plant canopy moisture is greater in the maps before the hurricane's landfall. After the hurricane's landfall, due to damage to the plant canopy, the plant canopy showed lower reflectance due to the lower wetness or moisture content (encircled by yellow in Figure 18.14b and d). This observation echoes what we saw in the TCT plots of Figure 18.10i and r.

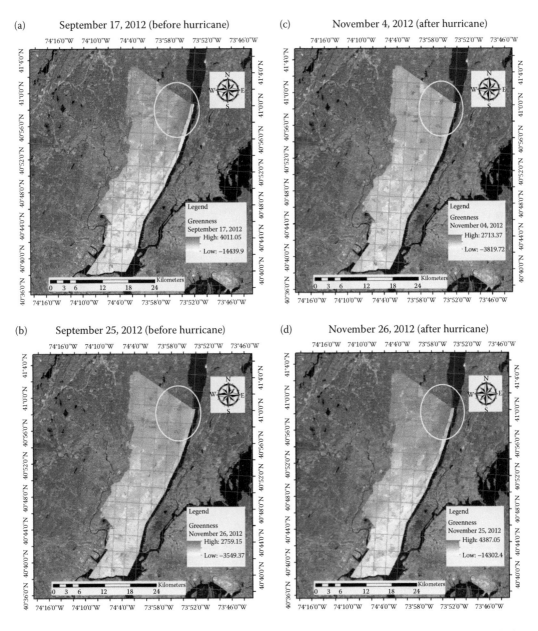

FIGURE 18.13 Greenness images of the Hackensack and Pascack watershed showing before and after hurricane landfall conditions. (a) and (b) represent the conditions before the hurricane's landfall while (c) and (d) represent the conditions after the hurricane's landfall.

18.7 DISPERSION ANALYSIS OF TCT VERSUS NDVI

The box and whisker plots depicted in Figure 18.15 show the comparison between the conditions of TCT plots (i.e., brightness, greenness, and wetness) in relation to NDVI for scenarios of before and after hurricane landfall. For the brightness versus NDVI box plots, there is significant difference in the values of brightness within the NDVI range of 0.00–0.20. The interquartile range of brightness in this NDVI range has a significant decrease in the case of after-landfall situation than in the case

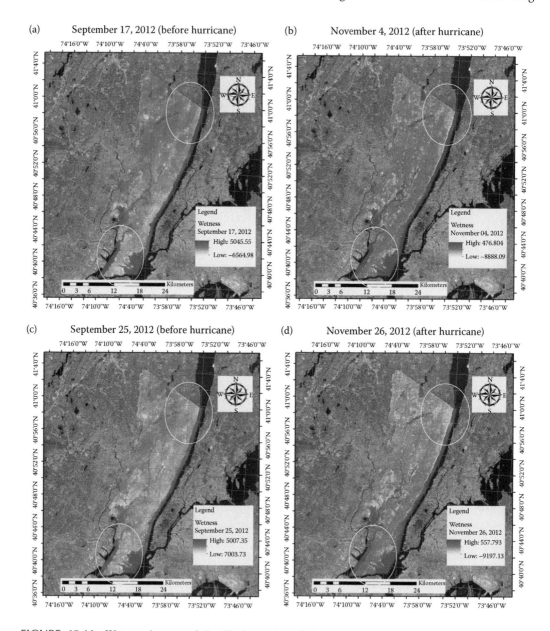

FIGURE 18.14 Wetness images of the Hackensack and Pascack watershed showing before and after hurricane landfall conditions. (a) and (b) represent the conditions before the hurricane's landfall while (c) and (d) represent the conditions after the hurricane's landfall.

of before-landfall situation (Figure 18.15a,b). The median value is also quite low after hurricane landfall, although for the NDVI ranges 0.21–0.40 and 0.41–0.60, the interquartile ranges and the median values are lower for scenarios of after-landfall than before-landfall. There are no values in the after-landfall scenario for NDVI range of 0.61–0.80 in the after-landfall plot, although there are values in the before-landfall scenario. This is an indication of lower vegetative index associated with loss of green vegetation due to landfall. For greenness in relation to NDVI, the overall situation is that there is a decrease in the interquartile ranges as well as the median values (Figure 18.15c,d). In the case of wetness versus NDVI values, there is not much significant change in the interquartile range as well as median in the case of the after-landfall values in the NDVI ranges, although there

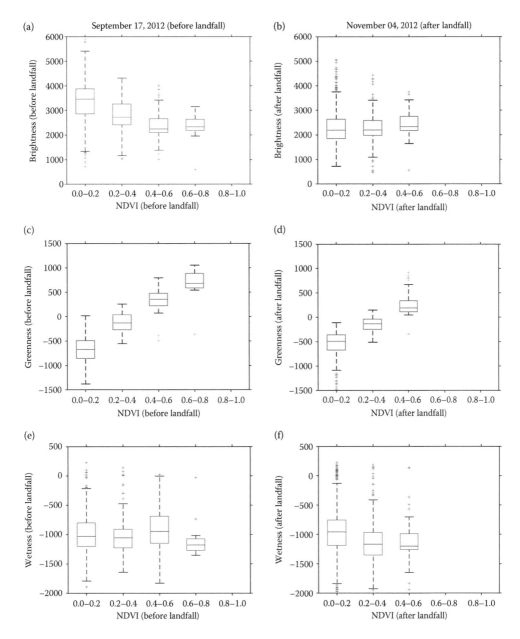

FIGURE 18.15 Box and whisker plots depicting the correlation between NDVI and Tasseled Cap Transformations before and after hurricane landfall. (a) Brightness vs. NDVI (before landfall); (b) brightness vs. NDVI (after landfall); (c) greenness vs. NDVI (before landfall); (d) greenness vs. NDVI (after landfall); (e) wetness vs. NDVI (before landfall); (f) wetness vs. NDVI (after landfall).

is a slight increase in median value in the NDVI range of 0.00–0.20 (Figure 18.15e and f). All these changes conform to the dispersion phenomenon noticed in the TCTs depicted before.

The TCT plots help us see the qualitative significance. However, it is important to back up the qualitative significance with quantitative analysis and prove the numerical significance. For this purpose, the t-test is used. To conduct the t-test, all 538,338 samples, that is pixels values, were collected for brightness, greenness, and wetness plots for before and after the landfall scenarios. The before-landfall date was chosen as September 17, 2012, and after-landfall date was November

04, 2012 that is, five days after hurricane landfall. The results of the t-test are tabulated in Tables 18.9 through 18.11. Both standard deviation and coefficient of variation are used for the dispersion assessment. The formulas for standard deviation and coefficient of variation are listed below.

- Standard deviation:

$$s = \sqrt{\frac{1}{N-1} \sum_{i=1}^{N} (x_i - \overline{x})^2}$$

where $\{x_1, x_2,...,x_N\}$ is the sample, s is the standard deviation, and $\overline{x}$ is the mean of the sample. The denominator $N - 1$ is the number of degrees of freedom in the vector $\{x_1 - \overline{x},...,x_N - \overline{x}\}$.

- Coefficient of variation:

$$CV = \frac{s}{\overline{x}}$$

TABLE 18.9

t-Test Results for Dataset Pairs of Before- and After-Landfall Scenarios for Brightness

	Brightness	
t-Test parameters	September 17, 2012	November 04, 2012
Mean	2585.25	2088.71
Standard deviation	936.51	790.10
Coefficient of variation	36.22%	37.83%
Pearson correlation	0.56	
Df (Degree of freedom)	538,338	
Alpha	0.05	
p-value	0.00	
t-value	445.62	
t-critical	1.96	

TABLE 18.10

t-Test Results for Dataset Pairs of Before- and After-Landfall Scenarios for Greenness

	Greenness	
t-Test parameters	September 17, 2012	November 04, 2012
Mean	602.45	322.27
Standard deviation	364.12	258.21
Coefficient of variation	60.44%	80.12%
Pearson correlation	0.01	
Df (Degree of freedom)	538,338	
Alpha	0.05	
p-value	0.01	
t-value	331.26	
t-critical	1.96	

TABLE 18.11

t-Test Results for Dataset Pairs of Before- and After-Landfall Scenarios for Wetness

t-Test parameters	Wetness	
	September 17, 2012	November 04, 2012
Mean	−877.56	−997.12
Standard deviation	449.76	545.98
Coefficient of variation	51.25%	54.76%
Pearson correlation	0.64	
Df (Degree of freedom)	538,338	
Alpha	0.05	
p-value	0.00	
t-value	204.11	
t-critical	1.96	

The t-test was conducted on the null hypothesis that the mean difference of the two datasets is zero with alpha value of 0.05. Alpha value is also referred to significance level, that is, the probability of rejecting the null hypothesis when the null hypothesis is true. The t values for brightness, greenness, and wetness are 445.62, 331.26, and 204.11, respectively. All the t values are presented in their absolute values. The t-critical value for all three cases is 1.96. The t-critical values are determined using the alpha value and the degree of freedom values. Since $t > $ t-critical and $p < 0.05$, the null hypothesis is rejected, that is, there is significant difference between the means of the two conditions. After calculations, we can surmise that there is a 19.21% change in pixel values for brightness; 46.51% change in greenness, and 13.47% difference in wetness when compared to pixel values of the before-landfall scenario. These significant differences in values conform to the tasseled cap plots that after hurricane landfall, there was significant change in the landscape which contributed to the dispersion of pixels, that is, change in the values of pixels. The Pearson correlation coefficient for the three transformations are also quite low, indicating a significant difference in the data pairs.

The standard deviation and coefficient of variation are effective statistical tools to compare between two datasets. For brightness, the standard deviation is 936.51 for the before-landfall and 790.10 for the after-landfall scenario. The standard deviation is lower in the case of the after-landfall than the before-landfall scenario. This indicates that the pixel values of the before-landfall scenario are more clustered around their mean than those of the after-landfall scenario. Yet the coefficient of variation is used to measure the spread or dispersion of data; the higher the value, the greater the dispersion. In this study, the coefficient of variation is 36.22% for the before-landfall scenario and 37.83% for the after-landfall scenario. This indicates that the pixel values for the after-landfall scenario are more dispersed than these of the-before landfall scenario. This conforms to the dispersion phenomenon shown in the tasseled cap plots. The same case is applicable for greenness and wetness as well. The coefficient of variation is significantly higher in case of the after-landfall scenario than these in the before-landfall scenario for greenness. This indicates a greater dispersion of pixel values for the after-landfall scenario than that of the before-landfall scenario, which indicates a significant change in the vegetation cover of the watershed. All of these phenomena prove that there was dispersion of data to some extent in all three transformation cases as evidenced in the TCT plots in Figure 18.10 driven by the hurricane landfall.

18.8   SUMMARY

In this chapter, a data fusion technique was adopted to generate fused images with better temporal and spatial resolution in support of the two biophysical models (NDVI and TCT) for multitemporal change detection of LULC over the selected coastal watershed under the Hurricane Sandy impact in October 2012. It shows that NDVI is not as capable as TCT of portraying a systematic response of the hurricane impact via the before and after scenarios. TCT plots can reveal a wealth of information in regard to how the landscape changed and when the recovery appeared. Further processing of the direct comparison of brightness, greenness, and wetness maps against the corresponding false color maps help in identifying where the change occurred as well as echoing the TCT plots interactively.

REFERENCES

Adam, E., Mutanga, O., Odindi, J., and Abdel-Rahman, E. M., 2014. Land-use/cover classification in a heterogeneous coastal landscape using RapidEye imagery: evaluating the performance of random forest and support vector machines classifiers. *International Journal of Remote Sensing*, 35, 3440–3458.

Anderson, G. L., Hanson, J. D., and Haas, R. H., 1993. Evaluating Landsat Thematic Mapper derived vegetation indices for estimating above-ground biomass on semiarid rangelands. *Remote Sensing of Environment*, 45, 165–175.

Ban, Y., Hu, H., and Rangel, I. M., 2010. Fusion of Quickbird MS and RADARSAT SAR data for urban land-cover mapping: Object-based and knowledge-based approach. *International Journal of Remote Sensing*, 31, 1391–1410.

Blake, E. S., Kimberlain, T. B., Berg, R. J., Cangialosi, J. P., and Beven II, J. L., 2013. Tropical cyclone report: Hurricane sandy. *National Hurricane Center*, 12, 1–10.

Carlson, T. N. and Ripley, D. A., 1997. On the relation between NDVI, fractional vegetation cover, and leaf area index. *Remote sensing of Environment*, 62, 241–252.

Chandrasekar, K., Sesha Sai, M. V. R., Roy, P. S., and Dwevedi, R. S., 2010. Land Surface Water Index (LSWI) response to rainfall and NDVI using the MODIS Vegetation Index product. *International Journal of Remote Sensing*, 31, 3987–4005.

Chang, N. B., Bai, K. X., Imen, S., Chen, C. F., and Gao, W., 2016. Multi-sensor satellite image fusion, networking, and cloud removal for all-weather environmental monitoring. *IEEE Systems Journal*, 1–17.

Chang, N. B., Vannah, B. W., Yang, Y. J., and Elovitz, M., 2014. Integrated data fusion and mining techniques for monitoring total organic carbon concentrations in a lake. *International Journal of Remote Sensing*, 35, 1064–1093.

Crist, E. P. and Cicone, R. C., 1984a. Application of the tasseled cap concept to simulated thematic mapper data. *Photogrammetric Engineering and Remote Sensing*, 50, 343–352.

Crist, E. P. and Cicone, R. C., 1984a. A Physically-Based Transformation of Thematic Mapper Data—The TM Tasseled Cap. *IEEE Transactions on Geoscience and Remote Sensing*, 22, 256–263.

Crist, E. P. and Kauth, R. J., 1986. The tasseled Cap De-Mystified. *Photogrammetric Engineering and Remote Sensing*, 52, 81–86.

Fauvel, M., Chanussot, J., and Benediktsson, J. A., 2006. Decision Fusion for the Classification of Urban Remote Sensing Images. *IEEE Transactions on Geoscience and Remote Sensing*, 44, 2828–2838.

Friedl, M. A., McIver, D. K., Hodges, J. C. F., Zhang, X. Y., Muchoney, D., Strahler, A.H., Woodcock, C. E., Gopal, S., Schneider, A., Cooper, A. et al., 2002. Global land cover mapping from MODIS: Algorithms and early results. *Remote Sensing of Environment*, 83, 287–302.

Fu, B. and Burgher, I., 2015. Riparian vegetation NDVI dynamics and its relationship with climate, surface water and groundwater. *Journal of Arid Environments*, 113, 59–68.

Gao, F., Masek, J., Schwaller, M., and Hall, F., 2006. On the blending of the Landsat and MODIS surface reflectance: Predicting daily Landsat surface reflectance. *IEEE Transactions on Geoscience and Remote Sensing*, 44, 2207–2218.

Gong, P., Wang, J., Yu, L., Zhao, Y. C., Zhao, Y. Y., Liang, L., Niu, Z. G., Huang, X. M., Fu, H. H., Liu, S., et al., 2013. Finer resolution observation and monitoring of global land cover: First mapping results with Landsat TM and ETM+ data. *International Journal of Remote Sensing*, 34, 2607–2654.

Hilker, T., Wulder, M., Coops, N. C., Linke, J., McDermid, G., Masek, J. G., Gao, F., and White, J. C., 2009. A new data fusion model for high spatial- and temporal-resolution mapping of forest disturbance based on Landsat and MODIS. *Remote Sensing of Environment*, 113, 1613–1627.

Holm, A. M., Burnside, D. G., and Mitchell, A. A., 1987. The development of a system for monitoring trend in range condition in the arid shrublands of Western Australia. *The Rangeland Journal*, 9, 14–20.

Huang, B. and Song, H., 2012. Spatiotemporal reflectance fusion via sparse representation. *IEEE Transactions on Geoscience and Remote Sensing*, 50, 3707–3716.

Huang, C., Yang, W. L., Homer, C., and Zylstra, G., 2002. Derivation of a tasseled cap transformation based on Landsat 7 at-satellite reflectance. Raytheon ITSS, USGS EROS Data Center.

Huang, B., Wang, J., Song, H., Fu, D., and Wong, K., 2013. Generating high spatiotemporal resolution land surface temperature for urban heat island monitoring. *IEEE Geoscience and Remote Sensing Letters*, 10, 1011–1015.

Jia, K., Liang, S., Wei, X., Yao, Y., Su, Y., Jiang, B., and Wang, X., 2014. Land cover classification of Landsat data with phenological features extracted from time series MODIS NDVI data. *Remote Sensing*, 6, 11518–11532.

Kauth, R. J. and Thomas, G. S., 1976. The tasseled Cap—A Graphic Description of the Spectral-Temporal Development of Agricultural Crops as Seen by LANDSAT. In: *Proceedings of the Symposium on Machine Processing of Remotely Sensed Data*. Purdue University of West Lafayette, Indiana, 4B-41 to 4B-51.

Liu, D. and Li, J., 2016. Data field modeling and spectral-spatial feature fusion for hyperspectral data classification. *Sensors*, 16, 2146.

Lukina, E. V., Stone, M. L., and Raun, W. R., 1999. Estimating vegetation coverage in wheat using digital images. *Journal of Plant Nutrition*, 22, 341–350.

Man, Q., Dong, P., and Guo, H., 2015. Pixel- and feature-level fusion of hyperspectral and lidar data for urban land-use classification. *International Journal of Remote Sensing*, 36, 1618–1644.

Mkhabela, M. S., Bullock, P., Raj, S., Wang, S., and Yang, Y., 2011. Crop yield forecasting on the Canadian Prairies using MODIS NDVI data. *Agricultural and Forest Meteorology*, 151, 385–393.

National Climatic Data Center (NCDC), 2012. Massachusetts Event Reports for October 29–30, 2012. National Climatic Data Center. National Oceanic and Atmospheric Administration. 2013. Retrieved May 15, 2017.

National Oceanic and Atmospheric Administration (NOAA), 2012. http://www.nhc.noaa.gov/data/tcr/summary_atlc_2012.pdf, accessed by May 2017.

Penuelas, J., Filella, I., and Gamon, J. A., 1995. Assessment of photosynthetic radiation-use efficiency with spectral reflectance. *New Phytologist*, 131, 291–296.

Pettorelli, N., Vik, J. O., Mysterud, A., Gaillard, J. M., Tucker, C. J., and Stenseth, N. C., 2005. Using the satellite-derived NDVI to assess ecological responses to environmental change. *Trends in Ecology & Evolution*, 20, 503–510.

Prasad, A. K., Chai, L., Singh, R. P., and Kafatos, M., 2006. Crop yield estimation model for Iowa using remote sensing and surface parameters. *International Journal of Applied Earth Observation and Geoinformation*, 8, 26–33.

Quarmby, N. A., Milnes, M., Hindle, T. L., and Silleos, N., 1993. The use of multi-temporal NDVI measurements from AVHRR data for crop yield estimation and prediction. *International Journal of Remote Sensing*, 14, 199–210.

Ran, Y. H., Li, X., and Lu, L., 2009. China land cover classification at 1 km spatial resolution based on a multi-source data fusion approach. *Advances in Earth Science*, 24, 192–203.

Rao, Y., Zhu, X., Chen, J., and Wang, J., 2015. An improved method for producing high spatial-resolution NDVI time series datasets with multi-temporal MODIS NDVI data and Landsat TM/ETM+ images. *Remote Sensing*, 7, 7865–7891.

Rouse Jr, J., Haas, R. H., Schell, J. A., and Deering, D. W., 1974. Monitoring vegetation systems in the Great Plains with ERTS. In: *Third Earth Resources Technology Satellite-1 Symposium- Volume I: Technical Presentations*, 309, NASA, Washington, D.C.

Scanlon, T. M., Albertson, J. D., Caylor, K. K., and Williams, C. A., 2002. Determining land surface fractional cover from NDVI and rainfall time series for a savanna ecosystem. *Remote Sensing of Environment*, 82, 376–388.

Shi, W., Zhu, C., Tian, Y., and Nichol, J., 2005. Wavelet-based image fusion and assessment. *International Journal of Applied Earth Observation and Geoinformation*, 6, 241–251.

Shimoni, M., Borghys, D., Heremans, R., Perneel, C., and Acheroy, M., 2009. Fusion of PolSAR and PolInSAR data for land cover classification. *International Journal of Applied Earth Observation and Geoinformation*, 11, 169–180.

Singh, K. K., Vogler, J. B., Shoemaker, D. A., and Meentemeyer, R. K., 2012. LiDAR-Landsat data fusion for large-area assessment of urban land cover: Balancing spatial resolution, data volume and mapping accuracy. *ISPRS Journal of Photogrammetry and Remote Sensing*, 74, 110–121.

Solaiman, B., Pierce, L. E., and Ulaby, F. T., 1999. Multisensor data fusion using fuzzy concepts: application to land-cover classification using ERS-1/JERS-1 SAR composites. *IEEE Transactions on Geoscience and Remote Sensing*, 37, 1316–1326.

Vermote, E. F., Kotchenova, S. Y., and Ray, J. P., 2008. MODIS surface reflectance user's guide. MODIS Land Surface Reflectance Science Computing Facility, Version, 1.

Vorovencii, I., 2007. Use of the "Tasseled Cap" transformation for the interpretation of satellite images. *Cadastre J. RevCAD*, 7, 75–82.

Wang, Q., Adiku, S., Tenhunen, J., and Granier, A., 2005. On the relationship of NDVI with leaf area index in a deciduous forest site. *Remote Sensing of Environment*, 94, 244–255.

Watkins, T., 2005. The Tasseled Cap transformation in remote sensing. available online http://www.sjsu.edu/faculty/watkins/tassel.htm, accessed by June 14, 2017.

Weng, Q., Fu, P., and Gao, F., 2014. Generating daily land surface temperature at Landsat resolution by fusing Landsat and MODIS data. *Remote Sensing of Environment*, 145, 55–67.

Wu, M.-Q., Wang, J., Niu, Z., Zhao, Y.-Q., and Wang, C.-Y., 2012. A model for spatial and temporal data fusion. *Journal of Infrared and Millimeter Waves*, 31, 80–84.

Xiao, Y., Jiang, Q., Wang, B., and Cui, C., 2016. Object-oriented fusion of RADARSAT-2 polarimetric synthetic aperture radar and HJ-1A multispectral data for land-cover classification. *Journal of Applied Remote Sensing*, 10, 026021.

Yengoh, G. T., Dent, D., Olsson, L., Tengberg, A. E., and Tucker, C. J., 2014. The use of the normalized difference vegetation index (NDVI) to assess land degradation at multiple scales: A review of the current status, future trends, and practical considerations. *Lund University Center for Sustainability Studies (LUCSUS), and the Scientific and Technical Advisory Panel of the Global Environment Facility (STAP/GEF)*.

Zhang, W., Li, A., Jin, H., Bian, J., Zhang, Z., Lei, G., Qin, Z., and Huang, C., 2013. An enhanced spatial and temporal data fusion model for fusing Landsat and MODIS surface reflectance to generate high temporal Landsat-like data. *Remote Sensing*, 5, 5346–5368.

Zhang, B., Zhang, L., Xie, D., Yin, X., Liu, C., and Liu, G., 2015. Application of synthetic NDVI time series blended from Landsat and MODIS data for grassland biomass estimation. *Remote Sensing*, 8, 10.

Zhang, F., Zhu, X., and Liu, D., 2014. Blending MODIS and Landsat images for urban flood mapping. *International Journal of Remote Sensing*, 35, 3237–3253.

Zhu, X., Chen, J., Gao, F., Chen, X., and Masek, J. G., 2010. An enhanced spatial and temporal adaptive reflectance fusion model for complex heterogeneous regions. *Remote Sensing of Environment*, 114, 2610–2623.

Multisensor Data Merging and Reconstruction for Estimating PM$_{2.5}$ Concentrations in a Metropolitan Region

19.1 INTRODUCTION

Fine particulate matter (PM$_{2.5}$) has long been considered a critical air pollutant due to its devastating impacts on air quality, atmospheric visibility, public health, and even climate change (Cheng et al., 2008a,b,c; Zheng et al., 2015). As a tiny particulate with an aerodynamic diameter less than 2.5 µm, PM$_{2.5}$ can penetrate deep into the respiratory system, which may in turn induce numerous diseases causing significant mortality because of the complex composition of inorganic salts (heavy metal), organic matter (hydrocarbons), and microbes (viruses and/or bacteria). Epidemiologic studies indicate that long-term exposure to high concentrations of PM$_{2.5}$ is associated with Lung Cancer (LC), Chronic Obstructive Pulmonary Disease (COPD), Ischemic Heart Disease (IHD), and Acute Lower Respiratory Infection (ALRI) (Dockery et al., 1993; Lim et al., 2012; Arnold, 2014; Burnett et al., 2014; Lelieveld et al., 2015). McDonnell et al. (2000) and Burnett et al. (2014) found that the PM$_{2.5}$ concentration is more highly related to mortality in non-smokers than that of PM$_{10}$ across different regions. Raaschou-Nielsen et al. (2013) demonstrated that a wide range of exposure levels of PM$_{2.5}$ is associated with a high risk of lung cancer for non-smokers with healthy habits in 17 European cohorts.

The daily risk of exposure to PM$_{2.5}$ individually might be moderate, but this exposure might be acute to the population who has a significant healthcare burden, such as the elderly. Recent studies have found that the changing meteorological episodes, landscape environment, land use characteristics, and geographical background correlate well with PM$_{2.5}$ concentrations (Wallace and Kanaroglou, 2007; Schaap et al., 2009). Although the current PM$_{2.5}$ monitoring stations enable us to provide ground-level measurements with high accuracy, such a sampling method is still costly and, most significantly, cannot provide observations over a wide range of area simultaneously, especially in some hot spots without any monitoring stations. These drawbacks significantly limit air quality management and epidemiologic studies of PM$_{2.5}$, both of which require finer scale PM$_{2.5}$ data with varying spatial and temporal resolution. By taking advantage of satellite-derived products of Aerosol Optical Depth (AOD) to estimate PM$_{2.5}$ concentrations, we are enabled to investigate PM$_{2.5}$ impacts over a large area with finer spatial and temporal resolution in concert with ground-based observations.

Since the 1980s, satellite-derived AOD products have become available due to the implementation of various satellite sensors for environmental applications, such as the Moderate Resolution Imaging Spectroradiometer (MODIS) (Lyapustin et al., 2011; Hsu et al., 2013; Levy et al., 2013; Mei et al., 2013a), the Ozone Monitoring Instrument (OMI) (Grey et al., 2006; Curier et al., 2009; Thomas et al., 2009; Mei et al., 2013b), the Multi-angle Imaging SpectroRadiometer (MISR) (Liu et al., 2007; Wallace and Kanaroglou, 2007), and the Visible Infrared Imager Radiometer Suite (VIIRS) (Von Hoyningen-Huene et al., 2003, 2011). Some general specifications of sensors and their relevant products are summarized in Table 19.1. It is clear that AOD products derived from distinct sensors may not agree well with each other due to the different wavelengths, viewing geometries,

TABLE 19.1

Summary of Instruments with Aerosol Products

Instrument	Platform	Mission Period	Nadir Resolution (km²)	Temporal Resolution	Spectral Resolution (μm)	Aerosol Products
AVHRR	TIROS-N	1978–1981	8*8	Daily	0.58–12.5	AAI[a], AOD
	NOAA-7	1978–1985				
	NOAA-15	1998–				
TOMS	Nimbus-7	1978–1994	50*50	Daily	0.98–1.02	AOD, aerosol type
	TOMS-EP	1996–2006				
SeaWIFS	SeaStar	1997–2010	1.1*1.1	Daily	0.41–0.86	AOD, α[b]
GOME	ERS-2	1995–2003	320*40	Daily	0.24–0.79	AAI, AOD
AATSR-2	ERS-2	1995–2003	1*1	Daily	0.55–1.2 (7γ)	AOD
POLDER-1/2	ADEOS I	1996–1997	18*18	15 days	0.44–0.91 (8γ)	AOD, α
	ADEOS II	2003-2003				
AATSR	Envisat	2002–2007	1*1	Daily	0.55–1.12 (4γ) 1.6–1.2 (3γ)	AOD, α
MERIS	Envisat	2002–2007	1.2*1.2	35 days	0.39–1.04 (15γ)	AOD, α
SEVIRI	MSG 8,9	2004–2007	3*3	Daily	0.81 or 0.55	AOD
	MSG10	2012–				
MODIS	Terra	2000–	10*10 and 3*3	Daily	0.4–14.4 (36 γ)	AOD, α, η[c]
	Aqua	2002–				
MISR	Terra	2000–	17.6*17.6	16 days	0.44–0.86	AOD
SCIAMACHY	Envisat	2002–	60*30	12 days	0.23–2.3	AOD, aerosol type
OMI	Aura	2004–	24*13	Daily	0.27–0.5	AOD
POLDER-3	PARASOL	2005–	6*6	15 days	0.44–0.91 (8γ)	AOD, α
GOME-2	MetOp-A	2006–	80*40	Daily	0.24–0.79	AOD, SSA[d]
	MetOp-B	2012–				
CALOIP	CALOIP	2006–	40*40	–	0.53–1.06	Aerosol profiles
OMPS	NPP	2011–	50*50	Daily	0.33–0.39	AOD, AAI,
VIIRS	NOAA	2013–	0.75*0.75 and 6*6	Daily	0.3–14	AOD, SSA

[a] Absorption Aerosol Index.
[b] Ångstrom exponent.
[c] Fine aerosol fraction.
[d] Single Scattering Albedo.

and retrieval algorithms applied for the derivation of AOD (Kokhanovsky et al., 2007). Often, single space-borne optical sensors cannot provide a spatiotemporally complete AOD data set due to various impacts such as cloud. These constraints motivate the synergistic use of multisensor AOD products through different methods of data merging or data fusion.

In order to fill in significant data gaps and improve the prediction accuracy of AOD products, Nirala (2008) developed a method to yield more spatiotemporally complete AOD products with optimal quality by spatially merging satellite AOD products from multiple satellite sensors. Several later studies attempted to merge the AOD data through the following two primary methods. One method is to employ the spatio-geometric analysis using topological, geometric, or geographic properties. These methods include, but are not limited to, the optimum interpolation (Yu et al., 2003; Xue et al., 2014), the empirical orthogonal functions (Liu et al., 2005), the linear or second-order polynomial functions (Mélin et al., 2007), the arithmetic and weighted average (Gupta et al., 2008), the least squares estimation (Guo et al., 2013), the Maximum Likelihood Estimation (MLE) (Nirala, 2008; Xu et al., 2015), the Universal Kriging (UK) method (Chatterjee et al., 2010; Li et al.,

TABLE 19.2

Summary of the Data Merging Methods for AOD Products

Name of Method	Data Source	Reference
UK	MODIS, OMI	Chatterjee et al. (2010)
MLE	MODIS, MIRS	Nirala (2008)
Geostatistical inverse model	MODIS, MIRS	Wang et al. (2013)
BME	MODIS, SeaWiFs	Tang et al. (2016)
SSDF	MODIS, MIRS	Nguyen et al. (2012)
Optimum interpolation	MOIDIS, MIRS	Yu et al. (2003)
Empirical orthogonal functions	MODIS, GOCART	Liu et al. (2005)
Second-order polynomial functions	MODIS, SeaWiFS	Mélin et al. (2007)
Least squares estimation	MODIS, MIRS	Guo et al. (2013)

UK: Universal Kriging; MLE: Maximum Likelihood Estimate; BME: Bayesian Maximum Entropy method; SSDF: Spatial Statistical Data Fusion.

2014), geostatistical inverse modeling (Wang et al., 2013), and the Spatial Statistical Data Fusion (SSDF) method (Nguyen et al., 2012; Puttaswamy et al., 2013). The other method is to perform a few nonlinear systems analyses making use of probability and machine learning theories, such as the Bayesian Maximum Entropy (BME) method (Beckerman et al., 2013) and the Artificial Neural Network (ANN) method (Wu et al., 2011, 2012; Guo et al., 2014; Di et al., 2016). The contemporary data merging methods applied for the generation of AOD products with two satellite sensors simultaneously are summarized in Table 19.2.

Research on satellite-derived AOD data with spatial and temporal variations began in the mid-1970s. In the past two decades, numerous studies attempted to estimate $PM_{2.5}$ concentrations by using satellite-derived AOD products with different resolutions. Many of them were directed to investigate the relation between $PM_{2.5}$ and AOD after the launch of the two Earth Observing System satellites—Terra (1999) and Aqua (2002). Wang and Christopher (2003) initiated the use of the MODIS AOD data for the prediction of ground-level $PM_{2.5}$ concentrations through linear correlations. Since then, encouraged by the precision improvement of satellite products, complex nonlinear estimations have become popular with the inclusion of meteorological parameters, land use data, population, and other environmental parameters. The quantitative relation between AOD and $PM_{2.5}$ has greatly been enhanced with better retrieval algorithms.

In general, there are three types of methods used in association with $PM_{2.5}$ retrieval algorithms, including (1) statistical models (Liu et al., 2005; van Donkelaar et al., 2006), such as the Linear Correlation (LC) (Wang and Christopher, 2003), the Multiple Linear Regression (MLR) (Liu et al., 2005), the Geographically Weighted Regression (GWR) model (Song et al., 2015), the linear Mixed-Effect Model (MEM) (Lee et al., 2012), the Generalized Additive Model (GAM) (Liu et al., 2009), some remote sensing formulas (Li et al., 2014; Zhang and Li, 2015), and the complex statistical downscaling method accounting for varied spatiotemporal relationships between AOD and $PM_{2.5}$ in a linear regression framework (Chang et al., 2014); (2) chemical transport models (Liu et al., 2004, 2007; van Donkelaar et al., 2006, 2010; Hystad et al., 2012; Lee et al., 2012); and (3) machine learning methods (Wu et al., 2011, 2012; Guo et al., 2014; Di et al., 2016). The commonly used methods and their accuracies in satellite remote sensing for the estimation of AOD and the retrieval of $PM_{2.5}$ concentrations are summarized in Table 19.3.

This chapter emphasizes the data merging of satellite AOD products collected by the MODIS-Terra and MODIS-Aqua by making use of the Modified Quantile-Quantile Adjustment (MQQA) algorithm (Bai et al., 2016b). In addition, AERONET (AErosol RObotic NETwork) program is a federation of ground-based remote sensing aerosol networks established by the National

TABLE 19.3

The Commonly Used Models and Parameters in Satellite Remote Sensing for AOD Estimation and Retrieval of $PM_{2.5}$ Concentrations

Methods	Parameters	Accuracy (R^2) and Source
LC	Relative Humidity (RH), Planetary Boundary Layer Height (PBLH)	0.47 (Wang et al., 2010)
MLR	Temperature, RH, wind speed, precipitation, air pressure, roads, population, elevation, land use data	0.82 (Fang et al., 2016)
GWR	Temperature, wind speed, sulfur dioxide, nitrogen dioxide, ozone, and carbon monoxide	0.69 (Song et al., 2015)
LUR	Temperature, RH, wind speed, PBLH	0.74 (Song et al., 2014)
MEM	RH, PBLH, wind speed, land use data	0.67 (Ma et al., 2016b)
BEYSIAN	Temperature, RH, PBLH, elevation, land use data	0.78 (Lv et al., 2016)
ANN	Temperature, RH, precipitation, wind speed, wind direction, PBLH, and time	0.43 (Wu et al., 2012)
TSM	PBLH, wind speed, RH, air pressure, precipitation, fire point data	0.73/0.79 (Ma et al., 2016a)
OLS + GWR	DAP, DTEMP	(Zou et al., 2016)

LC: Linear Correlations; MLR: Multiple Linear Regression; GWR: Geographically Weighted Regression; LUR: Land Use Regression; MEM: linear Mixed-Effect Models; ANN: Artificial Neural Network; TSM: Multi-stage models; OLS: Ordinary Least Squares model; RH: Relative humidity; PBLH: Planetary Boundary Layer Height; DTEMP: absolute Deviation of TEMPerature; DAP: Dust Area Percentage.

Aeronautics and Space Administration (NASA) and PHOtométrie pour le Traitement Opérationnel de Normalisation Satellitaire; Univ. of Lille 1, CNES, and CNRS-INSU (PHOTONS) (Figure 19.1). It provides continuous cloud-screened observations of spectral AOD, precipitable water, and inversion aerosol products in diverse aerosol regimes, and can be used as the ground truth to verify the merged remote sensing AOD products. For further verification, ground-based AOD measurements recorded by instruments from AERONET were used in our case study. Overall, this study aims to: (1) compare the accuracy of data merging of two satellite AOD products against the AERONET AOD measurements, and (2) examine the feasibility of predicting $PM_{2.5}$ concentrations using linear or nonlinear retrieval algorithms in a metropolitan region against the ground-based measurements collected from a few local monitoring stations managed by the United States Environmental Protection Agency (US-EPA).

19.2 AOD PRODUCTS AND RETRIEVAL ALGORITHMS

19.2.1 AOD Products

There exists a variety of methods for the retrieval of AOD, and the most popular two are deep blue and dark target methods. They produce different AOD values with significant degrees of uncertainty. Several studies have been conducted for inter-comparisons of sensors to determine the reasons for discrepancies in terms of cloud screening, method of calibration, and surface contribution (King et al., 1999). There are several common issues of uncertainties embedded in AOD products that deserve our attention, as described below.

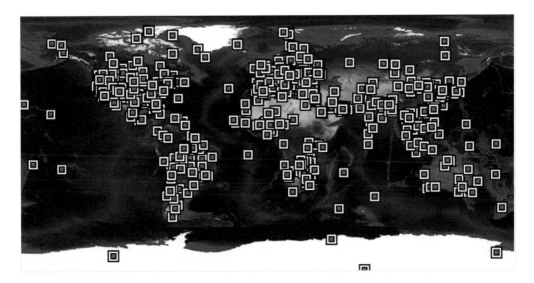

FIGURE 19.1   AERONET ground stations worldwide. (NASA, https://aeronet.gsfc.nasa.gov/.)

- *Cloud screening*: Two different cloud screening approaches in retrieving AOD data from one sensor can cause significant biases, and imperfect cloud screening can be one of the largest sources of errors (Mishchenko et al., 1999; Kacenelenbogen et al., 2011). Cloud screening is the most critical step in retrieving AOD, and the success of AOD retrieval strongly depends on eliminating inaccurate pixels (Kaufman et al., 2005). Moreover, studies have been conducted to compare the estimated AOD data from various satellites with different cloud masking algorithms. A consistency assessment of AOD data derived from MODIS and OMI indicated that the bias between these two sensors in cloud free conditions is relatively low, whereas the bias is higher in cloudy conditions (Ahn et al., 2008).

- *Calibration*: The other main factor introducing uncertainties in different satellite products is radiance calibration (Ignatov, 2002). Imperfect calibration is known as the primary source of uncertainty in AOD retrievals (Mishchenko et al., 1999). A sensitivity analysis on AOD derived from the Advanced Very-High-Resolution Radiometer (AVHRR) shows that calibration can change a retrieved AOD value by about 40% over the open ocean environment (Geogdzhayev et al., 2002). MODIS has an on board calibration system to guarantee prediction accuracy (Li et al., 2009). To reduce the uncertainties among sensors, SNO (Simultaneous Nadir Overpass) was developed. This method works by relying on inter-satellite radiometer calibration. SNO provides an ideal scenario for inter-calibration of radiometers between two satellites (Cao et al., 2004). Therefore, other instruments can cross-calibrate with MODIS to produce more accurate reflectance in their overlap period. It is significant that the value of AOD derived from AVHRR increased after calibrating AVHRR data using MODIS through the SNO method (Cao et al., 2004). As AOD values of MODIS are higher than those derived from the AVHRR, this inter-calibration could reduce cross mission biases between MODIS and AVHRR (Mishchenko et al., 2007).

- *Surface contribution*: Surface reflectance contribution to the radiances received by satellite sensors is often larger than that of aerosol. Therefore, inaccurate determination of surface reflectance contribution can result in significant biases in aerosol retrievals (Drury et al., 2008). Separating surface reflectance from atmospheric reflectance is mainly dependent on the instrument. Over the ocean area, surface reflectance in the MODIS is nearly zero at a specific wavelength, which may be used to generate quality-assured AOD products (Remer et al., 2005).

19.2.2   AOD Retrieval Algorithms

As mentioned above, the dark target algorithm and the deep blue algorithm are the two most commonly used algorithms applied for the derivation of AOD products associated with several satellites. It is worthwhile to extend our vision from a holistic viewpoint to the level of detail in the context of AOD algorithms. Here, AOD algorithms associated with MODIS instruments will be specifically introduced.

The MODIS instrument is an essential payload deployed on board Terra (viewing Earth from north to south across the equator locally at 10:30 am, descending) and Aqua (viewing Earth from south to north across the equator locally at 1:30 pm, ascending) satellites. It provides data in 36 spectral bands ranging from 0.4 to 14.4 μm on a daily basis. MODIS products are available in three groups of land, ocean, and atmosphere, in which AOD is distributed as one of the MODIS atmospheric products. NASA has developed three AOD retrieval algorithms based on three kinds of surface types. The dark target algorithm is performed to retrieve AOD from MODIS over ocean and vegetated or dark-soiled land. The dark target algorithm over these two areas is generally independent, however. These two retrieval algorithms for ocean and vegetated or dark-soiled land were developed before the Terra was launched and the primary description of these algorithms are explained in Tanré et al. (1997). On the other hand, MODIS uses the deep blue algorithm to retrieve AOD over bright areas—arid or semi-arid areas. Overall, all total column aerosol retrieval algorithm must remove the surface reflectance signal properly in order to determine the aerosol signal. Dark target and deep blue retrieval algorithms have different ways of accounting for the reflectance signal coming from the land surface. According to NASA (NASA, 2017), dark target uses a set of ratios and relationships between the 0.47, 0.67, and 2.1 μm channels to account for the surface signal (NASA, 2017). This method works best over dark vegetated targets and does not work over bright land surfaces (NASA, 2017). Deep blue uses maps and libraries of surface reflectance in the blue channels to account for the surface signal as well as spectral reflectance ratios over bright land (NASA, 2017). An overview of these three algorithms is included below for comparison.

- *AOD over vegetated land (dark target algorithm)*: The MODIS AOD retrieval algorithm over dark-soiled and vegetated land is based on the dark target method. It is assumed that surface reflectance over dense vegetated and dark soiled land in the visible channels consists of blue (0.47 μm) and red (0.66 μm), which are correlated with surface reflectance in SWIR (2.12 μm) (Kaufman et al., 1997). The MODIS AOD retrieval algorithm uses the Look-Up Table (LUT) to find the best condition for simulating the observed surface reflectance. It uses SWIR and visible channels simultaneously to retrieve AOD and surface reflectance.
- *AOD over ocean (dark target algorithm)*: The AOD retrieval algorithm over ocean is described in detail by Tanré et al. (1997) and Remer et al. (2005). AOD data over ocean are retrieved based on the pre-computed LUT for various aerosol parameters. The AOD derived from LUT is compared with the observed radiation. This algorithm assumes that aerosol property is a combination of a fine and a coarse aerosol mode with a suitable weighting factor (Equation 19.1)

$$\rho_\gamma^{LUT} = \eta\rho_\gamma^f + (1-\eta)\rho_\gamma^c \tag{19.1}$$

where ρ_γ^{LUT} is the average reflectance of a pure fine (ρ_γ^f) and a pure coarse (ρ_γ^c) mode of the reflectance with a weighting factor of η.

- *AOD over arid and semi-arid land (deep blue algorithm)*: Surface reflectance over arid, semi-arid, and urban areas is bright in red and near infrared wavelengths whereas darker in blue channel over these areas, and hence surface reflectance contribution to the observed radiance by satellite is larger than in vegetated areas. Hsu et al. (2004) developed the deep blue algorithm to retrieve AOD over bright surfaces. A global database of surface reflectance is generated on $0.1° \times 0.1°$ grid cells. In order to simulate the radiance received by the satellite, LUT is constructed based on the following Equation 19.2:

$$R(\mu, \mu_0, \Phi) = R_0(\mu, \mu_0, \Phi) + \frac{T A_s}{1 - s A_s} \qquad (19.2)$$

where R is radiance at the top of the atmosphere ($W \cdot sr^{-1} \cdot m^{-2}$), and μ, μ_0, and Φ are the cosine of the solar zenith angle, the cosine of the view zenith angle, and the relative azimuth angle, respectively. R_0 shows the path radiance ($W \cdot sr^{-1} \cdot m^{-2}$), T is the transmission function, A_s is the Lambertin reflectance, and s represents the spherical albedo of the atmosphere. The deep blue algorithm retrieves AOD only for cloud free pixels. The surface reflectances for these pixels are specified based on the database in three wavelengths (0.412, 0.490, and 0.67 μm). The appropriation of AOD is determined by matching the LUT radiances and the observed radiance by the maximum likelihood method.

19.3 CHALLENGES IN MERGING OF AOD PRODUCTS

Aside from MODIS, many other satellite sensors, such as OMI and VIIRS-NPP, provide AOD products as well. Hence, AOD products from these multiple sensors can be synergistically merged to increase spatiotemporal coverage for various applications. The AOD retrieval algorithm for VIIRS uses wavelengths of 0.67 μm or greater while MODIS uses wavelengths of 0.55 μm (Jackson et al., 2013). Yet, the 550 nm is an important wavelength which has been routinely used in global climate modeling analyses and ground-level AERONET AOD data. Remer et al. (2005) found that 550 nm AOD products performed better than those derived from other wavelengths. However, aerosol properties are normally derived from two independent algorithms. The first algorithm is the near-UV algorithm that makes use of OMI observations in the 350–390 nm spectral region to retrieve information on the absorption capacity of tropospheric aerosols; OMI-derived information on aerosol absorption includes the UV Aerosol Index and absorption optical depth at 388 nm. Unlike the MODIS aerosol products, the seventeen wavelengths that the OMI aerosol products use do not include 550 nm. In principle, the wavelength chosen for this purpose tends to avoid the ozone absorption wavelength lower than 330 nm and remains within the special range of the instruments while trying to match other sensors (Stammes, 2002). This discrepancy hinders the exploration of OMI-derived information on aerosol absorption for spatial data merging. To overcome this barrier, we list four possible methods below for converting them to the target band of 550 nm:

- *The Ångstrom exponent method (Angstrom, 1929)*: This method can be expressed as follows:

$$\tau(\lambda) = \beta \lambda^{-\alpha} \qquad (19.3)$$

where α is the Ångstrom exponent, ranging from 2.0 to zero (Kaufman et al., 1992). If the Ångstrom exponent is known, AOD at one wavelength can also be calculated from AOD at another wavelength (Li et al., 2012). The Ångstrom exponent can be calculated for AOD at UV, visible, and near-infrared spectral regions since it depends on wavelengths (Kaufman, 1993), yet further investigations are required.
- *Quadratic polynomial interpolation method*: This method interpolates the OMI 442, 483.5, and 342.5 nm channel AOD to 550 nm AOD (Li et al., 2014). The equation of quadratic polynomial interpolation is shown as follows.

$$\ln(\tau_\lambda) = a_0 + a_1 \ln \lambda + a_2 (\ln \lambda)^2 \qquad (19.4)$$

where τ_λ is the value of AOD at 550 nm, and $a_i(i = 0,1,2)$ are unknown coefficients. This method is more accurate than the first method (Eck et al., 1999; Li et al., 2014). This is mainly because the Ångstrom wavelength index method relies on Junge distribution, which does not hold in real world cases (King and Byrne, 1976).

- *The modified quantile-quantile adjustment (MQQA) method*: MQQA is used to perform spatial or temporal data merging given the overlapped time periods or spatial extent of several satellite sensors after the correction of associated cross mission biases (Bai et al., 2016b). With the aid of MQQA, the discrepancies of cross mission bias can be removed without using any *in situ* measurement. One of the advantages of MQQA is its independence on the wavelength distribution of each individual sensor. Taking MODIS-Aqua, MODIS-Terra, and OMI into account, the MODIS-Aqua AOD product can be selected as the baseline, and the OMI AOD and MODIS-Terra as the complementary products. Then the MQQA may be applied to identify the discrepancy between MODIS 550 nm AOD and the seventeen wavebands of OMI AOD toward getting the target product at designated wavelengths. However, the prediction accuracy requires further investigation given the need of band conversion between OMI AOD and MODIS-Aqua AOD product.

- *CTM—GOCART model*: Chemical transport models (CTMs), such as the well-known aerosol transport model, GOCART (the Goddard Chemistry, Aerosol, Radiation, and Transport model), have emerged as important mergers to fill in observational gaps by assimilating various observational data sets (e.g., Zhang et al., 2008; Colarco et al., 2010). Because of their physically meaningful frameworks, such methods seem to be more accurate. However, several studies found that such methods were considered less reliable, especially at 550 nm, due to the increase in the dependence of aerosols, although it may still function as expected (e.g., Torres et al., 2007).

19.4   STUDY FRAMEWORK AND METHODOLOGY

19.4.1   STUDY AREA

The Atlanta metropolitan area in Georgia, United States of America (USA), is selected as the study region. The Atlanta metropolitan covers a vast area (latitude range from 31° N to 35° N and longitude range from 86° W to 81° W) (Figure 19.2). With an estimated 2016 population of 5.7 million and a total area of 21,694 km², the Atlanta metropolitan area covers several counties, in which Fulton and DeKalb Counties are the major two. The petrochemical and electronic industries are well developed in this area.

In order to assess the prediction accuracy of the merged AOD, pairwise comparisons between the merged AOD and the MODIS AOD against the AERONET AOD will be conducted to deepen understanding in regard to the quality of data merging. There are two AERONET stations located in Atlanta, namely, Tech stations (84.4° W/33.78° N) and Yorkville stations (85.05° W/33.93° N) (Figure 19.2).

Meanwhile, there are 16 PM$_{2.5}$ monitoring stations managed by the US-EPA, which have been measuring both general population exposure in urban and suburban settings and environmental backgrounds. The geographical locations of these monitoring stations and their associated monitoring time periods are summarized in Table 19.4, while the spatial distribution of these monitoring stations for air quality monitoring is shown in Figure 19.2.

19.4.2   DATA SOURCES

There are three types of data used in this study, which are listed in Table 19.5:

- *Satellite AOD*: For illustration, MODIS AOD data from both Terra and Aqua satellites were collected. In this study, the MODIS Level 2 Aerosol Products, namely, MOD04 and MYD04, were acquired for the period of 2006–2016. Specifically, the

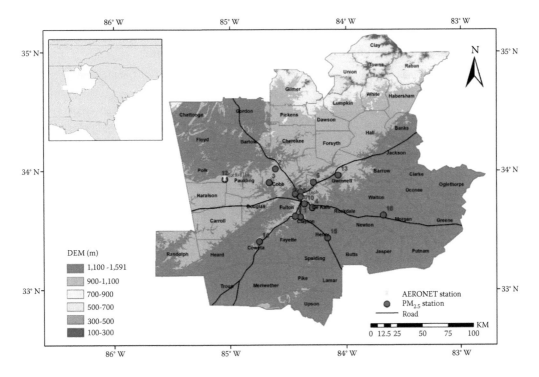

FIGURE 19.2 The study area of Atlanta and locations of air quality monitoring stations.

TABLE 19.4

The Ground PM$_{2.5}$ Stations in the Atlanta Metropolitan Area of Georgia, USA

County	Station Number	Station ID	Longitude/ Latitude	Data Availability
Clayton	1*	130630091	84.38W/33.61N	1999–
Cobb	2*	130670003	84.61W/34.02N	1999–
	3	130670004	84.66W/33.90N	2003–2012
DeKalb	4*	130890002	84.29W/33.69N	1999–
	5	130892001	84.28W/33.90N	1999–2012
	6	131210032	84.39W/33.82N	1999–2012
	7*	131210039	84.44W/33.80N	1999–
Fulton	8	131210048	84.40W/33.78N	2006–2008
	9	131211001	84.44W/33.62N	1999–2001
	10*	131210055	84.36W/33.72N	2005–
	11	131210056	84.39W/33.79N	2015–
Paulding	12*	132230003	85.04W/33.93N	1999–
Gwinnett	13*	131350002	84.07W/33.96N	2000–
Coweta	14*	130770002	84.75W/33.40N	2003–
Henry	15*	131510002	84.16W/33.43N	2003–
Walton	16	132970001	83.68W/33.63N	2005–2008

Note: Only Stations Marked with Asterisks were Used in PM$_{2.5}$ Modeling.

TABLE 19.5

Data Products Used for Estimating PM$_{2.5}$ Concentrations in Atlanta

Parameters	Temporal/Spatial	Data Provider/Instruments	Data Sources
Ground-level PM$_{2.5}$	Daily, 9 stations	US-EPA	https://www.epa.gov/outdoor-air-quality-data/download-daily-data
AOD	Daily, 10 km	MODIS-Terra MODIS-Aqua	https://ladsweb.modaps.eosdis.nasa.gov
Meteorological data	Daily, 0.125°	ECMWF	http://apps.ecmwf.int/datasets/

Optical_Depth_Land_And_Ocean of MOD04 and MYD04 derived at 550 nm were used for the derivation of PM$_{2.5}$.

- *Ground-level PM$_{2.5}$ concentration*: Daily mean PM$_{2.5}$ concentration for the period of 1999–2016 were acquired from the U.S. EPA (Figure 19.2, highlighted by red circles). In the past decade, the monitoring stations increased from 8 stations in 1999 to 14 stations in 2007 due to serious air pollution at that time. Yet, the total number of the stations was reduced to 9 stations in 2013 due to improved air quality conditions. Here, only PM$_{2.5}$ concentrations recorded at these 9 stations after 2013 were used in PM$_{2.5}$ modeling.
- *Meteorological data*: The ERA-Interim reanalysis fields, including 2-meter temperature (TEMP), relative humidity (RH), 10-meter U wind speed (WS_E), 10-meter V wind speed (WS_N), and planetary boundary layer height (PBLH), were acquired from the European Centre for Medium Range Weather Forecasts (ECMWF). The analyses field of each product at 18:00 (UTC) over Atlanta was downloaded, with a spatial resolution of 0.125° longitude × 0.125° latitude on a daily basis.

19.4.3 Methodology

In this study, AOD products derived from MODIS-Terra and MODIS-Aqua during 2013–2016 were merged for demonstration. Before merging, both AOD products were resampled to have a spatial resolution of 10 km × 10 km at each grid cell. In contrast, the spatial resolution of meteorological data is 0.125° × 0.125°. All these data sets were projected onto the WGS-84 coordinate system to ensure geographic collocation. The schematic flowchart of this study is depicted in Figure 19.3, which mainly involves the following procedure. (1) AOD data from MODIS-Aqua and MODIS-Terra are processed by using various image pre-processing approaches, including quality control with quality flags, resampling and re-projection, prior to data merging. (2) These pre-processed AOD products are merged to improve spatiotemporal coverage by making use of MQQA. More specifically, based on the limited AOD overlaps between two sensors during 2013–2016, the MQAA method is applied to perform cross mission bias correction in order to merge two AOD products spatially. (3) After data merging, the SMIR method (Chang et al., 2015) is applied to the merged AOD data set in order to further fill in data gaps remaining therein. (4) Once a suite of AOD data sets with relatively complete spatial coverage in some episodes becomes available, the surface PM$_{2.5}$ concentrations can be estimated with the aid of MLR and backward propagation (BP)-ANN tools. (5) Finally, the ground-based measurements of PM$_{2.5}$ concentration are used to assess the prediction accuracy based on a set of statistical indexes. The case study can be described in detail below.

1. *AOD Merging and Gap Filling*: Toward AOD merging, the MQQA method was used instead of simple linear fitting aiming to remove cross mission bias between these two AOD products. The detailed steps of MQQA can be found in Chapter 15. Before data merging practices, one of the two AOD products must be chosen as the baseline, while

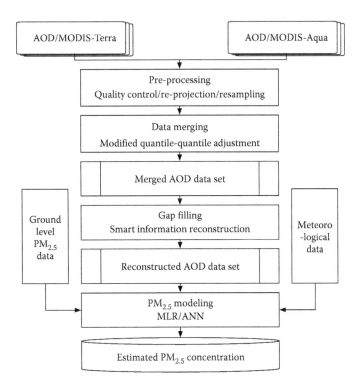

FIGURE 19.3 The schematic flowchart for the estimation of $PM_{2.5}$ concentration.

the other is to be used as complementary. In other words, the complementary data need to be mapped onto the baseline level so as to merge them together. In practice, several screening and ranking criteria were used toward such a goal. Some indicators such as monthly coverage ratio (*CR*) average and mean absolute bias (*MAB*) can be applied for the final determination as to which satellite should be the baseline (Bai et al., 2016a). Whereas *CR* indicates the spatial coverage of AOD products over the study area, MAB implies the degree of deviation of satellite AOD from the collocated ground-based observations.

As shown in Figure 19.4, when comparing those two satellite sensors with each other, the Terra sensor has higher monthly *CR* average than that of Aqua, particularly in winter seasons. Compared to the AOD products derived from each individual sensor the merged AOD product a between Terra and Aqua observations gain a larger monthly *CR* averages, benefiting from complementary observations, which is evidenced by the fact that CR values of AT are larger than that of Aqua or Terra alone. Thus, Terra has obvious priority over Aqua to be selected as the baseline sensor. In addition, comparisons of two satellite AOD data sets with the AERONET AOD show that choosing MODIS-Terra as the baseline sensor can get better agreement, despite the slightly larger *rmse* (Figure 19.5). Therefore, AOD product derived from MODIS-Terra should be chosen as the baseline and that of MODIS-Aqua as complementary.

After data merging, the SMIR method was further applied for filling in data gaps present in the merged AOD data set. The ultimate goal is to improve the spatial coverage based on the merged AOD data set. More details related to the SMIR method can be found in Chapter 12. In practical use of SMIR, long-term historical time series of the relevant data set (AOD here) are required. Because the merged AOD data set is created on the basis of MODIS-Terra AOD, the MODIS-Terra AOD data set should be used to create the referenced time series in the MQQA (see Figure 11.2).

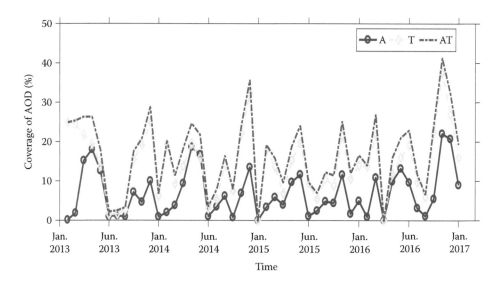

FIGURE 19.4 Average monthly coverage ratio during the 2013–2016 time period for the possible combinations of these sensors. The combinations of sensors are identified by the letters associated with each individual sensor (AT: Aqua + Terra).

2. *PM$_{2.5}$ modeling*: A myriad of empirical models for exploring the relationship between PM$_{2.5}$ and AOD are available for the estimation of ground-level PM$_{2.5}$ concentrations. In this study, two typical models were chosen for demonstration, as mentioned above. One is a statistical model, and the other is a machine learning model. They are introduced below.

   • *Multiple linear regression model*: Satellite-derived AOD products provide useful perspectives on the global air pollution, especially the surface PM$_{2.5}$ concentration (Wang and Christopher, 2003), by using linear correlation (Chu et al., 2003). The MLR model can be expressed as a function of PM$_{2.5}$ in terms of several dependent variables, including TEMP, RH, AOD, PBLH, Wind Direction (WD), and Wind Speed (WS), as shown below.

$$PM_{2.5} = \beta_0 + \beta_1 x_1 + \ldots + \beta_n x_n \tag{19.5}$$

   where, β_0 is the intercept and β_i are regression coefficients; $x_1, x_2, \ldots x_n$ are dependent variables such as TEMP, WS (WS_E and WS_N), WD, AOD, and PBLH, respectively. The function can be well solved in a least square manner.

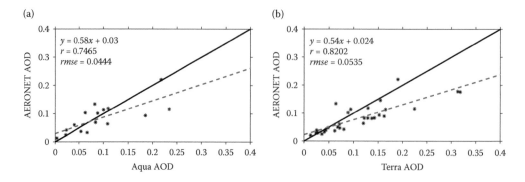

FIGURE 19.5 Comparisons of AOD between satellite- and AERONET-based data sets. (a) MODIS-Aqua and (b) MODIS-Terra.

TABLE 19.6

Samples Size for Model Fitting (Training) and Validation of the Proposed BP-ANN Model for the Generalization of PM$_{2.5}$-AOD Relationship

	The Whole Year	Spring	Summer	Fall	Winter
Total sample	2701	680	302	930	789
Training sample	2161	544	242	750	631
Validation sample	540	136	60	180	158

- *Artificial neural networks model*: A backward propagation-ANN (BP-ANN) model commonly has three layers of nodes: the input layer, the hidden layer, and the output layer. In this study, there are six nodes in the input layer and one node in the output layer. The input data in the input layer include AOD, PBLH, RH, TEMP, WS, and WD. Empirically, the appropriate number of nodes in the hidden layer ranges from n to 2n + u, where n is the number of nodes in the input layer and u is the number of nodes in the output layer. Thus, the number of nodes in the hidden layer may vary from six to 13 in the BP-ANN model. The transfer function is generally non-linear in the hidden layer and linear in the output layer. In this study, all data were simulated in a feed-forward network by using the Levenberg-Marquardt learning algorithm, and the number of nodes in the hidden layer was given as 13. In learning process, 70% of the pair-wised samples were used as the training data set, with the remaining 30% for validation (Table 19.6).

19.4.4 PERFORMANCE EVALUATION

The root-mean-square error (*rmse*) and correlation coefficient (*r*) were used here as statistical measures to evaluate the prediction accuracy. The *rmse* and *r* between observed and predicted PM$_{2.5}$ concentrations can be calculated by:

$$rmse = \sqrt{\frac{\sum_{i=1}^{N}(\widehat{y}_i - y_i)^2}{N}} \tag{19.6}$$

$$r = \frac{\sum_{i=1}^{N}(y_{oi} - \bar{y}_o)(y_{pi} - \bar{y}_p)}{\sqrt{\sum_{i=1}^{N}(y_{oi} - \bar{y}_o)^2 \sum_{i=1}^{N}(y_{pi} - \bar{y}_p)^2}} \tag{19.7}$$

where $\widehat{y}_i$ and y_i are the simulated and observed PM$_{2.5}$ data, respectively. A large value of *r* (close to 1) and a small value of *rmse* (around 0) indicate good predicting performance.

19.5 RESULTS

19.5.1 VARIABILITY OF PM$_{2.5}$ CONCENTRATIONS

Based on the long-term ground-level PM$_{2.5}$ data, we chose five stations out of the 16 in Atlanta and used them as a whole (data values are averaged over these five stations) to analyze long-term PM$_{2.5}$ concentrations (1999–2016). As shown in Figure 19.6, PM$_{2.5}$ concentrations had a prominent decreasing trend on a yearly basis in Atlanta from 1999 to 2016, in turn suggesting the improvement of air quality. Because of this improvement, the US EPA reduced the number of monitoring stations, keeping only nine stations active at present in 2017. The monthly average shown in Figure 19.7 may

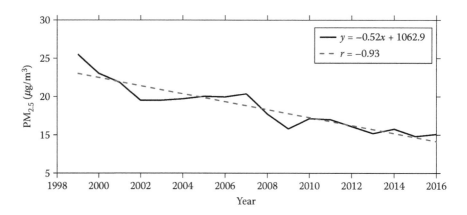

FIGURE 19.6   Linear trend of annual mean PM$_{2.5}$ concentrations in Atlanta during 1999–2016.

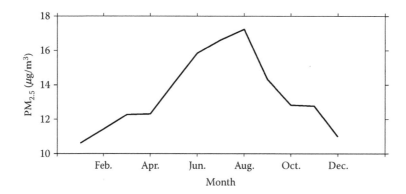

FIGURE 19.7   Variability of monthly averaged PM$_{2.5}$ concentrations in Atlanta during 1999–2016.

provide certain insight into the seasonality effect of PM$_{2.5}$ concentrations observed therein. It is clear that PM$_{2.5}$ concentrations are higher in summer, with a maximum shown in August (17.43 µg/m³). In contrast, low concentrations of PM$_{2.5}$ are present mainly in winter, with a minimum of 10.87 µg/m³ on average in January. According to the trend analysis, we chose the following four days at which few clouds were observed as our episodes for data merging of the two types of AOD products and for prediction analysis of ground-level PM$_{2.5}$ concentrations. They are: 20160217 (winter), 20160417 (spring), 20160701 (summer), 20160907 (fall), and the associated DOY (day-of-year) for these four days are 048, 108, 183, and 251, respectively.

19.5.2   Data Merging of AOD Products

Data merging aims to increase the spatial coverage of AOD products by mitigating the cloud impacts over the study area. In light of the niche of the MQQA method, AOD products before and after data merging may be compared based on the AOD maps side by side. This may provide visual evidence regarding how the individual satellite sensor viewed the study region and what the final merged image looks like based on MODIS-Aqua, MODIS-Terra, and merged AOD data.

In addition, seasonal AOD variations over the four selected days may be compared to one another to discover more about the seasonality effects. In winter, it is evident that several pixels with higher AOD values were observed in the downtown area (Figure 19.8); in spring, hot spots obviously popped up over the whole downtown region (Figure 19.9); in the summer, hot spots showed up from

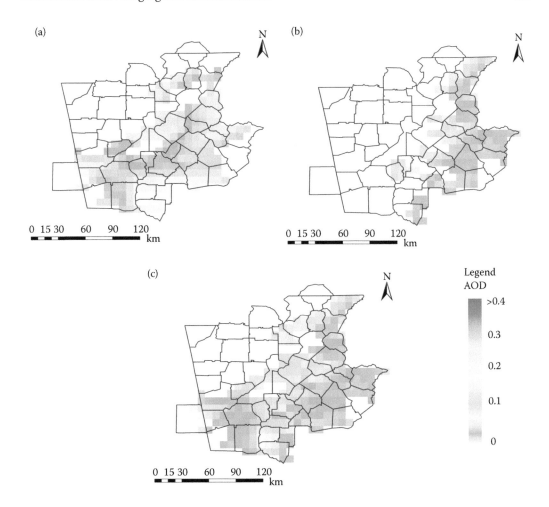

FIGURE 19.8   AOD maps derived from (a) MODIS-Aqua, (b) MODIS-Terra, and (c) merged AOD data set on 20160217 (winter time).

the downtown area all the way to the suburban regions, probably due to wind effects (Figure 19.10); and in fall, some hot spots remained in the downtown area and the northwestern suburban region (Figure 19.11).

To evaluate the accuracy of the merged AOD product, AOD retrievals on 20161114 were used as an example to justify the possible accuracy improvement after data merging. For a fair comparison, AOD data derived from MODIS-Terra were used as the baseline to compare with that of Aqua and the bias corrected Aqua data. Specifically, spatially collocated Terra and Aqua pixels with valid data values on 20161114 were first extracted and compared (Figure 19.12a). Then, those Aqua-based AOD data values may be mapped onto the Terra level by making use of MQQA, and such bias corrected data are thereby termed as Aqua_corrected AOD data (Figure 19.12b). As shown in Figure 19.12, it is noticeable that the corrected AOD data agree better with that of Terra than those before correction, because it shows higher correlation (r) and lower *rmse*.

Overall, even with the aid of data merging, it is still incapable of creating a spatially complete AOD image over the study area. From Figures 19.8 through 19.11, it is clear that the AOD map for the winter time had many more data gaps (value missing pixels) than those in other seasons. Restoring these data gaps can be made possible with the aid of interpolation/extrapolation methods or memory effects. The SMIR method via time-space-spectrum continuum is worthwhile to apply for bridging

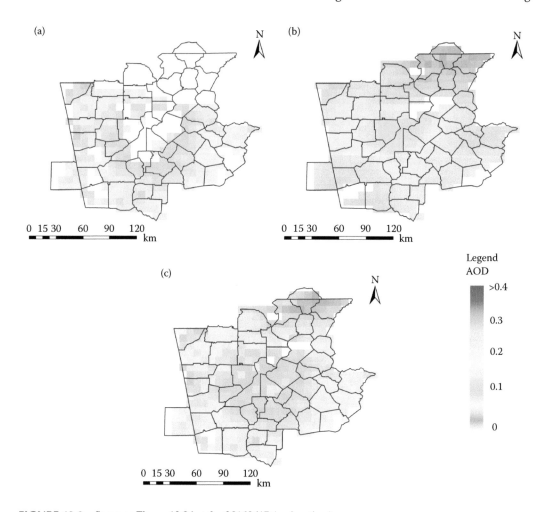

FIGURE 19.9   Same as Figure 19.8 but for 20160417 (spring time).

such a gap to recover the missing pixels via memory effects (see Chapter 12) (Chang et al., 2015). Results of the reconstructed AOD by making use of SMIR will be described in section 19.5.4.

19.5.3   PM$_{2.5}$ CONCENTRATION MODELING AND MAPPING

Since PM$_{2.5}$ has very strong effects on the diffraction of light in the atmospheric environment, its concentration thus has a close relationship with AOD which is related to the characterization of the extinct ability of the atmosphere. The larger the concentration of PM$_{2.5}$, the higher the ability of extinction. While the pure relation between PM$_{2.5}$ and AOD is hard to quantify, it is essential to understand the correlation of PM$_{2.5}$ with AOD and meteorological data (Table 19.7). Obviously, the proper estimations of ground-level PM$_{2.5}$ concentrations need to consider several meteorological variables, such as TEMP, WS, WD, RH, and PBLH, to formulate a useful model, such as an ANN model (Figure 19.13).

   Overall, PM$_{2.5}$ concentrations have different levels of correlation with the six potent dependent factors in different seasons. The seasonality effects are obvious according to Table 19.7. For example, PM$_{2.5}$ concentrations are negatively correlated with WS and WD except in the summer season, while PM$_{2.5}$ concentrations have a very strong positive correlation with TEMP, PBLH, and AOD in all seasons. Yet, PM$_{2.5}$ has strong negative and positive correlations with WD in summer and winter,

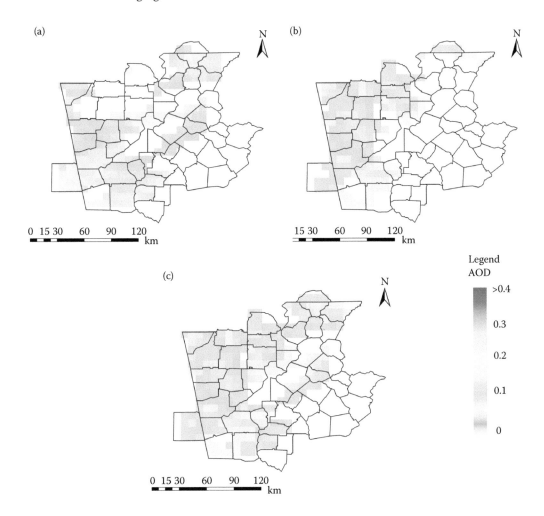

FIGURE 19.10 Same as Figure 19.8 but for 20160701 (summer time).

respectively. Comparatively, the correlation between $PM_{2.5}$ and RH is not strong in winter, as is the case between $PM_{2.5}$ and AOD in summer. These evidences strongly suggest the need to estimate the ground-level $PM_{2.5}$ concentrations in different seasons separately.

- *Comparisons of MLR model with BP-ANN model*: It is necessary to compare the MLR model and the BP-ANN model on the same basis in terms of the prediction of ground-level $PM_{2.5}$ concentrations. Such comparisons can be arranged in two categories. One is based on a seasonal scale and the other is based on an annual scale. Regardless of the category, 70% of the data were randomly selected to train each model and the remaining 30% were used for validation. The results in Table 19.8 show that the values of r and $rmse$ of the trained MLR model (testing) are 0.48 and 3.96, respectively. On the other hand, the values of r and $rmse$ of the trained ANN model are 0.53 and 0.25, respectively. The $rmse$ value of the tested ANN model is much better than that of the MLR model similar. On a seasonal basis, predicted values from MLR agree better with the ground truth in the spring due to a larger correlation and lower $rmse$ values. Similarly, the seasonal ANN model for summer has the highest r value and a relatively lower $rmse$ value. Even though the proposed ANN model has good performance in the stage of model training, all seasons' testing data's $rmse$ is bigger than the training data; this overfitting phenomenon is obvious given the discrepancies of

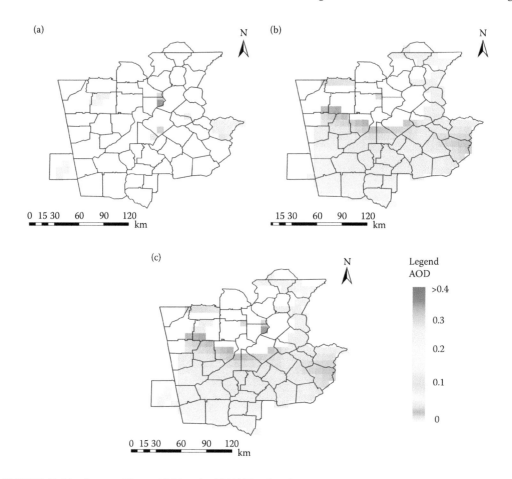

FIGURE 19.11   Same as Figure 19.8 but for 20160907 (fall time).

the r values between the training and the testing stages. Scatter plots between the observed and the estimated PM$_{2.5}$ concentrations are shown in Figures 19.14 and 19.15. Overall, the performance of the MLR model is generally worse than that of the ANN model.

* *PM$_{2.5}$ mapping*: As demonstrated above, the BP-ANN model shows better accuracy in generalizing complex AOD-PM$_{2.5}$ relationships. Thus, the well trained ANN models were used to estimate ground-level PM$_{2.5}$ concentrations for each season (Figure 19.16). For

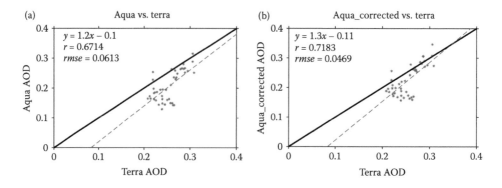

FIGURE 19.12   Comparisons of (a) MODIS-Aqua and MODIS-Terra, (b) MODIS-Terra and merged AOD data on 20161114.

TABLE 19.7

Correlation Coefficients between PM$_{2.5}$, Meteorological Factors, and AOD

PM$_{2.5}$	TEMP	RH	WS	WD	PBLH	AOD
Yearly	0.543**	0.110**	−0.206**	−0.170**	0.494**	0.389**
Spring	0.462**	0.059	−0.222**	−0.343**	0.213*	0.258**
Summer	0.289**	−0.355**	0.152	0.314**	0.388**	0.148
Fall	0.523**	0.266**	−0.275	−0.211**	0.512**	0.421**
Winter	0.354**	−0.053	−0.066**	−0.351**	−0.229**	0.058

* $p < 0.05$, ** $p < 0.001$.

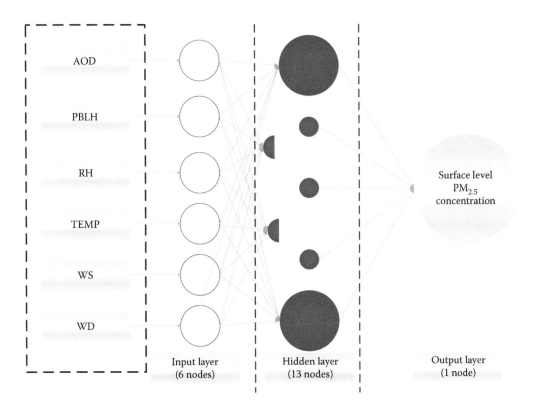

FIGURE 19.13 An illustrative structure of the proposed BP-ANN model for estimating PM$_{2.5}$ concentration.

demonstration, the following four dates including February 17, 2016 (winter), April 17 2016 (spring), July 1 2016 (summer), and September 7 2016 (fall), were used. They are noted as 20160217, 20160417, 20160701, and 20160907 in the following figures (Figures 19.16, 19.18, and 19.19). It shows that the spatial distributions of estimated PM$_{2.5}$ concentrations at the four dates have similar spatial distributions to AOD. In general, the higher the AOD level, the larger the PM$_{2.5}$ concentrations. PM$_{2.5}$ pollution in the downtown area of Atlanta is more serious than that in rural areas, which may be caused by emissions from factories or automobiles in Atlanta, Georgia, USA. In addition to the inclusion of AOD, the results also suggest that we need to integrate a variety of factors, such as meteorological factors, to better explain the observed PM$_{2.5}$ variations.

TABLE 19.8

Comparisons of Model Learning Accuracy between MLR and ANN in Estimating PM$_{2.5}$ Concentrations

| | MLR | | | | ANN | | | |
| | Training | | Testing | | Training | | Testing | |
	r	*rmse*	*r*	*rmse*	*r*	*rmse*	*r*	*rmse*
Annual	0.4906	4.2044	0.4843	3.9590	0.5348	0.1537	0.5276	0.2532
Spring	0.5659	3.4184	0.5082	3.5333	0.6338	0.3062	0.5590	1.0360
Summer	0.5148	3.2879	0.4866	3.2970	0.6182	0.2132	0.5744	0.6524
Fall	0.5020	4.8108	0.5096	5.2316	0.6007	0.1685	0.4939	0.3212
Winter	0.4367	3.4152	0.4724	3.2485	0.5505	0.2266	0.4930	0.5539

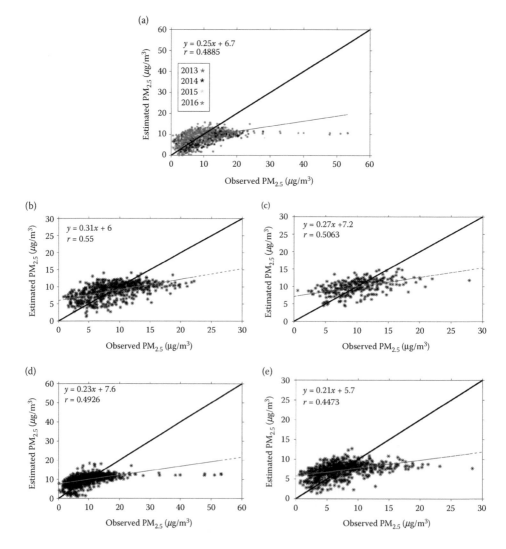

FIGURE 19.14 Comparisons between the observed and the estimated PM$_{2.5}$ concentrations derived from the MLR model. (a) Annual, (b) spring, (c) summer, (d) fall, and (e) winter.

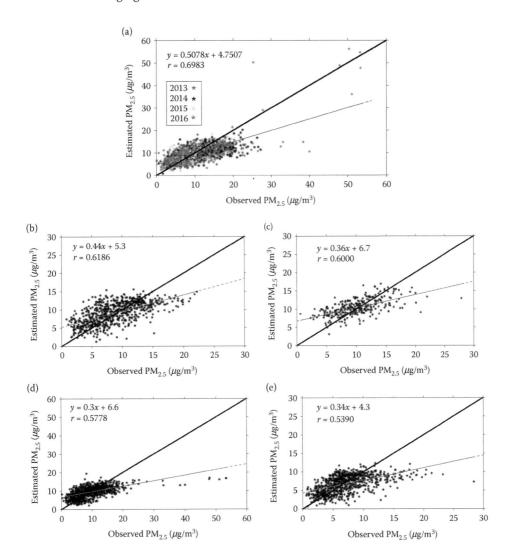

FIGURE 19.15 Same as Figure 19.14 but for the ANN model.

19.5.4 Gap Filling Using SMIR Method

SMIR method was first brought into satellite image processing field to reconstruct the cloud contaminated data by Chang et al. (2015). Compared with the conventional methods for cloudy pixel recovery, it performs better because the recovery process is driven by the similarity principles with a memory effect over spatial, spectral, and temporal domains simultaneously. In this study, SMIR was used to restore the value missing satellite AOD pixels. For demonstration, the merged AOD images associated with the four selected dates to represent the four seasons in 2016 (shown in Figure 19.16) were chosen to demonstrate how SMIR can improve the CR of these merged AOD temporally and spatially. In Figure 19.17, it is indicative that based on the data in 2013–2016, time period monthly CR averages of the merged data derived from MQQA is much lower than that of the recovered data derived from both MQQA and SMIR. This confirms the incremental progress made by the inclusion of SMIR albeit some missing data in 2014. Following the same logic, the side-by-side comparison of AOD spatial distribution shown in Figure 19.18 reveals that the incremental progress of CR after the inclusion of SMIR is phenomenal on these four selected dates.

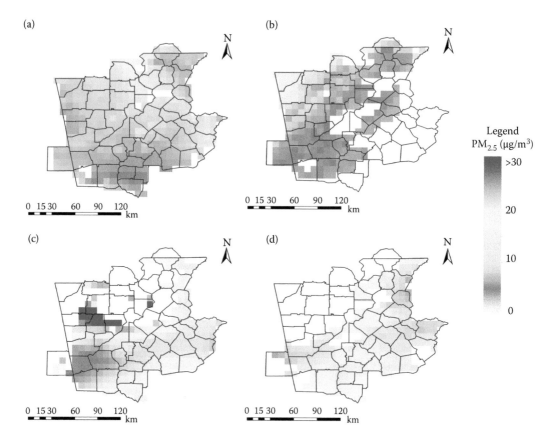

FIGURE 19.16   Spatial distribution of estimated PM$_{2.5}$ concentrations at day of (a) 20160417 (spring), (b) 20160701 (summer), (c) 20160907 (fall), and (d) 20160217 (winter).

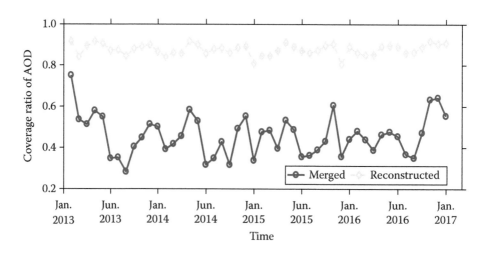

FIGURE 19.17   Monthly CR averages during the 2013–2016 time period for the merged data (MQQA) and reconstructed data (SMIR).

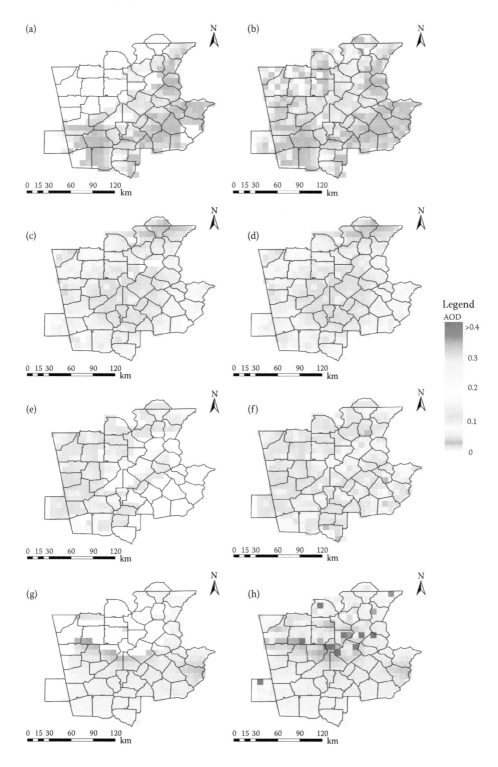

FIGURE 19.18 Spatial distribution of AOD to show the incremental progress of CR in relation to merged and recovered AOD. (a) 20160217 (MQQA), (b) 20160217 (MQQA + SMIR), (c) 20160417 (MQQA), (d) 20160417 (MQQA + SMIR), (e) 20160701 (MQQA), (f) 20160701 (MQQA + SMIR), (g) 20160907 (MQQA), and (h) 20160907 (MQQA + SMIR).

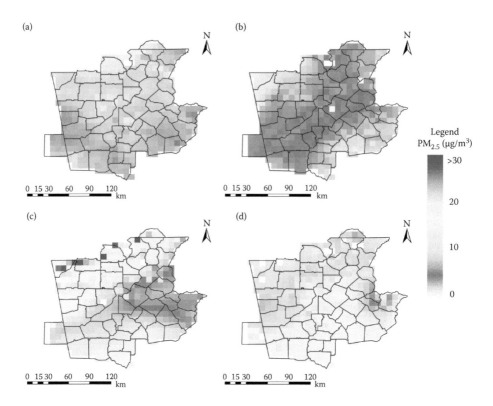

FIGURE 19.19 Spatial distribution of estimated $PM_{2.5}$ concentrations by using SMIR AOD data at day of (a) 20160417 (spring), (b) 20160701 (summer), (c) 20160907 (fall), and (d) 20160217 (winter).

To increase the application potential, we used the same validated BP-ANN model in above to estimate the ground level $PM_{2.5}$ concentrations again based on the results after the inclusion of SMIR. The maps shown in Figure 19.19 indicate that the spatial distributions of the estimated $PM_{2.5}$ concentrations on the four dates of 20160217, 20160417, 20160701, and 20160907 have similar spatial patterns to that shown in Figure 19.16, but have much more delicate and complete spatial coverage of the estimated $PM_{2.5}$ concentrations relative to the original ones before gap filling.

19.6 APPLICATION POTENTIAL FOR PUBLIC HEALTH STUDIES

As confirmed by many epidemiological studies on $PM_{2.5}$ exposure experiments, $PM_{2.5}$ did contribute to mortality (Dockery et al., 1993; Pope and Dockery, 2006). Lelieveld et al. (2015) assessed the contribution of $PM_{2.5}$ to mortality on a global scale and Wang et al. (2017) studied the long-term trends in $PM_{2.5}$-related mortality on the Northern Hemisphere, both of which proved that $PM_{2.5}$ links to mortality. Other scholars studied the effects of $PM_{2.5}$ with different demographic groups, such as traffic police (Novotna et al., 2007), taxi drivers (Pfeifer et al., 1999), women and children (Stieb et al., 2012), and found that $PM_{2.5}$ has a significant relationship with human health. The contents of this chapter can be linked to explore the public health impact of $PM_{2.5}$ (i.e., LC, COPD, IHD, and ALRI).

Mehta et al. (2013) used the Bayesian approach for meta-analysis to review and identify the relevance between $PM_{2.5}$ concentrations and ALRI in children. They found that per $10\,\mu g/m^3$ increase in annual average $PM_{2.5}$ concentrations there is an increase in ALRI occurrence risk of 1.12 in children. Pope et al. (2006) studied the relationship of long-term exposure to $PM_{2.5}$ concentrations by using the Cox model, which showed that mortality with IHD will increase by 8%–18% with a $PM_{2.5}$ concentration increase of 10 μg/m³. In 2016, Yu and Chien (2016) investigated the non-linear concentration response associations between fine particulate matters and respiratory diseases in Taipei (Taiwan). They found that higher $PM_{2.5}$ concentrations can cause IHD through a combination of autonomic, hemostatic, inflammatory, and vascular endothelial disturbances with consequent changes in cardiac and vascular function. There are many ways to link remote sensing studies for air quality assessment to epidemiological studies as more remote sensing images and retrieval algorithms will be available in the future.

19.7 SUMMARY

In this chapter, some commonly used models for the derivation of AOD data and methods for the retrieval of $PM_{2.5}$ concentrations have been reviewed and applied for a case study in Atlanta, Georgia, USA. It shows that the MQQA algorithm can work well given the variability of the atmospheric environment. SMIR may still help for recovering the information of value-mixing pixels in the future to facilitate the estimation of $PM_{2.5}$ concentrations. Overall, the UN-Habitat (2012) predicted that more than 70% of the world's population will live in cities by 2050, and more urbanization effects will trigger even worse air pollution issues and severe human health concerns. The type of research in this chapter will be in acute need for both air quality management and public health assessment in the future.

REFERENCES

Ahn, C., Torres, O., and Bhartia, P. K., 2008. Comparison of ozone monitoring instrument UV aerosol products with Aqua/Moderate Resolution Imaging Spectroradiometer and Multiangle Imaging Spectroradiometer observations in 2006. *Journal of Geophysical Research: Atmospheres*, 113, D16S27.

Angstrom, A., 1929. On the atmospheric transmission of sun radiation and on dust in the air. *Geografiska Annaler*, 11, 156–166.

Arnold, C., 2014. Disease burdens associated with PM2.5 exposure: how a new model provided global estimates. *Environmental Health Perspectives*, 122, A111.

Bai, K., Chang, N. B., and Chen, C. F., 2016a. Spectral information adaptation and synthesis scheme for merging cross-mission consistent ocean color reflectance observations from MODIS and VIIRS. *IEEE Transactions on Geoscience and Remote Sensing*, 54, 311–329.

Bai, K., Chang, N. B., Yu, H., and Gao, W., 2016b. Statistical bias corrections for creating coherent total ozone records with OMI and OMPS observations. *Remote Sensing of Environment*, 182, 150–168.

Beckerman, B. S., Jerrett, M., Martin, R. V., Van Donkelaar, A., Ross, Z., and Burnett, R. T., 2013. Application of the deletion/substitution/addition algorithm to selecting land use regression models for interpolating air pollution measurements in California. *Atmospheric Environment*, 77, 172–177.

Burnett, R. T., Arden Pope, C., Ezzati, M., Olives, C., Lim, S. S., Mehta, S., Shin, H. H., et al., 2014. An integrated risk function for estimating the global burden of disease attributable to ambient fine particulate matter exposure. *Environmental Health Perspectives*, 122, 397–403.

Cao, C., Weinreb, M., and Xu, H., 2004. Predicting simultaneous nadir overpasses among polar-orbiting meteorological satellites for the intersatellite calibration of radiometers. *Journal of Atmospheric and Oceanic Technology*, 21, 537–542.

Chang, H. H., Hu, X., and Liu, Y., 2014. Calibrating MODIS aerosol optical depth for predicting daily PM2.5 concentrations via statistical downscaling. *Journal of Exposure Science & Environmental Epidemiology*, 24, 398–404.

Chang, N. B., Bai, K., and Chen, C. F., 2015. Smart information reconstruction via time-space-spectrum continuum for cloud removal in satellite images. *IEEE Journal of Selected Topics in Applied Earth Observations and Remote Sensing*, 99, 1–19.

Chatterjee, A., Michalak, A. M., Kahn, R. A., Paradise, S. R., Braverman, A. J., and Miller, C. E., 2010. A geostatistical data fusion technique for merging remote sensing and ground-based observations of aerosol optical thickness. *Journal of Geophysical Research*, 115, D20207.

Cheng, Y. F., Heintzenberg, J., Wehner, B., Wu, Z. J., Su, H., Hu, M., and Mao, J. T., 2008a. Traffic restrictions in Beijing during the Sino-African Summit 2006: aerosol size distribution and visibility compared to long-term in situ observations. *Atmospheric Chemistry and Physics*, 8, 7583–7594.

Cheng, Y., Wiedensohler, A., Eichler, H., Heintzenberg, J., Tesche, M., Ansmann, A., Wendisch, M., Su, H., Althausen, D., and Herrmann, H., 2008b. Relative humidity dependence of aerosol optical properties and direct radiative forcing in the surface boundary layer at Xinken in Pearl River Delta of China: An observation based numerical study. *Atmospheric Environment*, 42, 6373–6397.

Cheng, Y. F., Wiedensohler, A., Eichler, H., Su, H., Gnauk, T., Brüggemann, E., Herrmann, H., et al., 2008c. Aerosol optical properties and related chemical apportionment at Xinken in Pearl River Delta of China. *Atmospheric Environment*, 42, 6351–6372.

Chu, D. A., Kaufman, Y. J., Zibordi, G., Chern, J. D., Mao, J., Li, C., and Holben, B. N., 2003. Global monitoring of air pollution over land from the Earth Observing System-Terra Moderate Resolution Imaging Spectroradiometer (MODIS). *Journal of Geophysical Research*, 108, 1–18.

Colarco, P., da Silva, A., Chin, M., and Diehl, T., 2010. Online simulations of global aerosol distributions in the NASA GEOS-4 model and comparisons to satellite and ground-based aerosol optical depth. *Journal of Geophysical Research: Atmospheres*, 115, D14207.

Curier, L., de Leeuw, G., Kolmonen, P., Sundstrom, A.-M., Sogacheva, L., and Bennouna, Y., 2009. Aerosol retrieval over land using the (A)ATSR dual-view algorithm, In: *Satellite Aerosol Remote Sensing over Land*. 135–159.

Di, Q., Kloog, I., Koutrakis, P., Lyapustin, A., Wang, Y., and Schwartz, J., 2016. Assessing PM2.5 Exposures with High Spatiotemporal Resolution across the Continental United States. *Environmental Science & Technology*, 50, 4712–4721.

Dockery, D. W., Pope, C. A., Xu, X., Spengler, J. D., Ware, J. H., Fay, M. E., Speizer, F. E., 1993. An association between air pollution and mortality in six US cities. *The New England Journal of Medicine*, 329, 1753–1759.

Drury, E., Jacob, D. J., Wang, J., Spurr, R. J., and Chance, K., 2008. Improved algorithm for MODIS satellite retrievals of aerosol optical depths over western North America. *Journal of Geophysical Research: Atmospheres*, 113, D16204.

Eck, T. F., Reid, J. S., Smirnov, A., Neill, N. T. O., Slutsker, I., and Kinne, S., 1999. Wavelength dependence of the optical depth of biomass burning, urban, and desert dust aerosols. *Journal of Geophysical Research*, 104, 31333–31349.

Fang, X., Zou, B., Liu, X., Sternberg, T., and Zhai, L., 2016. Satellite-based ground PM2.5 estimation using timely structure adaptive modeling. *Remote Sensing of Environment*, 186, 152–163.

Geogdzhayev, I. V., Mishchenko, M. I., Rossow, W. B., Cairns, B., and Lacis, A. A., 2002. Global two-channel AVHRR retrievals of aerosol properties over the ocean for the period of NOAA-9 observations and preliminary retrievals using NOAA-7 and NOAA-11 data. *Journal of the Atmospheric Sciences*, 59, 262–278.

Grey, W. M. F., North, P. R. J., Los, S. O., and Mitchell, R. M., 2006. Aerosol optical depth and land surface reflectance from multi-angle AATSR measurements: global validation and inter-sensor comparisons. *IEEE Transactions on Geoscience and Remote Sensing*, 44, 2184–2197.

Guo, Y., Feng, N., Christopher, S. A., Kang, P., Zhan, F. B., and Hong, S., 2014. Satellite remote sensing of fine particulate matter (PM2.5) air quality over Beijing using MODIS. *International Journal of Remote Sensing*, 35, 6522–6544.

Guo, J., Gu, X., Yu, T., Cheng, T., Chen, H., and Xie, D., 2013. Trend analysis of the aerosol optical depth over China using fusion of MODIS and MISR aerosol products via adaptive weighted estimate algorithm. In: Proceedings Volume 8866, Earth Observing Systems XVIII, 88661X.

Gupta, P., Patadia, F., and Christopher, S. A., 2008. Multisensor data product fusion for aerosol research. *IEEE Transactions on Geoscience and Remote Sensing*, 46, 1407–1415.

Hsu, N. C., Tsay, S.-C., King, M. D., and Herman, J. R., 2004. Aerosol Properties Over Bright-Reflecting Source Regions. *IEEE Transactions on Geoscience and Remote Sensing*, 42, 557–569.

Hsu, N. C., Jeong, M.-J., Bettenhausen, C., Sayer, A. M., Hansell, R., Seftor, C. S., Huang, J., and Tsay, S.-C., 2013. Enhanced deep blue aerosol retrieval algorithm: the second generation, *Journal of Geophysical Research: Atmospheres*, 118, 9296–9315.

Hystad, P., Demers, P. A., Johnson, K. C., Brook, J., van Donkelaar, A., Lamsal, L., Martin, R., and Brauer, M., 2012. Spatiotemporal air pollution exposure assessment for a Canadian population-based lung cancer case-control study. *Environmental Health*, 11, 22.

Ignatov, A., 2002. Sensitivity and information content of aerosol retrievals from the Advanced Very High Resolution Radiometer: radiometric factors. *Applied Optics*, 41, 991–1011.

Jackson, J. M., Liu, H., Laszlo, I., Kondragunta, S., Remer, L. A., Huang, J., and Huang, H. C., 2013. Suomi-NPP VIIRS aerosol algorithms and data products. *Journal of Geophysical Research: Atmospheres*, 118, 12,673–12,689.

Kacenelenbogen, M., Vaughan, M., Redemann, J., Hoff, R., Rogers, R., Ferrare, R., Russell, P., Hostetler, C., Hair, J., and Holben, B., 2011. An accuracy assessment of the CALIOP/CALIPSO version 2/version 3 daytime aerosol extinction product based on a detailed multi-sensor, multi-platform case study. *Atmospheric Chemistry and Physics*, 11, 3981.

Kaufman, Y. J., 1993. Aerosol optical thickness and atmospheric path radiance. *Journal of Geophysical Research*, 98, 2677–2692.

Kaufman, Y. J., Remer, L. A., Tanré, D., Li, R.-R., Kleidman, R., Mattoo, S., Levy, R. C., Eck, T. F., Holben, B. N., and Ichoku, C., 2005. A critical examination of the residual cloud contamination and diurnal sampling effects on MODIS estimates of aerosol over ocean. *IEEE Transactions on Geoscience and Remote Sensing*, 43, 2886–2897.

Kaufman, Y. J., Setzer, A., Ward, D., Tanfe, D., Holben, B. N., and Menzel, P., Pereira, M. C., and Rasmussen, R., 1992. Biomass burning airborne and spaceborne experiment in the Amazonas (BASE-A). *Journal of Geophysical Research*, 97, 14,581–14,599.

Kaufman, Y. J., Wald, A. E., Remer, L. A., Gao, B.-C., Li, R.-R., and Flynn, L., 1997. The MODIS 2.1-/spl mu/m channel-correlation with visible reflectance for use in remote sensing of aerosol. *IEEE Transactions on Geoscience and Remote Sensing*, 35, 1286–1298.

King, M. D. and Byrne, D. M., 1976. A method for inferring total ozone content from the spectral variation of total optical depth obtained with a solar radiometer. *Journal of the Atmospheric Sciences*, 33, 2242–2251.

King, M. D., Kaufman, Y. J., Tanré, D., and Nakajima, T., 1999. Remote sensing of tropospheric aerosols from space: Past, present, and future. *Bulletin of the American Meteorological Society*, 80, 2229–2259.

Kokhanovsky, A. A., Breon, F.-M., Cacciari, A., Carboni, E., and Diner D., 2007. Aerosol remote sensing over land: A comparison of satellite retrievals using different algorithms and instruments. *Atmospheric Research*, 85, 372–394.

Lee, H. J., Coull, B. A., Bell, M. L., and Koutrakis, P., 2012. Use of satellite-based aerosol optical depth and spatial clustering to predict ambient PM2.5 concentrations. *Environmental Research*, 118, 8–15.

Lelieveld, J., Evans, J. S., Fnais, M., Giannadaki, D., and Pozzer, A., 2015. The contribution of outdoor air pollution sources to premature mortality on a global scale. *Nature*, 525, 367–371.

Levy, R. C., Mattoo, S., Munchak, L. A., Remer, L. A., Sayer, A. M., Patadia, F., and Hsu, N. C., 2013. The collection 6 MODIS aerosol products over land and ocean. *Atmospheric Measurement Techniques*, 6, 2989–3034.

Li, Q., Li, C. and Mao, J., 2012. Evaluation of atmospheric aerosol optical depth products at ultraviolet bands derived from MODIS products. *Aerosol Science and Technology*, 46, 1025–1034.

Li, L., Shi, R., Zhang, L., Zhang, J., and Gao, W., 2014. The Data Fusion of Aerosol Optical Thickness Using Universal Kriging and Stepwise Regression in East China. In: *Proceedings of SPIE 9221, Remote Sensing and Modeling of Ecosystems for Sustainability XI*, 922112 (2014/10/08), 1–11.

Li, Z., Zhao, X., Kahn, R., Mishchenko, M., Remer, L., Lee, K.-H., Wang, M., Laszlo, I., Nakajima, T., and Maring, H., 2009. Uncertainties in Satellite Remote Sensing of Aerosols and Impact on Monitoring its Long-term Trend: a Review and Perspective, In: *Annales Geophysicae*. Copernicus GmbH, 2755–2770.

Lim, S. S., Vos, T., Flaxman, A. D., Danaei, G., Shibuya, K., Adair-Rohani, H., AlMazroa, M. A., Amann, M., Anderson, H. R., and Andrews, K. G., 2012. A comparative risk assessment of burden of disease and injury attributable to 67 risk factors and risk factor clusters in 21 regions, 1990–2010: A systematic analysis for the global burden of disease study. *Lancet*, 380, 2224–2260.

Liu, Y., Franklin, M., Kahn, R., and Koutrakis, P., 2007. Using aerosol optical thickness to predict ground-level PM2.5 concentrations in the St. Louis area: A comparison between MISR and MODIS. *Remote Sensing of Environment*, 107, 33–44.

Liu, Y., Park, R. J., Jacob, D. J., Li, Q., Kilaru, V., and Sarnat, J. A., 2004. Mapping annual mean ground-level PM2.5 concentrations using Multiangle Imaging Spectroradiometer aerosol optical thickness over the contiguous United States. *Journal of Geophysical Research: Atmospheres*, 109, 1–10.

Liu, Y., Paciorek, C. J., and Koutrakis, P., 2009. Estimating regional spatial and temporal variability of PM2.5 concentrations using satellite data, meteorology, and land use information. *Environmental Health Perspectives*, 117, 886–892.

Liu, H., Pinker, R. T., and Holben, B. N., 2005. A global view of aerosols from merged transport models, satellite, and ground observations. *Journal of Geophysical Research*, 110, D10S15.

Lv, B., Hu, Y., Chang, H. H., Russell, A. G., Cai, J., Xu, B., and Bai, Y., 2016. Daily estimation of ground-level PM2.5 concentrations at 4 km resolution over Beijing-Tianjin-Hebei by fusing MODIS AOD and ground observations. *Science of The Total Environment*, 580, 235–244.

Lyapustin, A., Wang, Y., Laszlo, I., Kahn, R., Korkin, S., Remer, L., Levy, R., and Reid, J. S., 2011. Multiangle implementation of atmospheric correction (MAIAC): 2. Aerosol algorithm. *Journal of Geophysical Research: Atmospheres*, 116, D03211.

Ma, Z., Hu, X., Sayer, A. M., Levy, R., Zhang, Q., Xue, Y., Tong, S., Bi, J., Huang, L., and Liu, Y., 2016a. Satellite-based spatiotemporal trends in PM2.5 concentrations: China, 2004–2013. *Environmental Health Perspectives*, 124, 184–192.

Ma, Z., Liu, Y., Zhao, Q., Liu, M., Zhou, Y., and Bi, J., 2016b. Satellite-derived high resolution PM2.5 concentrations in Yangtze River Delta Region of China using improved linear mixed effects model. *Atmospheric Environment*, 133, 156–164.

McDonnell, W. F., Ishikawa, N., Petersen, F. F., Chen, H., and Abbey, D. E., 2000. Relationships of mortality with the fine and coarse fractions of long-term ambient PM 10 concentrations in nonsmokers. *Journal of Exposure Analysis and Environmental Epidemiology*, 10, 427–436.

Mehta, S., Shin, H., Burnett, R., North, T., and Cohen, A. J., 2013. Ambient particulate air pollution and acute lower respiratory infections: A systematic review and implications for estimating the global burden of disease. *Air Quality, Atmosphere and Health*, 6, 69–83.

Mei, L. L., Xue, Y., de Leeuw, G., von Hoyningen-Huene, W., Kokhanovsky, A. A., Istomina, L., Guang, J., and Burrows, J. P., 2013a. Aerosol optical depth retrieval in the Arctic region using MODIS data over snow. *Remote Sensing of Environment*, 128, 234–245.

Mei, L. L., Xue, Y., Kokhanovsky, A. A., von Hoyningen-Huene, W., Istomina, L., de Leeuw, G., Burrows, J. P., and Guang, J., 2013b. Aerosol optical depth retrieval over snow using AATSR data. *International Journal of Remote Sensing*, 34, 5030–5041.

Mélin, F., Zibordi, G., and Djavidnia, S., 2007. Development and validation of a technique for merging satellite derived aerosol optical depth from SeaWiFS and MODIS. *Remote Sensing of Environment*, 108, 436–450.

Mishchenko, M. I., Geogdzhayev, I. V., Cairns, B., Rossow, W. B., and Lacis, A. A., 1999. Aerosol retrievals over the ocean by use of channels 1 and 2 AVHRR data: sensitivity analysis and preliminary results. *Applied Optics*, 38, 7325–7341.

Mishchenko, M. I., Geogdzhayev, I. V., Cairns, B., Carlson, B. E., Chowdhary, J., Lacis, A. A., Liu, L., Rossow, W. B., and Travis, L. D., 2007. Past, present, and future of global aerosol climatologies derived from satellite observations: A perspective. *Journal of Quantitative Spectroscopy and Radiative Transfer*, 106, 325–347.

NASA. 2017. What is the difference between dark target and deep blue? https://darktarget.gsfc.nasa.gov/content/what-difference-between-dark-target-and-deep-blue, accessed by Dec. 2 2017.

Nguyen, H., Cressie, N., and Braverman, A., 2012. Spatial statistical data fusion for remote sensing applications. *Journal of the American Statistical Association*, 107, 1004–1018.

Nirala, M., 2008. Technical note: Multi-sensor data fusion of aerosol optical thickness. *International Journal of Remote Sensing*, 29, 2127–2136.

Novotna, B., Topinka, J., Solansky, I., Chvatalova, I., Lnenickova, Z., and Sram, R. J., 2007. Impact of air pollution and genotype variability on DNA damage in Prague policemen. *Toxicology Letters*, 172, 37–47.

Pfeifer, G. D., Harrison, R. M., and Lynam, D. R., 1999. Personal exposures to airborne metals in London taxi drivers and office workers in 1995 and 1996. *Science of The Total Environment*, 235, 253–260.

Pope, C. A. and Dockery, D. W., 2006. Health Effects of Fine Particulate Air Pollution: Lines that Connect. *Journal of the Air & Waste Management Association*, 56, 709–742.

Pope, C. A., Muhlestein, J. B., May, H. T., Renlund, D. G., Anderson, J. L., and Horne, B. D., 2006. Ischemic heart disease events triggered by short-term exposure to fine particulate air pollution. *Circulation*, 114, 2443–2448.

Puttaswamy, S. J., Nguyen, H. M., Braverman, A., Hu, X., and Liu, Y., 2013. Statistical data fusion of multi-sensor AOD over the continental United States. *Geocarto International*, 29, 48–64.

Raaschou-nielsen, O., Andersen, Z. J., Beelen, R., Samoli, E., Stafoggia, M., Weinmayr, G., Hoffmann, B., and Fischer, P., 2013. Air pollution and lung cancer incidence in 17 European cohorts: prospective analyses from the European Study of Cohorts for Air Pollution Effects (ESCAPE). *The Lancet Oncology*, 14, 813–822.

Remer, L. A., Kaufman, Y., Tanré, D., Mattoo, S., Chu, D., Martins, J. V., Li, R.-R., Ichoku, C., Levy, R., and Kleidman, R., 2005. The MODIS aerosol algorithm, products, and validation. *Journal of the Atmospheric Sciences*, 62, 947–973.

Schaap, M., Apituley, A., Timmermans, R. M. A., Koelemeijer, R. B. A., and de Leeuw, G., 2009. Exploring the relation between aerosol optical depth and PM2.5 at Cabauw, the Netherlands. *Atmospheric Chemistry and Physics*, 9, 909–925.

Song, W., Jia, H., Huang, J., and Zhang, Y., 2014. A satellite-based geographically weighted regression model for regional PM2.5 estimation over the Pearl River Delta region in China. *Remote Sensing of Environment*, 154, 1–7.

Song, Y. Z., Yang, H. L., Peng, J. H., Song, Y. R., Sun, Q., and Li, Y., 2015. Estimating PM2.5 concentrations in Xi'an City using a generalized additive model with multi-source monitoring data. *PLoS One*, 10, e0142149.

Stammes, P., 2002. OMI Algorithm Theoretical Basis Document Volume III Clouds, Aerosols, and Surface UV Irradiance. Atbdomi03 III, 1–114.

Stieb, D. M., Chen, L., Eshoul, M., and Judek, S., 2012. Ambient air pollution, birth weight and preterm birth: A systematic review and meta-analysis. *Environmental Research*, 117, 100–111.

Tang, Q., Bo, Y., and Zhu, Y., 2016. Spatiotemporal fusion of multiplesatellite aerosol optical depth (AOD) products using Bayesian maximum entropy method. *Journal of Geophysical Research: Atmospheres*, 121, 4034–4048.

Tanré, D., Kaufman, Y., Herman, M., and Mattoo, S., 1997. Remote sensing of aerosol properties over oceans using the MODIS/EOS spectral radiances. *Journal of Geophysical Research: Atmospheres*, 102, 16971–16988.

Thomas, G. E., Carboni, E., Sayer, A. M., Poulsen, C. A., Siddans, R., and Grainger, R. G., 2009. Oxford-RAL Aerosol and Cloud (ORAC): aerosol retrievals from satellite radiometers, In: *Satellite Aerosol Remote Sensing over Land*. Praxis Publishing, Chichester, UK, 193–225.

Torres, O., Tanskanen, A., Veihelmann, B., Ahn, C., Braak, R., Bhartia, P. K., Veefkind, P., and Levelt, P., 2007. Aerosols and surface UV products form Ozone Monitoring Instrument observations: An overview. *Journal of Geophysical Research: Atmospheres*, 112, 1–14.

UN-Habitat, 2012. State of the World's Cities 2012/2013. https://sustainabledevelopment.un.org/content/documents/745habitat.pdf accessed by May 2017.

van Donkelaar, A., Martin, R. V., Brauer, M., Kahn, R., Levy, R., Verduzco, C., and Villeneuve, P. J., 2010. Global estimates of ambient fine particulate matter concentrations from satellite-based aerosol optical depth: Development and application. *Environmental Health Perspectives*, 118, 847–855.

van Donkelaar, A., Martin, R. V., and Park, R. J., 2006. Estimating ground-level PM2.5 using aerosol optical depth determined from satellite remote sensing. *Journal of Geophysical Research: Atmospheres*, 111, 1–10.

von Hoyningen-Huene, W., Freitag, M., and Burrows, J. P., 2003. Retrieval of aerosol optical thickness over land surface from top-of-atmosphere radiance, *Journal of Geophysical Research*, 108, 4260.

von Hoyningen-Huene, W., Yoon, J., Vountas, M., Istomina, L. G., Rohen, G., Dinter, T., Kokhanovsky, A. A., and Burrows, J. P., 2011. Retrieval of spectral aerosol optical thickness over land using ocean color sensors MERIS and SeaWiFS. *Atmospheric Measurement Techniques*, 4, 151–171.

Wallace, J. and Kanaroglou, P., 2007. An investigation of air pollution in southern Ontario, Canada, with MODIS and MISR aerosol data. In: *2007 IEEE International Geoscience and Remote Sensing Symposium*, 4311–4314.

Wang, J., Browna, D. G., and Hammerling, D., 2013. Geostatistical inverse modeling for super-resolution mapping of continuous spatial processes. *Remote Sensing of Environment*, 139, 205–215.

Wang, J. and Christopher, S., 2003. Intercomparison between satellite-derived aerosol optical thickness and PM2.5 mass: Implications for air quality studies. *Geophysical Research Letters*, 30, 2095.

Wang, J., Xing, J., Mathur, R., Pleim, J. E., Wang, S., Hogrefe, C., Gan, C., Wong, D. C., and Hao, J., 2017. Historical Trends in Pm2.5-Related Premature Mortality During 1990–2010 Across the Northern Hemisphere. *Environmental Health Perspectives*, 125, 400–408.

Wang, Z., Chen, L., Tao, J., Zhang, Y., and Su, L., 2010. Satellite-based estimation of regional particulate matter (PM) in Beijing using vertical-and-RH correcting method. *Remote Sensing of Environment*, 114, 50–63.

Wu, Y., Guo, J., Zhang, X., and Li, X., 2011. Correlation between PM concentrations and Aerosol Optical Depth in eastern China based on BP neural networks. In: *2011 IEEE International Geoscience and Remote Sensing Symposium*, 3308–3311.

Wu, Y., Guo, J., Zhang, X., Tian, X., and Zhang, J., 2012. Science of the Total Environment Synergy of satellite and ground based observations in estimation of particulate matter in eastern China. *Science of The Total Environment*, 433, 20–30.

Xu, H., Guang, J., Xue, Y., de Leeuw, G., Che, Y. H., Guo, J., He, X. W., and Wang, T. K., 2015. A consistent aerosol optical depth (AOD) dataset over mainland China by integration of several AOD products. *Atmospheric Environment*, 114, 48–56.

Xue, Y., Xu, H., Guang, J., Mei, L., Guo, J., Li, C., Mikusauskas, R., and He, X., 2014. Observation of an agricultural biomass burning in central and east China usingmerged aerosol optical depth data from multiple satellite missions. *International Journal of Remote Sensing*, 35, 5971–5983.

Yu, H.-L. and Chien, L.-C., 2016. Short-term population-based non-linear concentration-response associations between fine particulate matter and respiratory diseases in Taipei (Taiwan): a spatiotemporal analysis. *Journal of Exposure Science and Environmental Epidemiology*, 26, 197–206.

Yu, H. B., Dickinson, R. E., Chin, M., Kaufman, Y. J., Holben, B. N., Geogdzhayev, I. V., and Mishchenko, M. I., 2003. Annual cycle of global distributions of aerosol optical depth from integration of MODIS retrievals and GOCART model simulations. *Journal of Geophysical Research*, 108, 4128.

Zhang, J., Reid, J., Westphal, D., Baker, N. and Hyer, E., 2008. A system for operational aerosol optical depth data assimilation over global oceans. *Journal of Geophysical Research*, 113, D10208.

Zhang, Y. and Li, Z., 2015. Remote sensing of atmospheric fine particulate matter (PM2.5) mass concentration near the ground from satellite observation. *Remote Sensing of Environment*, 160, 252–262.

Zheng, G. J., Duan, F. K., Su, H., Ma, Y. L., Cheng, Y., Zheng, B., Zhang, Q., et al., 2015. Exploring the severe winter haze in Beijing: the impact of synoptic weather, regional transport and heterogeneous reactions. *Atmospheric Chemistry and Physics*, 15, 2969–2983.

Zou, B., Pu, Q., Bilal, M., Weng, Q., Zhai, L., and Nichol, J. E., 2016. High-resolution satellite mapping of fine particulates based on geographically weighted regression. *IEEE Geoscience and Remote Sensing Letters*, 13, 495–499.

20 Conclusions

20.1 INTRODUCTION

Multisensor data fusion and feature extraction with traditional or machine learning algorithms is a rapidly evolving research area that requires interdisciplinary knowledge in image processing, signal processing, artificial intelligence, probability and statistics, information engineering, system engineering, and so on. (Luo et al., 2002). This book has identified, appraised, and synthesized research principles, guidelines, and contemporary research topics regarding remote sensing data fusion and machine learning in individual chapters with respect to both quantitative and qualitative characteristics. However, there are still a myriad of challenges standing at the forefront that deserve more attention. These challenges are tied to the changes in spatial complexity embedded in the Earth system driven by evolving connectivity patterns from the integration of natural systems, semi-natural systems, and the built environment to cross-scale interactions in time and space among them (Chang et al., 2010, 2012, 2013). These challenges may be further complicated by possible structural changes in the study domain which could induce new connectivity patterns and cross-scale interactions that have no historical precedence. Consequently, optimizing the synergistic effects of sensors, platforms, and models in order to provide decision makers and stakeholders with timely decision support concepts, tools, and projections is regarded as a critical mission for the future (NCAR, 2002; NSF, 2003).

Once integrated data fusion and machine learning are successful, the powerful and innovative earth observatories can trigger more scientific breakthroughs and tell us about the water, energy, and carbon cycles, as well as the land use and land cover changes in relation to earth system processes. This can lead to the exploration of coupling effects among biosphere, geosphere, hydrosphere, lithosphere, and atmosphere and the corresponding feedback mechanisms intertwined in between. This series of efforts is helpful for entailing how scale affects processes and elucidating what are the scientific implications from multitemporal change detection. To achieve these research goals, relevant sensors, platforms, image processing and feature extraction algorithms, knowledge discovery tools, and content-based time series mapping/visualization techniques must be smoothly woven together to collect, analyze, fuse, synthesize, and visualize a myriad of remote sensing data sets that are more spatially, spectrally, and temporally comprehensive. This chapter aims to highlight the relevance of challenges and perspectives with particular attention paid to wrapping up the intertwined streamlines of this book.

20.2 CHALLENGES

20.2.1 DATA SCIENCE AND BIG DATA ANALYTICS

Earth big data collected from various space-borne platforms, air-borne platforms, ground-based stations, and social media has become available over various spectral, spatial, and temporal domains. With the recent advancement of data science from computation to storage, to transmission, to sharing, however, the assessment of the large amount of dynamic sensor observations for specific times and locations can be made easier (Chen et al., 2007). A few recent challenges when developing explorative algorithms to improve information retrieval from big remote sensing data are summarized below:

- *Sparse Remote Sensing Data*: Sparse signal processing is a common challenge in remote sensing from the multispectral to the hyperspectral to the microwave spectrum. The sparsity of signals embedded in a remote sensing image may appear in: (1) pan-sharpening,

(2) hyperspectral unmixing, (3) hyperspectral resolution enhancement, (4) Tomographic Synthetic Aperture Radar (TomoSAR) inversion issues, and (5) SAR imaging. By using the compressive sensing theory, it is possible to achieve higher resolution or reduce the required number of samples, given that the resolution requirement is known, when handling sparsity exploitation issues.

- *Non-Local Filtering Methods*: Noise reduction is a common challenge in remote sensing data processing. Several classical local filters, such as low pass filter and high pass filter, as well as look-processing for SAR and InSAR data, are often used. However, these methods always reduce the spatial resolution. This challenge calls for developing advanced non-local approaches that take advantage of the high degree of redundancy of natural images when handling data fusion between multispectral remote sensing images with different resolutions, InSAR filtering, and hyperspectral image denoising.

- *Robust Estimation in Feature Extraction:* Remote sensing image processing techniques for feature extraction often encounter many outliers and unmodeled noise, making the development of robust estimators a challenging task. Innovative robust estimation may help with: (1) the hyperspectral nonlinear unmixing method, (2) minimization of semantic gaps at the sub-pixel level, (3) dimensionality reduction in hyperspectral images, (4) land deformation monitoring using interferometric SAR data stacks, and (5) object reconstruction from TomoSAR point clouds.

- *Advanced Image or Data Fusion Techniques*: Image or data fusion by pairing optical and SAR images at the decision level are meritorious given that SAR may penetrate the cloud if the cloudy pixel reconstruction is not successful. Innovative pairing may include: (1) optical and SAR images, (2) optical and InSAR images, and (3) optical and PolInSAR images.

- *Deep and Fast Learning:* Deep learning algorithms such as convolutional neural networks and deep believe networks have become a series of emerging tools in the machine learning field that may help in feature extraction when making use of the nature of big remote sensing data. Fast learning, such as extreme learning machine (Huang et al., 2006), may largely reduce processing time while retaining reasonable prediction accuracy. All of them can be applied for: (1) newly modified convolutional neural networks and recurrent networks for multispectral and hyperspectral image processing, (2) innovative network architectures for data or image fusion of SAR and hyperspectral data or hyperspectral and multispectral data for complex remote sensing analyses, (3) data fusion of remote sensing satellite data and social media information with text retrieval algorithms, and (4) novel integration between deep and fast learning algorithms.

- *Cluster and Classifier Ensemble*: The employment of ensemble learning algorithms for both supervised and unsupervised classification may elevate feature extraction techniques to a new level. There are newly developed learning ensemble algorithms in the artificial intelligence field which include, but are not limited to: (1) consensus clustering for unsupervised classification, (2) ensemble learning or classifier ensemble for supervised classification, and (3) collaborative classification of SAR, hyperspectral, and visible images with deep learning algorithms.

- *Evolutionary Computation and High-Performance Computing*: Facing the big data era, large scale computational infrastructure is of much necessity. There are newly developed computational facilities and equipment that may come to help when dealing with data science and big data analytics, and they include, but are not limited to: (1) new evolutionary computation methods, (2) new image compression methods, (3) cloud computing networks, (4) graphics processing units, and (5) parallel computing units.

Hence, with the aid of the latest development of evolutionary computation and high-performance computing technologies, further data science ideas in sparsity exploitation, robust estimation, deep and fast learning algorithms, advanced image or data fusion techniques, cluster and classifier ensemble, and non-local filtering methods may be advanced with an unprecedented pace. It will certainly lead to enhanced image processing techniques and minimized semantic gaps for hyperspectral, multispectral, and SAR images directly or indirectly, improving data fusion/merging technologies collectively or individually. For example, pairing suitable optical and SAR sensors, such as ERS-2 (30 m) and Landsat-7 TM (30 m) fusion (Alparone et al., 2004), TerraSAR-X (1 m/3 m/18.5 m) and QuickBird MS fusion (0.65 m) (Amarsaikhan et al., 2010), and TerraSAR X (1 m/3 m/18.5 m) and ALOS MS (2.5 m) fusion (Bao et al., 2012) may be a challenge in demanding missions. In the future, it will be possible to pair Sentinel-1A (SLC-IW) SAR data (VV polarization) and Sentinel-2A (MSI) optical data and test the prediction accuracy for more complicated studies. However, in a regional study, such a space-borne effort may be supported by including ad-hoc air-borne remote sensing efforts that may be operated without regard to the cloud impact with the aid of sparsity exploitation, robust estimation, deep and fast learning algorithms, cluster and classifier ensemble, and non-local filtering methods. It is even possible that these remote sensing data sets may be stored in a public data service center to offer data and sensor planning service to end users (Chen et al., 2007). This progress may certainly result in "sensor web" and "model web" for end users to select mission-oriented sensors and models for specific decision-making.

These advancements may be signified and magnified by advanced visualization techniques in computer vision engineering such as virtual reality and actual reality interactive installations. Such advancements may play a critical role in knowledge discovery by creating a myriad of big data analytics and enriching decision science while smoothing out complex feature extraction and content-based mapping with visualization. This will lead to better understanding of the relationships between the components of the Earth system at different scales, making them seamlessly linked with each other. For a large multidisciplinary initiative which emphasizes contemporary data-driven research challenges from a strongly applied science perspective, a holistic effort may be geared toward targeting the interactive themes of environment, energy, food, health, ecology, and society as well as exploring causal effects among factors and constraints.

20.2.2 ENVIRONMENTAL SENSING

Environmental sensing is broadly defined as incorporating the span of different remote sensing technologies from satellite-based earth observation through air-borne remote sensing (either unmanned or manned aerial vehicles) to *in situ* sensor networks. Both satellite-based, air-borne, and *in-situ* sensors are important "sensor-web" components, as they provide spatiotemporal information as a function of scene element optical properties (Bradley et al., 2010). For example, weighted data fusion for unmanned aerial vehicle 3D mapping with camera and line laser scanner can be used to supplement space-borne remote sensing for urban surveillance (Jutzi et al., 2014). Recently, CubeSats have become part of a trend toward an increasingly diverse set of platforms with different sensors for understanding the rapidly changing environment such as: disaster management, emergency response, national security, and so on. This trend has great potential to avoid cloud impact in optical remote sensing with the aid of data fusion and data merging technologies. It can be foreseen that more private sectors will develop the capability to implement large-scale constellation missions utilizing CubeSats or CubeSat-derived technologies and the highly anticipated subsequent evolutionary development (National Academies of Sciences, Engineering, and Medicine, 2016). Future research directions of multisensor fusion technology include microsensors, smart sensors, and adaptive fusion techniques (Luo et al., 2002). Combinations of all the possible sensors and platforms with complementary capacity may support the range of environmental applications from monitoring human-produced environmental change to quantifying fundamental earth system processes with sustainability implications. Such combinations may be extended, especially to explore the interface of environmental sensing and data science in relation to the hydrosphere, lithosphere, biosphere, and atmosphere for multiscale sensing, monitoring, and modeling analyses.

Recently, quantum entanglement, a physical phenomenon that provides remote sensing an advanced basis for the future, has been well studied. Quantum entanglement occurs when pairs or groups of particles interact with one another in such a way that the quantum state of each particle cannot be described independently of the others. This is true even when the particles are divided by a large distance. Quantum sensing is thus deemed a new quantum technology that employs quantum coherence properties for ultrasensitive detection, which is especially useful for sensing weak signals. A quantum sensor is a device that exploits quantum correlations, such as quantum entanglement, to achieve a sensitivity or resolution that is better than can be achieved using only classical systems (Kapale et al., 2005). Quantum remote sensing, quantum sensors, and quantum sources have become hot topics in research. For example, infrared sensing technology has a central role to play in addressing 21st century global challenges in environmental sensing, and infrared imaging and sensing with the single-photon setting has been studied recently as a new quantum remote sensing technology (European Union, 2016). This type of new technology may deeply affect future environmental sensing (Han, 2014; Bi and Zhang, 2015).

20.2.3   Environmental Modeling

Modeling unobservable processes is by no means an easy task because the estimation of parameters relies on the ability to accurately relate them to quantities that can be measured or observed (LaDeau et al., 2011). With uncertainties, limited data on one particular temporal or spatial scale gives only partial information regarding the underlying process for which a model is attempting to communicate. Obtaining the best possible description of a process by combining all available data with the dynamics of a model is beneficial. Several techniques have been proposed and applied in various studies to bridge the gap between data and models which include data assimilation, data-model synthesis, and data-model fusion. They are delineated briefly below.

- Data assimilation applies model structure and ranges of parameter values as prior information using a Bayesian framework to represent the current state of knowledge. Global optimization techniques are then used to update parameters and state variables of a model based on information contained in multiple, heterogeneous data sets that describe the past and current states (Niu et al., 2014). While physically-based models can be useful for studying a wealth of earth system processes, parameters and state variables collected and updated by remote sensing or other means may account for an earth system process in a dynamic way.
- Data-model synthesis infers the dynamics of internal state variables and earth system processes that are not directly observable via models while combining multiple types of data operating on different scales and of different types (Dietze et al., 2013). The goal of data-model synthesis is to obtain the best possible description of a model by combining all available data with the dynamics (i.e., a known process or spatiotemporal distribution) of a model such that data are needed to run models, evaluate the dynamic patterns, and improve models.
- Data-model fusion seeks to apply techniques to minimize the difference between observations of a system state and modeled state variables (Renzullo et al., 2008) such that an integration of both model parameter estimation and data assimilation techniques can be applied to better understand a considered system (Wang et al., 2009). Data-model fusion is applied to improve the performance of a model by not only optimizing or refining the values of unknown parameters and initial conditions, but also improving the predictive capacity of a model by constraining the model by periodically updating data spatially and temporally to minimize the possible gaps driven by numerical modeling steps present in data assimilation or data-model synthesis in a cognitive way (Guiot et al., 2014).

For example, Chen et al. (2009) claimed that the ensemble Kalman filter (EnKF) and the ensemble square-root Kalman filter (EnSKF) were suitable for coastal ocean and estuarine modeling efforts within coastal ocean systems reflecting dynamics which vary significantly in the temporal and spatial domain (Madsen and Cañizares, 1999). Although EnKF may require extensive computational time, it is advantageous for applications on complex first-principle physics-based models due to its flexibility. The challenge, however, rests on tracking highly dynamic state variables, such as wave heights and coastal wind speed/direction, which require using data-model fusion in which remote sensing may be of help. In this context, the Ocean Surface Topography Mission on the Jason-2 or Jason-1 satellite may provide sea level observation that may serve as a state variable (i.e., wave height) in the context of data-model fusion. With big data available, therefore, proper investigations of synergies through various data-model fusion scenarios are deemed possible and feasible. The next-generation multiscale modeling system may be created via a smooth data-model fusion to form a highly adaptive context-aware intelligence system. This may be achieved by proper integration or coupling of land, air, and water modeling with the aid of data science for managing and providing next generation research capability in a form of comprehensive data-driven modeling analysis from system science perspectives. To achieve such a bold agenda, there is a need to emphasize cross-disciplinary research addressing global environmental challenges through methodological advances in data science, earth science, ecological science, environmental sciences, and sustainability science. In the future, different data collection schemes (i.e., air-borne, space-borne, ground-based stations, and social media) in support of multisensor data-model fusion may be considered to improve measurement and/or mapping accuracy to better model parameters and retrieve system state for the intelligent multiscale modeling analysis.

20.3   FUTURE PERSPECTIVES AND ACTUALIZATION

20.3.1   CONTEMPORARY RESEARCH TOPICS

Remote sensing practitioners generally understand the optical properties of electromagnetic radiation, the system design of sensors and platforms, the basic image processing algorithms, and the pertinent remote sensing products available for land, air, and water applications. The contemporary research topics may be categorized by the following nine areas of importance: (1) efficient and effective cloudy image reconstruction and enhancement with both microwave and optical remote sensing techniques with the aid of non-local concepts via statistical similarity and machine learning, (2) simultaneous integration between data fusion and cloudy image reconstruction, (3) hierarchical processes of cross-mission data merging and data fusion maintaining good signal-to-noise ratios with the aid of flexible satellite constellations, (4) development of data mining, machine learning or big data analytics techniques to aid in multisensor data fusion and feature extraction at pixel-level, feature-level, and decision-level, (5) integrated data fusion and machine learning algorithms with the aid of learning ensemble techniques for better content-based time series retrieval and domain adaptation, (6) decision support systems with the aid of "sensor web" (Figure 20.1) and "model web" (Figure 20.2) to expand the data collection capacity and multi-scale modeling platforms leading to create new cyberinfrastructure facilities such as new geo-portals to provide various new remote sensing products, (7) exploration of minute environmental changes at urban or local scales with Tomographic Synthetic Aperture Radar (TomoSAR), solving various types of problems by using multiple SAR images acquired from slightly different viewing angles, (8) linking the space-borne and air-borne remote sensing sensors/platforms to the ground-based sensor networks sustained by the Internet of Things, wireless transmission, and wireless charge to form the cloud-based data and services hub and cyber-physical systems for environmental monitoring and decision support for emergency response and regular management, and (9) the use of "sensor web" and "model web" in support of large scale and interactive data-model fusion and visualization.

20.3.2   REMOTE SENSING EDUCATION

Big data technologies depend on data storage, data search and transmission, and data sharing. With abundant data collected from these mission-oriented earth observation, the identification

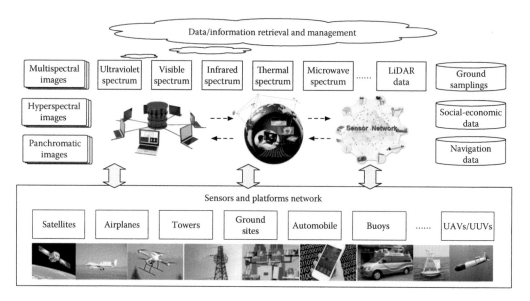

FIGURE 20.1   Sensor web architecture proposed.

of comparative advantages via content-based classification and context-based retrieval system for e-learning may be applied to potentially exhibit information and discover knowledge via different big data analytics (Chang et al., 2010, 2012, 2013). This may trigger a collaborative learning approach intent on achieving a multidisciplinary data science goal and tackling methodological challenges in answering key environmental questions.

Along this line, a growing challenge could be tied to how image processing, data assimilation, data mining, machine learning, and information retrieval techniques can be coordinated and

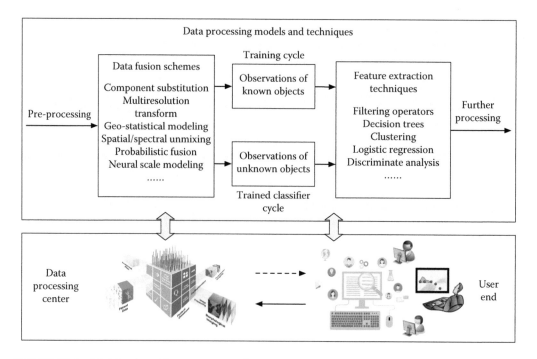

FIGURE 20.2   Model web architecture proposed.

developed for knowledge discovery in web-based e-learning systems as a final destination of remote sensing education. This movement will certainly help signify and magnify the core value of multisensor data fusion and machine learning for decision making in remote sensing education.

20.4 SUMMARY

This chapter presents challenges in terms of big data and data analytics, environmental sensing, and environmental modeling to support multisensor remote sensing data fusion and machine learning for environmental monitoring and Earth observation based on the paradigm of multisensor data fusion at different fusion levels. The chapter also highlights nine essential research topics that are worthwhile to consider from a highly applied perspective with an insightful footnote for both research and education. Overall, this book will serve as a foundation for future strategic planning of this scientifically stimulating and professionally rewarding realm of remote sensing study.

REFERENCES

Alparone, L., Facheris, L., Baronti, S., Garzelli, A., and Nencini, F., 2004. Fusion of multispectral and SAR images by intensity modulation. In: *Proceedings of the 7th International Conference on Information Fusion*, 637–643.

Amarsaikhan, D., Blotevogel, H. H., van Genderen, J. L., Ganzorig, M., Gantuya, R., and Nergui, B., 2010. Fusing high-resolution SAR and optical imagery for improved urban land cover study and classification. *International Journal of Image and Data Fusion*, 1, 83–97.

Bao, C., Huang, G., and Yang, S., 2012. Application of fusion with SAR and optical images in land use classification based on SVM. International archives of the photogrammetry. In: *Remote Sensing and Spatial Information Sciences, Volume XXXIX-B1, XXII ISPRS Congress*, 25 August–01 September, Melbourne, Australia.

Bi, S. and Zhang, Y., 2015. The study of quantum remote sensing principle prototype. In: *Proceedings of SPIE 9524, International Conference on Optical and Photonic Engineering (icOPEN 2015)*, Singapore, April 14, 2015, 95241F.

Bradley, E. S., Toomey, M. P., Still, C. J., and Roberts, D. A., 2010. Multi-Scale sensor fusion with an online application: Integrating GOES, MODIS, and webcam imagery for environmental monitoring. *IEEE Journal of Selected Topics in Applied Earth Observations and Remote Sensing*, 3, 497–506.

Chang, N. B., Han, M., Yao, W., Xu, S. G., and Chen, L. C., 2010. Change detection of land use and land cover in a fast-growing urban region with SPOT-5 images and partial Lanczos extreme learning machine. *Journal of Applied Remote Sensing*, 4, 043551.

Chang, N. B., Xuan, Z., and Yang, J., 2013. Exploring spatiotemporal patterns of nutrient concentrations in a coastal bay with MODIS images and machine learning models. *Remote Sensing of Environment*, 134, 100–110.

Chang, N. B., Yang, J., Daranpob, A., Jin, K. R., and James, T., 2012. Spatiotemporal pattern validation of chlorophyll-a concentrations in Lake Okeechobee, Florida using a comparative MODIS image mining approach. *International Journal of Remote Sensing*, 33, 2233–2260.

Chen, N., Di, L., Yu, G., Gong, J., and Wei, Y., 2007. Use of RIM-based CSW with sensor observation services for registry and discovery of remote-sensing observations. *Computing Geoscience*, 35, 360–372.

Chen, C., Malanotte-Rizzoli, P., Wei, J., Beardsley, R. C., Lai, Z., Xue, P., Lyu, S., Xu, Q., Qi, J., and Cowles, G. W., 2009. Application and comparison of Kalman filters for coastal ocean problems: An experiment with FVCOM. *Journal of Geophysical Research*, 114, C05011.

Dietze, M. C., Lebauer, D. S., and Kooper, R., 2013. On improving the communication between models and data. *Plant Cell Environment*, 36, 1575–1585.

European Union, 2016. Periodic Reporting for period 1 – IRIS (Infrared imaging and sensing: the single-photon frontier), http://cordis.europa.eu/result/rcn/198065_en.html, accessed by June 2017.

Guiot, J., Boucher, E., and Gea-Izquierdo, G., 2014. Process models and model-data fusion in dendroecology. *Frontiers in Ecology and Evolution*, 2, 1–12.

Han, J. H., 2014. Photon pair sources for quantum information networks and remote sensing systems. In: *12th International Conference on Optical Internet 2014 (COIN)*, Jeju, 1–2.

Huang, G. B., Zhu, Q. Y., and Siew, C. K., 2006. Extreme learning machine: Theory and applications. *Neurocomputing*, 70, 489–501.

Jutzi, B., Weinmann, M., and Meidow, J., 2014. Weighted data fusion for UAV-borne 3D mapping with camera and line laser scanner. *International Journal of Image and Data Fusion*, 5, 226–243.

Kapale, K. T., Didomenico, L. D., Lee, H., Kok, P., and Dowling, J. P., 2005. Quantum interferometric sensors. *Concepts of Physics*, Vol. II, 225–240.

LaDeau, S. L., Glass, G. E., Hobbs, N. T., Latimer, A., and Ostfeld, R. S., 2011. Data-model fusion to better understand emerging pathogens and improve infectious disease forecasting. *Ecological Applications*, 21, 1443–1460.

Luo, R. C., Yih, C. C., and Su, K. L., 2002. Multisensor fusion and integration: Approaches, applications, and future research directions. *IEEE Sensors Journal*, 2, 107–119.

Madsen, H. and Cañizares, R., 1999. Comparison of extended and ensemble Kalman filters for data assimilation in coastal area modelling. *International Journal for Numerical Methods in Fluids*, 31, 961–981.

National Academies of Sciences, Engineering, and Medicine, 2016. *Achieving Science with CubeSats: Thinking Inside the Box*. The National Academies Press, Washington, DC.

National Center for Atmospheric Research (NCAR), 2002. Cyberinfrastructure for Environmental Research and Education, Boulder, CO, USA.

National Science Foundation (NSF), 2003. Complex Environmental Systems: Synthesis for Earth, Life, and Society in the 21st Century. NSF Environmental Cyberinfrastructure Report, Washington DC, USA.

Niu, S., Luo, Y., Dietze, M. C., Keenan, T. F., Shi, Z., Li, J., and Chapin, III., F. S., 2014. The role of data assimilation n predictive ecology. *Ecosphere*, 5, 65.

Renzullo, L. J., Barrett, D. J., Marks, A. S., Hill, M. J., Guerschman, P. J., Mu, Q., and Running, S. W., 2008. Multi-sensor model-data fusion for estimation of hydrologic and energy flux parameters. *Remote Sensing of Environment*, 112, 1306–1319.

Wang, Y.-P., Trudinger, C. M., and Enting, I. G., 2009. A review of applications of model–data fusion to studies of terrestrial carbon fluxes at different scales. *Agricultural and Forest Meteorology*, 149, 1829–1842.

Index

T - #0166 - 111024 - C528 - 254/178/24 - PB - 9780367571979 - Gloss Lamination